33rd International Conference on Physics of Semiconductors (ICPS 2016)

Journal of Physics: Conference Series Volume 864

Beijing, China
31 July – 5 August 2016

ISBN: 978-1-5108-4720-0
ISSN: 1742-6588

Printed from e-media with permission by:

Curran Associates, Inc.
57 Morehouse Lane
Red Hook, NY 12571

Some format issues inherent in the e-media version may also appear in this print version.

Printed by Curran Associates, Inc. (2017)

Institute of Physics
Dirac House, Temple Back
Bristol BS1 6BE UK

Phone: 44 1 17 929 7481
Fax: 44 1 17 920 0979

techtracking@iop.org

Additional copies of this publication are available from:

Curran Associates, Inc.
57 Morehouse Lane
Red Hook, NY 12571 USA
Phone: 845-758-0400
Fax: 845-758-2633
Email: curran@proceedings.com
Web: www.proceedings.com

Table of contents

Volume 864

33rd International Conference on the Physics of Semiconductors
31 July to 5 August 2016, Beijing, China

Accepted papers received: 15 June 2017
Published online: 15 August 2017

Preface

Material growth, structural properties and characterization, phonons

Wide-bandgap semiconductors

Narrow-bandgap semiconductors

Carbon: nanotubes and graphene

2D Materials beyond graphene

Topological states of matter, topological Insulators and Weyl semimetals

Transport in heterostructures

Quantum Hall effects

Spintronics and spin phenomena

Electron devices and applications

Optical properties, optoelectronics, solar cells

Quantum optics, nanophotonics

Other topics in semiconductor physics and devices

IOP Conf. Series: Journal of Physics: Conf. Series **864** (2017) 011001 doi:10.1088/1742-6596/864/1/011001

Preface to the Proceedings of the 33rd International Conference on the Physics of Semiconductors, Beijing, 2016

Shaoyun Huang[1], Yingjie Xing[1], Yang Ji[2], Dapeng Yu[3], and Hongqi Xu[1]

[1]Beijing Key Laboratory of Quantum Devices, Key Laboratory for the Physics and Chemistry of Nanodevices and Department of Electronics, Peking University, Beijing 100871, China
[2]SKLSM, Institute of Semiconductors, Chinese Academy of Sciences, Beijing 100083, China
[3]State Key Laboratory for Mesoscopic Physics, Department of Physics, Peking University, Beijing 100871, China

From July 31st to August 5th, 2016, the 33rd International Conference on the Physics of Semiconductors (ICPS 2016) was held in Beijing, China, with a great success. The International Conference on the Physics of Semiconductors began in the 1950's and is a premier biennial meeting for reporting all aspects of semiconductor physics including electronic, structural, optical, magnetic and transport properties. Reflecting the state of the art developments in semiconductor physics, ICPS 2016 served as an international forum for scholars, researchers, and specialists across the globe to discuss future research directions and technological advancements. The main topics of ICPS 2016 included:

- Material growth, structural properties and characterization, phonons
- Wide-bandgap semiconductors
- Narrow-bandgap semiconductors
- Carbon: nanotubes and graphene
- 2D Materials beyond graphene
- Organic semiconductors
- Topological states of matter, topological Insulators and Weyl semimetals
- Transport in heterostructures
- Quantum Hall effects
- Spintronics and spin phenomena
- Electron devices and applications
- Optical properties, optoelectronics, solar cells
- Quantum optics, nanophotonics
- Quantum information
- Other topics in semiconductor physics and devices
- Special topic: Majorana fermions in solid state

The conference consists of a Nobel symposium, five plenary sessions, 72 oral session, three poster sessions, and an IUPAP Young Scientist Prize winner session. The Nobel symposium was featured with three Nobel laureate lectures and a technology lecture on the emerging semiconductor industries of China, and a panel discussion by Nobel laureates Klaus von Klitzing, Andre Geim, Hiroshi Amano and Shuji Nakamura, and Senior Vice President of SMIC (a major China chip company) Min-Hwa Chi. The plenary sessions included lectures by Yasuhiko Arakawa, David D. Awschalom, Jiangfeng Du, Andre Geim, Tony Heinz, Charles Marcus, Sang Il Seok, Michelle Y. Simmons, and Ali Yazdani. The oral sessions included 72 invited talks and 287 contributed talks. The poster sessions hosted 792 contributed presentations. In the IUPAP Young Scientist Prize winner session, the prize winners Dr. Samuel D. Stranks and Dr. Kimberly Dick Thelander presented plenary lectures on their prize-winning achievements. The conference was concluded by a best student poster award session, in which outstanding poster presentations by graduate students were honored

In the proceedings, we selected eighty-four presentations made in ICPS 2016 to provide a good overview of the topics covered in the conference. All the papers in this collection have undergone the same peer-review process as regular journal articles. The organizers of ICPS 2016 are particularly grateful to the Institute of Physics for allowing us to publish these papers in Journal of Physics: Conference Series. We thank all the authors for taking extra efforts to convert their valuable presentations into the presented journal articles and thus making their contributions available to a broad audience in semiconductor physics and related fields. We hope that the readers will find the results presented in these papers interesting and scientifically stimulating.

IOP Conf. Series: Journal of Physics: Conf. Series **864** (2017) 011001 doi:10.1088/1742-6596/864/1/011001

Address at the Opening Ceremony of the 33rd International Congress on the Physics of Semiconductors

Professor Chunli BAI, President, Chinese Academy of Sciences

Committees of the Conference, Distinguished guests, Dear colleagues and friends, Ladies and Gentlemen, it gives me great pleasure to join you in the beautiful city of Beijing at the special time of wonderful August. On behalf of the Chinese Academy of Science, let me extend warm congratulations on the opening of the 33rd International Conference on the Physics of Semiconductors. I want to welcome all guests who have traveled from afar to Beijing and I also express my sincere thanks to the committees of the conference.

Semiconductor physics and technologies have been the driving force and fundamental blocks to shape the modern society, and the ICPS is the premier meeting for reporting important advances in all related aspects. Since its founding in 1950, the biennial conference presents invaluable milestones and common memories to us. It is delighted to me personally to be here, among you prestigious experts and scientists, to continue the legendary tradition. At the same time, I want to take this opportunity to give you, ladies and gentlemen, a general introduction about CAS.

CAS was established on ruins in 1949, almost one year before the first ICPS. After decades of struggling and developing, today, CAS has become the linchpin of China's drive to explore and harness high technology and the natural sciences for the benefit of China and the world. Comprising a comprehensive research and development network, a merit-based academic society and a system of higher education. CAS now is consisting of 104 research institutes, five universities & supporting organizations, 12 branches and is functioning as the national scientific think tank and academic governing body, providing advisory and appraisal services on issues stemming from the national economy, social development, and science and technology progress. Based on the number of papers published by the Nature Publishing Group, the Chinese Academy of Science has ranked 1st among research institutions in the world according to the Nature Publishing Index elaborated by NPG in 2014 and 2015.

Ladies and Gentlemen, semiconductor physics and technologies remain core battlefields in CAS strategies. Date back to 1960, CAS has established special institute to improve semiconductor research and development in China, and contributed outstanding scientists such as Prof. Kun Huang et al. to the semiconductor physics society. As the initiative of CAS innovation 2020 in 2011, CAS will keep seeking to improve the academy's capacity for innovation and to make a significant contribution to the development of emerging industries related to semiconductors. To fulfill the purpose, CAS will set up more collaborative groups, initiate more international science programmes, and implement more international talent programmes to recruit more senior scientists and talented young scientists. The more the better, just as Prof. P. W. Anderson says: "More is different."

Ladies and gentlemen, please pay your continuing attentions to CAS. Let's push semiconductors physics and technologies forward together. And finally I wish you an unforgettable and prefect experience here.

Thank you!

IOP Conf. Series: Journal of Physics: Conf. Series **864** (2017) 011001 doi:10.1088/1742-6596/864/1/011001

Welcome Address at the Thirty-Third International Conference on the Physics of Semiconductors (ICPS2016)

Jianhua LIN, President of Peking University (August 1, 2016, Beijing)

Good morning! Committees of the Conference, Distinguished guests, Ladies and gentlemen, please allow me to begin by, as President of Peking University, offering my great congratulations on the opening of the THIRTY-THIRD INTERNATIONAL CONFERENCE ON THE PHYSICS OF SEMICONDUCTORS (ICPS2016), and expressing my sincere welcome to scientists and students who travelled from home and aboard to make ICPS2016 a memorable academic feast.

I would like to extend my special gratitude to the distinguished scholars, who attended the Nobel Symposium on Sunday July 31st, Nobel Laureates: Professor Klaus von Klitzing, Professor Andre Geim, Professor Hiroshi Amano, and Professor Shuji Nakamura.

Peking University, founded in 1898, was the China's first national multidisciplinary university. Now, Peking University is a key comprehensive university of China, located near to the Yuanmingyuan Garden and the Summer Palace.

The University is proud of its outstanding faculties, including 53 members of the Chinese Academy of Sciences (CAS), 7 members of the Chinese Academy of Engineering (CAE), and 14 members of the Third World Academy of Sciences (TWAS). It has effectively combined top-level research on a wide range of sciences and technologies with the training of personnel with a high level of knowledge and professional skill as demanded by the China's fast development and modernization. It strives not only for improvements in education and research, but also for the promotion of interaction and mutual influence among various disciplines.

Thus, Peking University has become a center for education and research, embracing diverse branches of learning and exploration, such as basic and applied sciences, social sciences and the humanities, and sciences of medicine, management, and education. Its aim is to rank among the world's best universities in the near future.

Well known as the premier meeting for reporting all aspects of semiconductor physics, the ICPS2016 offers us an opportunity together to focus on current developments in semiconductor physics, condensed matter physics, materials science, and IT technologies. In addition to the inspiration day of Sunday, when the Nobel Symposium was successfully held, I hope you will have five most productive days of forefront, insightful and stimulating discussions. And also hope the ICPS2016 will be a great success not only as a stage to share knowledge and techniques but also as the beginning of a fruitful cooperation and friendship among young research fellows who will shape our future.

Thank you very much!

IOP Conf. Series: Journal of Physics: Conf. Series **864** (2017) 011001 doi:10.1088/1742-6596/864/1/011001

Welcome Address at the Thirty-Third International Conference on the Physics of Semiconductors (ICPS2016)

Hongqi Xu, Chair of ICPS2016 (August 1, 2016, Beijing)

President BAI Chunli, President YANG Wei, President LIN Jianhua, Distinguished guests, Ladies and gentlemen, once more, on behalf of the organizing committee, I, as the chair of ICPS2016, wish to welcome you all, especially those colleagues and young scientists who traveled far abroad, to Beijing and to China. I am very happy and excited to open the 33rd International Conference on the Physics of Semiconductors (ICPS2016). ICPS started in 1950 after two years since the discovery of the transistor effect. ICPS2016 is in its 66th year and for the second time in Beijing, in China, after 24 years of ICPS-21 in Beijing in 1992.

ICPS2016 was initiated in 2012 when ICPS-31 was held in Zurich, Switzerland after strong encouragements and supports received from the international community of the physics of semiconductors. What you may already and will experience in this week is the outcome of the joint effort of many people including you all at this opening session and those who could not join us. I would like to thank all people who have provide encouragements, supports, and helps, and have worked hard to make the program of this conference possible. Especially, I would like to thank you all for writing, submitting, packing, and coming over to Beijing to present your accomplishments.

The physics of semiconductors is a dynamic, fast developing, multidisciplinary field and has made strong influences on the current world industrial revolution, the society development, and the way we all live today. ICPS2016 is featured by Nobel Symposium lectures, plenary lectures, IUPAP C8 Young Scientist Prize winner lectures, invited talks, oral presentations and poster presentations, in all aspects of semiconductor physics. ICPS2016 will also strongly reflect the emerging directions of the development in semiconductor physics in recent years, including, e.g., topological states of matter, two-dimensional materials, perovskite photovoltaic materials, and quantum information technologies. I hope that in the five days ahead, you will enjoy ICPS2016 as a stage to share knowledge and techniques and to create fruitful cooperation and friendship among participants, especially among young scientists, from different countries and with different cultural backgrounds.

The scientific program you have received is full and tight, which means that you will have five hard, heavy working days ahead in Beijing. I would like to encourage you to take some time off to explore the natural, cultural, and historical elements of Beijing, an ancient capital city of China. In particular, I do encourage you to visit the Forbidden City, the Great Wall, The Summer Palace, The Temple of Heaven, The National Museum of China, and Hutong and Siheyuan (old streets and houses where Beijing's people have lived). However, I have to advise you to be prepared for very crowded people in all tourist spots in Beijing and traffic jams in the city; July and August are the hottest tourist seasons in Beijing.

Enjoy ICPS2016! Enjoy your stay in Beijing!

Now let me pass the floor to Nobel Laureate Andre Geim, Professor at University of Manchester, to chair the first plenary session of ICPS2016.

Thank you!

IOP Conf. Series: Journal of Physics: Conf. Series **864** (2017) 011001 doi:10.1088/1742-6596/864/1/011001

The 33rd International Conference on the Physics of Semiconductors (ICPS2016)

Honorary Chairs:

Houzhi Zheng	Institute of Semiconductors, CAS, China
Enge Wang	Peking University & CAS, China

Chair:

Hongqi Xu	Peking University, China

Co-chairs:

Qikun Xue	Tsinghua University, China
Shushen Li	Institute of Semiconductors, CAS, China

International Advisory Committee

Chair:

Bangfen Zhu	Tsinghua University, China

Members:

Hiroshi Amano	Nagoya University, Japan
Tsuneya Ando	Tokyo Institute of Technology, Japan
Wanda Andreoni	Ecole Polytechnique Fédérale de Lausanne, Switzerland
Yasuhiko Arakawa	The University of Tokyo, Japan
David Awschalom	University of Chicago, USA
Gérald Bastard	Laboratorire Pierre Algrain-ENS, France
Fabio Beltram	Scuola Normale Superiore, Italy
Hélène Bouchiat	CNRS, Université Paris-Sud, France
Luis Brey	Consejo Superior de Investigaciones Cientificas, Spain
Leonard Brillson	The Ohio State University, USA
Federico Capasso	Harvard, USA
Yves J. Chabal	University of Texas at Dallas, USA
Jim Chelikowsky	University of Texas at Austin, USA
Sankar Das Sarma	University of Maryland, USA
Benoît Deveaud	Ecole Polytechnique Fédérale de Lausanne, Switzerland
Laurence Eaves	University of Nottingham, UK
Klaus Ensslin	ETH Zurich, Switzerland
Leo Esaki	The Science and Technology Promotion Foundation of Ibaraki, Japan
Leonard C. Feldman	Rutgers University, USA
Alfred Forchel	University of Wurzburg, Germany
Alfonso Franciosi	University of Trieste, Italy
Takashi Fukui	Hokkaido University, Japan
Zizhao Gan	Peking University, China
Mordehai Heiblum	Weizmann Institute, Israel
Chennupati Jagadish	The Australian National University, Australia
Erik Janzén	Linkoping University, Sweden

Young Dong Kim	Kyung Hee University, Korea
Belita Koiller	Federal University of Rio de Janeiro, Brazil
Jorg Kotthaus	Ludwig-Maximilians-Universität Munich, Germany
Leo Kouwenhoven	TU Delft, the Netherlands
Seung Joo Lee	Dongguk University, Korea
Young Hee Lee	Sungkyunkwan University, Korea
Daniel Loss	University of Basel, Switzerland
Yuri E. Lozovik	Institute for Spectroscopy, Moscow, Russia
Charles Marcus	University of Copenhagen, Denmark
Bruce D. McCombe	University of Buffalo, USA
Laurens Molenkamp	Würzburg University, Germany
Robin J. Nicholas	University of Oxford, UK
Hideo Ohno	Tohoku University,Japan
Mikael Ostling	Royal Institute of Technology (KTH), Sweden
Hiroyuki Sakaki	Toyota Technological Institute, Japan
Lars Samuelson	Lund University, Sweden
David Seiler	National Institute of Standards and Technology, USA
Pascale Senellart	CNRS, France
Xuechu Shen	Institute of Technical Physics, CAS, China
Seigo Tarucha	The University of Tokyo, Japan
Thomas Tiedje	University of Victoria, Canada
Daniel C. Tsui	Peking University, China
Klaus von Klitzing	Max Planck Institute for Solid State Research, Germany
Felix von Oppen	FU Berlin, Germany
Jianbai Xia	Institute of Semiconductors, CAS, China
Lu Yu	Institute of Physics, CAS, China
Youdou Zheng	Nanjing University, China

Program Committee

Chair:

Kai Chang (Chair)	Institute of Semiconductors, CAS, China
Li Lu (Co-chair)	Institute of Physics, CAS, China
Xincheng Xie (Co-chair)	Peking University, China

Members:

Jacqueline Bloch	LPN, CNRS, France
Guido Burkard	University of Konstanz, Germany
Irina Buyanova	Linkoping University, Sweden
Xi Chen	Tsinghua University, China
Kee Joo Chang	Korea Advanced Institute of Science and Technology, Korea
Mei-Yin Chou	Georgia Institute of Technology
Zhong Fang	Institute of Physics, CAS, China

IOP Conf. Series: Journal of Physics: Conf. Series **864** (2017) 011001 doi:10.1088/1742-6596/864/1/011001

Jerome Faist	ETH Zurich, Switzerland
Gwendal Feve	LPA, ENS, France
Silvano De Franceschi	CEA Grenoble, France
Ulf Gennser	CNRS, France
Bernard Gil	Université Montpellier 2, France
David Goldhaber-Gordon	Stanford University, USA
Vladimir Gritsenko	Institute of Semiconductor Physics, Siberian Branch of RAS, Russia
Guoping Guo	University of Science and Technology of China, Hefei, China
Alex Hamilton	The University of New South Wales, Australia
Rolf Haug	University of Hannover, Germany
Amr Helmy	University of Toronto, Canada
Yoshiro Hirayama	Tohoku University, Japan
Kazuhiko Hirakawa	University of Tokyo, Japan
Ru Huang	Peking University, China
Thomas Ihn	ETH Zurich, Switzerland
Koji Ishibashi	RIKEN, Japan
Frank Jahnke	Bremen University, Germany
Berend T. Jonker	Naval Research Laboratory, USA
Thomas Jungwirth	Institute of Physics ASCR Prague, Czech Republic& University of Nottingham, United Kingdom
Shingo Katsumoto	University of Tokyo, Japan
Gil-Ho Kim	Sungkyunkwan University, Korea
Jari Kinaret	Chalmers University of Technology, Sweden
Donghan Lee	Chungnam National University, Korea
Aristide Lemaitre	LPN, CNRS, France
Renbao Liu	The Chinese University of Hong Kong, Hong Kong
Roman M. Lutchyn	University of California at Santa Barbara, USA
Jochen Mannhart	Max Planck Institute for Solid State Research, Germany
Yuval Oreg	Weizmann Institute of Science, Israel
Francois Peeters	Universiteit Antwerpen, Belgium
Ravi Pillarisetty	Intel Corporation, USA
Vladimir Pudalov	P.N. Lebedev Physical Institute RAS, Russia
John Robertson	University of Cambridge, United Kingdom
Andrew Sachrajda	National Research Council of Canada, Canada
Dezhen Shen	Changchun Institute of Optics, Fine Mechanics and Physics, CAS, China
Shun-Qing Shen	University of Hong Kong, Hong Kong
Yi Shi	Nanjing University, China
Glenn Solomon	NIST, Gaithesburg, USA
Young-Woo Son	Korea Institute for Advanced Study, Korea
Lucia Sorba	SNS, Italy

IOP Conf. Series: Journal of Physics: Conf. Series **864** (2017) 011001 doi:10.1088/1742-6596/864/1/011001

Patrick Soukiassian	Commissariat à l'Energie Atomique et aux Energies Alternatives, Saclay, France
Vladislav Timofeev	Institute of Solid State Physics, RAS, Russia
Chris G. Van de Walle	University of California at Santa Barbara, USA
Luis Vina	Universidad Autónoma de Madrid, Spain
Jiannong Wang	The Hong Kong University of Science and Technology, Hong Kong
Lars-Erik Wernersson	Lund University, Sweden
Hiroshi Yamaguchi	NTT BRL, Japan
Alfredo Levy Yeyati	Autonomous University of Madrid, Spain
Dapeng Yu	Peking University, China
Uli Zulicke	Victoria University of Wellington, New Zealand

IUPAP Commission on Semiconductors (C8)

Chair	Michael Thewalt	Canada
Vice-Chair	Belita Koiller	Brazil
Secretary	Rolf Haug	Germany
Member	Qi-Kun Xue	China
Member	Pascale Senellart	France
Member	Young Dong Kim	Korea
Member	Amalia Patanè	United Kingdom
Member	Yasuhiko Arakawa	Japan
Member	Robert Suris	Russian Federation
Member	Per-Olof Holtz	Sweden
Member	Anna Cavallini	Italy
Member	Jacek Kossut	Poland
Member	Uli Zülicke	New Zealand
Member	Alan MacDonald	United States
Assoc. Mem.	Joël Cibert	France
Assoc. Mem.	Thaddeus Ladd	United States
Assoc. Mem.	Vladimir Kulakovskii	Russia

Organizing Committee

Honorary Chairs:

Houzhi Zheng	Institute of Semiconductors, CAS, China
Enge Wang	Peking University & CAS, China

Chair:

Hongqi Xu	Peking University, China

Co-chairs:

Qikun Xue	Tsinghua University, China
Shushen Li	Institute of Semiconductors, CAS, China

Secretary-General:

DaPeng Yu	Peking University, China

Vice Secretary-General:

IOP Conf. Series: Journal of Physics: Conf. Series **864** (2017) 011001 doi:10.1088/1742-6596/864/1/011001

Yang Ji	Institute of Semiconductors, CAS, China
Yingjie Xing	Peking University, China
Shaoyun Huang	Peking University, China

Local Committee Members:

Hongqi Xu	Peking University
Qikun Xue	Tsinghua University
Shushen Li	Institute of Semiconductors, CAS
Kai Chang	Institute of Semiconductors, CAS
Li Lu	Institute of Physics, CAS
XinCheng Xie	Peking University
Yun Jiang	Peking University
Hui Cai	Peking University
Dapeng Yu	Peking University
Yang Ji	Institute of Semiconductors, CAS
Yingjie Xing	Peking University
Shaoyun Huang	Peking University
Jian Liu	Institute of Semiconductors, CAS
Ning Kang	Peking University
Jingyun Wang	Peking University
Linhui Ye	Peking University
Zhimin Liao	Peking University
Chengjie Xing	Peking University
Yan Gao	Peking University

Plenary Speakers

Yasuhiko Arakawa	University of Tokyo, Japan
David D. Awschalom	University of Chicago, USA
Jiangfeng Du	University of Science and Technology of China, China
Andre Geim	University of Manchester, UK
Tony Heinz	Stanford University & Columbia University, USA
Charles Marcus	University of Copenhagen, Denmark
Sang Il Seok	Ulsan National Institute of Science and Technology, Korea
Michelle Y. Simmons	University of New South Wales, Australia
Samuel D. Stranks	Massachusetts Institute of Technology, USA
Kimberly Dick Thelander	Lund University, Sweden
Ali Yazdani	Princeton University, USA

Invited Speakers

Marc-Alexander Assmann	Technische Universität Dortmund, Germany
Georgy Astakhov	University,of,Würzburg, Germany

Mete Atatüre	University of Cambridge, UK
Adrian Bachtold	Institute of Photonic Sciences, Spain
Christopher Bauerle	CNRS, Institute NEEL, France
Sagar Bhandari	Harvard University, USA
Hartmut Buhmann	University of Würzburg, Germany
Guillaume Cassabois	University of Montpellier, France
Kookrin Char	Seoul National University, Korea
Evgeny Chekhovich	University of Sheffield, UK
Weimin M. Chen	Linköping University, Sweden
Yulin Chen	Tsinghua University & ShanghaiTech University, China
Zhanghai Chen	Fudan University, China
Hyeonsik Cheong	Sogang University, Korea
Xi Dai	Institute of Physics, CAS, China
Tomasz Dietl	Institute of Physics, Polish Academy of Sciences, Poland
Hong Ding	Institute of Physics, CAS, China
Rui-rui Du	Rice University/Peking University, USA/China
A. S. Dzurak	University of New South Wales, Australia
Laurence Eaves	University of Nottingham, UK
Saskia F. Fischer	Humboldt-Universitaet zu Berlin, Germany
Akira Fujiwara	NTT Basic Research Laboratories, Japan
Attila Geresdi	TU Delft, the Netherlands
David Gershoni	Israel Institute of Technology, Israel
Katsushi Hashimoto	Tohoku University, Japan
Simon Huppert	University Paris Diderot, France
Chennupati Jagadish	Australian National University, Australia
Jinfeng Jia	Shanghai Jiaotong University, China
Masashi Kawasaki	RIKEN, Japan
Keun Su Kim	Pohang University of Science and Technology, Korea
Philip Kim	Harvard University, USA
Takis Kontos	Ecole Normale Supérieure, France
Sergey Krishtopenko	University of Montpellier, France
Loic Lanco	LPN, CNRS, France
Snezana Lazic	Universidad Autónoma de Madrid, Spain
Young Hee Lee	leeyoung@skku.edu
Xian-Bin Li	Jilin University, China
Xiaoqin Li	University of Texas at Austin, USA
Daniel Loss	University of Basel, Switzerland
Steve G. Louie	University of California at Berkeley, USA
Chao-Yang Lu	University of Science and Technology of China,

IOP Conf. Series: Journal of Physics: Conf. Series **864** (2017) 011001 doi:10.1088/1742-6596/864/1/011001

	China
Stefan Ludwig	Ludwig-Maximilians-University München, Germany
Anders Mikkelsen	Lund University, Sweden
Shuji Nakamura	UC Santa Barbara, USA
Junsaku Nitta	Tohoku University, Japan
Masahiro Nomura	The University of Tokyo, Japan
Akira Oiwa	Osaka University, Japan
Wei Pan	Sandia National Laboratories, USA
Jason Petta	Princeton University, USA
Frédéric Pierre	CNRS, France
Heike E. Riel	IBM Zurich Research Laboratory, Switzerland
Jay Deep Sau	University of Maryland, USA
Christian Schönenberger	University of Basel, Switzerland
Leigh Smith	University of Cincinnati, USA
Ady Stern	Weizmann Institute, Israel
Seigo Tarucha	University of Tokyo, Japan
Felix von Oppen	Freie University Berlin, Germany
Timo Wagner	Leibniz University, Germany
Jian Wang	Peking University, China
Xinran Wang	Nanjing University, China
Yayu Wang	Tsinghua University, China
Richard J. Warburton	University of Basel, Switzerland
Claude Weisbuch	UC Santa Barbra, USA
Dieter Weiss	University of Regensburg, Germany
Jonas Wiedenmann	University of Würzburg, Germany
Wang Yao	University of Hong Kong, China
Peide D Ye	Purdue University, USA
Yuanbo Zhang	Fudan University, China
Jianhua Zhao	Institute of Semiconductors, CAS, China
Oded Zilberberg	ETH Zurich, Switzerland
Valery Zwiller	Royal Institute of Technology (KTH), Sweden

Peer review statement

All papers published in this volume of *IOP Conference Series: Materials Science and Engineering* have been peer reviewed through processes administered by the proceedings Editors. Reviews were conducted by expert referees to the professional and scientific standards expected of a proceedings journal published by IOP Publishing.

InAs/GaSb core-shell nanowires grown on Si substrates by metal-organic chemical vapor deposition

Xianghai Ji [‡], **Xiaoguang Yang** [‡], **Wenna Du** [‡], **Huayong Pan**[§], **Shuai Luo** [‡], **Haiming Ji**[‡], **Hongqi Xu**[§] **and Tao Yang** [‡]

[‡]Institute of Semiconductors, Chinese Academy of Sciences, Beijing 100083, China
[§]Key Laboratory for the Physics and Chemistry of Nanodevices and Department of Electronics, Peking University, Beijing 100871, China

E-mail: tyang@semi.ac.cn

Abstract. We report the growth of InAs/GaSb core-shell heterostructure nanowires with smooth sidewalls on Si substrates using metal-organic chemical vapor deposition (MOCVD) with no assistance from foreign catalysts. Sb adatoms were observed to strongly influence the morphology of the GaSb shell. In particular, Ga droplets form on the nanowire tips when a relatively low TMSb flow rate is used, whereas the droplets are missing and the radial growth of the GaSb is enhanced due to a reduction in the diffusion length of the Ga adatoms when the TMSb flow rate is increased. Moreover, transmission electron microscopy measurements revealed that the GaSb shell coherently grew on the InAs core without any misfit dislocations.

1. Introduction

Among the III-V semiconductor materials, InAs and GaSb are direct band gap compounds with extremely high electron and hole mobility, respectively. In addition, a low lattice mismatch of 0.6% and a unique type-II-broken band alignment between InAs and GaSb make this heterostructure system attractive. Based on one-dimensional nanowire geometry, InAs/GaSb core-shell nanowires have been used in many applications such as tunneling-based devices [1], ambipolar field-effect transistors [2,3], research on electron-hole hybridization [4].

Thus far, InAs/GaSb or GaSb/InAs core-shell nanowires have been mainly grown with the assistance of Au catalysts [2-6]. However, the introduction of Au may degrade the electronic properties of nanowires [7]. Therefore, it is highly desirable to develop the growth of nanowires without the use of any foreign catalysts. In this paper, we present the growth of InAs/GaSb core-shell nanowires on Si(111) substrates using metal-organic chemical vapor deposition (MOCVD) with no assistance from foreign catalysts. In particular, the influence of Sb on the growth of the GaSb shell is investigated in detail. Finally, a homogeneous GaSb shell around the InAs core is achieved through careful control of the growth conditions.

2. Results and discussions

Figure 1a shows a representative SEM image of bare InAs nanowires. The nanowires were observed to be vertically grown on the substrate without tapering. To investigate the influence of Sb adatoms on the GaSb shell growth, a series of InAs/GaSb core-shell nanowires were grown using different TMSb flow rates of 1.3×10^{-6} mol/min, 2.6×10^{-6} mol/min and 3.9×10^{-6} mol/min, respectively. However, the TMGa flow rate was fixed at 0.25×10^{-6} mol/min. The growth time of the GaSb shell was 15 min. The majority of the nanowires in Figure 1b have Ga droplets on their tips. However, the situation changed when the TMSb flow rate was increased to 2.6×10^{-6} mol/min. The droplets were missing, and a uniform GaSb shell around the InAs core was achieved, as shown in Figure 1c. Moreover, the radial growth of GaSb was clearly observed to be enhanced when the TMSb flow rate was increased from 1.3×10^{-6} mol/min to 2.6×10^{-6} mol/min. When the TMSb flow rate was further increased to 3.9×10^{-6} mol/min, the sidewalls of the nanowires became rough.

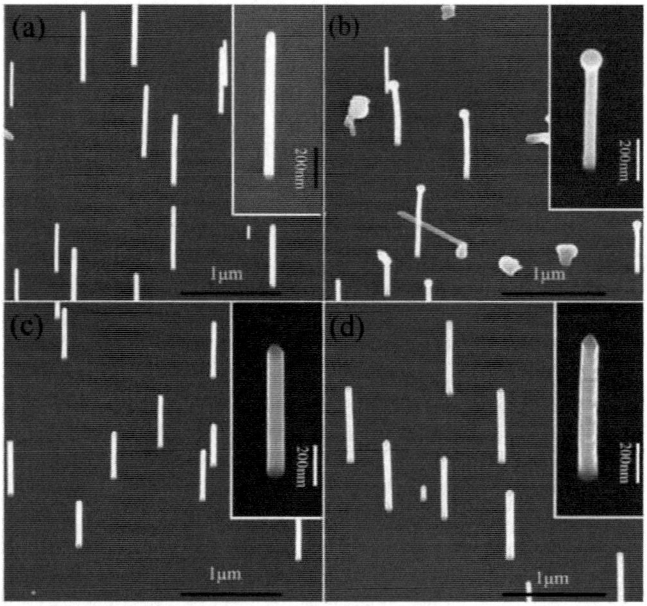

Figure 1. 30°-tilted SEM images of the bare InAs and InAs/GaSb core-shell nanowires: (a) InAs core nanowires, (b-d) InAs/GaSb core-shell nanowires with the TMSb flow rate of (b)1.3×10^{-6}mol/min, (c)2.6×10^{-6}mol/min, (d)3.9×10^{-6}mol/min. The TMGa flow rate was kept constant at 0.25×10^{-6}mol/min. Insets in (a-d) show higher-magnification SEM images.

Generally, Ga atoms are collected by the droplet via three pathways [8]: (i) direct impingement of Ga atoms on the droplet; (ii) adsorption of Ga atoms onto the nanowire side facets and subsequent diffusion to the nanowire tip; (iii) adsorption of Ga atoms onto the substrate surface and subsequent diffusion to the nanowire tip along the nanowire sidewalls. Because the growth temperature and the TMGa flow rate were unchanged for the growth of the GaSb shell in Figures 1b-d, the Ga atoms from pathway(i) should be the same for all cases. Additionally, we assume that the contribution in this manner is weak because of the small surface area of the droplet. In terms of the contribution from pathway(iii), Ga adatoms can hardly diffuse from the substrate surface to the tip of nanowires because of the relatively low growth temperature of the GaSb shell (440°C) [9]. Moreover, the GaSb shell without tapering in Figure 1c also indicates that the Ga adatoms that diffused from the substrate surface to the sidewalls of nanowires can be ignored. Therefore, the main contribution to the droplet should be from pathway(ii). Furthermore, we speculate that the missing of Ga droplets is a consequence of the reduction in the surface diffusion length of the Ga adatoms. Specifically, Sb, which has a heavier atomic mass and lower volatility than As and P, is easier to adsorb onto the nanowire side facets. To move, Ga adatoms must break their temporary bonds with the adsorbed Sb. Eventually, the surface mobility of the Ga adatoms on the nanowire side facets is suppressed kinetically, and the diffusion length of the Ga adatoms is gradually reduced under increasing TMSb flow rate. Therefore, the Ga droplets cannot be formed when the diffusion length of the Ga adatoms is reduced to a certain degree, resulting in enhancement of the radial growth of the GaSb via vapor-solid (VS) mechanism, as schematically illustrated in Figure 2. Additionally, when the TMSb flow rate was further increased to 3.9×10^{-6}mol/min, the diffusion length of the Ga adatoms was so short that the nucleation of GaSb was not homogeneous on the whole InAs sidewalls. Therefore, to obtain good morphology of the GaSb shell, the TMSb flow rate should be carefully controlled.

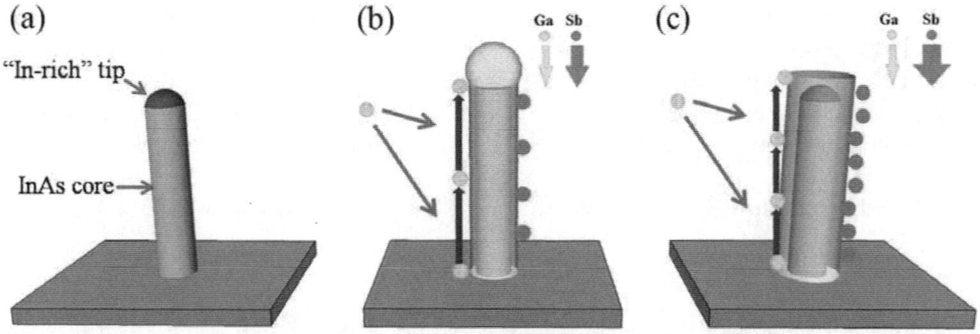

Figure 2. Schematic of the effect of Sb on the morphology of the InAs/GaSb core-shell nanowire:(a) bare InAs nanowire, (b) with a small amount of Sb, and (c) with relatively more Sb.

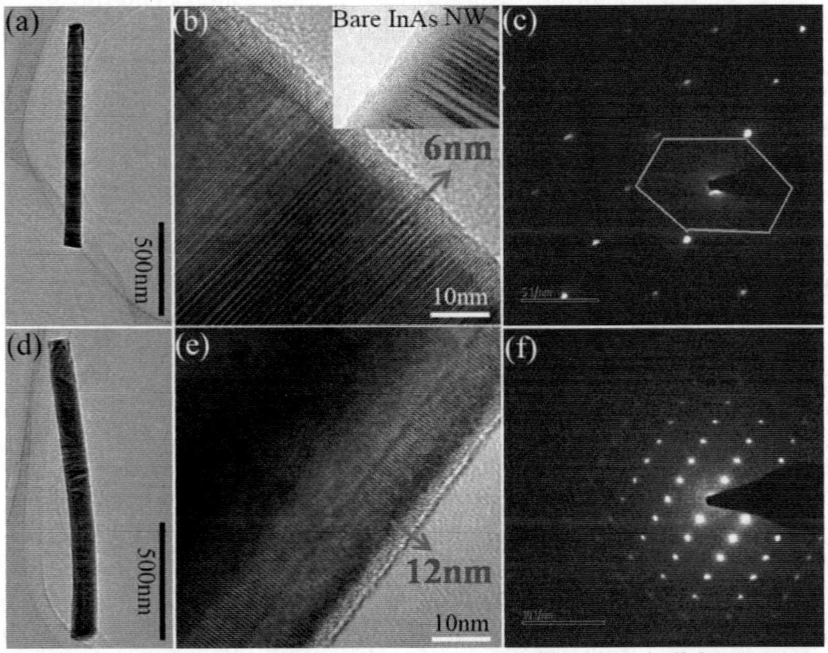

Figure 3. TEM images and SAED patterns of the InAs/GaSb core-shell heterostructure nanowires with the shell growth times of (a-c) 15 min, (d-f) 30 min, respectively. (a, d) Low-magnification TEM image of the corresponding nanowire. (b, c) High-resolution TEM image and corresponding SAED pattern of the nanowire shown in (a) viewed along the <110> zone axis. (e, f) High-resolution TEM image and corresponding SAED pattern of the nanowire shown in (d) viewed along the <211> zone axis. The inset in (b) shows a typical high-resolution TEM image of a bare InAs nanowire.

Detailed TEM investigations are shown in Figure 3. The TMSb and TMGa flow rates were of 2.6×10^{-6}mol/min and 0.25×10^{-6}mol/min, respectively; while the growth times of the GaSb shells investigated in Figures 3a-c and d-f were varied. Figure 3a shows a low-resolution TEM image of the nanowire with the shell growth time of 15 min. A corresponding high-resolution (HR) TEM image viewed from the <110> zone axis was shown in Figure 3b. A GaSb shell, approximately 6 nm in thickness, was observed to be radially grown around the InAs core. The nanowire is wurtzite (WZ)-

dominated and contains a large number of planar defects along its growth direction, which is very common in self-catalyzed InAs nanowires, as shown in the inset of Figure 3b. We speculate that these planar defects were recorded from the InAs core to the GaSb shell. In fact, such a phenomenon has been observed in core-shell nanowires of other material systems [10,11]. This behavior highlights the need to improve the crystal quality of the self-catalyzed InAs core nanowires.We note that, in Figures 3d and e, the TEM images were acquired from the <211> zone axis and the growth time of the GaSb shell was extended to 30 min. Planar defects cannot be observed from the <211> zone axis [12]. However, because the sidewalls of the InAs core are six {110}-family facets, this direction can always provide a clear contrast between the InAs core and the GaSb shell. Moreover, no Moiré fringes are visible in Figure 3e, which means that no plastic relaxation occurred. This lack of plastic relaxation is attributed to the relatively thin GaSb shell and the small lattice difference between the InAs and GaSb.

3. Conclusion

In summary, we have demonstrated the growth of InAs/GaSb core-shell heterostructure nanowires on Si substrates using MOCVD. The impact of Sb adatoms on the growth behavior of the GaSb shell is investigated in detail. Ga droplets form on the tips of nanowires but without visible axial GaSb growth under a relatively low TMSb flow rate. However, the droplets were missing and the radial growth of GaSb was enhanced with increasing the TMSb flow rate because of the reduction in the diffusion length of the Ga adatoms. Detailed TEM analyses revealed that the overgrowth of the GaSb shell was highly uniform and coherent with the InAs core without misfit dislocations. The results presented here may provide a new understanding of the growth of other antimonide nanowires. Furthermore, the as-grown InAs/GaSb core-shell nanowires on Si substrates may introduce new possibilities for applications in future nanowire-based devices

References

[1] Dey A W, Svensson J, Ek M, Lind E, Thelander C and Wernersson L E 2013 *Nano Lett.* **13** 5919
[2] Ganjipour B, Ek M, Mattias B B, Dick K A, Pistol M E, Wernersson L E and Thelander C 2012 *Appl. Phys. Lett.* **101** 103501
[3] Ganjipour B, Sepehri S, Dey A W, Tizno O, Borg B M, Dick K A, Samuelson L, Wernersson L E and Thelander C 2014 *Nanotechnology* **25** 425201
[4] Ganjipour B, Leijnse M, Samuelson L, Xu H Q and Thelander C 2015 *Phys. Rev. B* **91** 161301
[5] Jan G G, Martin L, Bahram G and Kimberly A D 2015 *Acs Nano* **9** 7033
[6] Namazi L, Nilsson M, Lehmann S, Thelander C and Dick K A 2015 *Nanoscale* **7** 10472
[7] Wang X Y, Du W N, Yang X G, Zhang X W and Yang T 2015 *J. Cryst. Growth* **426** 287
[8] Ye H, Yu Z Y, Kodambaka S and Shenoy V B 2012 *Appl. Phys. Lett.* **10** 263103
[9] Wu J, Borg B M, Jacobsson D, Dick K A and Wernersson L E 2013 *J. Cryst. Growth* **383** 158
[10] Ghalamestani S G, Munshi A M, Dheeraj D L, Fimland B O, Weman H and Dick K A 2013 *Nanotechnology* **24** 405601
[11] Kawaguchi K, Heurlin M, Lindgren D, Borgstrom M T, Ek M and Samuelson L 2011 *Appl. Phys. Lett.* **99** 131915
[12] Johansson J, Karlsson L S, Svensson C P T, Martensson T, Wacaser B A, Deppert K, Samuelson L and Seifert W 2006 *Nat. Mater.* **5** 574

Acknowledgments

This work was financially supported by the Ministry of Science and Technology of China (Grant No.2012CB932701) and National Natural Science Foundation of China (Grant No. 91433206).

Leakage currents mechanism in thin films of ferroelectric $Hf_{0.5}Zr_{0.5}O_2$

Damir R Islamov[1,2], A G Chernikova[3], M G Kozodaev[3], A M Markeev[3], T V Perevalov[1,2], V A Gritsenko[1,2] and O M Orlov[4]

[1] Rzhanov Institute of Semiconductor Physics SB RAS, Novosibirsk 630090, Russian Federation

[2] Novosibirsk State University, Novosibirsk 630090, Russian Federation

[3] Moscow Institute of Physics and Technology, Dolgoprudniy 141700, Russian Federation

[4] JSC Molecular Electronics Research Institute, Zelenograd, Moscow 124460, Russian Federation

E-mail: damir@isp.nsc.ru

Abstract. We study the charge transport mechanism in ferroelectric $Hf_{0.5}Zr_{0.5}O_2$ thin films. Transport properties of $Hf_{0.5}Zr_{0.5}O_2$ are described by phonon-assisted tunnelling between traps. Comparison with transport properties of amorphous $Hf_{0.5}Zr_{0.5}O_2$ demonstrates that the transport mechanism does not depend on the structure. The thermal and optical trap energies 1.25 eV and 2.5 eV, respectively, in $Hf_{0.5}Zr_{0.5}O_2$ were determined based on comparison of experimentally measured data on transport with simulations within phonon-assisted tunnelling between traps. We found that the trap density in ferroelectric $Hf_{0.5}Zr_{0.5}O_2$ is slightly less than one in amorphous $Hf_{0.5}Zr_{0.5}O_2$. A hypothesis that oxygen vacancies are responsible for the charge transport in $Hf_{0.5}Zr_{0.5}O_2$ is confirmed by *ab initio* simulation of electronic structure.

1. Introduction

Recently ferroelectricity was demonstrated in thin of the solid solution $Hf_{0.5}Zr_{0.5}O_2$ (HZO) [1–4]. Ferroelectricity in these films is associated with the ability to stabilize non-centrosymmetric orthorhombic phase $Pbc2_1$ [2]. These materials have a number of advantages over conventional ferroelectric regarding compatibility with technological processes used in microelectronics, and has already demonstrated their ability to provide very high element density. Taking into account advantages of the ferroelectric random access memory (non-volatile, high-speed performance, high number of switching cycles) the discovery of ferroelectric effect in hafnia-based materials gave an impetus for development of universal memory concept, which may lead to a significant breakthrough in the development of memory devices.

An unsolved problem in the way of development of FeRAM-based universal memory is the spontaneous depolarization of the active medium, which leads to short data storage time (retention) [5]. One of the possible reason of spontaneous depolarization is leakage currents through the ferroelectric thin films. High-κ HZO thin films ($\kappa > 16$) can be used as sub-gate dielectrics in MOSFET and FinFET transistors instead of SiO_2. Thus, it is very important to know charge transport mechanisms in thin dielectrics films to control the leakage currents for development of high quality memory devices. Purpose of the present work is to study charge transport mechanism in ferroelectric HZO.

IOP Conf. Series: Journal of Physics: Conf. Series **864** (2017) 012002 doi:10.1088/1742-6596/864/1/012002

2. Experiment details

Transport measurements were performed for TiN/HZO/Pt structures. Test structure were fabricated with ALD technique. The TiN layer (10 nm) was deposited on oxidized Si substrate. Then the 10-nm-thick HZO films we deposited at 240 °C from TEMAHf, TEMAZr, and H_2O precursors. The thickness and stoichiometry films were confirmed by laser ellipsometry and Rutherford backscattering spectroscopy. A part of samples was treated by rapid thermal annealing *RTA) 400 °C in N_2 atmosphere. The crystalline structures of the films were examined by symmetrical X-Ray diffraction (XRD) using ARL X'TRA tool (Thermo Scientific) utilizing Cu$K\alpha$ radiation. Top Pt electrodes were deposited through a shadow mask with round holes (area is $7.1 \times 10^4\ \mu m^2$) by electron beam evaporation on HZO films.

The DC current-voltage (I-V) curves were measured using Agilent B1500A Semiconductor device parameter analyser. Polarization-voltage (P-V) curves were obtained by integration of displacement current response in reply to the stepped triangle voltage. The leakage current was subtracted from the current response in order to eliminate the polarization and displacement current contributions.

The electronic structure of HZO was calculated within the spin polarized density functional theory using the *ab initio* simulation program package Quantum ESPRESSO with B3LYP functional. The oxygen vacancy (V_O) was generated by the removal of an O atom, followed by relaxation of remaining atoms in 96-atom supercell.

3. Results and discussions

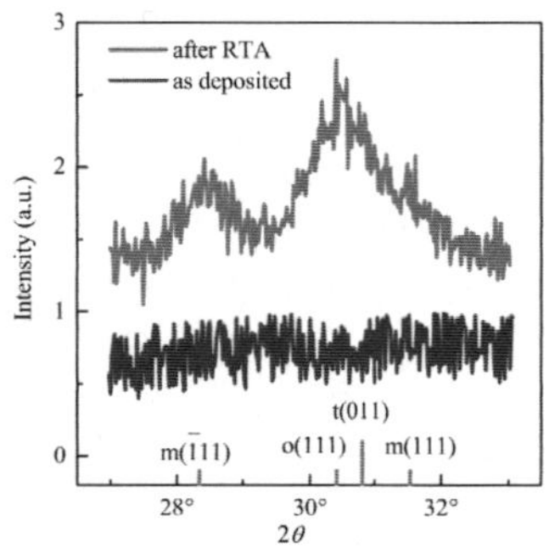

Figure 1. XRD spectra of as deposited and after RTA HZO.

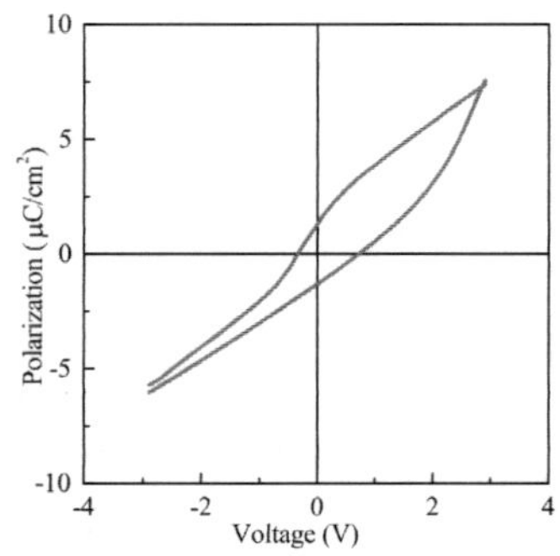

Figure 2. Polarization-voltage characteristics for TiN/f-HZO/Pt structures.

The XRD spectra of as deposited and annealed films are presented in Fig. 1. One can see that as deposited films have amorphous structure (a-HZO), while annealed films are polycrystalline with monoclinic, tetragonal, and orthorhombic phases. The presence of ferroelectricity of after-RTA-films is confirmed by observing characteristic hysteresis on the P-V plate (Fig. 2).

Figure 3 shows results of *ab initio* calculations performed for four types of oxygen vacancy in orthorhombic HZO. The calculated partial density of states (PDOS) of oxygen, hafnium and zirconium atoms for all types of neutral oxygen vacancies V_O in HZO are shown in Figures 4(a)–(d). The ZrZr=V_O–Hf and HfHf=V_O–Zr defects represent threefold oxygen vacancies with 2 Zr+Hf and 2 Hf+Zr metal atoms around, while HfHf=V_O=ZrZr and HfZr=V_O=HfZr defects represent fourfold oxygen vacancies with different spatial arrangement of Hf and Zr atoms around the vacancy. The schematic spatial configurations of the defects are shown in the insets of the Fig. 4(a)–(d)). One can see that the electronic structures of the different-type oxygen vacancies in HZO are similar: threefold and fourfold coordinated vacancies form filled states in the band gap of 3.4 eV and 2.7 eV above the valence band, respectively. The nearest metal atoms to

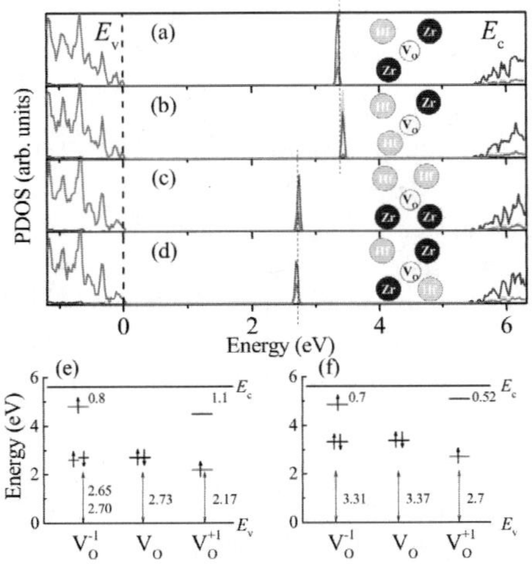

Figure 3. The calculated partial density of states (PDOS) of oxygen, hafnium and zirconium atoms for all types of neutral oxygen vacancy V_O in HZO: (a) $ZrZr{=}V_O{-}Hf$, (b) $HfHf{=}V_O{-}Zr$, (c) $HfHf{=}V_O{=}ZrZr$, (d) $HfZr{=}V_O{=}HfZr$. The energy levels of $HfHf{=}V_O{=}ZrZr$-type (e) and $HfHf{=}V_O{-}Zr$-type (f) oxygen vacancy in different charge states: $V_O{}^{-1}$ is negatively charged state, $V_O{}^{+1}$ is positively charged state. The spins of localized on the defect electrons are shown with arrows. The distance between energy levels and the top of the valence band (in eV) are shown by bilateral arrows. The reference point is the top of valence band E_v.

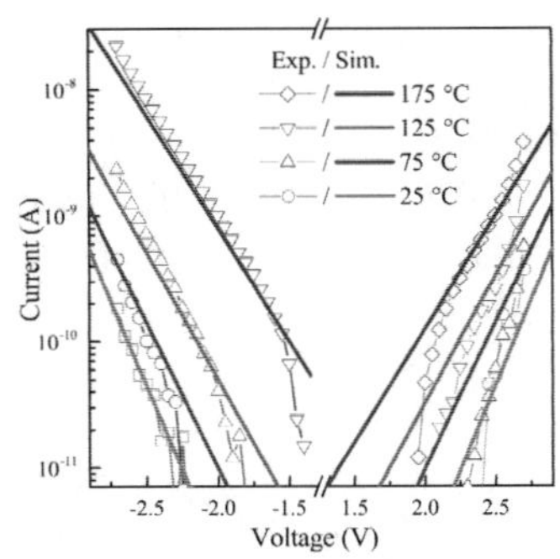

Figure 4. Experimental (characters) current-voltage characteristics and simulations (lines) for TiN/f-HZO/Pt structures at different temperatures.

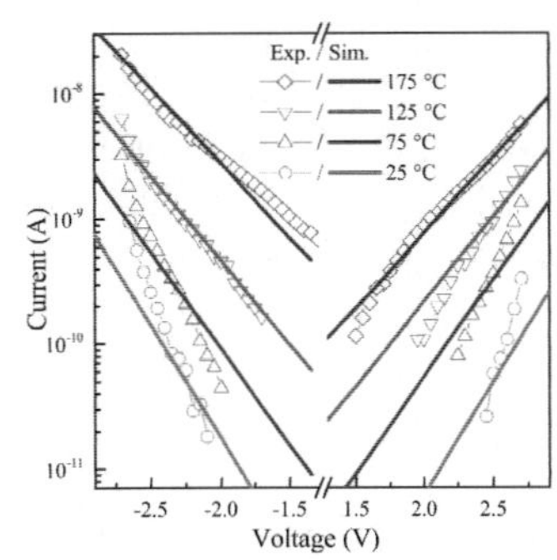

Figure 5. Experimental (characters) current-voltage characteristics and simulations (lines) for TiN/a-HZO/Pt structures at different temperatures.

the oxygen vacancy give the most contribution to the defect states as shown in Fig. 4(a)–(d). The next most important contribution to the defect state is given by the nearest to the oxygen vacancy atoms. Calculated Kohn-Sham energy levels for the oxygen vacancies in different charge states are shown in Figs. 4(e)–(f). We can see that a trapped on the defect electron has an energy of $\sim 1\,\mathrm{eV}$ below E_c. Trapping a hole leads to a descent of the energy level compared to neutral oxygen vacancy. Thus, we conclude that the defect states in the band gap are localized, and the oxygen vacancies can act as a trap for electrons, as well as for holes, being involved in the charge transport.

Despite that *ab initio* calculations predict amphoteric nature of traps, the capacitance-voltage measurements on trapped charge accumulation demonstrate that mostly holes are involving in transport processes [6]. Thus, further analysis we will perform in terms of monopolar hole conductivity.

Figure 4 shows experimental current-voltage characteristics of leakage current contribution in f-HZO at different temperatures T on semi-log I–V–T plate. The current depends on the voltage and temperature exponentially. The Frenkel model [7], can describe experimental I–V–T dependencies using underestimated value of frequency factor. Following by [6,8], we analysed experimentally measured I–V–T in terms of phonon-assisted tunnelling between traps [9]:

$$J = \frac{e}{s^2}\gamma(1-\gamma)\frac{\sqrt{2\pi}\hbar W}{m^* s^2\sqrt{(W_{\rm opt}-W_{\rm t})kT}}\exp\left(-\frac{W_{\rm opt}-W_{\rm t}}{2kT}\right)\exp\left(-\frac{2s\sqrt{2m^*W_{\rm t}}}{\hbar}\right)\sinh\left(\frac{eFs}{2kT}\right), \qquad (1)$$

where J is the current density, e is the elementary charge, s is mean distance between traps, γ is trap filling factor, $\hbar$ is the Planck constant, $W_{\rm t}$ is thermal trap energy, $W_{\rm opt}$ is optical trap energy, m^* is the effective mass, k is the Boltzmann constant, F is the electric field in the dielectric. One can see that the current-voltage characteristics (1) exhibits a maximum at $\gamma = 0.5$. Phonon-assisted tunnelling between traps (lines on Fig. 4) describes experimental I–V–T curves with very good quantitative agreement using thermal and optical trap energies $W_{\rm t} = 1.25\,{\rm eV}$ and $W_{\rm opt} = 2.5\,{\rm eV}$, respectively, $m^* = 0.21m_0$ (hereby m_0 is the free electron mass), trap density $N = s^{-3} = 3\times10^{19}\,{\rm cm}^{-3}$.

The same procedure was applied for MOS structures with amorphous HZO. Figure 5 represents comparison of experimentally measured current-voltage characteristics for a-HZO at different temperatures with calculated within (1). The best agreement of calculated curves with experimental data was got with parameter values: $W_{\rm t} = 1.25\,{\rm eV}$, $W_{\rm opt} = 2.5\,{\rm eV}$, $m^* = 0.3m_0$, $N = 1\times10^{20}\,{\rm cm}^{-3}$. It should be noted that obtained values of trap energies are the same as for amorphous a-$Hf_xZr_{1-x}O_y$, synthesized by different techniques [6].

Since thermal and optical energies for amorphous and ferroelectric HZO films do not depend on crystal phase, one can assume that the traps in different phases of these films have the same nature. Recently it was shown that the oxygen vacancies act as the charge traps in hafnium oxide [8]. At the same time, thermal and optical trap energies in HfO_2 have the same values. Taking into account, that the oxygen vacancies in HfO_2 and ZrO_2 have the same electronic structure, the fact that HZO is the solid solution of these oxides, one can assume that the oxygen vacancies act as the charge traps in a-HZO and f-HZO. Thus, we conclude that the leakage currents mechanism in HZO does not depend on crystal structure of HZO films, and it is phonon-assisted tunnelling between oxygen vacancies. The only difference is charge trap density, which is lower in f-HZO. This difference might be caused by RTA treatment during film preparation.

4. Summary

In conclusion, we demonstrated that the charge transport of amorphous and ferroelectric HZO films is described by phonon-assisted tunnelling between traps. The thermal and optical trap energies in HZO were revealed as 1.25 eV and 2.5 eV, respectively. It was concluded that the traps are oxygen vacancies. A hypothesis that oxygen vacancies are responsible for the charge transport in HZO is confirmed by electronic structure *ab initio* simulation.

Acknowledgments

The work was supported by the Ministry of Education and Science of the Russian Federation (project #RFMEFI57614X0065) (synthesis of HZO films) and the Russian Science Foundation (project #14-19-00192) (transport measurements and simulation).

References

[1] Müller J *et al* 2011 *Appl. Phys. Lett.* **99** 112901
[2] Müller J *et al* 2012 *Nano Lett.* **12**(8) 4318–4323
[3] Hyuk Park M *et al* 2013 *Appl. Phys. Lett.* **102** 112914
[4] Chernikova A *et al* 2015 *Microelectron. Eng.* **147** 15–18
[5] Cheng C H and Chin A 2014 *IEEE Electron Device Lett.* **35** 138–140
[6] Islamov D R *et al* 2015 *Appl. Phys. Lett.* **106**(10) 102906 (*Preprint* 1501.02370)
[7] Frenkel J 1938 *Tech. Phys. USSR* **5**(9) 685–695
[8] Islamov D R, Gritsenko V A, Cheng C H and Chin A 2014 *Appl. Phys. Lett.* **105** 222901 (*Preprint* 1409.6887)
[9] Nasyrov K A and Gritsenko V A 2011 *JETP* **112** 1026–1034 ISSN 1063-7761

IOP Conf. Series: Journal of Physics: Conf. Series **864** (2017) 012003 doi:10.1088/1742-6596/864/1/012003

Mechanism of charge transport of stress induced leakage current and trap nature in thermal oxide on silicon

Damir R Islamov[1,2], **V A Gritsenko**[1,2], **T V Perevalov**[1,2], **O M Orlov**[3] **and G Ya Krasnikov**[3]

[1] Rzhanov Institute of Semiconductor Physics SB RAS, Novosibirsk 630090, Russian Federation

[2] Novosibirsk State University, Novosibirsk 630090, Russian Federation

[3] JSC Molecular Electronics Research Institute, Zelenograd, Moscow 124460, Russian Federation

E-mail: `damir@isp.nsc.ru`

Abstract. We study the charge transport mechanism of electron via traps in thermal SiO_2 on silicon. Electron transport is limited by phonon-assisted tunnelling between traps. Charge flowing leads to oxygen vacancies generation, and the leakage current increases. Long-time annealing at high temperatures decreased the leakage current to initial values due to oxygen vacancies recombination with interstitial oxygen. Taking into account results of *ab initio* simulations, we found that the oxygen vacancies act as electron traps in SiO_2.

1. Introduction

Write/erase operations of flash memory with *poly*-Si-floating gates are driven by tunnelling injections of electrons and holes through thin 1.8–8.0 nm (tunnel) SiO_2 layer [1, 2]. The current through the tunnel oxide during erase/rewrite operations by applying high electric field $\sim$ 10 MV/cm in the oxide increases the conductivity of the layer at low electric fields $\sim$ 1 MV/cm. This is caused by the appearance of additional current components through the tunnel oxide compared to the current flowing in a strong electric field on the mechanism of Fowler-Nordheim [3]. This phenomenon is called Stress Induced Leakage Current, SILC. SILC limits reprogramming cycles number of the flash memory cells based on floating gate and the *poly*-silicon to 10^4–10^5. SILC also leads to accelerate the de-charging of the flash memory elements in the data storage mode, i.e., leads to memory characteristics degradation. In spite of SILC intensive study, the nature of this phenomenon is still a debatable question [2, 4–6]. The purposes of the present study are identification of the electron trap ionization and transport mechanisms in SiO_2 in external electric field and determination of the trap parameters (ionization energy, concentration).

2. Experiment details

SILC and transport measurements were performed for test structures based on FET transistors with n^+-type *poly*-Si floating gate, manufactured using the 180-nm design rule technology. The

p-Si substrate with n^+-type channel (n^+-Si$_p$) was used as the bottom contact. The n^+-Si floating gate was used as the top contact. The 7.5-nm-thick SiO_2 was used as tunnel layers between contacts. Test samples were formed by 16 parallel-connected n^+-Si/SiO$_2$/n^+-Si$_p$ capacitors with the total area of the *poly*-Si electrode $S = 8 \times 10^4\,\mu m^2$. Current-voltage ($J$–$V$) characteristics were measured at different temperatures 25–70°C before and after charge stress $\Sigma = (0.01$–$10)\,C/cm^2$. The silicon electrode resistivity is much less that the one of SiO_2 layer, and the most part of the electric field falls on the oxide layer.

The electronic structure of oxygen vacancy in α-SiO$_2$ was calculated within the spin polarized density functional theory using the *ab initio* simulation code QUANTUM ESPRESSO with B3LYP functional [7]. The oxygen vacancy was generated by the removal of an O atom, followed by relaxation of remaining atoms in 72-atom supercell. This method yields the α-SiO$_2$ band gap of 8.0 eV.

3. Results and discussions

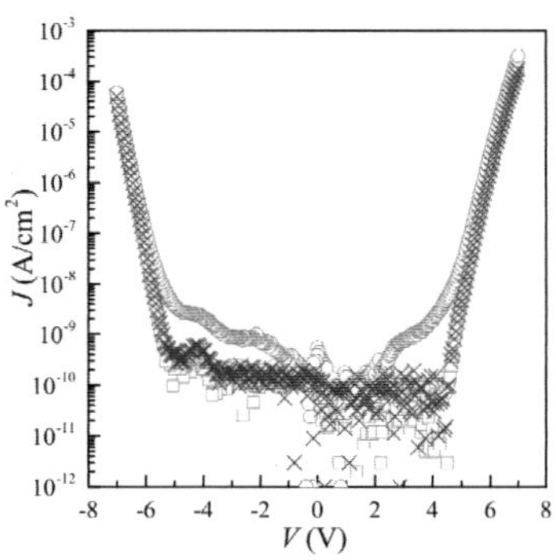

Figure 1. Experimental current-voltage characteristics of n^+-Si/SiO$_2$/n^+-Si$_p$ structures at room temperature (25°C). Boxes (□) present results of measurements for "as deposited" samples before stress and annealing, rounds (○) — after stress and before annealing, crosses (×) represent data for stressed samples after long time annealing.

Experimental current-voltage characteristics of the fresh (as deposited) SiO_2 are shown in Fig. 1 by boxes. The high-field ($|V| > 5\,V$, $F > 6\,MV/cm$) current through the structure was limited by Fowler-Nordheim tunnelling

$$J_{FN} = \frac{e^3F^2}{8\pi h\Phi_0}\exp\left(-\frac{8\pi\sqrt{2m^*}}{3ehF}\Phi_0^{3/2}\right), \tag{1}$$

where J_{FN} is the tunnel current density, e is the elementary charge, F is the electric field, h is the Planck constant, Φ_0 is the height of the triangular potential barrier for electrons at the Si/SiO$_2$ interface, m^* is the electron effective tunnelling mass [3]. According to photoemission measurements, the height of the triangular potential barrier for electrons at the Si/SiO$_2$ interface is $\Phi = 3.14\,eV$ [8]. In high fields at the Si/SiO$_2$ interface the effective barrier height decreases to $\Phi_0 = 2.9\,eV$ due to quantization of electron energy spectrum [9]. Taking this into account, we obtained the electron effective mass in SiO_2 $m^*/m_0 = 0.5 \pm 0.02$ from simulations (1) at both positive and negative biases on the *poly*-Si electrode. These results are in consistent with the literature data [9, 10]. Low-field ($|V| < 5\,V$, $F < 6\,MV/cm$) measured current values are determined by the sensitivity of the measuring devices and bulk properties of the substrate.

The experimental current-voltage characteristics after SILC ($10\,C/cm^2$) are shown in Fig. 1 by rounds. High-field current through SiO_2 layer is limited by Fowler-Nordheim tunnelling (1). Low-field current is SILC. The characters in Fig. 2 represent the experimental J–V characteristics of n^+-Si/SiO$_2$/n^+-Si$_p$ structures at room temperature after different stresses. The SILC currents ($2\,V < |V| < 5\,V$) grow with increasing of the stress, while the slope of J–V curves decreases. The temperature increase leads to growing SILC currents on 30–50%, but Fowler-Nordheim tunnelling remains the same level.

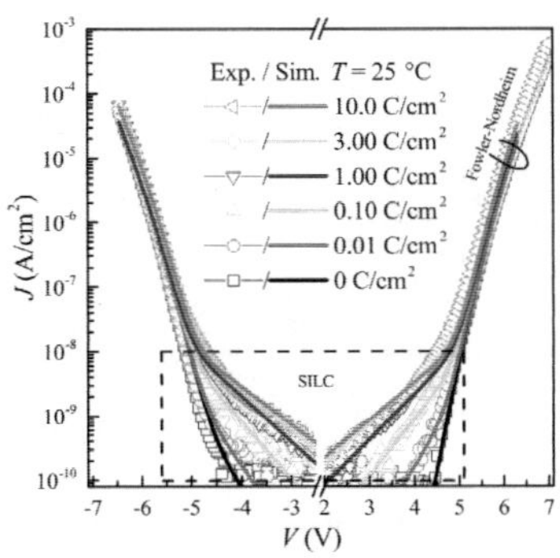

Figure 2. Current-voltage characteristics of n^+-Si/SiO$_2$/n^+-Si$_p$ structures at room temperature (25°C) before and after stress. Characters show experimental data; lines represent results of simulations. Inset: stress charge values.

Charge transport in dielectrics is determined by traps. When the trap density is high and the distance between them is short, trapped electrons (or holes) can tunnel between the neighbouring traps without ionisation to the conduction band [11]. Due to the external electric field electrons on the neighbouring traps have different energy levels, e.g. phonon emission and phonon absorption. According to phonon-assisted tunnelling model, the rate of such transitions is given as

$$P_{\mathrm{PAT}} = \frac{2\sqrt{\pi}\hbar W_{\mathrm{t}}}{m^* a^2 Q_0 \sqrt{kT}} \exp\left(-\frac{2a\sqrt{2m^* W_{\mathrm{t}}}}{\hbar}\right) \exp\left(-\frac{W_{\mathrm{opt}} - W_{\mathrm{t}}}{2kT}\right) \sinh\left(\frac{eFa}{2kT}\right). \quad (2)$$

where $\hbar = h/2\pi$, $Q_0 = \sqrt{2(W_{\mathrm{opt}} - W_{\mathrm{t}})}$, W_{opt} and W_{t} are thermal and optical trap energies, k in the Boltzmann constant, T is the temperature. In the static one-dimensional case, the charge transport through the dielectrics is described current-voltage characteristics, including Fowler-Nordheim tunnelling (1)

$$J = \frac{s}{S}\frac{e}{a^2}\frac{n_{\mathrm{t}}}{N}\left(1 - \frac{n_{\mathrm{t}}}{N}\right) P_{\mathrm{PAT}} + J_{\mathrm{FN}}. \quad (3)$$

Here J is the current density, n_{t} is the filled trap density, $N = a^{-3}$ is the trap density, Here we assume that the stress generated new defects on some part of test sample, creating a filament-like structures. The total area of the stressed part is s (SILC area). Recently it was show that in spite of high trap density, trapped charge density in dielectric does not exceed 5×10^{18} cm^{-3} due to strong electron-electron (Coulomb) interaction, while the charge carriers form Wigner-like glass hexagonal lattice [12–14].

Experimental data is described quantitatively using thermal and optical trap energies of $W_{\mathrm{t}} = 1.6$ eV and $W_{\mathrm{opt}} = 3.2$ eV. Obtained value of W_{t} is equal to a half of Stokes shift of photoluminescence near 4.4 eV exited by 7.6-eV-photons on oxygen vacancy in SiO$_2$ [15] and equal to calculated trap energy for trapped electrons and holes on oxygen vacancies. The W_{opt} is close to measured electron trap energy in SiO$_2$ [16]. Simulations of experimental current-voltage characteristics measures at 70°C gives the same trap energy values. These results confirm correctness of using the phonon-assisted tunnelling between traps model to describe SILC. The obtained values of trap densities and the squares of SILC area s depending on the total stress charge Σ are given in table 1. Charge trap density before stress was less than 10^{20} cm^{-3}. Charge stress caused arising of the trap density up to 7×10^{21} cm^{-3} depending on the stress. The SILC area increases with the stress up to 10 μm^2 (i.e. $\sim 3\,\mu$m $\times 3\,\mu$m). Note, that obtained SILC area in our experiments is less than 0.01% of total sample area.

After the charge stress some samples were treated by annealing at the temperature of 250°C during 120 hours. A comparison of the current-voltage characteristics measures for the "as deposited", after the stress and after annealing structures are shown in Fig. 1. One can see

Table 1. Obtained values of fitting parameters.

Σ (C/cm^2)	0	0.01	0.1	1	3	10
N (cm^{-3})	$\ll 10^{20}$	1×10^{21}	2×10^{21}	4×10^{21}	5×10^{21}	7×10^{21}
s (μm^2)	0	0.1	0.8	3	5	10

that J-V curves, after annealing, are almost identical to the "as deposited" control ones. I.e., long-time annealing leads to the electronic structure of the tunnel SiO_2, which is identical to one of the "as deposited" films. This means that the trap density decreases with annealing to initial values. Since charge stress leads to generatio of Frenkel pairs of oxygen vacancies and interstitial oxygen atoms, annealing leads to recombination of oxygen vacancies with interstitial oxygen. This phenomenon makes it possible to controllably suppress the leakage currents caused by the tunnel SiO_2 degradation due to the charge stress.

4. Summary

In conclusion, we demonstrated that charge transport via traps in thermal SiO_2 is phonon-assisted tunnelling between traps. This transport model in SILC is analytical and much easier than previously proposed multiphonon model of charge transport that requires complicate numerical calculations to describe transport in SILC. Furthermore, phonon-assisted tunnelling between traps model shows that oxygen vacancies are the charge traps in SiO_2, whereas recently proposed multiphonon model does not clarify the nature of the traps responsible for SILC [4,6]. Obtained thermal and optical trap energies are $W_t = 1.6\,\text{eV}$ and $W_{opt} = 3.2\,\text{eV}$, respectively. Charge flowing leads to oxygen vacancies generation, and the leakage current increases. Long-time annealing at high temperatures decreased the leakage current to initial values due to oxygen vacancies recombination with interstitial oxygen. Taking into account results of *ab initio* simulations we found that the oxygen vacancies act as electron traps in SiO_2.

Acknowledgments

This work was supported by the Russian Science Foundation, project #16-19-00002.

References

[1] Brown D and Brewer J F (eds) 1998 *Nonvolatile Semiconductor Memory Technology. A Comprehensive Guide to Understanding and Using NVSM Devices* (New York: IEEE Press)
[2] Blöchl P E and Stathis J H 1999 *Phys. Rev. Lett.* **83**(2) 372–375
[3] Fowler R H and Nordheim L 1928 *Proc. R. Soc. A* **119** 173–181 ISSN 0950-1207
[4] Endoh T, Chiba T, Sakuraba H, Lenski M and Masuoka F 1999 *J. Appl. Phys* **86**(4) 2095–2099
[5] Komiya K and Omura Y 2002 *J. Appl. Phys* **92**(5) 2593–2601
[6] Jiménez-Molinos F, Palma A, Gámiz F, Banqueri J and López-Villanueva J A 2001 *J. Appl. Phys* **90**(7) 3396–3404
[7] Giannozzi P *et al* 2009 *J. Phys. Condens. Matter* **21** 395502–395520
[8] Gritsenko V A, Mogil'nikov K P and Rzhanov A V 1978 *JETP Lett.* **27** 375–377
[9] Weinberg Z A 1982 *J. Appl. Phys* **53**(7) 5052–5056
[10] Nasyrov K A, Shaimeev S S, Gritsenko V A and Han J H 2009 *J. Appl. Phys* **105** 123709
[11] Nasyrov K A and Gritsenko V A 2011 *J. Appl. Phys* **109** 093705
[12] Shaimeev S S and Gritsenko V A 2011 *JETP* **112** 479–481 ISSN 1063-7761
[13] Shaimeev S S, Gritsenko V A and Wong H 2010 *Appl. Phys. Lett.* **96** 263510
[14] Gritsenko V A, Perevalov T V and Islamov D R 2016 *Phys. Rep.* **613** 1–20 ISSN 0370-1573
[15] Gritsenko V and Wong H 2011 *Crit. Rev. Solid State Mater. Sci.* **36** 129–147
[16] Yamabe K and Miura Y 1980 *J. Appl. Phys* **51** 6258–6264

Comparison of self-assisted VLS GaAs nanowires grown by MBE on Si (111) and GaAs (111)B substrates

Samatcha Vorathamrong[1], Somsak Panyakeow[1], Somchai Ratanathammaphan[1], Piyasan Praserthdam[2], and Chiraporn Thongyam[3]
[1]Semiconductor Device Research Laboratory, Department of Electrical Engineering
[2]Center of Excellence on Catalysis and Catalytic Reaction Engineering, Department of Chemical Engineering
[3]Center of Research and Technology Development, Mektec Manufacturing Ltd., Thailand
Chulalongkorn University, Bangkok 10330, Thailand. mesavor@gmail.com

Abstract In this work GaAs nanowires were grown by self-assisted growth method with completely identical growth parameters, such as growth temperature, growth time, Ga and As flux, on GaAs (111)B and Si (111) substrates using Molecular Beam Epitaxy (MBE). All samples were then characterized by Scanning Electron Microscope (SEM), Energy-dispersive X-ray spectroscopy (EDX), and X-ray Diffraction (XRD). The results from both substrates were compared in order to understand the effect of substrate type on nanowires.

1. Introduction

Nanowire growth method has been an interesting topic recently due to gaining popularity of using nanowires in electrical and optical applications [1][2][3]. First nanowire growth method was proposed by Wagner and Ellis in 1964 [4]. Their Au-assisted Vapor Liquid Solid (VLS) method utilizes Au droplets as seed particles to grow nanowires. This growth method is very well-known and convenient however it comes up with a problem of gold contamination which can cause negative effects to electrical and optical properties of nanowires. To eliminate the problems from Au contamination, Fontcuberta proposed self-assisted VLS method in 2008 [5]. Instead of Au, this method utilizes an element metal which is the composition of nanowire as seed particles, and ultra-thin SiO_2 layer deposited on substrate surface improves growth rate of nanowires by a reasonable margin. This has been one of the most developed technique to grow the compound semiconductor nanowires up to date. However as good as it may look, this method still possesses some crucial problems such as occurrence of parasitic growth on SiO_2 layer [6], and tilted or non-directional growth nanowires caused by non-uniformity thickness of SiO_2 layer [7]. All these issues prevent it from being an ideal nanowire growth method. Thus, we came up with the idea to grow GaAs nanowires directly on GaAs substrate. Since our method does not need Au and SiO_2 layer in the process, we believe that all previous problems can be overcome.

2. Experimental procedures

GaAs nanowires were carried out in COMPACT-21 Molecular Beam Epitaxy (MBE) system on n-doped GaAs (111) and Si (111) substrates. All substrates were free from any surface or chemical treatment and only prepared by degassing in pre-heat chamber at 400°C for 1 hours to get rid of the humidity and some contamination on surface before loaded into growth chamber. After loaded into the growth chamber, substrate temperature was ramped up to growth temperature of 450°C and remained constant for the whole process. To form seed particles for growth initiation, Ga droplets were deposited with constant flux pressure of 1.6×10^{-6} Torr for 40 seconds. Substrates were then exposed with Ga and As at flux pressure of 1.6×10^{-6} and 8×10^{-6} Torr respectively to grow nanowires. Growth process continued for 30 minutes before Ga supply was cut off to terminate the process. After growth finished, substrate temperature was ramped down to 240°C before As_4 supply was cut off to prevent damage of GaAs surface. After growth process finished, the samples were unloaded from MBE system and cut into pieces for characterization and analysis. The fabrication process of this experiment is displayed in Fig. 1.

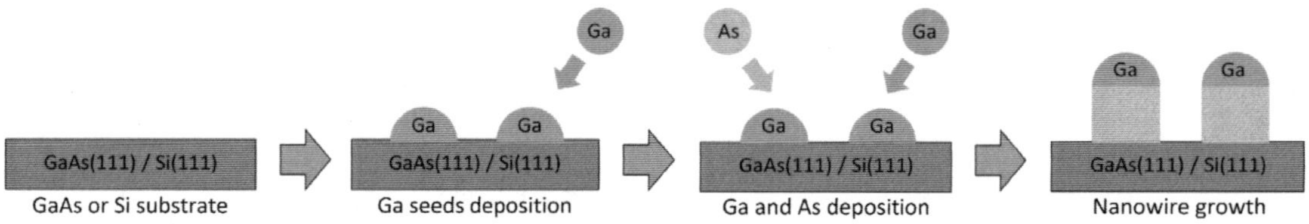

Fig.1 Diagram of GaAs nanowire growth process

3. Results and discussions

(1) Surface morphology

SEM images of the surface morphology at ×30,000 magnification show vast difference between nanowires grown on GaAs (111) and on Si (111) substrates as shown in Fig.2. On GaAs (111), nearly all nanowires grew vertically with only few tilted nanowires, no parasitic growth or bulk structure exists on substrate. While on Si (111) substrate, tilted nanowires and parasitic growth are clearly visible. This may result from non-uniformity thickness of SiO_2 layer which required to perform GaAs nanowire growth on Si substrate [8].

Fig.2 SEM images of GaAs nanowires grown on GaAs (111) substrate (a) and Si (111) substrate (b)

(2) Growth rate, density, vertical yield, and elemental composition

As clearly seen in Fig. 3, the average growth rate of nanowires grown on Si(111) is almost 3 times of nanowires grown on GaAs (111) substrate. This might result from lower sticking coefficient of Ga on SiO_2 surface that caused Ga atoms to have higher diffusion length and higher tendency to reach seed particles than on GaAs substrate [9]. However the nanowires on GaAs (111) substrate are much superior in density and vertical yield. The growth rate of nanowires correlates with density, shorter nanowires are likely to grow with higher density than longer ones and vice versa due to mass conservation [10]. For vertical yield, since our Si (111) substrate was completely free of any surface treatment, substrate surface was normally covered with native SiO_2 layer. As discussion in previous topic, non-uniformity thickness of the native SiO_2 layer can lead to tilted nanowires and parasitic growth [6]. Ga/As ratio is similar for nanowires grown on both substrates. This suggests that nanowires are GaAs nanowires with nearly stoichiometric composition.

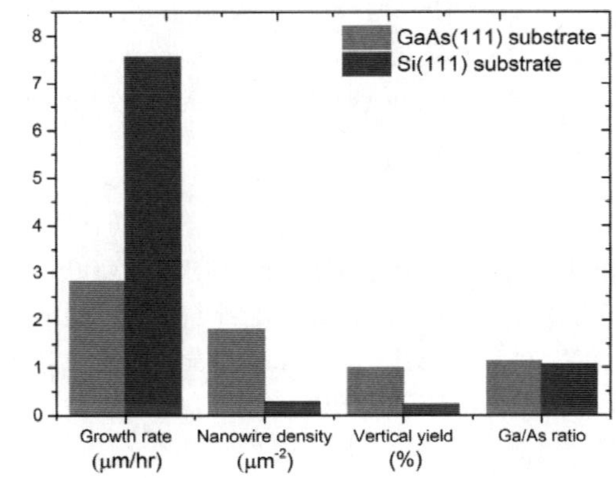

Fig.3 Chart of growth rate, density, vertical yield, and Ga/As ratio of nanowires grown on GaAs (111) substrate compared to grown on Si (111) substrate

(3) Crystal structure

XRD result of nanowires grown on GaAs (111) substrate shows the highest peak at 27.4° which corresponds to Zincblende (ZB) GaAs (111) structure. There is shoulder peak at 26.5° to 27.0°, this probably represent Wurtzite (WZ) GaAs structure. This implies that nanowires grown on GaAs (111) substrate have mixed structure of ZB and WZ GaAs.

For Si (111) substrate sample, the main peak at 28.6° is likely to result from Si (111) substrate. Shoulder peak exists at around the same position as ZB GaAs (111) peak. A slight peak shift and broad FWHM is likely to result from defects in crystal structure. There is no WZ GaAs peak exists for Si (111) substrate.

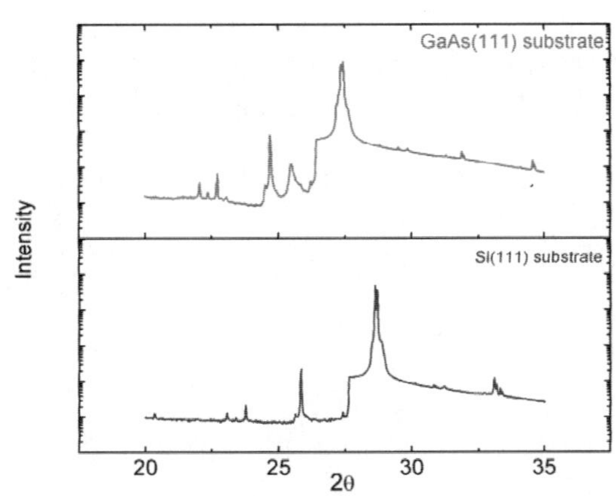

Fig.4 XRD scans of nanowire samples grown on GaAs (111) and Si (111) substrates

4. Conclusion

We succeeded in the growth of GaAs nanowires by self-assisted method on GaAs (111) and Si (111) substrates. The nanowire growth on both substrates shows the different properties. GaAs nanowires were grown on GaAs (111) possess higher density, and excellent vertical yield. While on Si (111), the nanowires were grown with higher growth rate, however there are many parasitic islands existing on substrate. XRD results show that nanowire sample grown using Si (111) substrate seem to be dominated with ZB GaAs structure whereas mixed structures of WZ and ZB GaAs are found on GaAs (111) substrate.

Acknowledgments This work is supported by the Ratchadapisek Sompoch Endowment Fund (2015), Chulalongkorn University (CU-58-056-EN), the 100th Anniversary Chulalongkorn University Fund for Doctoral Scholarship, National Science and Technology Development Agency, National Nanotechnology Center, Asian Office of Aerospace Research and Development (FA2386-14-1-4081) and Office of Naval Research Global.

References

[1] Cui Y, Zhong Z, Wang D, et al., NANO LETTERS 2003 Vol. 3, No. 2 149-152

[2] Duan X, Huang Y, Agarwal R, and Lieber C M, Nature 421, 241-5 16 January 2003

[3] Krogstrup P, Ingerslev Jorgensen H, Demichel O, et al. Nature Photonics 32, 2013

[4] Wagner R S, Ellis W C (1964). Appl. Phys. Lett. 4 (5): 89.

[5] Morral A, Columbo C, Abstreiter G, et al, ApplPhysLett 2008;92:063112.

[6] Breuer S, *"Molecular Beam Epitaxy of GaAs Nanowires and their Suitability for Optoelectronic Applications"*, Dissertation, 2011.

[7] Matteini F, Tütüncüo ğ lu G, et. al., Journal of Crystal Growth 404 (2014) 246–55.

[8] Dimakis E, Lähnemann J, Jahn U, Breuer S, Hilse M, Geelhaar L, and Riechert H (2011) Cryst. Growth Des. 11, 4001–08.

[9] Breuer S, Hilse M, Geelhaar L, Riechert H, Nucleation and growth of Au-assisted GaAs nanowires on GaAs(111)B and Si(111) in comparison, J. Crystal Growth 323, 311 (2011)

[10] Cao L, White J S, Park J S, Schuller J A, Bruce M, Nature Materials 8, 643-7 (2009)

Electronic structure of graphene/hexagonal boron nitride heterostructure revealed by NanoARPES

Chaoyu Chen[1], José Avila[1], Shuopei Wang[2], Rong Yang[2], Guangyu Zhang[2] & Maria C. Asensio[1*]

[1] ANTARES Beamline, Synchrotron SOLEIL & Université Paris-Saclay, L'Orme des Merisiers, Saint Aubin-BP 48, 91192 Gif sur Yvette Cedex, France.
[2] Beijing National Laboratory for Condensed Matter Physics and Institute of Physics, Chinese Academy of Sciences, Beijing 100190, China

E-mail: asensio@synchrotron-soleil.fr

Abstract. The ability to explore electronic structure and their role in determining material's macroscopic behaviour is essential to explain and engineer functions of material and device. Graphene/hexagonal boron nitride heterostructure (G/h-BN) has become a model system to study the emergent behaviour in 2D van der Waals heterostructure. Here by employing angle-resolved photoemission spectroscopy with sub-micrometric spatial resolution (NanoARPES), we give a full description on the electronic structure of G/h-BN, demonstrating the power of NanoARPES to detect the microscopic inhomogeneity of electronic structure for different materials.

1. Introduction

G/h-BN has become model system to study the emergent behaviour in 2D van der Waals heterostructure[1]. Previous ARPES measurements have shown its electronic band structure roughly[2, 3]. However, the lack of high-resolution (energy, momentum, space) is strangling the full understanding of G/h-BN' exotic behaviour. This is because that currently available high-quality G/h-BN on SiO_2/Si substrate[4, 5] are fabricated mainly by plasma-enhanced chemical vapor depositing (PECVD) or exfoliating graphene on mechanical cleaved h-BN, with typical domain size of several micrometer or sub-micrometer[6, 7, 8]. This scale is beyond the spatial resolution of conventional ARPES setups (typical $\sim 100\mu$m). In this work, by applying ARPES with sub-micrometer resolution (NanoARPES[9, 10]), we have been able to measure the comprehensive band structure and Fermi surface of high-quality PECVD-G/h-BN. We observe sharp features of valence bands of graphene Dirac cone and hBN, confirming that the G/h-BN system is atomically clean[11].

2. Experimental details

We first introduce the working principle of NanoARPES, as illustrated in Fig. 1a. Fresnel zone plate is used to focus the synchrotron radiation beam into nanometer size. Order sorting aperture can block the undesired orders of diffraction. The sample is mounted on a high-precision scanning stage with five degrees of freedom. Piezoscanners allow zero backlash and nanometric

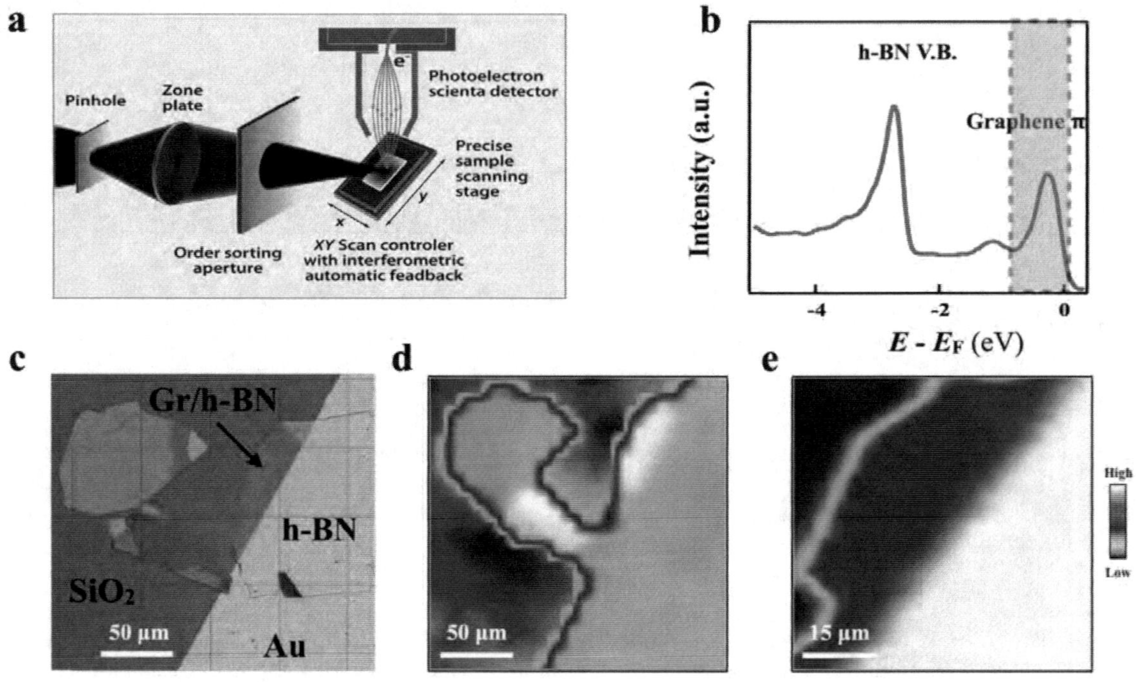

Figure 1. **a,** Working principle of NanoARPES. **b,** Valence band density of states of G/h-NB. The grey area indicates the energy window from which the valence band density of states images are integrated. **c,** Optical image of G/h-BN heterostructure. **d, e,** NanoARPES image of the valence band intensity.

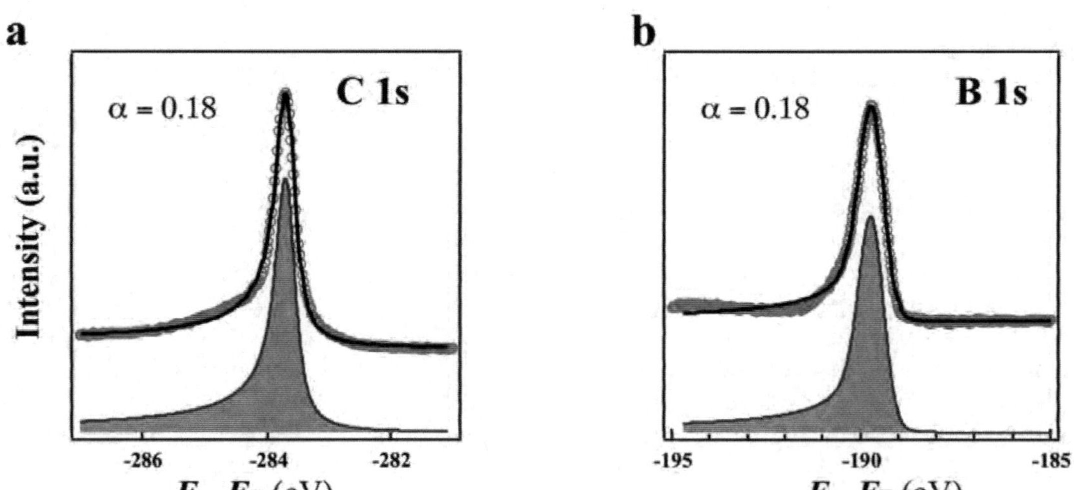

Figure 2. **a,** C 1s core level. **b,** B 1s core level. The core level peaks are fitting by asymmetric Lorentzian line shape (blue). Red circles are the original data. Black solid lines are fitted spectra.

IOP Conf. Series: Journal of Physics: Conf. Series **864** (2017) 012005 doi:10.1088/1742-6596/864/1/012005

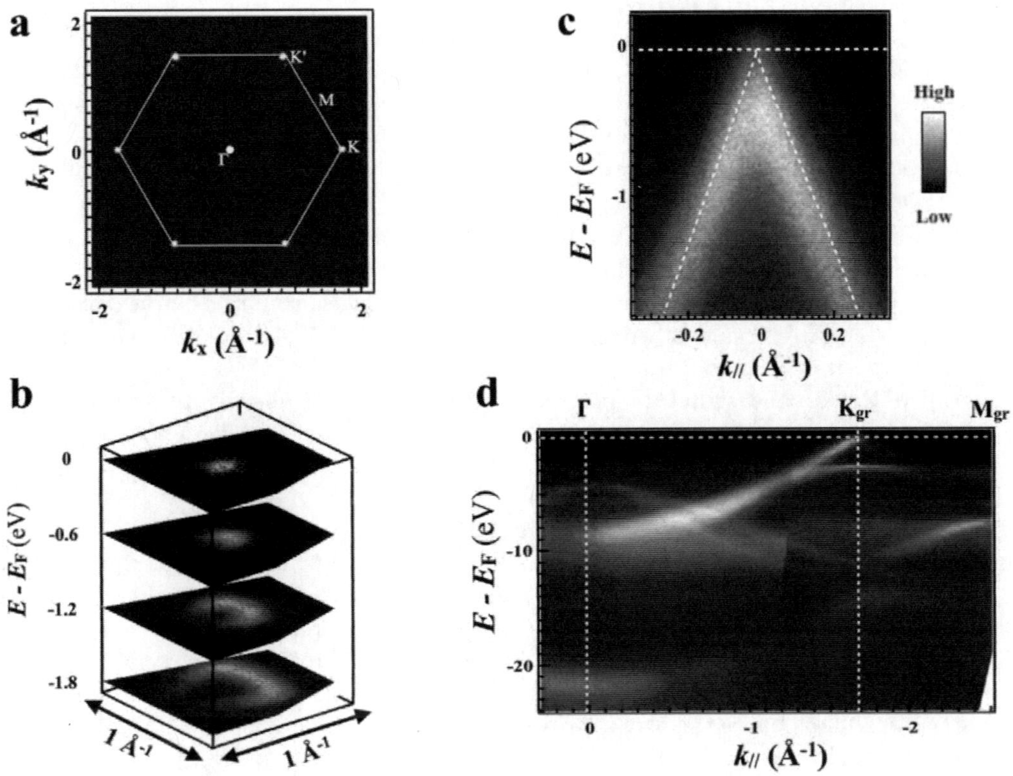

Figure 3. **a**, Symmetrised Fermi surface of G/h-BN . **b**, Spectral intensity distribution in E-k_x-k_y space. **c**, Detailed electronic structure of Dirac cone. **d**, Complete valence band structure of G/h-BN.

accuracy in the linear motion. Interferometers are used to control the simultaneous movement of sample and zone plate. By synchronised scanning of the sample along in-plane two directions and collecting the photoelectron intensity by energy analyser, images of the sample chemical composition (by element-specified core levels) and angle-resolved valence-band states can be obtained.

G/h-BN was prepared by PECVD growing graphene on exfoliated h-BN micro crystals. As described in the previous work[8], the growth of single crystalline monolayer graphene follows the van der Waals epitaxy mode, with zero crystallographic alignment angle respect to h-BN (Fig. 1c). SiO_2/Si was chosen as substrate for the convenience of transport measurement and also to avoid the possible interference in the valence band of G/h-BN from the d-bands of metals or transition metals. G/h-BN was grounded by Au contact for ARPES measurement (optical image in Fig. 1c).

As shown in Fig .1b, the spectra density of states of G/h-BN shows the peaks from h-BN valence band and graphene Dirac cone. By integrating the intensity within the selected energy window while scanning the sample along two in-plane direction, the spectral intensity distribution image can be generated, as shown in Fig. 1d and e. Compared with the optical image in Fig. 1c, it is clear that we can locate the G/h-BN sample (grounded) precisely with NanoARPES. From any position within the scanning range we can extract or remeasure the detailed electronic structure, including the core level spectra and valence band dispersion, as well as the Fermi surface, which will all be discussed in detail in the following.

3. Core level spectra of G/h-BN

Fig. 2 shows the core level spectra of C 1*s* and B 1*s*, measured by NanoARPES with phonon energy of 350 eV. Both panels present sharp core level peaks. For a more quantitative analysis, we have fitted the peaks by a procedure involving the deconvolution of the peak with a Doniach and Sunjic analysis[12], using a Lorentzian line shape. Both C 1*s* and B 1*s* show an asymmetry parameter of 0.18 and the fitting results are satisfactory with the usage of a linear background.

4. Fermi surface, Dirac cone and valence band structure

We show here, for the first time, the complete Fermi surface of G/h-BN heterostructure. As proved in Fig. 3a, the Fermi surface of G/h-BN originates from the Dirac cone of graphene, composing a standard hexagon, with the *K*-*K'* distance $\sim$1.7 $\AA^{-1}$. This clearly indicates that there is no structural change for graphene from the weak van der Waals interaction with h-BN. The detailed Dirac cone structure plotted in Fig. 3b and c clearly proved that there is no perturbation for the standard Dirac cone. The graphene is non-doped. The linear fitting for the Dirac cone shown in Fig. 3c gives a Fermi velocity $\sim$ 7.5 eV $\AA$ ($\sim$1.1 $\times$ 10^6 m s^{-1}), in agreement with reported value[13].

The complete band structure plotting in Fig. 3d clearly show the sharp features of well-defined dispersions for valence band of G/h-BN. Generally speaking, the G/h-BN valence band can be treated as the superposition of free graphene and h-BN valence bands, however with detailed but significant modification to be discussed elsewhere[14].

Up to now we have shown in detail that, by employing NanoARPES, we are able to measure the complete electronic structure, including the core levels and valence bands, for G/h-BN heterostructure. This paves the way for future exploration of microscopic novel heterostructures' electronic structure by NanoARPES.

5. Acknowledgments

The Synchrotron SOLEIL is supported by the Centre National de la Recherche Scientifique (CNRS) and the Commissariat à l'Energie Atomique et aux Energies Alternatives (CEA), France. G.Z. thanks the support from the National Basic Research Program of China (973 Program, grant No. 2013CB934500), the National Natural Science Foundation of China (NSFC, grant No. 61325021).

6. References

[1] Geim A K and Grigorieva I V 2013 *Nature* **499** 419-425
[2] Usachov D et al. 2010 *Physical Review B* **82** 075415
[3] Roth S et al. 2013 *Nano Letters* **13** 2668-2675
[4] Dean C R et al. 2010 *Nature Nanotechnology* **5** 722-726
[5] Gannett W et al. 2011 *Applied Physics Letters* **98** 242105
[6] Xue J et al. 2011 *Nature Materials* **10** 282-285
[7] Yankowitz M et al. 2012 *Nature Physics* **8** 382-386
[8] Yang W et al. 2013 *Nature Materials* **12** 792-797
[9] Avila J and Asensio M C. 2014 *Synchrotron Radiation News* **27** 24-30
[10] Chen C Y et al. 2017 *J. Phys.: Condens. Matter* **29** 183001
[11] Haigh S J et al. 2012 *Nature Materials* **11** 764-767
[12] Doniach S and Sunjic M. 1970 *J. Phys. C Solid State Phys* **3** 285-291
[13] Zhou S Y et al. *Nature Physics* **2** 595-599
[14] Chen C Y et al. arXiv:1707.00184.

IOP Conf. Series: Journal of Physics: Conf. Series **864** (2017) 012006 doi:10.1088/1742-6596/864/1/012006

Impacts of the annealing profile on AIC thin film solar cell characteristics fabricated by magnetron sputtering

Omid Shekoofa, Jian Wang, Yi Luo, De Jie Li, Changzheng Sun, Zhibiao Hao, Yanjun Han, Bing Xiong, Lai Wang, Lei Wang, and Hongtao Li

Tsinghua National Laboratory for Information Science and Technology, Department of Electronic Engineering, Tsinghua University, Beijing 100084, People's Republic of China

Corresponding author's email: wangjian@tsinghua.edu.cn

Abstract. Two different layer orders of p-Si/Al (as structure #1) and Al/p-Si (as structure #2) thin films were deposited on n-type mono-crystalline Silicon wafer and quartz glass by magnetron sputtering at 200 °C. The fabricated films were annealed at different temperatures for various durations. Then, they were analyzed by XRD, Raman and SEM methods. According to XRD results the largest grain size of both structures was smaller than 20 nm. The Raman spectra of the samples annealed at 1000 °C determined a crystallization ratio of 98% and 75% of structures #1 and #2 respectively. SEM images confirmed that the crystallization happened for structure #1 at lower temperatures than for structure #2. The impact of annealing on electrical and photovoltaic performance of the samples were studied after the fabrication of metal contact by sputtering of a few hundred nanometers of aluminium. The highest measured V_{oc} were 360 and 402 mV, and the best J_{sc} were 2.2 and 0.12 mA/cm^2, for structure #1 and #2 respectively.

1. Introduction

Aluminium inducing crystallization (AIC) is becoming a common method for crystallization and passivation of amorphous Silicon (a-Si) thin films which are used in fabricating low cost photovoltaic (PV) devices and Thin-Film Transistor (TFT) displays [1, 2]. Among different a-Si thin film growth methods like magnetron sputtering, PECVD, LPCVD or EBE; the magnetron sputtering has recently attracted more and more attentions because of its advantages such as being low cost and environment-friendly, and benefiting the high level of uniformity and scalability [3-7]. Poly-Si thin films has been fabricated on various types of substrates such as Si wafer [8], glass [2] and ceramic [9], with inducing different ratios of Al-to-Si for lowering the temperature and shortening the time of annealing [10]. High crystallization ratio is an essential criterion for the thin films which are fabricated for PV applications; either as emitter layer on Si wafers or in the p-i-n structure of thin film solar cells. However, just a few groups have reported the complete fabrication of poly-Si thin film solar cell by AIC and magnetron sputtering, and in majority of research the poly-Si films have been only grown as a primary step to create the so called "Seed Layer" for future growth of high quality films by PECVD or other methods [10, 11]. This paper investigated the AIC of a-Si throughout the fabrication of two simple solar cell structures completely by magnetron sputtering. They were fabricated and annealed at both low and high temperatures for various durations. Subsequently, their material properties were characterized and electrical parameters were measured. Finally, the resulted data were analysed and discussed in detail.

2. Experiments

The sputtering experiments were carried out by a MSP3200P magnetron sputtering machine. A p-type Si disk, with the purity of 99.9999% and resistivity of 1Ω.cm, and an Al disk, with purity of 99.99% , were used as the targets of Si and Al, respectively. All the films were deposited by DC magnetron sputtering power of 250 W, at 200 °C. The base vacuum pressure before sputtering was lower than 10^{-3} Pa. During the sputtering the pressure was raised to 1.2 Pa by flowing Ar inside the chamber at a rate of 40 sccm. The deposition rates of 0.5 nm/sec and 0.33 nm/sec were achieved for Al and Si respectively. Quartz glass and n-type Czochralski <100> mono-Si wafer with the thickness of 300 μm and typical resistivity of 20-30 Ω.cm, were used as the substrate. All AIC processes were done in vacuum ($<10^{-4}$ Pa) or N_2 atmosphere (~10 Pa). To measure the electrical parameters metal contacts were created on the front and back surfaces of the deposited thin films by sputtering of 150-250 nm of Al.

3. Results and Discussions

Two thin film structures are shown in the inset diagrams of figure 1. The first structure is an example of (1:1) ratio of Si/Al layers that typically has been reported as a typical ratio in AIC method for good poly-Si film fabrication [11]. The second one, which is usually called "Invert" structure [12], is an instance of highly unequal Al/Si layers' ratio (1:50) and it was investigated to see if such layer order and thin Al layer leads to good crystallization and PV property or not.

3.1. Material Properties

The XRD spectra are presented in figure 1. As shown in the left graph, the peak of Si <111> plane became observable after annealing at 550 °C or above. By extending the annealing time from 1.5 to 6 hours not only the peak intensity of Si <111> plane increased but also <200> and <311> peaks appeared in the XRD spectra. At 1000 °C these peaks became visible even for the annealing times as short as 2 hours. In the right-side graph, the peaks of poly-Si of structure #2 were observable only at the higher temperatures than 800 °C. The grain size of both structures were smaller than 20 nm.

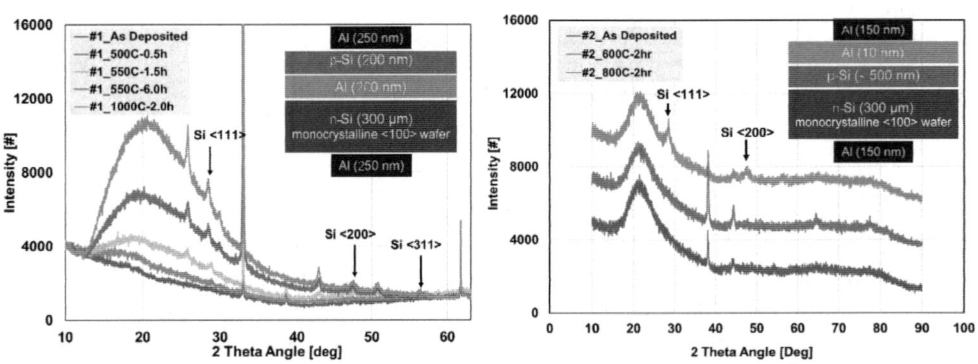

Figure 1. Si/Al/n-Si solar cell as structure #1 and Its XRD spectra (left), and Al/p-Si/n-Si solar cell as structure #2 with the XRD spectra of Al/p-Si sputtered on Glass

Following the XRD test, Raman spectra was measured to determine the crystallization ratio of the sputtered thin films. According to the Raman results of structure #1, shown in the left graph of figure 2, the crystallization ratio was not considerable for the annealing temperatures below than 550 °C. For the sample annealed at 550 °C for 6 hours, the crystallization ratio was about 44%. After annealing at 1000 °C for 2 hours, the poly-Si peak was very sharp and strong (shown by dash line) that didn't fit to the graph. The crystallization ratio for this sample was more than 98%. The Raman spectra of structure #2, right-side graph of figure 2, shows small peaks due to Si crystallization after annealing below 800 °C, while annealing at higher temperatures resulted in the stronger peaks. However, the crystallization ratio was less than 75% even after annealing at high temperatures up to 1000 °C.

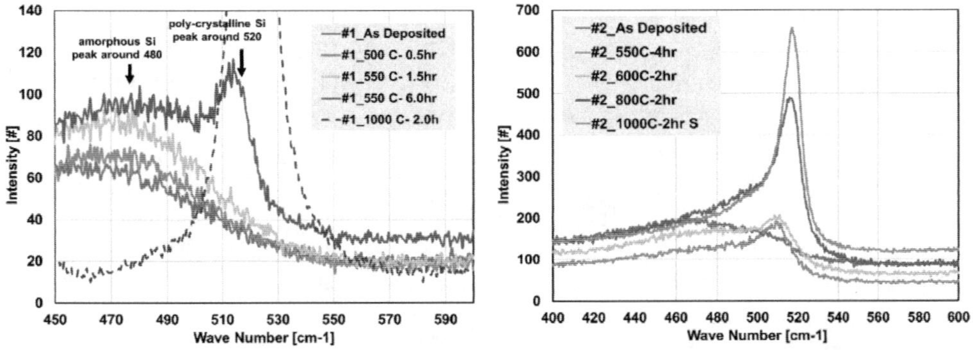

Figure 2. The Raman spectra of different samples of structure #1 (left) and #2 (right), after applying of different annealing profiles

Next part of the analysis was the study of thin films surface morphology by SEM images. The top row of figure 3 shows three different samples of structure #1. The left picture shows the smooth surface of the as-deposited sample, cover by a-Si. The middle one belongs to the sample annealed for 0.5 hour at 500 °C, wherein the surface roughness increased and some dark areas appeared. At 1000 °C, and due to the diffusion of Al into a-Si, the extra Al transferred to the top layer and formed an almost continuous layer on the top surface of the sample that prevented the film surface from receiving the light. In the second row, the as-deposited sample of structure #2 is displayed in the left-side, which expectedly was smooth. The middle picture shows that after annealing at 800 °C for 2 hours, dark poly-Si area was formed and surface roughness increased a little bit due to the diffusion of Al into the a-Si. The last picture belongs to the sample annealed at 1000 °C for 2 hours, wherein the phase-change obviously took place because the number and dimension of the dark areas increased.

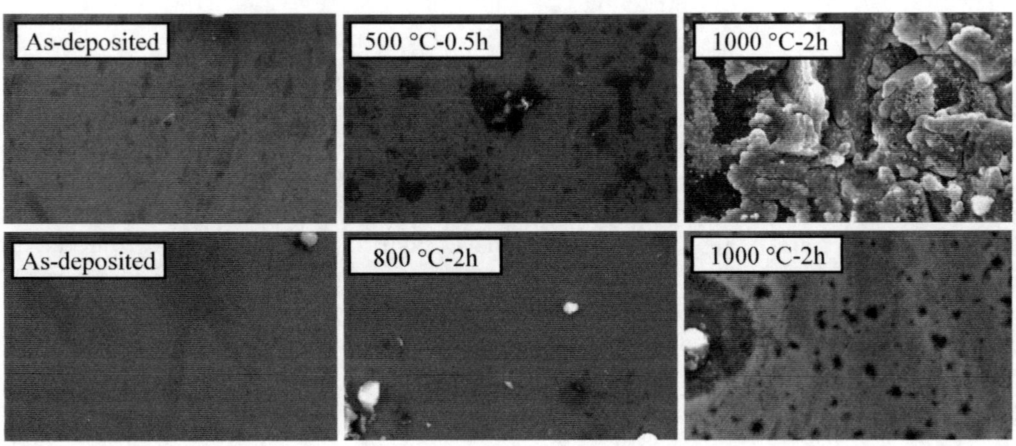

Figure 3. SEM images of structure #1 (top row) and structure #2 samples (bottom row)

3.2. *Electrical Properties*

Figure 4 shows the J-V curves of structure #1 samples in dark condition (left) and under the irradiance of 50 mW/cm^2 (right). All samples showed good rectification properties in dark condition. The reverse saturation current obviously reduced with longer annealing and at higher temperatures, while the corresponding threshold voltages increased. Furthermore, the PV performance of the samples under the illumination improved with increasing the annealing time and/or temperature, as shown in figure 4-right. The J-V curves of structure #2, shown in figure 5, are very different than of the structure #1. In dark condition (figure 5-left), the rectification took happen only after annealing at temperatures higher than 800 °C. Under the illumination (figure 5-right), the PV effect was very weak for all the samples annealed at temperatures lower than 600 °C and the voltage polarity of the solar cell was the inverse of the cell voltage annealed at higher temperatures, probably due to the metal-semiconductor junction formed at the top layer (besides the main p-n junction). Annealing with longer durations and at higher temperatures resulted in larger open circuit voltage (V_{oc}). However, the short circuit current density (J_{sc}) was reduced for the samples annealed at very high temperature for longer time than 1 hour. Finally, measurements in AM1.5 condition showed that structure #2 generated higher V_{oc} up to 402 mV but its J_{sc} was limited to 0.12 mA/cm^2, while structure #1 generated lower V_{oc} (about 360 mV) however its best measured J_{sc} was almost 20 times larger than of the structure #2 (about 2.2 mA/cm^2).

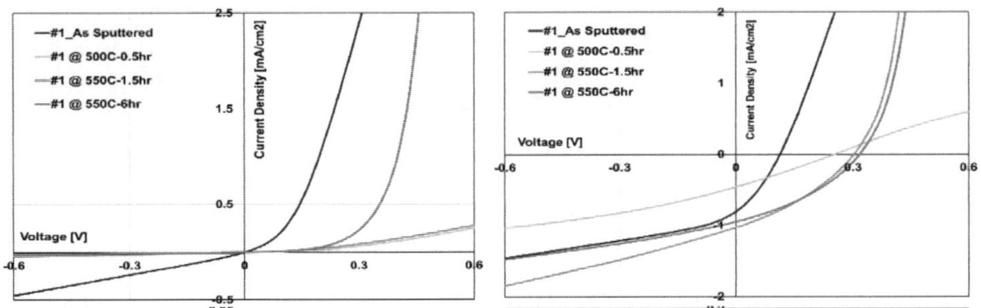

Figure 4. The J-V curves of structure #1 solar cell in dark (left) and under illumination of 50mW/cm^2 (right)

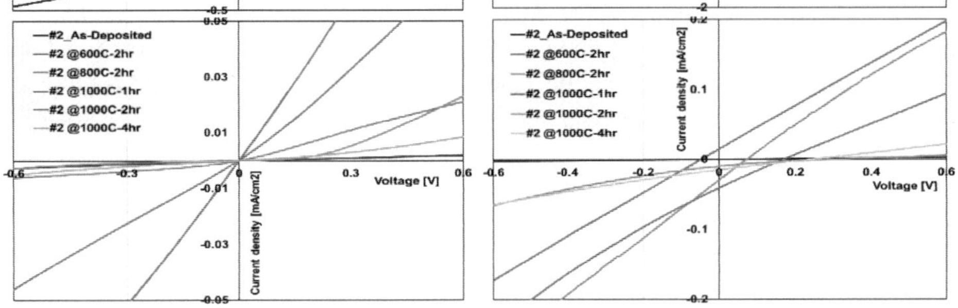

Figure 5. The J-V curves of structure #2 solar cell in dark (left) and under illumination of 50mW/cm^2 (right)

4. Conclusions

In this research we fabricated two different solar cell structures, i.e. (Al/p-Si) and (p-Si/Al) thin film layers on n-type Si wafer, by AIC and magnetron sputtering. Fabricated samples were annealed in an electrical annealing furnace at different temperatures and for various durations. The poly crystal films grain size, measured by XRD peak analysis, showed that the largest grain size is smaller than 20 nm in both structures. The crystallization ratio, determined by Raman peaks intensity measurement, was about 98% and 75% for structure #1 and #2 respectively. SEM images confirmed that the crystallization took place at higher temperatures for structure #2 (>600 °C), while it happened at 500 °C for structure #1. The I-V curve measurements showed the J_{sc} up to 2.2 mA/cm^2 for solar cells of structure #1 with a V_{oc} of 360 mV, whilst the best V_{oc} measured structure #2 was 402 mV with a J_{sc} of 0.12 mA/cm^2.

Acknowledgment
This work is supported by National Basic Research Program of China (no. 2015CB351900), National Natural Science Foundation of China (no. 61176059), and '863' High Technology Program of China (no. 2015AA017101), Tsinghua University Initiative Scientific Research Program (20131089364, 20161080068).

References

[1] G. Fortunato, 1997 Thin Solid Films 296.
[2] Haschke, J. et al., 2014 Solar Energy Materials and Solar Cells, 128, 190–197.
[3] Z. Tang, et al., 2009 Journal of Optoelectronics and Advanced Materials, 11, 1651–1655.
[4] G. Jin, et al., 2010, Progress in Photovoltaics Research and Application, 18, 582–589.
[5] G. J. Qi, et al., 2005 Surface & Coatings Technology, 198, 300–303.
[6] Jaeger, C., et al., 2010 Journal of Applied Physics, 108 (11), 113513.
[7] I. Safi, 2000, Surface and Coatings Technology 127 203-219
[8] Y. Ohmura, et al., 2001, Phys. B Condens. Matter, vol. 308–310, pp. 257–260, 2001.
[9] I. T. Gordon, et al., 2005 Thin Solid Films 487 113–117.
[10] Gestel, D. Van, et al., 2013 Sol. Energy Mater. Sol. Cells, 119, 261-270
[11] O. Nast and S. R. Wenham, 2000 J. Appl. Phys., vol. 88, no. 1, p. 124.
[12] K. Hiroshi, et al., 2009 Applied Physics Express vol. 2 (1) p. 015501.

Original method of GaN and InGaN quantum dots formation on (0001)AlN surface by ammonia molecular beam epitaxy

K S Zhuravlev[1], D V Gulyaev[1], I A Aleksandrov[1], T V Malin[1], V G Mansurov[1], Yu G Galitsyn[1], K.A. Konfederatova[1], Yen-Chun Chen[2] and Wen-Hao Chang[2]

[1] Rzhanov Institute of Semiconductor Physics, Siberian Branch of Russian Academy of Siences, Novosibirsk 630090, Russia
[2] Department of Electrophysics, National Chiao Tung University, Hsinchu 30010, Taiwan

E-mail: mansurov@isp.nsc.ru

Abstract. We report original method of formation Ga(In)N/AlN quantum dots with low density by ammonia MBE on the (0001)AlN surface by using a decomposition process of Ga(In)N thin layer. Low density of quantum dots have been obtained in the range 10^7-10^9 cm^{-2}. Single quantum dots photoluminescence lines corresponding to exciton and biexciton transitions were observed in micro-photoluminescence spectra.

1. Introduction

Formation of the wide band gap semiconductor GaN and InGaN quantum dots (QDs) in AlN matrices are important for fabrication of modern quantum devices such as single electron transistors, single and entangled photons sources, UV light emitting diodes, lasers and detectors, and etc. The GaN and InGaN QDs formation at the (0001)AlN surface usually implemented by using of the Stranski-Krastanov growth mode [1], or by using the droplet epitaxy technique, where at first Ga(In) droplets are deposited onto the surface, and then droplets are transformed into GaN (InGaN) 3D islands as result of the treatment in active nitrogen flux (like NH_3) at temperatures about 700 °C [2]. Usually density of QDs ensemble is quite high as 10^{10}-10^{11} cm^{-2} [1,3]. But for fabrication of single electron transistors or single photons sources the density of QDs of 10^8 cm^{-2} is required.

In the present work we propose an original method for the low density GaN (InGaN) QDs formation by using an Ga(In) wetting layer on the (0001)AlN surface [4] to form controllable thin two dimensional (2D) GaN layer and subsequent decomposition of this layer.

2. Experimental

Samples were grown by molecular-beam epitaxy (MBE) on 2 inch (0001)-oriented sapphire substrates in a Riber CBE-32 machine with ammonia as a source of active nitrogen. Formation process of the quantum dots were studied *in situ* by high energy electron diffraction (RHEED). The obtained structures with QDs were investigated by micro-photoluminescence (micro-PL). Photoluminescence was excited by continuous-wave HeCd laser with photon energy of 3.81 eV. The laser beam was focused in the spot with diameter of 1 μm on the sample by micro-objective. The PL was analyzed by

Acton SP-750i spectrometer with nitrogen-cooled CCD detector. A helium cryostat was used to control of the sample temperature.

3. Results and discussion

We have studied a two dimensional to three dimensional (2D-3D) growth mode transition during epitaxial growth of GaN (InGaN) layers with rate of 0.25 monolayers per second (ML/s) under ammonia flux (200 sccm) in wide temperature range (400-1000 °C) in order to determine suitable conditions for preparing of low density quantum dots ensemble. When the growth of GaN initiated, first the 2D growth mode was observed, that is evidenced by streaky RHEED pattern of the growing surface. Then the 3D transmission spots were appeared and it was found that the time interval from the growth initiation to transition point strongly depends on the surface temperature. The 3D GaN spot intensity behavior is shown in Figure 1. It should be noted that the 2D GaN growth without 2D-3D transition was observed at temperatures higher 750 °C. At temperatures higher ~870 °C the GaN growth rate drops because of decomposition of the GaN layer.

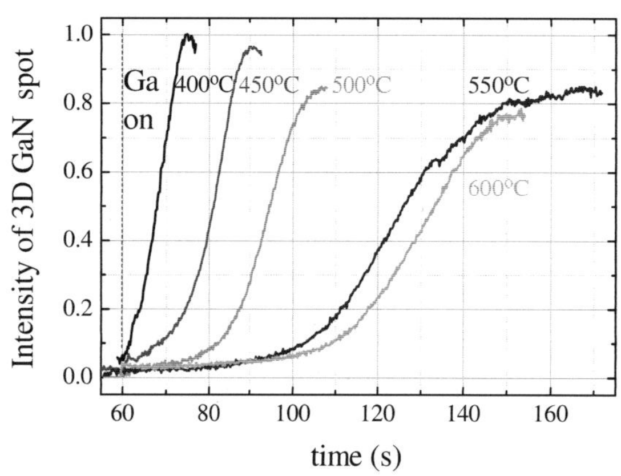

Figure 1. GaN 3D islands formation on (0001)AlN during epitaxial growth at different temperatures.

It is clear from the Figure 1 that the kinetic curves of 3D GaN islands (or GaN QDs) formation have a S-like character and the rate of this process decreases with increasing of temperature. It is evidence that the 3D GaN islands formation process is connected with a surface phase transition. The order parameter of the phase transition can be defined as densities difference between high density and low density phases of GaN QDs ensemble on the surface in close analogy with the van der Waals type of transitions from gas to liquid. However, this process is not fit to our goals because it results in high density of GaN QDs 10^9-10^{11} cm^{-2}.

We have developed an original method of GaN QDs formation. It is known that thermal decomposition of GaN occurs at temperatures higher 900 °C. The GaN decomposition is the multi-step process with appearing on the surface of GaN 3D islands [5].

Our QDs formation process consists of following stages. Three monolayers of Ga (or In+Ga) was deposited onto clean AlN surface at temperatures 500-600 °C. The RHEED pattern is quenched as a result of liquid Ga(In) layer formation on the surface. Then the surface was treated at the same temperature under ammonia flux (10 sccm) during of 2-5 min that leads to the Ga(In)N formation. Next stage is an increasing of temperature up to 1000 °C without ammonia flux. Appearing of GaN QDs was observed in temperature range of 900-930 °C. The RHEED patterns of the main stages of GaN QDs formation are shown in Figure 2.

Kinetics of GaN QDs formation is shown in Figure 3. The kinetic curve shape also demonstrates S-like character as in Figure 1, consequently this process should be treated as phase transition. In Figure 3 the μ denotes the chemical potential of quantum dot, ε – is the bonding energy of the QD with the surface, T – surface temperature and k is the Boltzmann constant. The value of μ increases with time because of increasing of entropy part of the chemical potential as the QDs are accumulated on the surface. The experimental kinetic curve of GaN QDs formation in fact represents a QDs density dependence on chemical potential μ, so this is an isotherm of the phase transition.

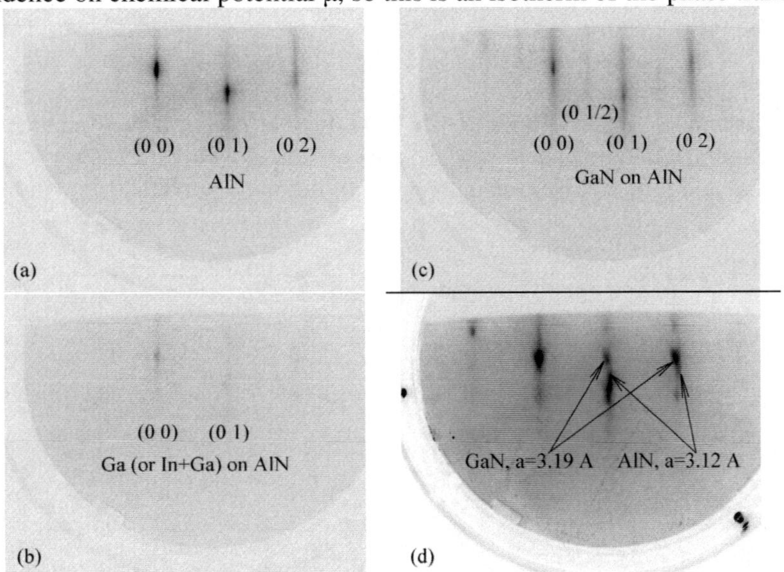

Figure 2. RHEED patterns observed at different stages of GaN QDs formation: (a) pattern of the initial AlN surface, (b) pattern appeared after Ga metal deposition, (c) pattern of the 2D GaN layer under ammonia flux, (d) pattern observed at point of 2D-3D transition during heating of the substrate from 540°C till 1000°C.

The experimental isotherm was described in the frame of lattice gas model for the first-order phase transitions. For calculations we have used the equation (6) on page 137 of the Reference [6]. Obviously, the experimental curve can not exactly follow for the discontinuous transition; nevertheless satisfactory agreement between theoretical and experimental curves is clearly seen.

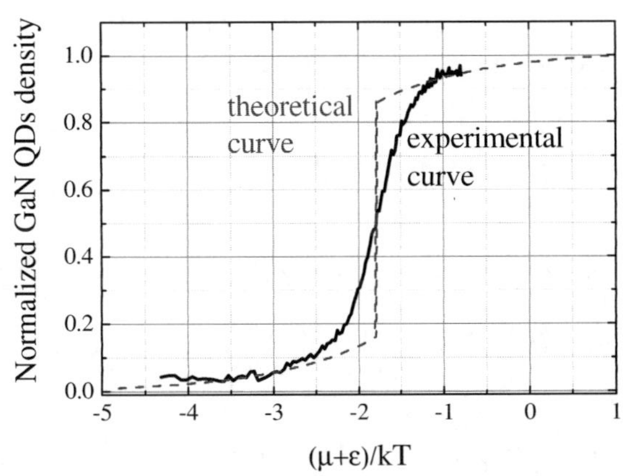

Figure 3. Experimental and theoretical isotherms of GaN QDs formation at T=900 °C.

Density of the GaN QDs was found about 10^8 cm^{-2}, as it confirmed by micro-photoluminescence (micro-PL) spectra obtained from single quantum dot. The sample for micro-PL measurements was fabricated by overgrowth of the prepared GaN QDs ensemble by AlN layer of 100 nm thick. Figure 4 shows micro-PL spectra of the GaN/AlN QDs at different excitation powers. The spectra contain two peaks at 401.2 nm and 403.7 nm with full width at half maximum of about 10 meV. The linewidth is similar to the observed in [7] for the single-QD transitions. The linewidth is larger than lifetime-determined linewidth of the excitonic transition in the single QD. We suppose that the broadening is caused by charge fluctuations on defects near the QDs. The first peak have superlinear power dependence and disappears at low excitation power. Power dependence of the intensity of the PL lines is well described by Poisson-distribution model [8]:

$$I(P) = \frac{I_0}{k!}\left(\frac{P}{P_0}\right)^k \exp\left(-\frac{P}{P_0}\right) \quad (1)$$

With k=1 (exciton) for the line at 403.7 nm and k=2 (biexciton) for the line at 401.2 nm. Thus we attribute the peak at 403.7 nm to the excitonic transition and the peak at 401.2 nm to the biexcitonic transition. Biexcitonic peak is higher in photon energy than the excitonic peak by 19 meV which is in agreement with typical biexcion binding energy in GaN/AlN QDs [9]. The QDs with sufficiently large size have negative biexciton binding energy because of electron and hole wavefunctions are separated by built-in electric field and the interaction between two electrons and two holes is higher than electron-hole interaction [10].

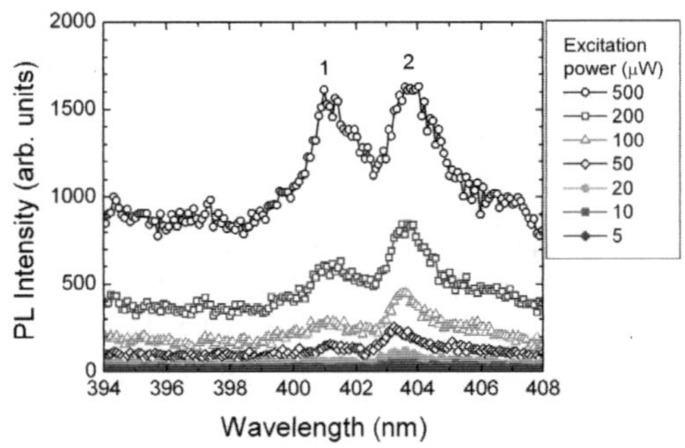

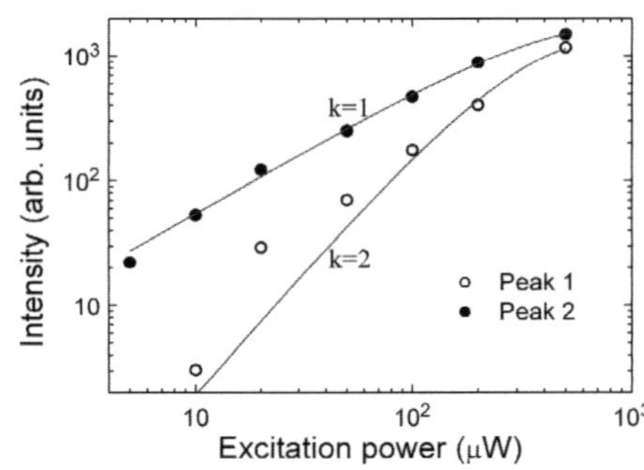

Figure 4. (a) Micro-photoluminescence spectra of GaN/AlN QD structure at different excitation powers. Sample temperature is 4 K. (b) Power dependence of integrated intensities of the PL lines (circles – experiment, solid lines – fitting by equation (1)).

4. Conclusion

Original method of Ga(In)N/AlN quantum dot formation by ammonia MBE have been developed. The method allows obtaining high-quality QDs with density in the range 10^7-10^9 cm^{-2}. PL lines corresponding to excitonic and biexcitonic transitions of single QDs were observed.

Acknowledgments

The authors gratefully acknowledge support from RFBR (grant # 14-02-92007).

References

[1] Daudin B, Widmann F, Feuillet G, Samson Y, Arlery M and Rouviere J L 1997 *Phys. Rev. B* **56** R7069
[2] Song H, Lee S-H, Jang E-S, Kim D-W, Navamathavan R, Kim J S and Lee Ch-R 2009 *J. Cryst. Growth* **311** 4418
[3] Damilano B, Brault J and Massies J 2015 *J. Appl. Phys.* **118** 024304
[4] Brown J S, Koblmüller G, Averbeck R, Riechert H and Speck J S 2006 *J. Vac. Sci. Technol. A* **24** 1979
[5] Zhang J et al. 2014 *Nanoscale Research Letters* **9** 341
[6] Galitsyn Yu G, Lyamkina A A, Moshchenko S P, Shamirzaev T S, Zhuravlev K S and Toropov A I 2012 Self-addembled Quantum Dots: From Stranski-Krastanov to Droplet Epitaxy *Self –Assembly of Nanostructures (The INFN Lectures vol III*) ed S Bellucci (London: Springer) chapter 3 p 137
[7] Kako S, Hoshino K, Iwamoto S, Ishida S and Arakawa Y 2004 *Appl. Phys. Lett.* **85** 64
[8] Grundmann M and Bimberg D 1997 *Phys. Rev. B* **55** 9740
[9] Simeonov D, Dussaigne A, Butte R and Grandjean N 2008 *Phys. Rev. B* 77 075306
[10] Amloy S, Yu K H, Karlsson K F, Farivar R, Andersson T G, and P. O. Holtz P O 2011 Appl. Phys. Lett. 99, 251903 (2011)

Improved planar perovskite solar cells using the modified hole transporting layer and solvent annealing

Shangzheng Pang, Chunfu Zhang#, Dazheng Chen, and Yue Hao
State Key Discipline Laboratory of Wide Band Gap Semiconductor Technology, School of Microelectronics, Xidian University, Xi'an 710071, China
#*Email: cfzhang@xidian.edu.cn*

Abstract

Organometal trihalide perovskites have been demonstrated as excellent light absorbers for high efficiency solar cell [1]. In this work, Glass/ITO/PEDOT:PSS/Perovskite/PCBM/Ag planar heterojunction structure was designed and investigated. For the reference cell, the efficiency of 12.5 % was achieved. An relative low J_{sc} was exhibited at the result, mainly owing to the low conductivity of the electric transport layer – PEDOT:PSS. By controlling the amount of DMSO doped in PEDOT:PSS, a superior device was obtained with the efficiency of 13.6% although the roughness of the PEDOT:PSS layer was also increasing with doping DMSO. Furthermore, another key factor for a perovskite solar cell is the high crystallization of the perovskite layer[2]. Solvent annealing was adopted to improve the crystallization in this work. With dropping IPA around the solar cell and annealing for 20 minute, a surprising characteristics of the device with the efficiency of 15%, V_{OC} of 0.99 V, Fill Factor of 67.9%, J_{SC} of 22.3mA/cm^2 was obtained.

1. Introduction

Due to the energy shortage, solar emery has acquired enormous attentions[3]. The growth of the perovskite solar cell as one new novel solar energy device is quite amazing. Only developed for a few years, the efficiency has exceeded 20%[4]. There are some possibilities that this type of device could replace the silicon solar cell for its low cost and simple in preparation[5].

In order to achieve high power conversation efficiency (PCE), the proper hole transport layer (HTL) with a high electrical conductivity and a high perovskite quality are required[6]. In this work, by controlling the amount of DMSO doped in PEDOT:PSS, the electric conductivity of PEDOT:PSS is improved and a superior device was obtained with the efficiency of 13.6%. Furthermore, solvent annealing was adopted to improve the crystallization in this work[7]. With dropping IPA around the fabricated solar cell and annealing for 20 minute, a surprising characteristic of the device with the efficiency of 15%, V_{OC} of 0.99 V, Fill Factor of 67.9%, J_{SC} of 22.3mA/cm^2 was obtained.

2. Experimental

Material and methods

ITO glass substrates are obtained from Zhuhai Kaivo Optoelectronic Technology Co.. Poly (3,4-ethylenedioxythiophene):poly(styrenesulfonate) (PEDOT:PSS Clevios P VP Al 4083) solution was acquired from Heraeus. Phenyl-C61-butyric acid methyl ester (PCBM) was acquired from American Dye Source. MAI (99.8%, Dyesol), PbI_2 (99.999%, ultra dry, Alfa), Dimethyl Sulphoxide

(99.8%, anhydrous, Sigma), Chlorobenzene (99.8%, anhydrous, Sigma), Toluene (99.8%, anhydrous, Sigma), γ-Butyrolactone (99.8%, anhydrous, Aladdin) were used as received.

Device Fabrication and characterization

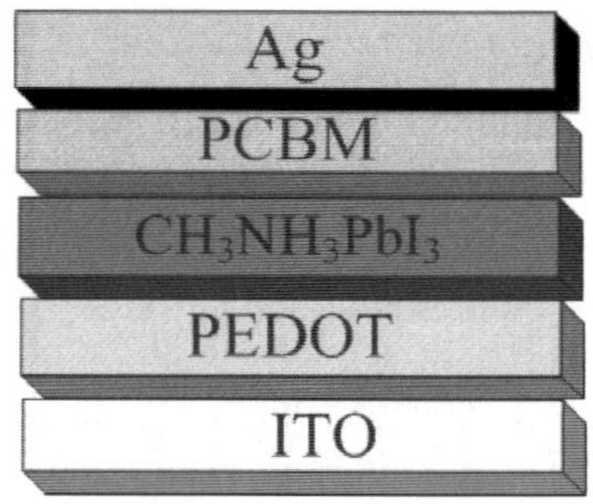

Fig. 1. Schematic of perovskite solar cell.

The fabricated device structure is shown in Fig. 1 and the corresponding energy diagram is shown in Fig. 2. The perovskite solar cell is based on a planar heterojunction structure, with HTL of PEDOT:PSS deposited on the transparent conductor ITO layer. The fabrication procedures are briefly stated as the following. Frist, the ITO glass substrates were cleaned with detergent, DI water, acetone, alcohol for 15 mins, respectively. And then they were treated by O_2 plasma for 15 minutes before the device fabrication. A layer of PEDOT:PSS was spun coated on ITO substrate at 7000 rpm for 40 seconds. Then, they were annealed at 150℃ for 20 minutes. The substrates were transferred into a glove box filled with nitrogen. $CH_3NH_3PbI_3$ perovskite films were prepared by spin coating a solution containing 1.4M PbI_2 and 1.35M CH_3NH_3I in co-solvent of DMSO:GBL (3:7 vol. ratio). The spin-coater was started at a rotation speed of 1000 rpm for 15 seconds and 5000 rpm for another 25 seconds. 350μl toluene was added quickly at 40 seconds after the start of spin coating process. The perovskite films were then annealed at 100℃ for 20 minutes. 20 mg/ml chlorobenzene solution of PCBM was spin-coated at 2000 rpm for 40 seconds. Finally, the films were transferred to a metal evaporation chamber, 100 nm thick Ag contacts were deposited under high vacuum ($< 4\times10^{-4}$ Pa).

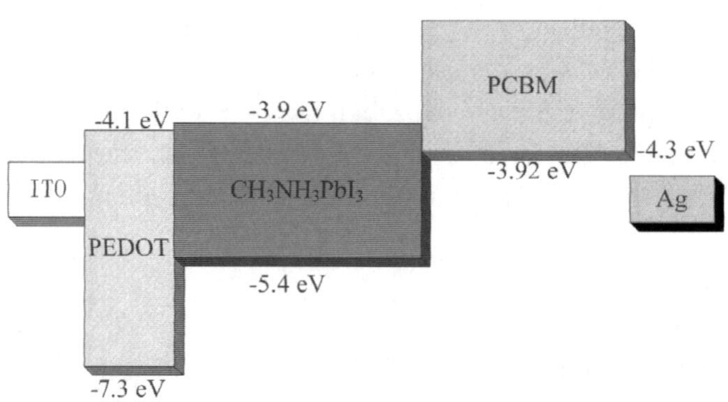

Fig. 2. Band alignment of Glass/ITO/PEDOT/Perovskite/PCBM/Ag.

3. Results and discussion

The I-V characteristics of the devices are shown in Fig. 3. As is shown, the reference device without any HTL treatment and solvent annealing, the device had a low PCE of 12.5%. Considering the weak electric conductivity of the HTL, it's necessary to dope the PEDOT:PSS. Dimethyl Sulphoxide (DMSO) was one ideal doping material for PEDOT:PSS[8]. By controlling the doping ratio of DMSO in PEDOT:PSS (1%, 2%, 5%, and 10%), we could find that PCE of the perovskite solar cells is continuously increased until the doping dose reaches 5%. Beyond this doping ratio, PCE of the perovskite solar cells began to decrease. Thus the optimal doping ratio of DMSO in PEDOT:PSS is 5% and at this time, the highest device PCE was 13.6%. Doping the DMSO in the PEDOT:PSS could improve the electric conductivity, so that at the beginning PCE was improved[9]. When the doping dose reached one critical point, the electric conductivity reached the best. When increasing the DMSO further would dilute the concentration of PEDOT:PSS, therefore the PCE droped[10].

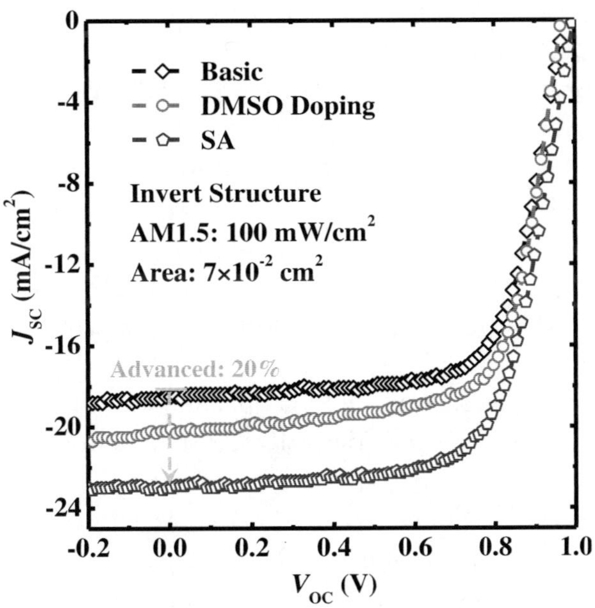

Fig. 3.Current densityvesusvoltage (JSC–VOC) characteristics of devices withundoped, DMSO doping PEDOT and annealed DMSO doping PEDOT. basic structure in black line the efficient of 12.5 % was achieved. By controlling the amount of DMSO doped in PEDOT,(red line) a superior device was obtained with Efficiency of 13.6%, With dropping IPA around the DMSO dopping solar cell and annealing for 20 minute,a surprising characteristics of the device with the efficient of 15%(blue line),VOC : 0.99 V,Fill Factor: 67.9%,JSC: 22.3mA/cm2

By using solvent annealing further, PCE acquired obvious enhancement. From the Fig. 3, we could find that PCE rise from 13.6% to 15%. When solvent annealing was used to treat on the perovskite, the PCE enhancement originates from the improved perovskite layer quality. The solvent was mixed by DMF and IPA (with ratio as 5:95). In the annealing process, DMF would evaporate and reach the perovskite layer surface. The DMF vapor could dissolve the perovskite, so the perovskite was in the dynamic equilibrium between crystallization and dissolution. At the start, perovskite crystallization wasn't perfect, crystalline grain were in different size, and the perovskite layer was not flat enough. When the device was not treated by solvent annealing, its PCE is low. After solvent annealing, the layer becomes more flat[11], and has an excellent contact with HTL layer PEDOT:PSS and ETL PCBM. All of these produce the highest PCE of 15%.

4. Conclusion

In conclusion, by controlling the amount of DMSO doped in PEDOT:PSS, a superior device was obtained with the efficiency of 13.6%. Furthermore, another key factor for a perovskite solar cell is the high crystallization of the perovskite layer. Solvent annealing was adopted to improve the crystallization in this work. With dropping IPA around the solar cell and annealing for 20 minute, a surprising characteristic of the device with the efficiency of 15%, V_{OC} of 0.99 V, Fill Factor of 67.9%, J_{SC} of 22.3mA/cm^2 was obtained.

Reference

[1]Della Gaspera E, Peng Y, Hou Q, et al. Ultra-Thin High Efficiency Semitransparent Perovskite Solar Cells Nano Energy 2015, 13, 249–257[J].

[2]G.E. Eperon, D. Bryant, J. Troughton, S.D. Stranks, M.B. Johnston,T. Watson, D.A. Worsley, H.J. Snaith, J. Phys. Chem. Lett. 6 (2015)129–138.

[3]J. Liu, Y. Wu, C. Qin, X. Yang, T. Yasuda, A. Islam, K. Zhang,W. Peng, W. Chen, L. Han, Energy Environ. Sci. 7 (2014)2963–2967.

[4]T. Schwab, S. Schubert, S. Hofmann, M. Fröbel, C. Fuchs,M. Thomschke, L. Müller-Meskamp, K. Leo, M.C. Gather, Adv.Opt. Mater. 1 (2013) 707–713.

[5]N.P. Sergeant, A. Hadipour, B. Niesen, D. Cheyns, P. Heremans,P. Peumans, B.P. Rand, Adv. Mater. 24 (2012) 728–732.

[6]Dai X, Zhang Y, Shen H, et al. Working from Both Sides: Composite Metallic Semitransparent Top Electrode for High Performance Perovskite Solar Cells[J]. ACS applied materials & interfaces, 2016, 8(7): 4523-4531.

[7]Suzuki A, Kida T, Oku T. Fabrication and Characterization of Perovskite-based Solar Cells using Lead Phthalocyanine as Hole Transport Layer[C]//International Conference on Sustainable Energy & Environmental Sciences (SEES). Proceedings. Global Science and Technology Forum, 2015: 85.

[8]Tong P, Cui Y, Hao Y, et al. Improving Performance of Organic Solar Cells with PEG-coated Gold Nanorods Doped in the Active Layer[C]//Asia Communications and Photonics Conference. Optical Society of America, 2015: ASu1A. 4.

[9]Yang J, Siempelkamp B D, Liu D, et al. Investigation of CH3NH3PbI3 degradation rates and mechanisms in controlled humidity environments using in situ techniques[J]. ACS nano, 2015, 9(2): 1955-1963.

[10]Si H, Liao Q, Zhang Z, et al. An innovative design of perovskite solar cells with Al 2 O 3 inserting at ZnO/perovskite interface for improving the performance and stability[J]. Nano Energy, 2016, 22: 223-231.

[11]Meng L, You J, Guo T F, et al. Recent Advances in the Inverted Planar Structure of Perovskite Solar Cells[J]. Accounts of chemical research, 2015.

IOP Conf. Series: Journal of Physics: Conf. Series **864** (2017) 012009 doi:10.1088/1742-6596/864/1/012009

Magnetic and ferroelectric properties of sol – gel synthesized rhombohedral phase $AlFeO_3$ nanoparticles

Mandar M. Shirolkar, Ming Li, Haiqian Wang

Hefei National Laboratory for Physical Sciences at Microscale, University of Science and Technology of China, Hefei 230026, People's Republic of China
hqwang@ustc.edu.cn

Abstract. We report the preparation of a rhombohedral phase (space group $R\bar{3}c$) $AlFeO_3$ nanoparticles (NPs) and its magnetic and ferroelectric properties. The high purity rhombohedral phase AFO NPs of an average size 50 nm were prepared by sol – gel route followed by annealing at 700 °C. The NPs show antiferromagnetic nature at room temperature and near room temperature, weak ferroelectric properties. The coexistence of magnetic and ferroelectric features indicate that AFO NPs can be used as a prospective lead free multiferroic material for the spintronics applications.

1. Introduction

Recently, the perovskite type oxide ABO_3 materials exhibiting ferro or anti – ferromagnetic and ferroelectric ordering, known as multiferroics have gained significant scientific and technological attention [1-4]. In the multiferroic materials, the electric and magnetic properties are mutually controlled, which enable its use in potential device applications such as sensors, data storage devices, etc.[3-6] . In this context, the search for single-phase magnetoelectric materials, with appropriate properties for practical applications is a significant challenge for physicists and material scientists. Furthermore, most of the single-phase magnetoelectric materials are produced from heavy metal such as lead (Pb) [7, 8]. Therefore, for environmental and public health reasons, the researchers and manufacturers of devices that usage ferroelectric and/or magnetoelectric materials are increasingly interested in reducing or even, in extreme cases, completely eliminating the heavy metals from these materials i.e., in producing lead – free materials [7].

$AlFeO_3$ (AFO) is a lead – free ABO_3 type perovskite material, which exists in two different phases, orthorhombic and rhmbohedral [9]. The orthorhombic AFO (O – AFO) retains collinear ferrimagnetic structure with Nèel temperature (T_N) in the range 210 – 250 K [9-11]. The magnetic ordering in O – AFO is due to the metal – oxygen – metal antiferromagnetic superexchange interactions [11]. In O – AFO, the Al and Fe atoms, respectively occupy the A and B positions of the standard ABO_3 perovskite. Thus, Fe atom is having six-fold coordination inside the oxygen octahedra of perovskite [9, 12]. The preparation and multiferroic properties of O - AFO bulk particles and thin films have been reported by several researchers [9-11, 13]. However, there are no experimental reports on the preparation and physicochemical properties of the rhombohedral AFO (R – AFO) particles. Herein, we report the preparation of high purity R – AFO nanoparticles and its structural, magnetic and ferroelectric properties.

2. Experimental

2.1. Preparation of $AlFeO_3$ nanoparticles

AFO nanoparticles were prepared by the chemical route followed by the annealing. Briefly, the reaction steps are as follows. All the precursors were of analytical grade and used without further purifications. The initial precursor preparation was carried out at room temperature. The precursors, $Al(NO_3)_3.9H_2O$ and $Fe(NO_3)_3.9H_2O$ in equilimolar ratio were dissolved in 20 ml Milli-Q water. The mixture was kept under constant stirring condition and 10 ml NH_4OH (25 wt % concentration) was added dropwise to above mixture. The addition of NH_4OH, turns the mixture dark red in color. The reaction was further kept under constant stirring conditions for 1 hour. The small amount of acetone was added to above reaction to precipitate the precursor required for $AlFeO_3$ nanoparticles preparation. The obtained precursor was further washed with Milli – Q water and dried at 80°C. The obtained powder was further annealed at 700°C for 1 hour.

2.2. Characterizations

The as prepared nanoparticles were characterized by several characterization techniques. The X – ray diffraction (XRD) of the sample was obtained using PANALYTICAL X'Pert Pro (Cu – Kα radiation source with λ ≈ 1.5406 Å). The Rietveld refinement of XRD pattern was carried out using Fullprof (Version July 2016). The transmission electron microscopy (TEM) of the nanoparticles was carried out using JEOL JEM – ARM200F microscope. The magnetic measurements (M – H and dc magnetization) on the nanoparticles were performed using a magnetic property measurement system (MPMS) from MPMS 5, Quantum Design Inc. San Diego, CA equipped with a 7 T superconducting magnet. The ferroelectric characteristic of the nanoparticles was studied at room temperature. For the measurement, sample in the pellet form was prepared using a conventional axial cold pressed technique followed by sintering at 500°C for 30 minutes (pellet diameter: 5 mm, electrode material: silver paste). The measurements were carried out using aixACCT TF Analyzer 2000 FE at a 1 kHz ac frequency.

3. Results and discussion

The phase purity of the sample was determined by XRD (see figure 1(a)). The measurement confirm the $R\bar{3}c$ phase of nanoparticles [10]. The lattice parameters were observed to be a = 5.030 Å and c = 13.350 Å, which are nearly close to the reported values for the rhombohedral phase of AFO particles [9].

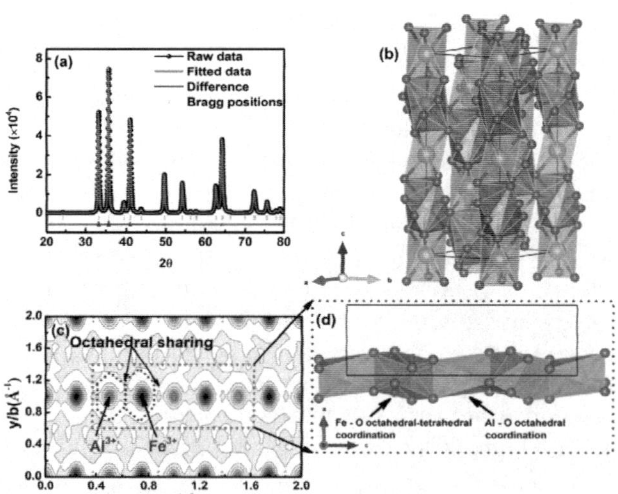

Figure 1. (a) Rietveld refined XRD pattern of AFO nanoparticles. (b) Unit cell of $AlFeO_3$ nanoparticles representing FeO_6 polyhedral and AlO_6 octahedral coordination. (c) and (d) shows 2D electron density plot of unit cell in the (104) direction and 3D representation of polyhydra (Fe – O) – octahedra (Al – O) sharing with reference to electron density map with in $AlFeO_3$ unit cell respectively.

IOP Conf. Series: Journal of Physics: Conf. Series **864** (2017) 012009 doi:10.1088/1742-6596/864/1/012009

The refinement also shows that Fe – O coordination forms distorted octahedra and appears to be polyhedra, while Al – O coordination observed to be octahedra (see figure 1(b) and (c)). The polyhedra – octahedra coordination was observed to be either face sharing or edge sharing with this the unit cell (see figure 1(d)).

The particle size was determined by TEM. The nanoparticles were observed to be a nearly spherical in shape with an average size ≈ 50 nm (see figure 2(a) and (b)).

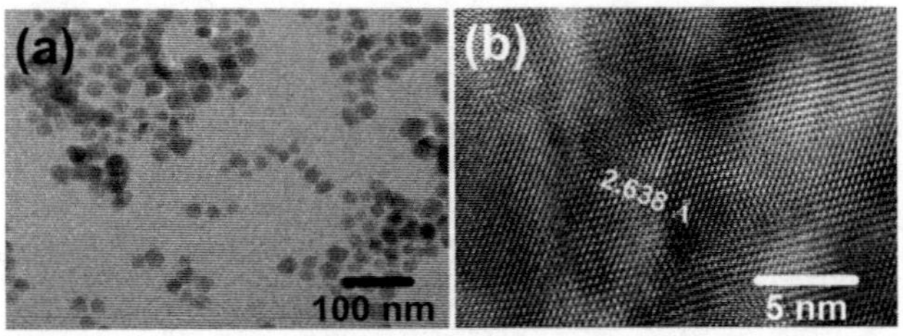

Figure 2. (a) and (b) shows TEM and HRTEM micrographs of $AlFeO_3$ nanoparticles respectively.

Figure 3(a) shows magnetic measurements on AFO nanoparticles. The magnetic measurements indicate that at room temperature (300 K) AFO nanoparticles exhibit linear hysteresis loop indicating antiferromagnetic behavior of the nanoparticles.

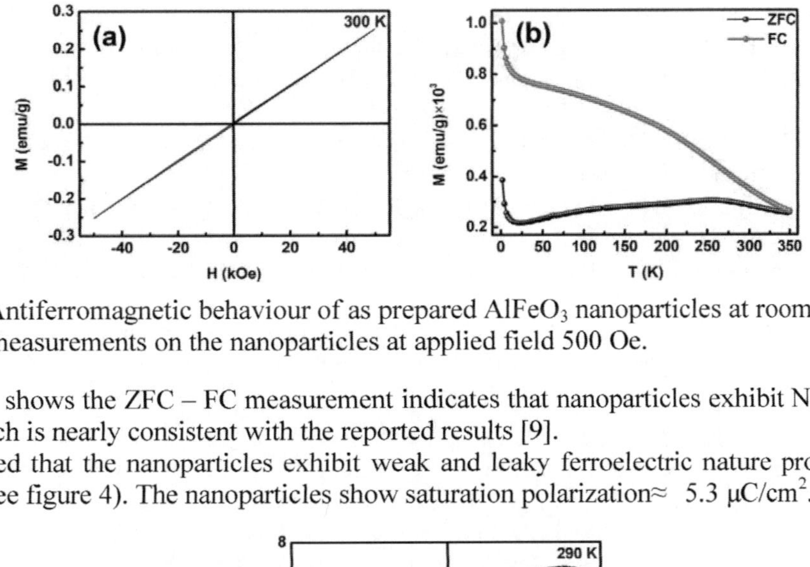

Figure 3. (a) Antiferromagnetic behaviour of as prepared $AlFeO_3$ nanoparticles at room temperature, (b) ZFC – Fc measurements on the nanoparticles at applied field 500 Oe.

Figure 3(b) shows the ZFC – FC measurement indicates that nanoparticles exhibit Nèel temperature ≈ 290 K, which is nearly consistent with the reported results [9].

We observed that the nanoparticles exhibit weak and leaky ferroelectric nature proximity to Nèel temperature (see figure 4). The nanoparticles show saturation polarization≈ 5.3 μC/cm^2.

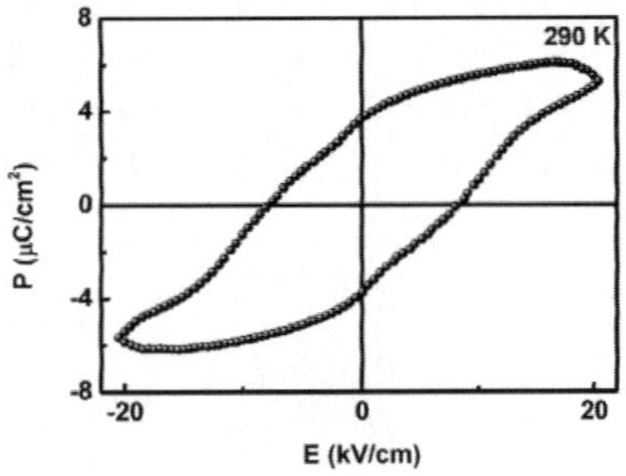

Figure 4. Ferroelectric hysteresis loop acquired on the as prepared $AlFeO_3$ nanoparticles.

4. Conclusions

In summary, the perovskite rhombohedral phase $AlFeO_3$ nanoparticles of an average size 50 nm were prepared by chemical route. The magnetic and ferroelectric properties were explored on these as prepared nanoparticles. The results presented here shed some light on multiferroic behavior of AFO nanoparticles, which can be a prospective candidate for fabrication of future device applications.

Acknowledgments
This work is supported by the National Natural Science Foundation of China (Grant No.: 21427804) and China Postdoctoral Foundation Grant (Grant No.: BH2340000067).

References

[1] Catalan G and Scott J F 2009 Physics and Applications of Bismuth Ferrite *Advanced Materials* **21** 2463-85
[2] Hill N A 2000 Why Are There so Few Magnetic Ferroelectrics? *The Journal of Physical Chemistry B* **104** 6694-709
[3] Ma J, Hu J, Li Z and Nan C W 2011 Recent progress in multiferroic magnetoelectric composites: from bulk to thin films *Adv Mater* **23** 1062-87
[4] Shirolkar M M, Hao C, Dong X, Guo T, Zhang L, Li M and Wang H 2014 Tunable multiferroic and bistable/complementary resistive switching properties of dilutely Li-doped $BiFeO_3$ nanoparticles: an effect of aliovalent substitution *Nanoscale* **6** 4735-44
[5] Rao C N, Sundaresan A and Saha R 2012 Multiferroic and Magnetoelectric Oxides: The Emerging Scenario *J Phys Chem Lett* **3** 2237-46
[6] Tokura Y and Seki S 2010 Multiferroics with spiral spin orders *Adv Mater* **22** 1554-65
[7] Edri E, Kirmayer S, Cahen D and Hodes G 2013 High Open-Circuit Voltage Solar Cells Based on Organic-Inorganic Lead Bromide Perovskite *J Phys Chem Lett* **4** 897-902
[8] Oishi Y, Tsukamoto E, Shimoda M, Takamuku T, Narita T and Era M 2013 A novel preparation method of lead-based layered perovskite Langmuir films with a negligible amount of $PbBr_2$ *New Journal of Chemistry* **37** 568
[9] Saha R, Shireen A, Shirodkar S N, Singh M S, Waghmare U V, Sundaresan A and Rao C N 2011 Phase transitions of $AlFeO_3$ and $GaFeO_3$ from the chiral orthorhombic ($Pna2_1$) structure to the rhombohedral ($\overline{R}3c$) structure *Inorg Chem* **50** 9527-32
[10] Saha R, Shireen A, Shirodkar S N, Waghmare U V, Sundaresan A and Rao C N R 2012 Multiferroic and magnetoelectric nature of $GaFeO_3$, $AlFeO_3$ and related oxides *Solid State Communications* **152** 1964-8
[11] Shireen A, Saha R, Mandal P, Sundaresan A and Rao C N R 2011 Multiferroic and magnetodielectric properties of the $Al_{1-x}Ga_xFeO_3$ family of oxides *J. Mater. Chem.* **21** 57-9
[12] Saha R, Sundaresan A and Rao C N R 2014 Novel features of multiferroic and magnetoelectric ferrites and chromites exhibiting magnetically driven ferroelectricity *Mater. Horiz.* **1** 20-31
[13] Hamasaki Y, Shimizu T, Taniguchi H, Taniyama T, Yasui S and Itoh M 2014 Epitaxial growth of metastable multiferroic $AlFeO_3$ film on $SrTiO_3$ (111) substrate *Applied Physics Letters* **104** 082906

MBE growth of nanowires using colloidal Ag nanoparticles

A D Bouravleuv[1-3], I V Ilkiv[1,4], R R Reznik[1], I V Shtrom[1-4], A I Khrebtov[1], Yu B Samsonenko[1,3], I P Soshnikov[1-3], G E Cirlin[1-4], H Lipsanen[5]

[1]St. Petersburg Academic University RAS, St. Petersburg 194021, Russia
[2]Ioffe Institute RAS, St. Petersburg 194021, Russia
[3]Institute for Analytical Instrumentation RAS, St. Petersburg 190103, Russia
[4]St.Petersburg State University, St.Petersburg 199034, Russia
[5]Aalto university, Espoo FI-02150, Finland

e-mail: bour@mail.ioffe.ru

Abstract. Ag colloidal nanoparticles are used as a catalyst for growth of GaAs nanowires by the molecular beam epitaxy on the Si(111) and GaAs(111)B substrate surfaces. The scanning electron microscopy measurements revealed that the nanowire formation occurs in different ways on different substrates, but the parameters of the synthesized nanowires open great prospects for their further use.

1. Introduction

Semiconductor nanowires (NWs) are considered to be the one of the most promising materials for the realization of new nano- and optoelectronic devices. Nowadays, the majority of semiconductor compounds can be grown in a shape of these quasi one-dimensional nanostructures. Moreover, they can be formed on semiconductor substrates with large lattice mismatch, which opens up great technological advantages and opportunities. It is well known that the surface location of individual semiconductor NWs can be predefined by the location of catalytic droplets. Up to now, the NW synthesis via a well-established vapour–liquid–solid (VLS) mechanism has been carried out using different metals as a catalyst for their growth [1], since the most common catalytic metal, such as gold, could result in the incorporation of its atoms into NWs with solubility exceeding the solubility limit of the bulk materials [2-4]. The existence of uncontrolled doping can seriously limit the number of possible NW applications, because gold can introduce deep level traps the band gap and act as a recombination center for charge carriers. Therefore, the other technological approaches, for example self-catalyzed [5] or template-assisted growth of NWs [6], and the use of alternative metals, e.g. Ag, which creates relatively shallow energy levels, are of particular interest. The silver has been already used for the growth of different NWs, such as In_2O_3 [7], Si [8], InSb [9], Ge [10], InAs [11], InP [12], $In_xGa_{1-x}As$ [13] and GaAs [14]. The metal catalytic droplets can be formed on the semiconductor substrate surface by means of different methods. One of the most simple among them, which is the thermal annealing of preliminary deposited metal thin films, does not allow us to control the surface density and size distribution of metal droplets created by their decomposition. Furthermore it cannot provide ultra low surface density of the droplets, which is of importance for some NW applications and their investigations. Other techniques based on different kind of beam lithography are quite time consuming. Alternatively, metal nanoparticle colloids, which have gained widespread acceptance in different areas due to their properties, are very attractive to be used for NW synthesis [15]. They, especially gold nanoparticle colloids, have been widely used as a catalyst for the growth mostly by different CVD methods [16].

Here we report on Ag and Au colloidal nanoparticle assisted MBE growth of GaAs NWs on Si and GaAs substrates.

2. Experiments

The solid-source Compact 21 TM Riber MBE system equipped with additional high vacuum connected chamber for gold deposition, which allows to perform the annealing of the samples at the temperatures up to 950°C, was used for the growth experiments. The growth of GaAs NWs was carried out on epi-ready Si(111) and GaAs(111)B substrates. We used commercially available gold and silver nanoparticle colloids with diameters of 40 and 20 nm, correspondingly, produced by BBI Company. The deposition of the colloidal nanoparticles was carried out after the substrate cleaning in deionized water and ethanol.

Prior to growth, the samples with deposited nanoparticles were loaded into additional chamber for the thermal annealing, which was carried out for 5 minutes at 850°C and 630°C for Si and GaAs substrates, correspondingly. Then, the temperature was decreased and the samples were transferred to the main chamber equipped with Ga effusion cell and As cracking cell in high vacuum conditions. After reaching the growth temperature of 580°C, Ga and As shutters were simultaneously opened. The growth rate was corresponded to 1 ML/s of 2D layers. The total NW growth time was equal to 15 minutes. The reflection high-energy electron diffraction (RHEED) method was used to control in situ the processes occurring on the substrate surface. RHEED patterns demonstrated the beginning NW nucleation appeared after 20 seconds of the growth. According to the RHEED patterns GaAs NWs had mixed wurzite/zincblende phases.

The morphological properties of the samples have been studied using scanning electron microscopy (SEM) Zeiss SUPRA 25.

3. Results

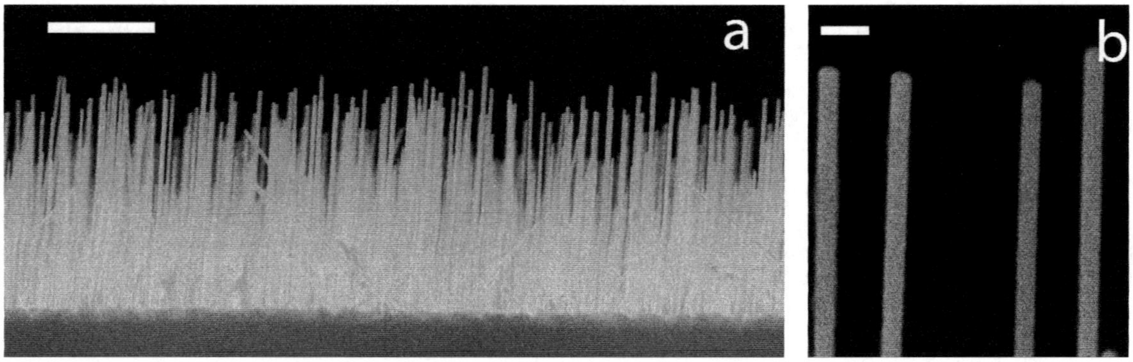

Figure 1. SEM images of GaAs NWs grown on the Si(111) substrates using Au (a) and Ag (b) colloidal nanoparticles. Scale bar corresponds to 2 μm (a) and 200 nm (b).

The GaAs NWs obtained by Au and Ag nanoparticle assisted MBE growth on the Si(111) substrates are shown in Figure 1. The results obtained demonstrate that arrays of vertical GaAs NWs can be grown using colloidal nanoparticles by MBE. The NWs grown on Si using Au nanoparticles have higher surface density may be due to the higher concentration of Au compare to Ag nanoparticles inside colloidal solution (see Fig. 1 a). Their average NW length is about 4.3 μm, whereas NW obtained using Ag nanoparticles were a bit smaller. They are of 3.2 μm in length. At the same time their diameters are quite close to each other, despite to different sizes of Au and Ag nanoparticles, which were 40 and 20 nm, correspondingly. The average diameter reasonably exceeds those values, since it is about 90 nm. This question requires an additional clarification. But it can be connected with different solubility of Ga atoms inside Au and Ag catalytic droplets. It should be noted that NWs were grown perpendicular to the substrate surface, i.e. they were grown along <111> crystallographic direction. As it was expected, they have quite small size distributions and they are do not change their shape along the growth axe. This opens up good perspectives for their implementations.

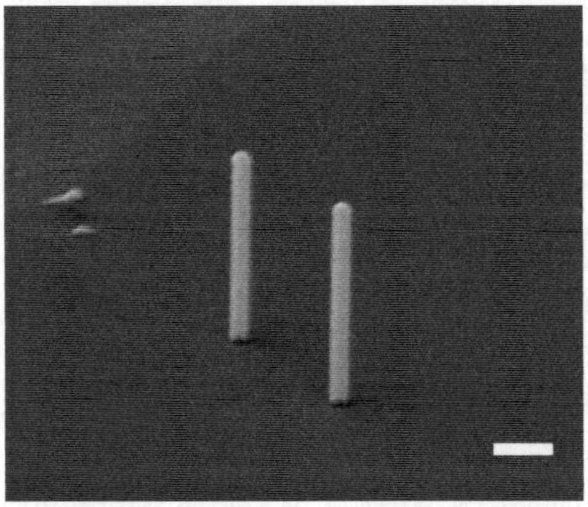

Figure 2. SEM image of GaAs NWs grown on the GaAs(111)B substrate using Ag colloidal nanoparticles as a catalyst for their growth. Scale bar corresponds to 200 nm.

The use of GaAs substrate for the Ag-assisted MBE synthesis of GaAs has also led to good results. The SEM investigation of the samples obtained shows that NWs obtained were also straight, as one can see in Figure 2. They were growth along <111> direction. But their average diameter (55 nm), as well as their length (about 0.6 μm) was smaller compare to the parameters demonstrated by NWs grown on Si substrates.

4. Conclusions

In summary, the GaAs NWs have been obtained for the first time by molecular beam epitaxy using Ag colloidal nanoparticles as a growth catalyst on the Si(111) and GaAs(111) substrate surface. The NWs obtained are straight and have small size distribution. It was found that despite similar growth condition, the formation of NWs on different semiconductor substrates could result in significant differences in the NW parameters.

Acknowledgments

This work was supported by the grant of the Ministry of Education and Science of Russian Federation (agreement № 14.613.21.0044, unique project identifier RFMEFI61315X0044). I.V. Shtrom and I.V. Ilkiv are grateful for the financial support provided by St.Petersburg State University (grant № 11.37.210.2016).

References

[1] Dick K A and Caroff P 2014 *Nanoscale* **6** 3006
[2] Perea D E, Allen J E, May S J, Wessels B W, Seidman D N and Lauhon L J 2006 *Nano Lett.* **6** 181
[3] Allen J E, Hemesath E R, Perea D E, Lensch-Falk J L, Li Z Y, Yin F, Gass M H, Wang P, Bleloch A L, Palmer R E and Lauhon L J 2008 *Nature Nanotech.* **3**, 168
[4] Bar-Sadan M, Barthel J, Shtrikman H and Houben L 2012 *Nano lett* **12**, 2352
[5] Samsonenko Yu B, Cirlin G E, Khrebtov A I, Bouravleuv A D, Polyakov N K, Ulin V P, Dubrovskii V G, Werner P 2011 *Semiconductors* **45** 431
[6] Ikejiri K, Sato T, Yoshida H, Hiruma K, Motohisa J, Hara S and Fukui T 2008 *Nanotechnology* **19** 265604
[7] Zhang J, Qing X, Jiang F, Dai Z 2003 *Chem. Phys. Lett.* **371** 311
[8] Wittemann J V, Münchgesang W, Senz S and Schmidt V J 2010 *Appl. Phys. Lett.* **107** 096105
[9] Vogel A T, de Boor J, Becker M, Wittemann J V, Mensah S L, Werner P and Schmidt V 2011 *Nanotechnology* **22** 015605
[10] Lotty O, Biswas S, Ghoshal T, Glynn C, O' Dwyer C, Petkov N, Morris M A and Holmes J D 2013 *J. Mater. Chem. C* **1**, 4450
[11] Pan D, Fu M, Yu X, Wang X, Zhu L, Nie S, Wang S, Chen Q, Xiong P, von Molnár S, and Zhao J 2014 *Nano Lett.* **14**, 1214
[12] Huang K, Zhang Z, Zhou Q, Liu L, Zhang X, Kang M, Zhao F, Lu X, Gao X and Liu J 2015 *Nanotechnology* **26** 255706
[13] Sarkar K, Palit M, Banerji P, Chattopadhyay S, Halder N N, Biswas P, Nagabhusana B and Chowdhurya S 2015 *CrystEngComm* 10.1039/c5ce01565k
[14] Lindberg C, Whiticar A, Dick K A, Sköld N, Nygård J and Bolinsson J 2016 *Nano Lett* **16**, 2181
[15] Gomes U P, Ercolani D, Zannier V, Beltram F and Sorba L 2015 *Semicond. Sci. Technol.* **30** 115012
[16] Tchernycheva M, Rigutti L, Jacopin G, de Luna Bugallo A, Lavenus P, Julien F H, M.Timofeeva, Bouravleuv A D, Cirlin G E, Dhaka V, Lipsanen H, Largeau L 2012 *Nanotechnology* **23**, 265402

High Quality Transferable AlN Thin Film by PLD

Heng Li, Xiaoli Lu*, Xin Li, Jincheng Zhang and Yue Hao

Wide Band Gap Semiconductor Technology Discipline State Key Laboratory, School of Microelectronics, Xidian University, Xi'an 710071, China
*Email: xllu@xidian.edu.cn

Abstract. AlN thin film was epitaxial grown on c-plane sapphire substrate by pulsed laser deposition. To reduce structural defects from largely lattice mismatched substrate, MgO or ZnO buffer layer was inserted between AlN and sapphire. Crystal structure and surface morphology of as prepared AlN were characterized by XRD, AFM, and SEM. It was found that buffer layers significantly improve crystalline quality of AlN, especially using ZnO. Furthermore, a general and steady wet chemical process was developed to selectively etch away ZnO layer, so that high quality free-standing AlN thin film was obtained. This film could be transferred onto any other host substrates such as Si, quartz, etc. Moreover, with no clamping effect from the substrate, the as-prepared free-standing AlN thin films may find potential applications in high sensitivity piezoelectric devices, flexible wearable detectors and so on.

1. Introduction

Aluminium nitride（AlN）is a wide bandgap semiconductor with a variety of outstanding physical and chemical properties, which make it one of the most promising next-generation semiconductor materials for several applications. With its wide bandgap of 6.2 eV[1], high electrical resistivity, high temperature stability, and high breakdown voltage, AlN can be used to fabricate power devices. Thin films of AlN are commonly exploited to manufacture optoelectronic devices covering wavelength range from visible spectral to deep ultra-violet. Moreover AlN exhibits high sound and piezoelectricity that make it suitable candidate for resonators and surface acoustic wave devices [2]. And AlN is a suitable buffer layer not only for GaN but also for SiC with similar crystal structure and lattice mismatch as little as 2% [3].

Since high quality single crystal AlN had yet been successfully fabricated, several deposition techniques were used to prepare thin films of AlN, for instance, magnetron sputtering, MOCVD, MBE and PLD. Compared to those reactive deposition techniques，PLD is a physical process with low cost and high efficiency, so it remains a challenge to enhance the activity of N element and improve the quality [4]. It was reported that introducing NH_3 or N plasma during the deposition can result in better epitaxial quality of AlN [5]. In this paper, N components ablated by laser were found remaining active in high vacuum, based on this result, quality of AlN film was further improved by inserting MgO or ZnO buffer layer. More importantly, the buffer layer can be selectively etched in solution and AlN film was lifted off from original substrate and transferred onto any other substrate, which provided a versatile way to fabricate high quality free-standing semiconductor nanosheets on large scale with various potential applications.

2. Experiment

Epitaxial AlN films were deposited through a standard PLD process. Single crystalline c-plane sapphire was used as substrate and temperature was up to 600℃ during deposition. Laser energy density varied from 1 J/cm^2 to 3 J/cm^2 for the pre-deposition of MgO or ZnO and deposition of AlN. After finishing deposition, N_2 gas was introduced into the chamber and the films cooled down to room temperature.

MgO or ZnO inserted between AlN and sapphire substrate also acted a sacrificial layer to transfer AlN film onto any substrate. PMMA photoresist was spread on AlN surface and after baked on the hot plate, the photoresist turned to be a firm and stable protector. Then these multilayers were immerged into $(NH_4)_2SO_4$ solution and the MgO or ZnO was selectively etched away. Free-standing AlN film

IOP Conf. Series: Journal of Physics: Conf. Series **864** (2017) 012011 doi:10.1088/1742-6596/864/1/012011

was lifted off from original substrate and transferred onto Si or transparent quartz, while acetone was used to dissolve PMMA. This process was graphically illustrated in Figure 1.

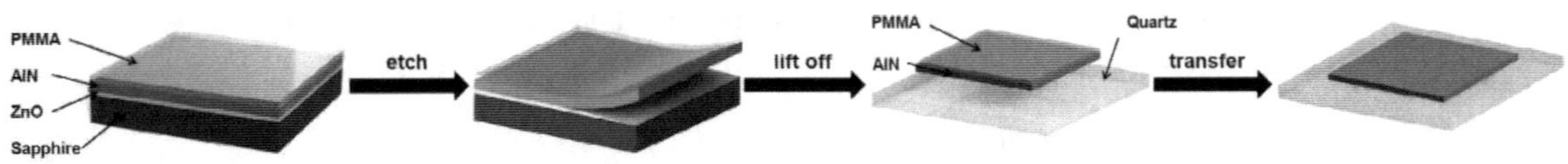

Figure 1. Etching and transfer process of AlN film

3. Results and discussion

3.1 Multilayer structure

As prepared MgO or ZnO buffer layers between AlN film and sapphire substrate were carefully characterized. Figure 2a showed the cross section of AlN thin films with MgO layers on c-plane sapphire. Clear boundaries across the section were used to measure different thicknesses of layers on the interface. The deposition rates was evaluated as 200 nm AlN of 5000 pulses, 80 nm MgO of 5000 pulses, and 90 nm ZnO of 3000 pulses approximately. The thickness data were further verified by fitting results of ellipsometer. And from the top view of AlN film as Figure 2a, the surface morphology is extremely smooth with very low density of particulates.

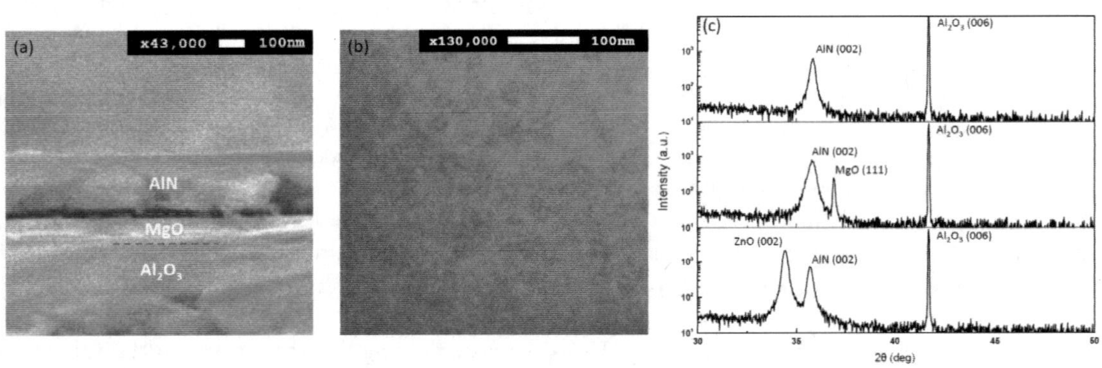

Figure 2. (a) SEM cross section, (b) top view of AlN/MgO/sapphire multilayer structure, (c) XRD 2θ-ω scans of different multilayer structures

Figure 2c was 2θ-ω scan result of as prepared multilayer samples by X-ray diffraction (XRD). The diffraction pattern showed that all the AlN films from different structure were highly c-axis oriented grown. And the MgO peak located at 35.9° was indexed to (111) orientation, and ZnO at 34.4° was of (002) orientation [6-7].

3. 2 Improvement of AlN with buffer layer

The MgO or ZnO layer of the multilayer structures was used to improve the quality of epitaxial AlN film as a buffer layer. Further researches have been carried out to reveal this. The full-width at half maximum (FWHM) of the ω-scans (rocking curves) by XRD could be used to describe the crystal quality of as deposited AlN films [8]. According to Tab.1, AlN film on ZnO buffer layer showed the narrowest FWHM of 0.74°, which indicating the highest degree of epitaxial quality, compared to the films on MgO buffer layer or bare sapphire substrate.

IOP Conf. Series: Journal of Physics: Conf. Series **864** (2017) 012011 doi:10.1088/1742-6596/864/1/012011

Table 1. FWHM of AlN films

	AlN (002) on Al_2O_3 (006)	AlN (002) on MgO (111)	AlN (002) on ZnO (002)
mismatch	13.2%	4.4%	4.1%
FWHM	2.01°	1.02°	0.74°

Atomic force microscopy (AFM) was used to study the surface morphology of AlN films. A smooth surface is an essential requirement for various applications, for example, to provide a suitable substrate for epitaxial growth of GaN [9]. As shown in Figure 3, the direct grown AlN film without buffer layer showed a relatively rough surface with root mean square (RMS) roughness of 2.699 nm. By pre-depositing MgO buffer layer, the RMS roughness was significantly reduced to 0.5642 nm as shown in Figure 3b. And ZnO buffer layer can further improve the surface morphology of AlN films with RMS roughness as low as 0.2684 nm as shown in Figure 3c.

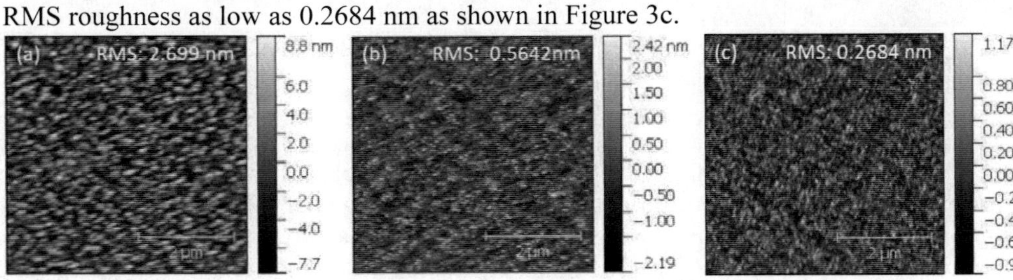

Figure 3. AFM images of AlN films surfaces (a) AlN/Sapphire (b)AlN/MgO/Sapphire (c)AlN/ZnO/Sapphire

3. 3 Transferred AlN thin films

After the etching and transfer process of AlN thin film, XRD and SEM measurement were firstly carried out to characterize the quality of transferred AlN film on Pt coated Si substrate. As the XRD result shown in Figure 4a, only AlN (002) peak was found, indicating a c-axis crystal orientation AlN as same as the original one. Figure 4b and 4c were the top view of transferred AlN film by SEM. Smooth and uniform AlN nanosheets were found in large scale up to tens of micrometers, and the close-up SEM image in large magnification revealed a similar surface topography to the surface in Figure 2b. All these results confirmed that crystal structure and surface morphology of free-standing AlN films were maintained, and the process of etching and transfer was steady and highly reliable.

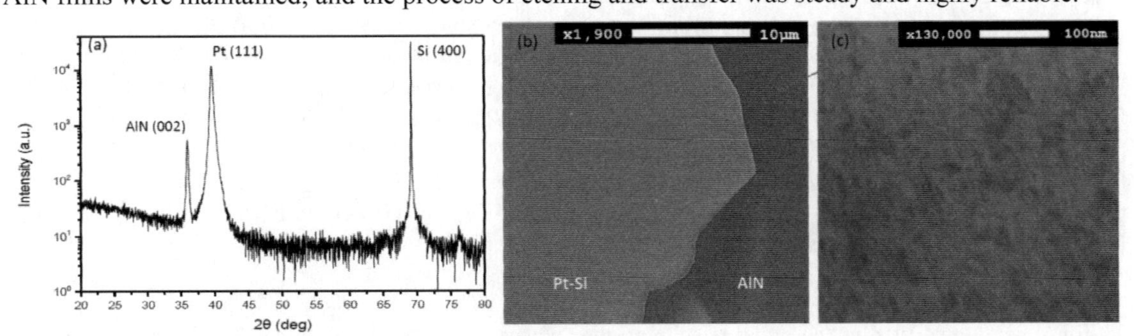

Figure 4. Characterization of AlN film transferred on Pt-Si (a) 2θ-ω result from XRD (b) SEM image in large scale (c) large magnification SEM image of a part

Optical transmittance spectra was also used to investigate the bandgap of transferred AlN film on full wavelength transparent quartz. Using the method proposed by Swanepoel, the absorption coefficient was calculated therefore the bandgap could be obtained from intercept of the plot [10]. As

IOP Conf. Series: Journal of Physics: Conf. Series **864** (2017) 012011 doi:10.1088/1742-6596/864/1/012011

shown in Figure 5, light transmittance remained above 80%, until strong absorption started at about 250 nm wavelength. The bandgap of transferred AlN film was then calculated from this spectra data to be nearly 6.0 eV, which is very close to the theory value 6.2 eV [11]. This result revealed that the transferred AlN film on transparent quartz may find promising applications in ultraviolet photoelectric field, as well as in other fields on respective substrate

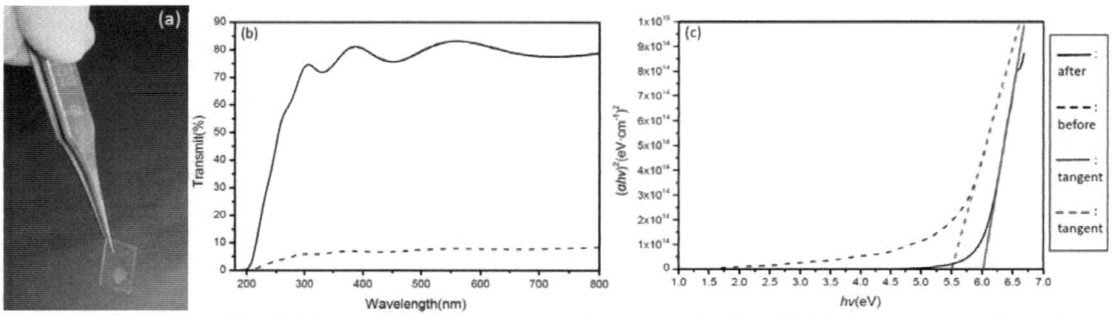

Figure 5. (a) photo of AlN film transferred on quartz (b) transmit of AlN film before transfer and after transfer (c) fitting of AlN optical bandgap

4. Conclusion

MgO or ZnO film was epitaxial growth on c-plane sapphire substrate as a buffer layer to improve the quality of PLD deposited AlN film. Through XRD, AFM, and SEM measurements, ZnO was found to be a better choice compared with MgO, since its similar crystal structure and less lattice mismatch towards AlN. And then ZnO was selectively etched away in $(NH_4)_2SO_4$ solution, releasing AlN layer from the original sapphire substrate. As obtained free-standing AlN film could be transferred onto any substrate. Researches of the films on Pt-Si and quartz substrate revealed that the AlN remained chemical stable, morphology steady and performance outstanding. This high reliable transferable AlN thin film provides a variety of promising applications, for instance, heteroepitaxial growth of GaN on any substrate, flexible and wearable ultraviolet photo detectors, clamping free SAW devices and so on.

References

[1] Zhongqi Shi, Yongyong Zou, Ruifeng Jing *Functional Materials Letters* (Vol. 8, No. 2 2015)
[2] Hullavarad S S, Vispute, RD, Nagaraj *47th Electronic Materials Conference* (JUN 2005)
[3] Duan Huantao, Hao Yue, Zhang Jincheng *Journal of Semiconductors* (SEP 2009)
[4] A basillais, r Benzerga, h Sanchez *Applied Physics* (A80, 2005)
[5] Tokumoto Y, Sato Y, Yamamoto *International Symposium on Characterization of Real Materials and Real Processing by Transmission Electron Microscopy* (JAN 26-27, 2005)
[6] Jin Bock Lee, Myung Ho Lee, Chang Kyun Park, Jin Seok Park *Thin Solid Films* (JUN, 2008)
[7] K Noda, Y Terai, N Miura, H Udono, Y Fujiwara, *Phys Proced* (23.5, 2012)
[8] D G Zhao, D S Jiang, L L Wu, L C Le, L Li *Alloys Compd* (544 94, 2012)
[9] L Vergara, J Olivares, and E Iborra *Thin Solid Films* (515, 2006)
[10] WeiFan Hsu, HuiLing Kao, ZihPing Lin *Journal of Crystal Growth* (FEB 2016)
[11] Huan He, Lirong Huang, Min Xiao *Mater Electron* (24, 2013)

Acknowledgments

The authors acknowledge the support of National Natural Science Foundation of China (Grant Nos. 51202176, 51572211).

Doping effect in Si Nanocrystals/SiO_2 multilayers

Dongke Li, Peng Lu, Minqing Qian, Jun Xu*, Wei Li, Kunji Chen
School of Electronic Science and Engineering, National Laboratory of Solid State Microstructures, Collaborative Innovation Center of Advanced Microstructures, Nanjing University, Nanjing, 210093, China.

*E-mail: junxu@nju.edu.cn

Abstract. Doping in Si nano-crystals (Si NCs) is currently a great challenge to develop the high-performance nano-devices. Here, we fabricated P-doped Si NCs by PECVD in Si/SiO_2 multilayered structures after annealing at high temperature. It is demonstrated experimentally that P dopants can be incorporated into Si NCs substitutionally. Furthermore, we found that the photoluminescence properties of Si NCs can be drastically modified by P doping in Si NCs/SiO_2 multilayers with the ultra-small size. A subband light emission centred at 1200nm is observed, which can be ascribed to the radiative recombination via the P-induced deep level in the gap of the Si NCs. Interestingly, it is also found that the subband emission can be enhanced obviously by B and P co-doping. Our results suggest that doping plays an important role in the electronic structures and optoelectronic characteristics of Si NCs, which provide a new route to realize the Si NCs-based optical and electronical devices.

1. Introduction

Doping in Si nano-crystals (Si NCs) is a fundamental issue in order to develop the next generation nano- and opto-electronic devices, such as nonvolatile memories[1], light emitting devices[2] and solar cells[3]. However, the theoretical calculation implies that doping in Si NCs is quite difficult since the formation energy related to the impurities is higher than that in mono-crystalline case. So the impurities tend to be expelled from the inner sites of Si NCs which is so-called self-purification effect[4]. In spite of these theoretical difficulties, several experimental studies have reported the doping effect in Si NCs[5-6]. It was reported that the dopant atoms location, the active doping concentration as well as the influences of dopants on the physical properties of Si NCs is quite different from the doping in bulk Si, and the doping mechanism is still unclear[7-8].

In the present work, doped Si QDs/SiO_2 multilayers were fabricated in conventional PECVD system after high temperature annealing. The doping behaviors in Si NCs with various doping concentrations were systematically studied by Raman, TEM and low temperature ESR techniques. It was found that photoluminescence properties of the ultra-small size Si NCs can be drastically modified by impurity doping and the subband emission was observed. Moreover, the subband emission can be improved by co-doping.

2. Experiment

The P-doped amorphous-Si/SiO_2 stacked multilayers were deposited on p-type Si wafers (1–3 Ω/cm) and quartz substrates by a PECVD system with the radio frequency of 13.56 MHz. The a-Si:H layer was deposited by using SiH_4 with a flow rate of 5 sccm. P doping or B/P co-doping was realized through adding PH_3 or PH_3+B_2H_6 into chamber during the a-Si layer deposition process. The actual content of

PH_3 and B_2H_6 are 1% by H_2 dilution. The deposition time and flow rate of PH_3 and B_2H_6 were varied to obtain different layer thickness and doping concentrations. Subsequently, the in-situ plasma oxidation was performed in oxygen ambient for 90s with a fixed O_2 flow rate. The deposition and oxidation process was alternatively repeated for times to get stacked multilayered structures. During the whole deposition process, the RF power was set at 50W and the substrate temperature was fixed at 250 °C. Following, the as-deposited samples were first dehydrogenated at 450 °C and then annealed at various high temperatures (800-1000 °C) for 1h under N_2 protection to form doped Si NCs.

The microstructure was revealed by HRTEM using a Tecnai G2 F20 electron microscope operating at 200 kV. The low temperature X-band ESR spectra were obtained by Bruker EMX 10/12$^+$. The photoluminescence spectra were obtained by using liquid N_2 cooled InGaAs detector and the excitation source is a 325 nm He-Cd laser with 30 mW power.

3. Results and discussion

3.1. The microstructure of HRTEM. The cross-sectional TEM image of the Si NCs/SiO_2 multilayers after 1000 °C high temperature annealing is shown in Fig.1, and the inset is the magnified image of a single Si NC. The thickness of Si layer and SiO_2 layer are about 7.2 nm and 3.7 nm, respectively. In the Si layer, the dot-shaped Si NCs are formed after 1000 °C annealing, and they are uniformly dispersed in the Si layer. As shown in the inset, the crystalline structure can be identified and the diameter of the NCs is about 6.8 nm, which is consistent with the thickness of the initial a-Si layer. The lattice fringes of the single Si QDs is about 0.31 nm, which is corresponding to (111) plans of Si.

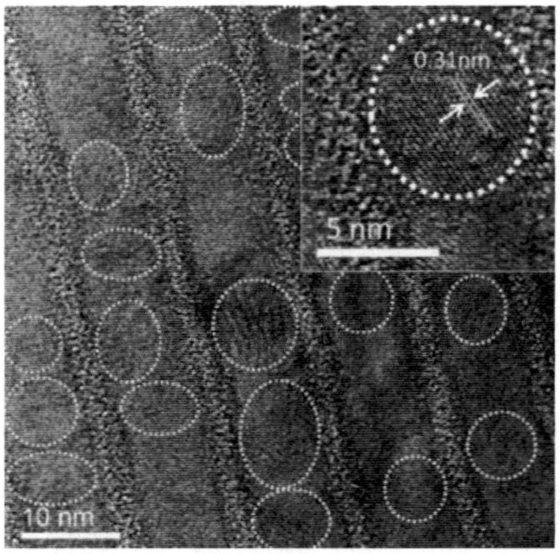

Figure 1. Cross-sectional HRTEM image of P doped Si NCs after 1000°C annealing.

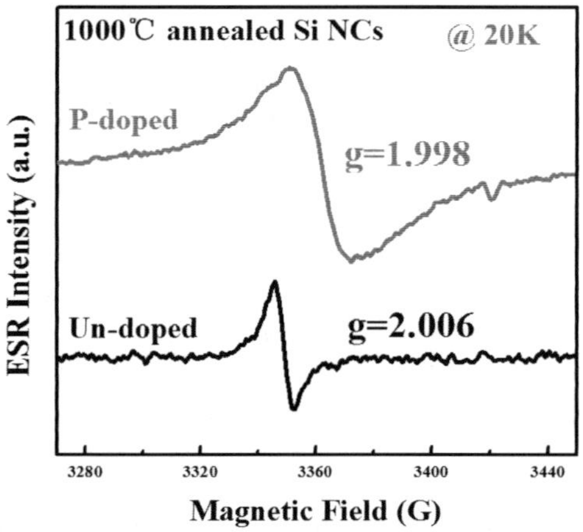

Figure 2. 20K low temperature X-band ESR spectra for un-doped and P doped Si NCs/SiO_2 multilayers with different P concentration after 1000 °C annealing.

3.2. The low temperature ESR for dopant location. The low temperature ESR can help us to explore the hyperfine microstructure of NCs and analyse the dopant location. Fig.2 is the X-band ESR spectra of Si/SiO_2 multilayers after 1000 °C annealing measured at 20 K with and without P doping. As shown in Fig.2, the ESR signal with $g = 2.006$ is detected for un-doped sample, which indicates the existence of dangling bonds in Si NCs-based films. However, when P atoms are introduced into Si NCs, the signal with $g \approx 1.998$ due to conduction electrons (CEs) is observed[9]. The CEs signal indicate some P dopants locate at the substitutional sites inside the Si NCs. When the P concentration is gradually increased, the g value remains at 1.998 for the CEs, and the linewidth of ESR spectra broadens, which prove more P atoms become the active atoms to provide free electrons.

3.3. The subband PL for doped Si NCs. For the ultra-small sized Si NCs, a radiative deep level will be induced in the bandgap by P impurities[10]. It is found that the 2 nm sized P-doped Si NCs/SiO_2 multilayers exhibit a broad subband emission band centered at 1.03 eV (1200 nm) after high temperature annealing. More interesting, we find the enhanced emission intensity with the similar PL spectra for B/P co-doped Si NCs. The mechanism for B/P co-doping luminescence behaviors is unclear at the present stage and the further studied is needed.

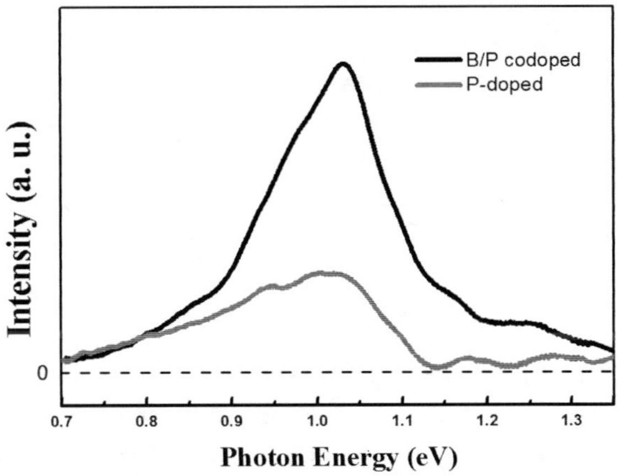

Figure 3. PL spectra for the P doped and B/P co-doped Si NCs/SiO2 multilayers

4. Conclusion

We have fabricated the a-Si/SiO_2 multilayers by PECVD and obtained the P doped and B/P co-doped Si NCs/SiO_2 multilayers after high temperature annealing. The ESR measurements reveal that P atoms can enter the Si NCs at the substitutional sites. For the 2nm ultra-small size Si NCs, P doping can induce a deep energy level and result in a subband light emission centred at 1.03 eV. The subband emission intensity can be enhanced in B/P co-doped Si NCs. Our results suggest that doping plays an important role in the electronic structures and optoelectronic characteristics of Si NCs, which provide a new route to realize the Si NCs-based devices.

Acknowledgments

This work was supported by "973 program" (2013CB632101) and NSFC (No.11274155) and "333 project" of Jiangsu Province (BRA2015284).

References

[1] Lien Y C *et al* 2012 *Appl. Phys. Lett.* **100** 143501
[2] Mu W W *et al* 2014 *IEEE J. Sel. Top.Quant.* **20** 8200106
[3] Xu J *et al* 2014 *Part. Part. Syst. Charact.* **31** 459
[4] Dalpian G M *et al* 2006 *Phys. Rev. Lett.* **96** 226802
[5] Pi X D *et al* 2008 *Appl. Phys. Lett.* **92** 123102
[6] Lu P *et al* 2016 *Sci. Rep.* **6** 22888
[7] Zhou J *et al* 2009 *Opt. Express* **17** 156
[8] Priolo F *et al* 2014 *Nat. Nanotechnol.* **9** 19
[9] Müller J *et al* 1999 *Phys. Rev. B* **60** 11666
[10] Ni Z Y *et al* 2014 *Phys. Rev. B* **89** 035312

IOP Conf. Series: Journal of Physics: Conf. Series **864** (2017) 012013 doi:10.1088/1742-6596/864/1/012013

Growth of InAs NWs with controlled morphology by CVD

Y S Huang [1], M Li, J Wang*[1], Y Xing[1] and H Q Xu*[12]
[1]Key Laboratory for the Physics and Chemistry of Nanodevices and Department of Electronics, Peking University, Beijing 100871, China
[2]Division of Solid-State Physics, Lund University, Box 118, Lund S-221 00, Sweden
* Corresponding authors. Electronic mails: jywang00@pku.edu.cn; hqxu@pku.edu.cn

Abstract: We report on the growth of single crystal InAs NWs on Si/SiO_x substrates by chemical vapor deposition (CVD). By adjusting growth parameters, the diameters, morphology, length and the proportion of superlattice ZB InAs NWs (NWs) can be controlled on a Si/SiO_x substrate. Our work provides a low-cost route to grow and phase-engineer single crystal InAs NWs for a wide range of potential applications.

1 Introduction

III-V InAs NWs have attracted much attention as a direct bandgap semiconductor with a small bandgap, a small effective mass and a high room-temperature electron mobility. Several novel functional devices have been realized with InAs NWs. Applications of InAs NWs in the future call for an effective control of their morphology and crystal structure during their growth.[1-3] Compared with other epitaxial methods, there are only limited reports of the effects of growth parameter on the morphology and structure of InAs NW.[4-7] While the understanding of the effect of the growth parameters is essential to achieve optimal control growth in CVD. This article demonstrate that the morphology, and the proportion of superlattice ZB InAs NWs can be effectively controlled on a Si/SiO_x substrate using growth parameters in CVD.

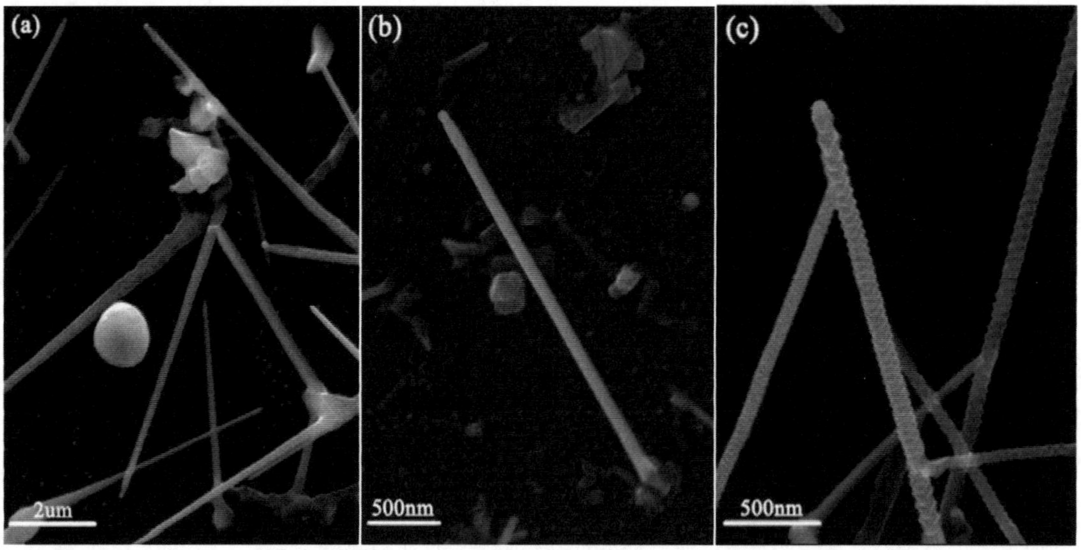

Fig. 1 the three morphology of InAs NWs (a) the taperd InAs NWs, (b) uniform InAs NWs, (c) twin superlattice NWs

2 Experimental details

Si/SiOx substrate was cleaned with ethanol, acetone and deionized (DI) water and then immersed in mercaptan-ethanol solution. Gold colloid of 5nm diameters were dispersed on the substrate. InAs powder (purity 99.999%) was placed at the spot to be heating up to 800 °C in a tube furnace (Lindberg

TF55035KC). The growth substrate was placed near the end of the tube where the temperature was around 390 to 640 °C during NW's growth. During the growth, the flow rate of H_2 carrier gas varied from 60 to 120 sccm (standard cubic centimeters per minutes). The pressure of the growth chamber was kept at 150 Torr. The growth time was about 40 minutes. The products was characterized using a field-emission scanning electron microscope (SEM, FEI Quant 600F, operated at 20 KV).

3 Results and discussion

The growth window of InAs NWs was found to within 390 ◦C and 640 ◦C. There are three kinds of InAs morphology, see Fig. 1, (a) the taperd InAs NWs, (b) uniform InAs NWs, (c) twin superlattice NWs. The structure and the composition of these InAs NWs has been reported in detail in Ref. 8. All the NWs are of zincblende structure. The InAs NW morphology as function of both the growth temperature and the flux rate of carrier gas was shown in Fig. 2. The tapering extent was exressed by color in Fig. 2 and the percentage represent the proportion of twin superlattice NW. Generally speaking, the InAs NWs grown on substrates are tapered at medium temperatures and become uniform at the lower temperature and higher temperature coupled with high flux rate. Radial growth is a thermally activated process, that is, kinetically limited. [9, 10] With the growth temperatures increasing, adatom diffusion rate and length increases and diffusing adatoms are more likely to incorporate into NW sidewalls and cause raxial growth by vapour-solid mechanism. While as the growth temperature increased further, the diffusion length increases and more adatom are able to diffuse long enough distances to the interface of Au nanoparticle and nanowire, this facilitating their incorporation at the NW tip instead of the NW sidewalls.

Twin-plane formation is less energetically favoured compared with ordinary plane nucleation and

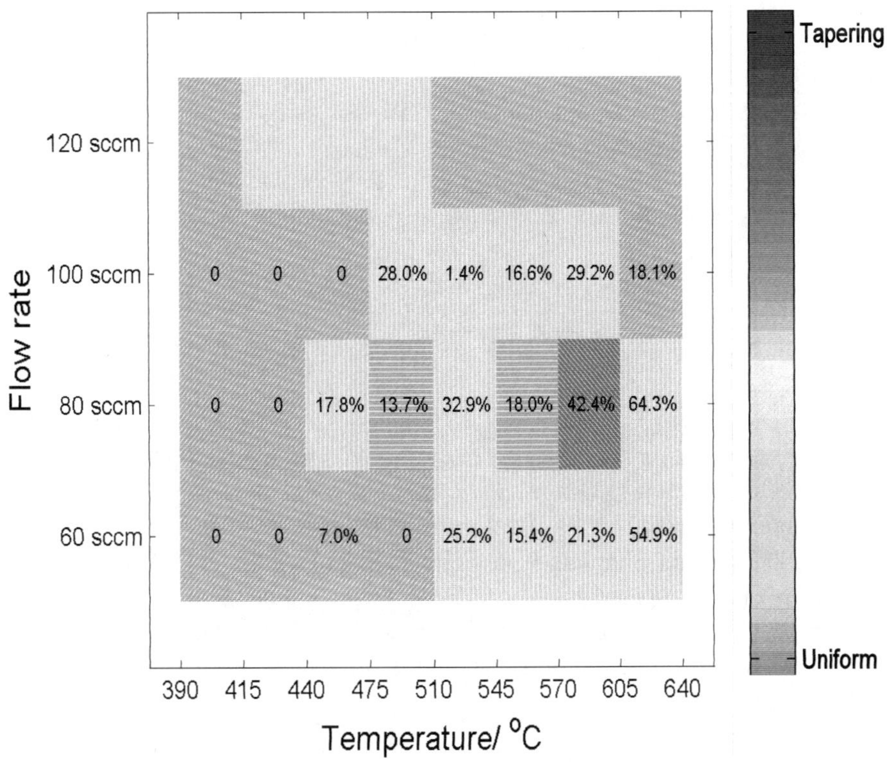

Fig. 2 InAs NW morphology as function of both the growth temperature and the flux rate of carrier gas

thus larger supersaturation is required. High flux rate, and fast growth rate promotes twin formation.[11-14] and therefore the proportion of superlattice NWs increase with increasing flux rate of carrier gas at certain growing temperature. As the flux rate increased further, the proportion of twin superlattice NWs then decrease. It is thought both As and In species reduce surface and interface tensions throughout growth[15], and this may be responsible for the reduction in twin superlattice when As and In precursor flow rates are high temperature and high flux rate. In conclusion, there are two opposing mechanisms that contribute to this result.

In summary, we have showed that the morphology and crystal structure can be controlled in the CVD growth of InAs NWs on Si/SiO_x substrates. The lower growth temperature has significant advantages, namely the minimization of undesirable radial growth, elimination of twinning defects and tapering. The higher growth temperature has similar advantages but has the shortcoming, the maximum of radial growth and the dimeter distribution.

This work was supported by the National Basic Research Program of China (Nos 2012CB932700, 2012CB932701, and 2012CB932703), the National Natural Science Foundation of China (91221202, 91421303, and 61321001). HQX also acknowledges financial support from the Swedish Research Council (VR).

REFERENCES

[1] Dayeh S A, Soci C, Bao X-Y, Wang D L 2009 Advances in the synthesis of InAs and GaAs nanowires for electronic applications *Nano Today* **4** 347

[2] Thelander C, Caroff P, Plissard S, Dey AW, Dick K A 2011 Effects of crystal phase mixing on the electrical properties of InAs NWs *Nano Lett.* **11** 2424

[3] Schroer M D ,Petta J R 2010 Correlating the nanostructure and electronic properties of InAs NWs *Nano Lett.* **10** 1618

[4] Park H D, Prokes S M, Cammarata R C 2005 Growth of epitaxial InAs NWs in a simple closed system *Appl. Phys. Lett.* **87** 063110

[5] He M Q, Fahmi M M E, Mohammad S N, Jacobs R N, Salamanca-Riba L, Felt F, Jah M, Sharma A, Lakins D 2003 InAs NWs and whiskers grown by reaction of indium with GaAs *Appl. Phys. Lett.* **82** 3749

[6] Ford A C 2008 Synthesis, contact printing, and device characterization of Ni-catalyzed, crystalline InAs NWs *Nano Research* **1** 32

[7] Park H D, Prokes S M, Twigg M E, Cammarata R C, Gaillot A -C 2006 Si-assisted growth of InAs NWs *Applied Physics Letters* **89** 223125

[8] Li M, Wang J Y, Li K, Xing Y J, Xu H Q 2015 Growth of InAs nanowires with the morphology and crystal structure controlled by carrier gasflow rate *J. Crys. Growth* **430** 87

[9] Johansson J, Svensson C P T, Mårtensson T, Samuelson L, Seifert W 2005 Mass transport model for semiconductor nanwire growth *J. Phys. Chem. B* **109** 13567

[10] Borgström M, Deppert K, Samuelson L, Seifert W 2004 Size- and shape-controlled GaAs nano-whiskers grown by MOVPE: a growth study *J. Cryst. Growth* **260** 18

[11] Hurle D T J 1995 A mechanism for twin formation during Czochralski and encapsulated vertical Bridgman growth of III-V compound semiconductors *J. Cryst. Growth* **147** 239 and Hurle D T J, Rudolph P 2004 A brief history of defect formation, segregation, faceting, and twinning in melt-grown semiconductors *J. Cryst. Growth* **264** 550

[12] Johansson J, Karlsson L S, Dick K A, Bolinsson J, Wacaser B A, Deppert K, Samuelson L 2008 The structure of <111> B oriented GaP nanowires *J. Cryst. Growth* **310** 5102

[13] Mandl B, Stangl J, Mårtensson T, Mikkelsen A, Eriksson J, Karlsson L S, Bauer G, Samuelson L, Seifert W 2006 Au-Free epitaxial growth of InAs nanowires *Nano Lett.* **6** 1817

[14] Glas F, Harmand J C, Patriarche G 2007 Why does Wurtzite form in nanowires of III_V Zinc Blende semiconductors *Phys. ReV. Lett.* **99** 146101
[15] Copel M, Reuter M C, Kaxiras E, Tromp R M 1989 Surfactants in epitaxial-growth *Phys. ReV. Lett.* **63** 632

Optical Tailoring of RF Magnetron Sputtered ZnO Thin Films

A Hassan*[1], C Feng[1], S Riaz[2], S Naseem[2] and Y Jiang[1]

[1]Institute of Laser Engineering, Beijing University of Technology, 100124, Beijing, China

[2]Centre of Excellence in Solid State Physics, University of the Punjab, 54590, Lahore, Pakistan

Email:15alirao@gmail.com

Abstract. Pure and N_2 doped ZnO thin films of thickness ranging ~300-500nm with 5,10,15,20,25, and 50 sccm inflow ratios of N_2 are deposited on soda-lime glass by means of RF magnetron sputtering system, and observed the dependence of optical properties of ZnO by the function of doping with the help of spectroscopic ellipsometer. And found that the N_2-inflow highly affects the optical properties of ZnO thin films. Even the high transmittance of about 97% is achieved and absorbance graph also shows that slight variation in N_2 inflow affects the absorbance, which is maximum with in UV region. Optical conductivity of ZnO is also observed high with the increase of N_2 inflow. With 25sccm N_2 inflow conductivity rose to the maximum value of about$1.4 \times 10^7 \Omega^{-1}cm^{-1}$, with 15sccm inflow of N_2 conductivity value is $2.0 \times 10^6 \Omega^{-1}cm^{-1}$ in the visible region. This is a strong contribution towards next generation photovoltaic devices.

Keywords: TCO, ZnO, RF magnetron sputtering, spectroscopic ellipsometer, optoelectronic.

1. Introduction:

transparent conductive oxide thin films (TCO's) are currently gain much interest of researchers worldwide[1], because of their intense penetration in the field of optoelectronic devices[2], photovoltaic[3], and semiconductors lasers[4] and modern LED's[5]. The efficiency of transparent conductive oxides (TCO's) can be judged by examine their transmittance and resistivity, which is a core issue in this era. Transmittance and resistivity of different transparent conductive oxides (TCO's) materials (CdO[6], ITO[7], FTO[8], ZnO[9], SnO_2 and In_2O_3[10]) have been investigated and found that material having high transmittance with low value of resistivity is the need of time these days[1]. CdO regards first transparent conductive oxide (TCO) material been investigated with resistivity of $1.4 \times 10^{-4} \Omega cm$. However, the toxicity and narrow band gap of CdO limits its practical applications [6]. ZnO based transparent conductive oxides (TCO's) attracted much interest in this regards because of its multiple advantages over all other transparent conductive oxide (TCO's) materials, such as its wide and direct band gap of ~3.37eV, high value of binding energy ~60meV at room temperature and high transparency as well [9]. Thin films of ZnO based transparent conductive oxides (TCO's) are being prepared by different deposition techniques, such as spray pyrolysis [7], pulsed laser deposition [8], sol-gel method [6], molecular beam epitaxy [9], chemical vapor deposition [10].

Doping of different dopants (Al, N_2, Cu, K, Na, Sn, Pb and Bi) in ZnO based transparent conductive oxides is also intensively studied to enhance the optical properties of ZnO based transparent conductive oxides [11]. Among all other dopants N_2 is supposed to be the best candidate due to its quite similar radii with Zn atoms and with non-toxic and environment friendly behavior. The p-type behavior of doped ZnO thin films is a key issue in this regards, some researchers claims that, but it is not be considered as large scale production[12].

In current work, we investigated the optical properties of N_2-doped sputtered deposited ZnO thin films via RF magnetron sputtering system, and characterized them with the help of spectroscopic ellipsometer (M-2000FI).

2. Experimental Details:

N_2-doped ZnO thin films on glass substrate are deposited with different N_2-inflow ratios varies from 0-50sccm inflow. Before deposition, the glass substrate are cleaned firstly with detergent, and then given ultrasonic bath of acetone and IPA for 15-20mins, after that dry them with N_2.ZnO target of 99.99% purity is used. Sputtering was carried out in the environment of (Ar $+N_2$) at room temperature and 10^{-5} Pa of pressure. Distance between target and substrate surfaces was maintained 72mm. Film thickness and optical parameters were investigated using spectroscopic ellipsometer (M2000-FI.J.A Woollam, USA).

3. Results & Discussions:

Figure.1 (a) exhibits the transmittance spectra of 7 samples. All the samples show high

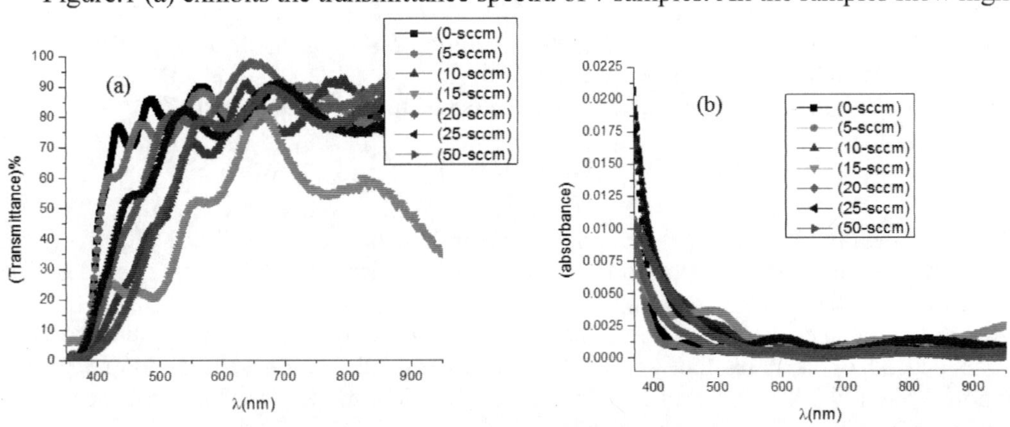

Figure 1: undoped & doped (a) transmittance spectra (b) and Absorbance spectra of ZnO thin films with different N_2 inflow ratio as a function of wavelength.

transparency of >80% in visible region which drops to 0% at 350nm wavelength and it can be seen that with the increase of N_2-inflow transmittance increases generally, with 20sccm flow of N_2 the maximum transmittance of about 97% is observed which is high value reported till now. The reason behind this can be the substitution of N_2 atoms in ZnO are in perfect ratio to overcome the lattice mismatches and defects. Clear variations in transmittance spectra with slight change of N_2 inflow reminds that doping highly affect the transmittance of ZnO, and with doping concentration one can easily control the transmittance of ZnO. This is quite easy and affordable way [13].

The coefficient of absorption "α" can be calculated using Eq.1

$$\alpha = -\frac{lnT}{d} \ldots\ldots\ldots\ldots (1)$$

"T" is transmittance and "d" denotes thickness which calculated by Eq (1) .The absorbance of ZnO thin films is also affected by N_2 doping.

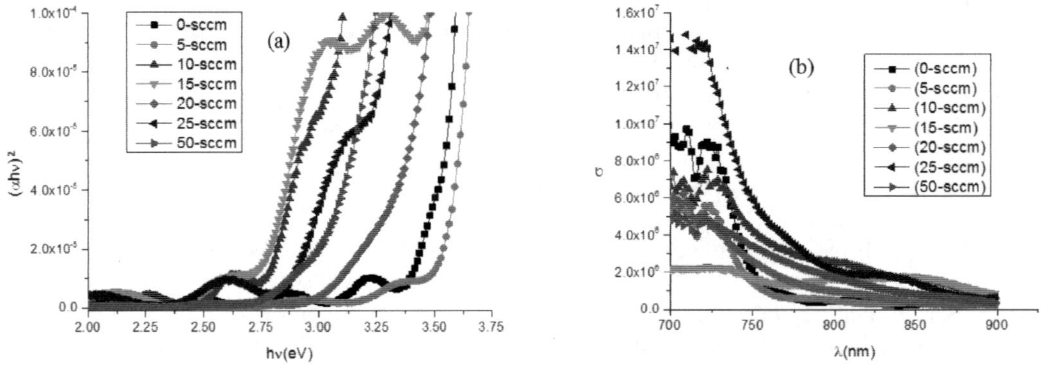

Figure 2: (a) (αhv) 2 vs photon energy plot of undoped & doped ZnO thin films (b) variation of optical conductivity vs wavelength with different inflow ratio of N_2.

Figure.1 (b) show that in UV region the absorbance is high and in visible spectrum it is found very low and goes to zero from infrared region and onward. A clear shift in absorbance edge is observed with the function of doping concentration, which shifts the absorbance edge into visible region. With 50sccm inflow of N_2 the absorbance edge falls on 440nm wavelength. And in pure ZnO thin films it falls on near 400nm wavelength. It concludes that the high doping of N_2 made ZnO thin films very good UV absorber. Which can be used as UV filters [13].

Band-gap of direct band-gap semiconductors can be evaluated by the following Eq.2

$$\alpha hv = A(hv - E_g)^{1/2} \ldots\ldots\ldots\ldots\ldots (2)$$

"A" is proportionality constant; "hv" is photon energy calculated in "eV" and "E_g" refers to the band-gap of the material. Linear exploration of the photon energy "hv" vs $(\alpha hv)^2$ also describes us about the band-gap of that specific material.

Figure.2 (a) shows the decreasing trend of band-gap with increase in the N_2-inflow ratio, the variation of band-gap depend upon many factors like film thickness, fabrication process, doping type and concentration. Present work is the strong evidence of theoretical models, which suggest the decrease of band-gap with increase in doping concentration.

The optical conductivity of any material is an important parameter in order to check its optoelectronic properties, because it gives the healthy information about the band structure of materials, more general it tells how the electronic state are disturbed when optical radiations strikes the surface of that material. It is related strongly with the well-known "moss Burstein effect". The following formula is helpful to calculate optical conductivity

$$\sigma = \frac{\alpha nc}{4\pi} \ldots\ldots\ldots\ldots\ldots (3)$$

Here "α" is coefficient of absorption, "n" is refractive index and "c" is speed of light in free space which is constant [13]. The optical conductivity of ZnO thin films is also analyzed and results are found in such a way with 25sccm N_2 inflow the conductivity rose to the maximum value of about $1.4 \times 10^7 \Omega^{-1}cm^{-1}$, with 15sccm inflow of N_2 the conductivity value is $2.0 \times 10^6 \Omega^{-1}cm^{-1}$ in the visible region after that it is decreasing exponentially in figure 2 (b).

The high value of optical conductivity in the visible region of electromagnetic spectrum with doping of N_2 made a great contribution to leads the future optoelectronic devices.

Conclusion:

In short, we examined that the optical properties of ZnO thin films can be perfectly tailored by introducing N_2 contents. The results show the precise control of doping concentration on each optical parameter. As the doping level increases the enhancement in optical parameters can be seen. And it is also observed that the doping of N_2 highly affects the optoelectronic properties of ZnO thin films. In fact, the transmittance of ZnO thin films increased up to 97%, while absorbance is also high in the UV region, and the increase in optical conductivity due to N_2 doping made ZnO thin films a perfect alternate of ITO thin films for high efficiency solar cells of modern era.

References

1. M. Berginc, A. Čampa, K. Vojisavljević, B. Malič, P. Panjan and M. Topič, Energy Procedia 84, 183-189 (2015).
2. D. W. Mohammed, R. Waddingham, A. J. Flewitt, K. A. Sierros, J. Bowen and S. N. Kukureka, Thin Solid Films 594, Part A, 197-204 (2015).
3. X. Wang, J. Huang, Z. Niu, X. Zhang, Y. Sun and C. Zhan, Tetrahedron 70 (32), 4726-4731 (2014).
4. J. S. Suelzer, A. Prasad, R. Ghosh and G. Vemuri, Optics Communications 370, 209-221 (2016).
5. K.-W. Kim, N.-J. Choi, K.-B. Kim, M. Kim and S.-N. Lee, Journal of Alloys and Compounds 666, 88-92 (2016).
6. S. Duman, G. Turgut, F. Ş. Özçelik and B. Gurbulak, Materials Letters 126, 232-235 (2014).
7. S. Marikkannu, M. Kashif, N. Sethupathy, V. S. Vidhya, S. Piraman, A. Ayeshamariam, M. Bououdina, N. M. Ahmed and M. Jayachandran, Materials Science in Semiconductor Processing 27, 562-568 (2014).
8. L.-j. Huang, N.-f. Ren, B.-j. Li and M. Zhou, Journal of Alloys and Compounds 617, 915-920 (2014).
9. D. Adolph, T. Tingberg and T. Ive, Journal of Crystal Growth 426, 129-134 (2015).
10. A. Vomiero, M. Ferroni, M. M. Natile, T. Fischer, R. Fiz, S. Mathur and G. Sberveglieri, Applied Surface Science 323, 59-64 (2014).
11. A. A. Othman, M. A. Ali, E. M. M. Ibrahim and M. A. Osman, Journal of Alloys and Compounds 683, 399-411 (2016).
12. J. J. Ortega, A. A. Ortiz-Hernández, J. Berumen-Torres, R. Escobar-Galindo, V. H. Méndez-García and J. J. Araiza, Materials Letters 181, 12-15 (2016).
13. V. K. Jayaraman, A. M. Álvarez, Y. M. Kuwabara, Y. Koudriavstev and M. d. l. l. Olvera Amador, Materials Science in Semiconductor Processing 47, 32-36 (2016).

IOP Conf. Series: Journal of Physics: Conf. Series **864** (2017) 012015 doi:10.1088/1742-6596/864/1/012015

PVT growth of AlN single crystals with the diameter from nano- to centi-meter level

H Wu, R Zheng, Y Guo, Z Sun

Key Laboratory of Optoelectronic Devices and Systems of Ministry of Education and Guangdong Province, Shenzhen University, Shenzhen 518060, China

Email: hlwu@szu.edu.cn

Abstract. Physical vapor transport (PVT) is the most successful and widely used approach for bulk aluminum nitride (AlN) single crystals. During the process of PVT growing AlN crystals, crucible materials, the growth setup, and the growth parameters (e.g., temperature distribution, growth pressure) are crucial. This work proposes a detailed study on the PVT growth of single AlN crystals with sizes ranging from nanometers to centimeters. AlN crystals with different sizes are grown by spontaneous nucleation. Furthermore, it discusses and contrasts the growth conditions and mechanisms of AlN crystals with different sizes. The structural and optical properties of the AlN crystals are also involved.

1. Introduction

Aluminum nitride (AlN), gallium nitride (GaN), silicon carbide (SiC), diamond, and other wide-band gap semiconductor materials are often referred to as third-generation semiconductor materials. Compared to first-generation and second-generation semiconductor materials, the third generation is characterized by high breakdown field strength, high thermal conductivity, high electron saturation velocity, and high resistance to radiation; they are suitable for the production of high-frequency, anti-radiation, large power, and high-temperature semiconductor devices. AlN, a typical representative of this group, has a widest direct band gap of 6.2 eV as well as excellent electrical, optical, mechanical, and piezoelectric properties. The properties make AlN as one of the best choices for electronic and optoelectronic applications. For example, AlN is a promising material for deep-ultraviolet (DUV) solid-state light-emitting diodes (LEDs) and laser diodes (LDs). AlN LEDs with a wavelength of 210 nanometers have been made successfully [1,2].

There is no natural occurrence of AlN as a mineral. Since AlN powder was first synthesized in 1862 by molten aluminum in nitride atmosphere, the first AlN single crystal (Φ0.03 mm×0.3 mm) was obtained in 1956 and AlN crystal bullets were grown in 1976. [3] Due to a very high melting point and a very large dissociation pressure at the melting point, AlN crystals are hardly grown from the melt at normal pressure condition. Over the past several decades, physical vapor transport (PVT – also named the sublimation-recondensation method) has been proved to be the most successful and widely used approach for bulk AlN crystals growth [4,5,6]. Now the maximum size of AlN single crystal grown in laboratory is about 1 inch in diameter [4,7]. Moreover, high quality AlN nanocrystals can be also obtained by PVT [8]. However, the contrastive research on the PVT growth of AlN crystals with different sizes is absent. The study of AlN crystal-growth technology remains a hot topic of solid state physics in the near future.

In this work, a detailed study was proposed on the PVT growth of AlN single crystals with size range from nano- to centi- metre. The growth conditions and mechanism of AlN crystals with different sizes were also discussed contrastively. Finally, attention was also paid to investigate the structural and optical properties of AlN crystals.

2. Experimental procedure

The growth experiments were performed by the PVT technique. The resistance-heated furnace was operated with a high purity nitrogen atmosphere (99.999%) at 90-120 kPa. AlN powder source usually begins to sublimate at the temperature of 1850 °C. [3] During the growth, the species gaseous of AlN or Al and N_2 from the source located at the hotter bottom of the crucible sublimated and re-condensate on the crucible lid where the temperature is relatively lower. This process is really that the overall reaction AlN(s) = Al(g) + $\frac{1}{2}N_2$ is run in the forward (sublimation) direction at the source, and in the reverse (recondensation) direction at the top of the crucible. To achieve high quality bulk AlN crystals with fast growth rates (>200 μm/h), high temperature are required [9,10]. The upper limit of growth temperature is about 2500 °C, when the Al vapor pressure reaches 1 atm [11]. The induction heating unit as shown as Figure 1 included a main heater and a top heater which heated on sublimation zone and crystalline zone, respectively. A vertical thermal gradient was established by the two heaters. The temperatures of two zones were measured and controlled separately by infrared two-color pyrometers on the body and the lid of the crucible. Growth experiments were carried out in a tungsten crucible. The high purity tungsten was also used for the resistance heater and the thermal shield. AlN polycrystals, which had been grown by

commercially available AlN powder, were employed as the AlN source and resublimed for 4 hours at 1850-1900 °C prior to the use for achieving the highest possible degree of purification.

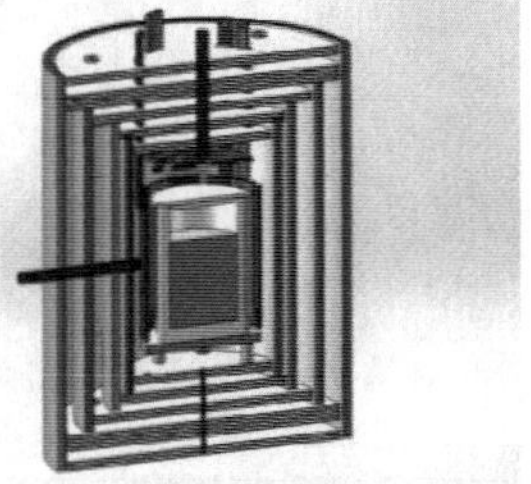

Figure 1 Scheme of the induction heating unit mainly composed of tungsten

The suitable growth temperature and the temperature gradient between the crystallization and the sublimation zone (i.e. the crucible lid and body) were investigated to restrain preferential growth AlN crystals along the c-axis orientation and controlled by the top heater and the main heater. However, a high temperature usually results in too many defects in product, since the thermal equilibrium concentration of point defects increases with rising growth temperature.[7,12] During the cooling, an auxiliary annealing process with optimal is employed to reduce the defects. X-ray diffraction (XRD) with Cu Kα radiation on a Philips x-ray diffractometer at 45 kV and 40 mA. Moreover, the Energy Dispersive Spectrometer (EDS) were employed to investigate the composition of AlN crystals

3. Results and discussion

The sublimation-recondensation process of AlN can be described by the simple reaction: AlN(s) = Al(g) + ½N_2 . In fact, the PVT growth of AlN crystal includes following five sub-processes: (a) sublimation of the AlN source, (b) mass transfer of the vapor species, (c) adsorption of the vapor species on the substrate surface, (d) diffusion and nucleation on the surface, (e) desorption [9]. The sublimation behavior of AlN, which is well described with experimentally and calculationally relevant curves, is mainly related to growth temperature (usually above 1850°C) and vapor pressure (80-120 kPa) [3,4]. The gas phase transport and the adsorption mechanisms of the vapor species has been also discussed intensively [9,10,11]. Diffusion is the main mass transport model for the growth species. Gaseous Al and N_2 are as the main species during the growth, while possible vapor species like metastable Al_xN (g) (x = 2,3,4) can be almost excluded [13]. The growth experiments are usually performed with a high purity nitrogen atmosphere at 80-120 kPa, which excesses multiple times above the partial pressure. However, a nitrogen molecule has the strong binding energy. Thus sticking coefficients of gaseous N_2 (α_{N2}), which are much less than one, are introduced to model correct during the description of the adsorption behavior and surface reactions [14]. Higher temperatures lead to enhance values of α_{N2} and surface diffusion of the adatoms, which are better for improving structural quality of the AlN crystals. The maximal temperature of AlN crystals PVT growth is lower than the decomposition point about 2430 °C at 100 kPa nitrogen [15].

Considering these factors, the growth rate of crystalline AlN, or V_G, can be expressed in Eq(1). [10,11,13]

$$V_G = \frac{M_{AlN}}{\rho_{AlN}} \frac{D_{Al0} P_0 \Delta H}{T_0^{1.8} R} \frac{\exp(\frac{\Delta S}{R} - \frac{\Delta H}{RT})}{P_T^{1.5} T^{1.2}} \frac{\Delta T}{\Delta d} = k \frac{\exp(\frac{\Delta S}{R} - \frac{\Delta H}{RT})}{P_T^{1.5} T^{1.2}} \frac{\Delta T}{\Delta d} \quad (1)$$

where k is a constant; ΔS and ΔH represent entropy and enthalpy of sublimation, respectively; R is the universal gas constant; PT is the pressure when the local gas phase temperature is T; δ is the distance from the source to the substrate surface; and ΔT is the temperature difference between the crystallization and sublimation zones. The equation shows that ΔT constitutes a key factor in the growth rate V_G, especially when T and P are determined. It is obtained that the common temperature range for the AlN source and the nucleation zone is about 2050-2320°C, which provides sufficient surface mobility and mass transport. [16]

During the early stages of growth, self-nucleation and the subsequent growth of millimeter-sized freestanding crystals dominate the growth process. It can be seen in Figure 3 that the AlN crystal pattern changes from nanometer-sized (in diameter) whiskers to micrometer-sized needles to millimeter-sized, perfect, well-faceted hexagonal prisms as the growth temperature increases from 1850 °C to 2200 °C. AlN crystal growth demonstrates a strong anisotropy and preferential behavior along the c-axis. However, this habit decreases as the temperature increases. With increasing temperature, polycrystalline AlN boules, shown in Figure 3, rather than single crystals are produced on the tungsten substrate because of the absence of seeds. The majority of crystallies are (0001) oriented, but some are twisted and tilted against each other by several degrees. This creates the opportunity to develop new techniques to grow bulk AlN crystals.

(a) (b) (c) (d)

Figure 2 SEM photos of AlN crystals with the diameter from (a) nano- to (d) milli-meter level

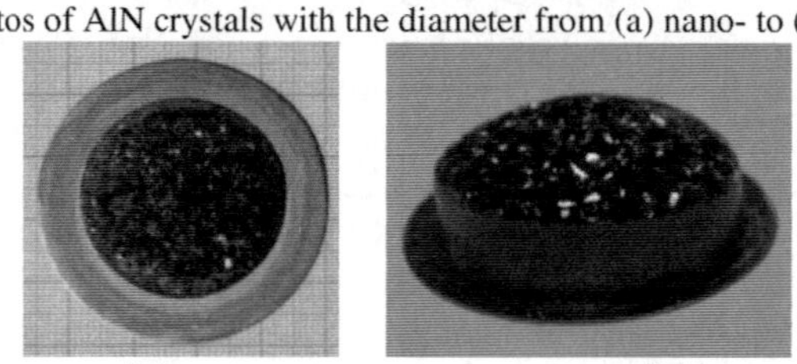

(a) (b)

Figure 3 Polycrystalline AlN boule, 28mm in diameter, and (a) 3mm thick and (b)16mm thick grown at 2250 °C

During growth, proper process conditions lead to low nucleation densities and high growth rates. Numerical simulations of the thermal field and supersaturation can determine the optimal parameters. The local supersaturation relative to thermal equilibrium can be defined as [17]

$$S = \frac{p_{Al}^2 \cdot p_{N_2}}{K(T)} - 1 \qquad (2)$$

where p_{Al} is the gaseous Al pressure, p_{N2} is the gaseous N_2 pressure, and K(T) is the equilibrium constant. Excessively low supersaturation leads to very low growth rates, and excessively high supersaturation results in defective densities and high-density nucleation points that cause the polycrystallization of AlN. Supersaturation values in the range of $0.25<S<0.3$ have proven to be suitable for growing separate freestanding AlN crystals with appreciable size and high structural quality [5]. At a constant ambient pressure in a given growth setup, the variable process parameters are the growth temperature and the temperature difference between crystallization and the sublimation zone. Due to the high vapor pressures of Al and N_2 at temperatures of 2050-2320°C, the temperature gradients must be maintained at 5-20 °C /cm.

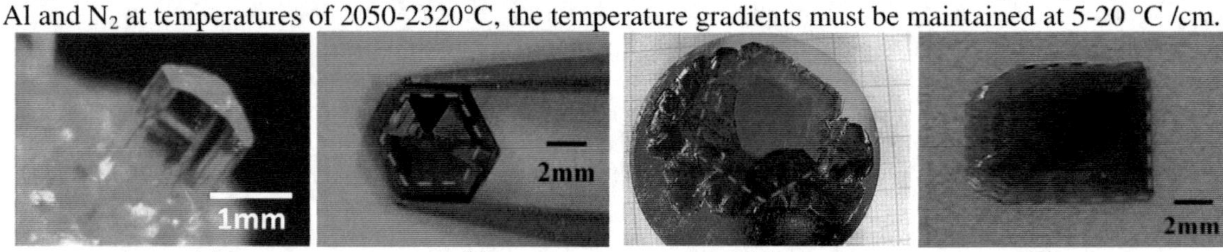

(a) (b) (c) (d)

Figure 4 Optical photos of AlN single crystal with the diameter from (a) milli- to (b) (c) (d)centi-meter level

The crystal growth was composed of three main sub-processes: (a) heating-up, (b) holding, and (c) cooling. According to the traditional growth model where the temperature of the crystallization zone is higher than that of the sublimation zone during the heating-up process, or $\Delta T<0$, the AlN polycrystals form on the tungsten lid, as shown in Figure 3. Hexagonal AlN single crystals in <0001> orientation form at a growth temperature of 2,250 °C, with a temperature difference or ΔT of positive 20-50 °C during sub-processes (a) and (c). Maintaining the proper values of ΔT before growth can help avoid too much crystallization on the substrate, where conditions are close to thermal equilibrium and only a few detached AlN single crystals grow, as shown in Figure 4 (b) (c), and recrystallize in sub-process (c). Transparent yellow or amber crystallites roughly grow in <0001> orientation, with excellent hexagonal morphology. By optimizing experimental parameters, a single AlN crystal has grown to a maximum diameter of 22 mm, as shown in the dash-dotted frame in Figure 4(c).

Strong spontaneous and piezoelectric polarization fields affect AlN crystals and other structures grown in the polar c-axis direction. AlN has an anisotropic emission pattern where light is barely emitted from the c-plane and is emitted preferentially from the a-plane. The a-plane emission intensity is estimated to be 25 times higher than that of the c-plane. [18] Therefore, controlling growth direction is the most effective and directive method to grow nonpolar plane AlN crystals. During sub-process (b) of AlN crystal growth, M-plane AlN single crystals can also form on the tungsten lid when the temperature difference is negative 10-40 °C. Reducing the values of ΔT and other parameters allows M-plane AlN crystals to grow easily. Figures 4 (b) and (d) show C-plane and M-plane AlN single crystals with ΔT values of 30 °C and 15 °C, and growth temperatures of 2200-2250 °C. The $\omega/2\theta$ XRD data of the sample (Figure 4 (d)) has only a sharp diffraction peak for wurtzite (100), indicating the growth of M-plane surface-oriented AlN crystals with a size of about 1cm. The full width at half maximum (FWHM) of the symmetric x-ray rocking curve (XRC) is about 300 arcsec. Figure 5 shows the cathodeluminescence (CL) spectra of AlN crystals from a low temperature of 8 K to room temperature or 300 K, over a narrow spectral range of 190-230nm. Free exciton (FX) emission peaks in the near band-edge emissions spectra shift from 206 nm to 210 nm as the temperature changes from 8 K to 300 K.

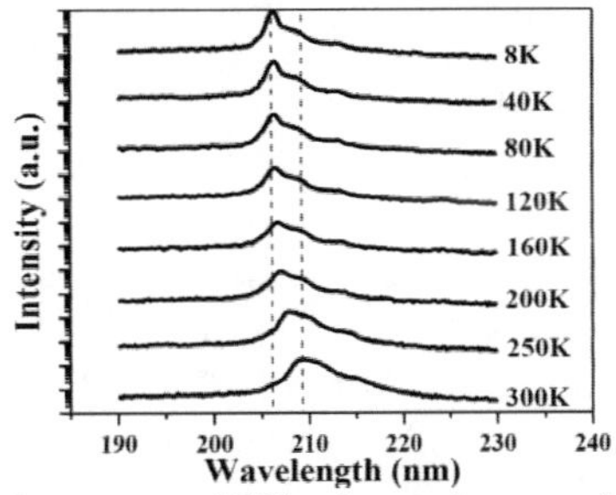

Figure 5 Near band-gap Cathodeluminescence (CL) of sample recorded at 8–300 K with 15 KV excitation

4. Conclusions

AlN single crystals with the diameter from nano- to centi-meter level have been grown by PVT. During the growth, the optimized temperature range for the AlN source and the nucleation zone, about 2050-2320°C. The size of AlN crystal increases as the temperature increases. The temperature difference between the crystallization and sublimation zones (ΔT) is crucial to grow AlN single crystal. Maintaining the proper values of ΔT before growth can help avoid too much crystallization on the substrate and also affect growth direction from C-plane to M-plane. With the CL spectra at 8–300 K with 15 KV excitation, AlN crystal FX emission peaks in the near band-edge emissions spectra shift from 206 nm to 210 nm.

Acknowledgement

This work was supported by the National Natural Science Foundation of China (Grant Nos. 11447029 and 61136001), and the Science & Technology Bureau of Shenzhen (Grant No. 20140418095735610).

References

[1] Taniyasu Y *et al* 2006 *Nature* **441** 325
[2] Taniyasu Y and Kasu M 2014 *Diam. Relat. Mater.* **29** 084002.1
[3] Slack G A and McNelly T F 1976 *J. Cryst. Growth* **34** 263
[4] Hartmann C *et al* 2014 Semicond. Sci. Tech. **29**, 3652
[5] Hartmann C *et al* 2008 Semicond. Sci. Technol. **244** 1780
[6] Lu P *et al* 2008 J. Cryst. Growth **310** 2464
[7] Raghothamachar B *et al* 2012 *Mater. Sci. Forum* **717–720** 1287
[8] Zheng W *et al* 2015 *Adv. Mater.* **27** 3971
[9] Hartmann C 2007 *Semicond. Sci. Technol.* **244** 1780
[10] Wu B and Zhang H 2004 *Int. J. Heat Mass Tran.* **47** 2989
[11] Segal A S *et al* 2000 *J. Cryst. Growth* **211** 68
[12] Huang F and Banfield J F 2005 *J. Am. Chem. Soc.* **127** 4523
[13] Li Y andBrenner D W 2006 *J. Appl. Phys.* **100** 084901
[14] Mueller S G *et al* 2009 *Phys. Status Solidi a* **206** 1153
[15] Adekore B T*et al* 2006 *J. Electron. Mater.* **35** 1104
[16] Epelbaum B M *et al* 2007 *J. Cryst. Growth* **305** 317
[17] Hartmann C *et al* 2013 *Japan. J. Appl. Phys.* **52** 08JA06
[18] Taniyasu Y and Kasu M 2010 *NTT Tech. Rev.* **8** 1

IOP Conf. Series: Journal of Physics: Conf. Series **864** (2017) 012016 doi:10.1088/1742-6596/864/1/012016

Modification of the yellow luminescence in gamma-ray irradiated GaN bulk single crystal

Y. Torita [1], **N. Nishikata** [1], **K. Kuriyama**[1], **K. Kushida**[2], **A. Kinomura**[3] **and Q. Xu**[3]

[1]College of Engineering and Research Center of Ion Beam Technology, Hosei University, Koganei, Tokyo 184-8584, Japan

[2]Department of Arts and Sciences, Osaka Kyoiku University, Kashiwara, Osaka 582-8582, Japan

[3]Research Reactor Institute, Kyoto University, Kumatori, Osaka 590-0494, Japan

E-mail: kuri@ionbeam.hosei.ac.jp

Abstract. We report the variation of the yellow luminescence (YL) in GaN bulk single crystals by gamma-ray irradiation. The crystals are irradiated at room temperature with gamma-rays of 1.17 and 1.33 MeV from a cobalt-60 source. Gamma-ray dose is 160 kGy. The resistivity varies from 30 Ωcm for an un-irradiated sample to 10^4 Ωcm for gamma-ray irradiated one. The nitrogen displacement in gamma-ray irradiated samples is observed by Rutherford backscattering channeling experiments using proton beam, suggesting the existence of the deep energy level relating to interstitial nitrogen atoms. The YL from the un-irradiated GaN with a peak at 557 nm (2.22 eV) is observed at around 440 nm to 800 nm, whereas that of the gamma-ray irradiated GaN shows a peak at 532 nm (2.33 eV) although the YL spectrum is almost overlapped with un-irradiated ones. Compton electrons emitted by the gamma-ray irradiation induce the shallow donor located at about 50 meV bellow the conduction band. This energy level is close to that of nitrogen vacancy. The modification of YL is attributed to a transition from the shallow donor induced by the gamma-ray irradiation to the native gallium vacancy.

1. Introduction

Examining the defects caused by various radiations to GaN under the space environment is important. In our previous study [1, 2], we reported that the energy levels relating to nitrogen vacancy (V_N) and gallium vacancy (V_{Ga}) were induced by neutron and proton irradiated GaN. The neutron irradiation has been used as the neutron transmutation doping into semiconductors such as GaAs [3], GaP [4], and GaN [5]. Atoms in semiconductors mainly transmute by a (n, γ) reaction. Therefore, to survey the radiation effect of gamma ray alone is meaningful. It is expected that Frenkel pairs on Ga and N sublattices are produced by the Compton electrons having a mean energy of about 700 keV [6]. In the present study, we report the variation of the yellow luminescence (YL) in GaN bulk single crystals by gamma-ray irradiation.

2. Experiments

GaN bulk single crystals with a thickness of 450 ± 50 μm were purchased from Furukawa Co. Ltd. The crystals were irradiated at room temperature with gamma-rays of 1.17 and 1.33 MeV from a cobalt-60 source of Kyoto University Research Reactor Institute. Samples were irradiated with an absorption dose rate of 1.771 KGy/h. Total gamma-ray dose was 160 kGy. The resistivity measured by van der Pauw method varied from 30 Ωcm for an un-irradiated sample to 10^4 Ωcm for gamma-ray irradiated one. Photoluminescence (PL) spectra were measured at 77 K using a He-Cd laser. To evaluate the lattice displacement, Rutherford backscattering spectroscopy (RBS) analysis was performed using the Van de Graaff accelerator of Hosei University. 1.5-MeV H^+ ions were used in the present experiment because the kinematic factor of N atoms for He-ion as a probing beam is very small. The experimental arrangement has already been reported elsewhere [7, 8]. The samples were mounted on a triple axis goniometer and the backscattered ions were collected by a silicon surface barrier detector placed at a scattering angle of $\theta = 150°$. The beam spot and the beam divergence were ~1.0 mm in diameter and ~0.03°, respectively.

3. Result & Discussion

Figure 1 (a) shows the PL emission spectra before and after gamma-ray irradiation. The band edge emission was observed at 356 nm (3.48 eV) in both un-irradiated and gamma-ray irradiated samples. The YL from the un-irradiated GaN with a peak at 557 nm (2.22 eV) is observed at around 440 nm to 800 nm (see Fig.1(b)), whereas that of the gamma-ray irradiated GaN shows a peak at 532 nm (2.33 eV) although the YL spectrum is almost overlapped with un-irradiated ones (see Fig.1(a)). Therefore, the modification of YL observed in the present study would be attributed to a transition from the shallow donor, which is induced by Compton electrons emitted by the gamma-ray irradiation, to a native V_{Ga}. Since the usual YL has been proposed as a transition from a shallow donor to V_{Ga} located at about 1.1 eV above the valence band [9], the gamma-ray induced shallow donor is located at about 50 meV below the conduction band. This energy level is close to a donor level (64 ± 10 meV [10]) relating to V_N. Figure 2 shows the YL model estimated from the photoluminescence at 77 K.

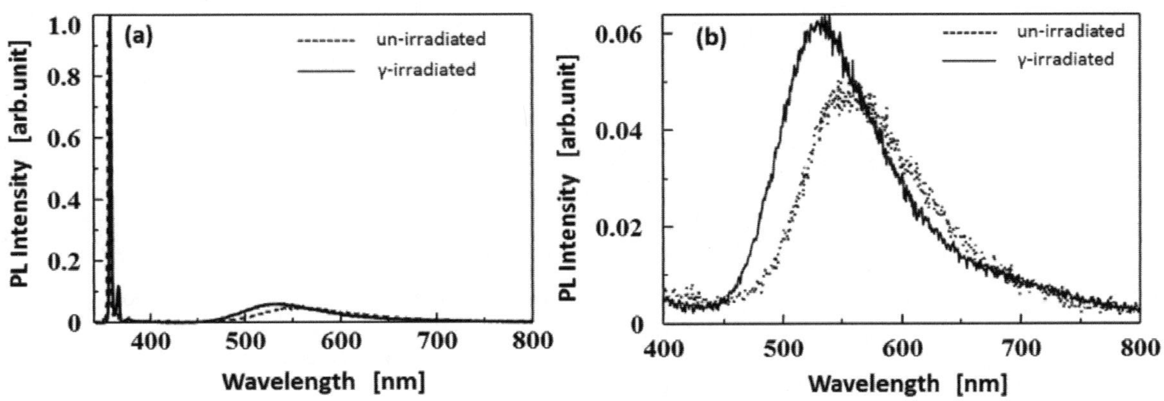

Figure 1. PL emission spectra before and after gamma-ray irradiation measured at 77 K. The PL intensity is normalized by the band edge emission. Fig.1(b) is the expansive spectra of YL.

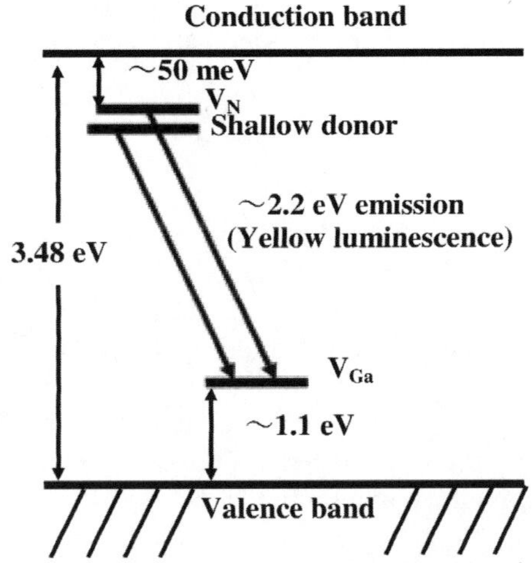

Figure 2. Yellow luminescence estimated from the photoluminescence. The gamma-ray induced shallow donor relating to nitrogen vacancy is located at about 50 meV below the conduction band.

Figure 3 shows typical random and aligned RBS spectra of as-irradiated and un-irradiated GaN bulk single crystal. The aligned spectra were obtained from scattering along the <0001> channeling directions. The aligned yield increases with increasing the depth within the crystal since the dechanneling probability increases with increasing the depth. Therefore, the minimum yield χ_{min} (the ratio between aligned and random) of Ga and N was evaluated using a width of about 20 channels (120 nm) behind the surface peak. Near the surface, the number of atoms displaced into channels in both Ga and N is approximately related to the minimum yield by [11,12]

$$N_D = N_{Ga\ or\ N}(\chi_{min} - \chi^0_{min}) / (1 - \chi^0_{min}), \qquad (1)$$

where $N_{Ga\ or\ N}$ is the Ga or N density (4.43×10^{22} cm^{-3}) in GaN, χ^0_{min} the minimum yield for the un-irradiated samples, and χ_{min} the observed minimum yield for irradiated ones. The values of χ_{min} for Ga were 1.5 % for un-irradiated and 2.3 % for gamma-ray irradiated samples. On the other hand, the values of χ_{min} for N were 7.7 % for un-irradiated and 9.5 % for gamma-ray irradiated samples. The number of displaced Ga and N atoms estimated using Eq. (1) were 3.5×10^{20} cm^{-3} and 8.7×10^{20} cm^{-3}, respectively. Although the disorder is recognized in Ga and N lattices, the displacement concentration of N atoms is about two times larger than that of Ga atoms. This suggests that N interstitial (N_i) exists in gamma-ray irradiated GaN bulk single crystal. This result is similar to that of the neutron irradiated GaN [13]. Since N_i atoms form the deep acceptor level at 960 meV below the bottom of the conduction band [13], the origin of the high resistivity after the gamma-ray irradiation is attributed to the carrier compensation effect due to the deep level of N_i.

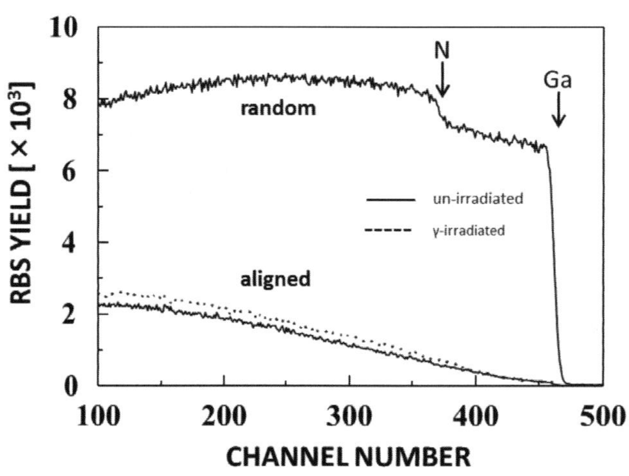

Figure 3. Random and aligned RBS spectra of un-irradiated and gamma-ray irradiated GaN bulk single crystal.

4. Conclusion

By gamma-ray irradiation, the resistivity of GaN bulk single crystals increased from 30 Ωcm to 10^4 Ωcm. The nitrogen displacement in gamma-ray irradiated samples was observed by Rutherford backscattering channeling experiments using proton beam. The high resistivity was attributed to the carrier compensation due to the deep acceptor level relating to interstitial nitrogen atoms. The peak energy of yellow luminescence after gamma-ray irradiation varied from 2.22 eV to 2.33 eV. The modification of YL was attributed to a transition from the shallow donor induced by the gamma-ray irradiation to the acceptor of native gallium vacancy. The shallow donor was related to nitrogen vacancy located at about 50 meV below the conduction band.

Acknowledgements

Part of this work was carried out under the Visiting Researchers Program of Kyoto University, Research Reactor Institute (KURRI).

References

[1] Kuriyama K, Ooi M, Onoue A, Kushida K, Okada M and Xu Q 2006 *Appl. Phys. Lett.* **88** 132109-1-3
[2] Nakamura T, Nishikata N, Kamioka K, Kuriyama K and Kushida K 2016 *Nucl. Instrum, Method Phys. Res.* **B 371** 251-253
[3] Satoh M, Kuriyama K and Kawakubo T 1990 *J. Appl. Phys* **67** 3542-3544
[4] Kuriyama K, Miyamoto Y, Koyama T, Ogawa O and Okada M 1999 *J. Appl. Phys.* **86** 2352-2354
[5] Kuriyama K, Tokumasu T, Takahashi J, Kondo H and Okada M 2002 *Appl. Phys. Lett.* **80** 3328-3330
[6] Emtsev VV, Davydov V Yu, Kozlovskii VV, Lundin VV, Poloskin DS, Smirnov AN, Shmidt NM, Usikov AS, Aderhold J, Klausing H, Mistele D, Rotter T, Stemmer J, Semchinova O and Graul J 2000 *Semicond. Sci. Technol.* 15, 73-78
[7] Kuriyama K, Satoh M and Kim C 1986 *Appl. Phys. Lett.* **48** 411-412
[8] Takahashi Y, Kuriyama K, Matsui Y and Kamijoh T 1993 *Appl. Phys. Lett.* **63** 1071-1073.
[9] Neugebauer J and Van de Walle CG 1996 *Appl. Phys. Lett.* **69** 503-505
[10] Look DC, Reynolds DC, Hemsky JW, Sizelove JR, Jones RL and R. J. Molnar RJ 1997 *Phys. Rev. Lett.* **79** 2273-2276
[11] Chu WK, Mayer JW and Nicolet AM, in *Backscattering Spectrometry* (Academic, New York, 1978), Chap. 8.
[12] Hollis MJ 1973 *Phys. Rev. B* **8** 931-935
[13] Nakamura T, Kamioka K, Kuriyama K, Kushida K, Xu Q, and Hasegawa M 2015 *Solid State Commun.* **205**, 1-3

IOP Conf. Series: Journal of Physics: Conf. Series **864** (2017) 012017 doi:10.1088/1742-6596/864/1/012017

Persistent photoconductivity in hydrogen ion-implanted $KNbO_3$ bulk single crystal

R. Tsuruoka[1], A. Shinkawa[1], T. Nishimura[1], C. Tanuma[1], K. Kuriyama[1] and K. Kushida[2]

[1]College of Engineering and Research Center of Ion Beam Technology, Hosei University, Koganei, Tokyo 184-8584, Japan
[2]Department of Arts and Sciences, Osaka Kyoiku University, Kashiwara, Osaka 582-8582, Japan

E-mail: kuri@ionbeam.hosei.ac.jp

Abstract. Persistent Photoconductivity (PPC) in hydorogen-ion implanted (001) oriented $KNbO_3$ bulk single crystals (perovskite structure at room temperature; ferroelectric with a band gap of 3.16 eV) is studied in air at room temperature to prevent the crystallinity degradation caused by the phase transition. Hydrogen is implanted into $KNbO_3$ bulk single crystals using the energy (the peak ion fluence) of 500 keV (5.0×10^{15} cm^{-2}). The resistivity varies from $\sim 10^8$ Ω/□ for an un-implanted $KNbO_3$ sample to 2.3×10^5 Ω/□ for as-implanted one. suggesting the presence of donors consisting of hydrogen interstitial and oxygen vacancy. The PPC is clearly observed with ultraviolet and blue LEDs illumination rather than green and infrared, suggesting the release of electrons from the metastable conductive state below the conduction band relating to the charge states of the oxygen vacancy as observed in electron irradiated ZnO.

1. Introduction

The ferroelectric field-effect transistor uses the surface conduction of insulating ferroelectric under the gate oxide. In the previous study, we artificially formed a surface conducting layer in $KNbO_3$ (perovskite structure, a=5.697, b=3.971, and c=5.722 Å at room temperature) by the hydrogen ion-implantation [1]. Persistent Photoconductivity (PPC) in electron irradiated ZnO bulk single crystal has been studied using light emitting diodes (LEDs) with various wavelengths above and below the band gap [2]. Irradiation induced oxygen vacancy (V_O) plays an important role in PPC. However, there are few reports on the PPC phenomena in ferroelectric material. The PPC has been observed in the layered perovskite $Nd_2Ti_3O_9$ [3] and the Niobate nanosheets [4] under the vacuum condition, suggesting the presence of the surface oxygen vacancy evidenced by the disappearance of PPC phenomenon in the ambient air. Lead zirconate titanate (PZT) has excellent properties, however, it contains harmful lead and, thus influence on the environment is concerned. $KNbO_3$ is expected as a substitute material for PZT. $KNbO_3$ is a unique dielectric material that has four phase transitions, namely rhombohedral ($T < -10$ °C), orthorhombic (-10 °C $< T <$ 225 °C), tetragonal (225 °C $< T <$ 435 °C), and cubic ($T >$ 435 °C) [5]. $KNbO_3$ shows ferroelectricity below 435 °C. However, this material cannot maintain single-crystallinity by the rapid heating and cooling. In the present study, we report the PPC observed in hydrogen-ion implanted $KNbO_3$ bulk single crystals at room temperature in air.

2. Experiments

The (001) oriented $KNbO_3$ bulk single crystals were purchased from SurfaceNet GmbH, Germany. The energy band gap of $KNbO_3$ estimated from optical absorption is 3.16 eV [1]. The hydrogen concentration near the surface of $KNbO_3$ evaluated by elastic recoil detection analysis (ERDA) using a 1.5 MeV $^4He^+$ beam was 5.1 x 10^{14} cm^{-2} for un-implanted sample [1]. Therefore, the residual hydrogen was recognized in un-implanted samples.

Hydrogen-ion implantation was performed using an energy of 500 keV (ion fluence: 5.0 x 10^{15} cm^{-2}). During implantation the samples were kept at room temperature. The resulting implantation depth and the hydrogen concentration were 3650 nm and 1.62 × 10^{20} cm^{-3}, respectively. The hydrogen concentration at around 60 nm from the surface evaluated by ERDA was 5.6 x 10^{14} cm^{-2} for as-implanted and 100 °C annealed samples [1]. In sheet resistance measurements using Van der Pauw technique, electrodes were fabricated using titanium/gold. The resistance varied from $>10^8$ Ω/□ for an un-implanted sample to 2.3 × 10^5 Ω/□ for as-implanted and 100 °C annealed ones. The decrease in resistance would be attributed to a donor related to the complex defects consisting of hydrogen interstitials and vacancy defect such as V_O generated by the hydrogen-ion implantation. This situation has been observed in the H-ion implanted ZnO [6].

For the PPC measurements, samples were kept at room temperature in air to maintain single-crystallinity by the rapid heating and cooling, and then illuminated with an ultraviolet-emitting diode (a peak wavelength of λ=400 nm) for the excitation above the energy band gap. The measurements for excitation below the band gap were also performed in comparison with the illumination above the energy gap, using blue (λ=465 nm), green (520 nm), and infrared (940 nm) emitting diodes.

3. Result & Discussion

Figure 1 shows the PPC of 100 °C annealed samples taken at room temperature in air for the illumination with various LEDs. Although twice on- and off-operations using the blue and green LEDs were tried, similar decay curve was observed. The photocurrent for the excitation above the energy band gap using a ultraviolet LED increases more in comparison with those below the band gap, showing the effective excitation from the valence band and/or the neutral oxygen vacancy (V_O^0) level to the conduction band. Furthermore, for the excitation below the band gap the photocurrent using the blue LED is larger than those using other LEDs. This suggests the presence of photo-responsible

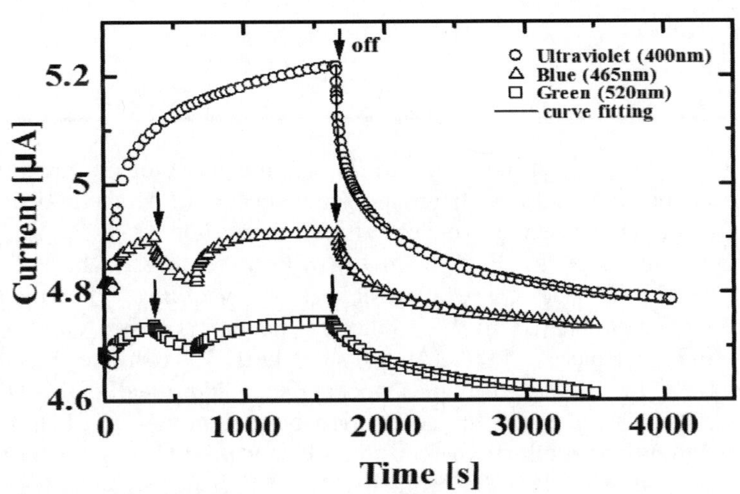

Figure 1. Persistent photoconductivity taken at room temperature in air for the illumination with various LEDs. The solid curves show the curve fitting with Eq.(1).

defects such as V_O for blue band region located at around 2.6 eV from the valence band. The decay kinetics of PPC seen here follow stretched exponential functions as described in previous works [7],

$$I_{PPC}(t) = I_{PPC}(0)\ \exp[-\ (t/\tau)^{\beta}],\ \ (\beta < 1), \qquad (1)$$

where $I_{PPC}(0)$ is the factor defined as the PPC build up level at the moment of light excitation being removed, τ is the PPC decay time constant, and β is the decay exponent. The solid curves in Fig.1 are the least squares fit of data with Eq.(1). Table 1 shows τ and β values for some LED illuminations. Table 2 also shows the percentage value of the PPC measured at 600 sec after cut off of the LED light. This indicates that the illumination due to LEDs above and near the band gap induces the effective PPC phenomenon.

Table 1. Decay time constant (τ) and decay exponent (β) for each LED estimated by eq.1
.

LED (wavelength; nm)	Decay time constant (τ; sec)	decay exponent (β)
Ultraviolet (365)	1.0×10^6	0.28
Blue (470)	1.2×10^8	0.247
Green (520)	4.8×10^7	0.279
Infrared (950)	1.7×10^5	0.48

Table 2. Percentage value of the PPC measured at 600 sec after cut off of the LED light.
.

LED (wavelength; nm)	Percentage value of the PPC (%)
Ultraviolet (365)	63.7
Blue (470)	56.3
Green (520)	41.4
Infrared (950)	14.4

The perovskite $KNbO_3$ ($K^{1+}Nb^{5+}O^{2-}{}_3$) structure has the K cation in 12-fold coordination surrounded by O anions, and the Nb cation in 6-fold coordination surrounded by an octahedron of O anions. PPC phenomenon in $KNbO_3$ would be expected to occur in the similar origin as PPC in ZnO [2]. According to the Lany and Zunger's model [8] for PPC in ZnO ($Zn^{2+}O^{2-}$) (see Fig.2), the $V_O{}^0$ creates a defect localized state (DLS) deep in the band gap, which localizes two electrons. When the two electrons at the DLS of $V_O{}^0$ are optically excited to $V_O{}^{2+}$, the Zn neighbors relax outwards and reach a configuration with a Zn-Zn distance of 4.0 Å. As a result, the DLS is shifted into the inside of the conduction band and the DLS of $V_O{}^{2+}$ becomes unoccupied. Additionally, the DLS spreads and resonates with the conduction band above the conduction band minimum (CBM). The resonance of the DLS with the conduction band results in the perturbed-host state (PHS) as a metastable conductive state (MCS) near the CBM (see Fig.2 (b)). Consequently, the PHS behaves as a shallow donor state [8]. Thus, the electrons in the PHS can be easily excited to the conduction band, leading to PPC. In $KNbO_3$, when the two electrons at the DLS of $V_O{}^0$ are optically excited to $V_O{}^{2+}$, the K and/or Nb neighbors would relax outwards and the DLS is shifted into the inside of the conduction band and the DLS of $V_O{}^{2+}$ becomes unoccupied. This lattice relaxation is similar to the situation that lattice distortions

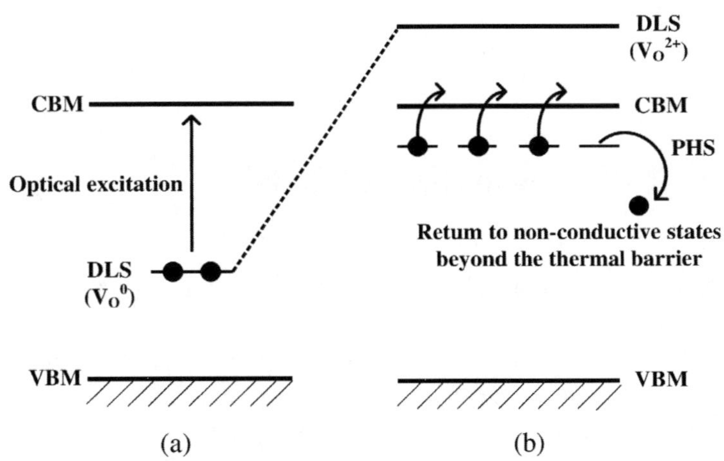

Figure 2. Schematic diagram based on Lany and Zunger's model [8]. When the two electrons at the defect localized state (DLS) of V_O^0 in the band gap are optically excited to V_O^{2+} (Fig.2 (a)), the resonance of the DLS with the conduction band results in the perturbed-host state (PHS) near the conduction band minimum (CBM) (Fig.2 (b)). The electrons in the PHS as a shallow donor state can be easily excited to the conduction band, leading to PPC.

increase in magnitude in going from the neutral to the (2+) charge state of V_O in perovskite $SrTiO_3$ [9]. As a result, electrons occupy the MCS near the CBM. In $KNbO_3$, at higher temperature such as room temperature, the thermal excitation to the CBM for the electrons in the PHS would be enhanced than that at lower temperature, causing a rapid decrease of PPC.

4. Conclusion

The Persistent Photoconductivity in hydrogen-ion implanted $KNbO_3$ bulk single crystals was observed in air at room temperature. Hydrogen was implanted into $KNbO_3$ bulk single crystals using the energy (the peak ion fluence) of 500 keV (5.0×10^{15} cm^{-2}). The resistivity varied from $>10^8$ Ω/□ for an un-implanted sample to 2.3×10^5 Ω/□ for as-implanted one, suggesting the presence of donors consisting of hydrogen interstitial and oxygen vacancy. The PPC was clearly observed with ultraviolet and blue LEDs illuminations rather than green and infrared ones. According to the Lany and Zunger's model for PPC in ZnO, the thermal excitation to the conduction band minimum for the electrons in the perturbed-host state near the conduction band minimum was enhanced at room temperature, causing a rapid decrease of PPC.

References

[1] Shinkawa A, Shibasaki Y, Nishimura T. Tanuma C, and Kuriyama K 2016 *Nucl. Instrum, Method Phys. Res.* **B 371** 283-285.
[2] Oga T, Izawa Y, Kuriyama K, Kushida K, and Xu Q 2011 *Solid State Commun.* **151** 1700-1703.
[3] Dulieu B, Bullot J, Wery J, Richard M, and Brohan L 1996 *Phys. Rev. B* **53** 10641-10650.
[4] Okamoto K, Sato H, Saruwatrai K, Tamura K, Kameda J, Kogure T, Umemura Y, and Yamagishi A 2007 *J. Phys. Chem. C* **111** 12827-12833.
[5] Shirane G, Newnham R, and Pepinsky R 1954, *Phys. Rev.* **96** 581-588.
[6] Kaida T, Kamioka K, Nishimura T, Kuriyama K, Kushida K, and Kinomura A 2015 *Nucl. Instrum, Method Phys. Res. B* **365** 171-174.
[7] Jiang HX and Lin JY 1990 *Phys. Rev. Lett.* 64 2547-2550.
[8] Lany S and Zunger A 2005 *Phys. Rev. B* **72** 035215-1-13.
[9] Buban JP, Iddir H, and Ogut S 2004 *Phys. Rev. B* **69** 180102-1-4.

High Temperature Annealing of MBE-grown Mg-doped GaN

S Contreras[1], L Konczewicz[1], H Peyre[1], S Juillaguet[1], M Al Khalfioui[2], S Matta[2], M Leroux[2], B Damilano[2] and J Brault[2]

[1] Laboratoire Charles Coulomb (L2C), UMR 5221 CNRS-Université de Montpellier, CC074, Montpellier, FR-34095, France
[2] CRHEA-CNRS, Centre de Recherche sur l'Hétéro-Epitaxie et ses Applications, CNRS, Rue B. Grégory, Valbonne, FR-06560, France

sylvie.contreras@umontpellier

Abstract. In this report, are shown the results of high temperature resistivity and Hall Effect studies of Mg-doped GaN epilayers. The samples studied were grown on (0001) (c-plane) sapphire by molecular beam epitaxy and 0.5 μm GaN:Mg layers have been achieved on low temperature buffers of GaN (30 nm) and AlN (150 nm). The experiments were carried out in the temperature range from 300 K up to 900 K. Up to about 870 K a typical thermally activated conduction process has been observed with the activation energy value E_A=215 meV. However, for higher temperatures, an annealing effect is observed in all the investigated samples. The increase of the free carrier concentration as a function of time leads to an irreversible decrease of sample resistivity of more than 60%.

1. Introduction

Nitride-based applications are very broad, from light emitting devices to high power transistors, and have continuously increased over the last decade, making gallium nitride and related compounds, the second semiconductor in volume after Si. The p-type doping of GaN and other wide-band-gap III-V nitrides is the key for all these applications. So far magnesium (Mg) is the only acceptor dopant which generates sufficient p-type conductivity in a reproducible manner. In the case of metal-organic vapor phase epitaxy (MOVPE), Mg acceptors are electrically inactive in as-grown material and an additional dopant activation procedure is required. It is generally accepted that this Mg activation consists in dissociation of Mg-H complexes [1]. Contrary to MOVPE, in the case of molecular beam epitaxy (MBE), Mg acceptors are electrically active in the as-grown material and no additional annealing procedure is necessary [2]. Neither annealing in nitrogen ambient furnace (at various temperatures between 600 and 850 °C) nor rapid thermal annealing at 950 °C for 2 min were found to have any discernible influence on the electrical properties of the films [3]. In this work, we present the results of high temperature resistivity and Hall Effect measurements on MBE-grown, Mg-doped GaN epilayers. The experiments have been carried out from 300 K to 900 K. As far as the temperature was lower than approximatively 870 K, a typical thermally activated p-type conduction process has been observed and the temperature dependence of free holes concentration was analyzed assuming a single Mg-acceptor impurity level. Moreover, at temperatures higher than 870 K, resistivity became time-dependent for all investigated samples. A more than 60% irreversible decrease of the sample resistivity as well as an increase of free hole concentration has occurred.

2. Experimental methods

Mg-doped GaN layers were grown on 2-inch c-plane sapphire substrates by MBE in a Riber 32 MBE system. Effusion cells were used for Ga and Mg and ammonia (NH3) as nitrogen source. Initially, the surface was nitrided with ammonia at ~ 950°C for 10 min. Then a GaN nucleation layer (30 nm) was grown at low temperature, followed by a 150 nm-thick AlN layer grown at 950°C, with a growth rate of 0.1 μm/h. This AlN layer was inserted as an insulating layer in order to get a very low residual concentration of unintentional donors before the growth of the p-type GaN layer. Finally, the Mg doped GaN layer was grown at a growth rate of 0.8 μm/h with a Mg flux varied between $(1.2 \div 2)x10^{-10}$ Torr corresponding to Mg atom concentrations between $(3 \div 5)x10^{19}$ cm^{-3}. The growth temperature was varied between 760°C and 780°C, in order to modify the hole concentration of the layers, the hole compensation being strongly temperature dependent [3]. The samples were cut into 5×5 mm^2 squares size with electrical contacts in the corners of the sample. The ohmic contacts were formed by the evaporation of Ti(30nm)/Al(180nm)/Ni(40nm)/Au(150nm) electrodes and subsequent annealing in N_2 ambient. The resistivity and Hall Effect measurements (magnetic field of 1 T) were performed using the van der Pauw method. During all the measurements, the current through the sample was kept adequately low to ensure ohmic conditions. For high temperature measurements and the in-situ electrical characterization during annealing the samples were bonded with gold-wires to a ceramic support. The sample holder was installed in a quartz tube inside a furnace to keep during the high temperature experiments a vacuum better then 10^{-6} bar. The temperature was measured with a platinum (Pt) resistor and stabilized with a precision better than 1 K. The room temperature electrical characteristics of the Mg-doped GaN layers under study are presented in Table 1. The preliminary studies of transport effects as a function of temperature for sample E, have demonstrated that as far as the temperature was lower than approximatively $T_{critical} = 870$ K, a typical thermally activated conduction processes takes place.

Table 1. Characteristics of the Mg-doped, GaN layers under study

Sample	ρ(300K) (Ωcm)	p_H(300K) (cm^{-3})	μ(300K) (cm^2/Vs)
A	3.1	$3.9x10^{17}$	5.2
B	7.7	$2.2x10^{17}$	3.7
C	1.3	$1.8x10^{18}$	2.6
D	3.9	$8.1x10^{17}$	2.0
E	1.2	$2.5\ x10^{18}$	2.0

However, at temperatures higher then $T_{critical}$, the electrical transport parameters were found to be time-dependent. For this reason the steps of the experimental procedure were as follows:

- the conductivity and Hall Effect were measured as a function of increasing temperature from 300 K up to approximately 870 K,
- when the time-dependent conductivity was observed, an annealing temperature T_{ann} was established and the sample resistivity ρ and Hall Effect were measured *vs* time during the isothermal annealing process,
- as soon as the ρ(t) values were practically saturated, the electrical parameters were once again measured as a function of decreasing temperature.

Figure 1 shows the resistivity ρ of samples A, B, C and D as a function of reciprocal of temperature. Below a critical temperature $T_{critical}$ a typical thermally activated conduction processes takes place for all samples. The same behavior was observed for the Hall Effect, $p_H(1/T)$ measurements. The analysis of this part of results allowed the determination of the ionization energy of Mg-acceptors as E_A=215 meV for sample A. For temperatures higher than approximatively 870 K, a time-dependent decrease of resistance takes place in all investigated samples. As an example the change of the resistivity of sample B with time is given in Figure 2. The annealing leads to the more than 60% decrease of the sample resistivity that practically saturates above three hours.

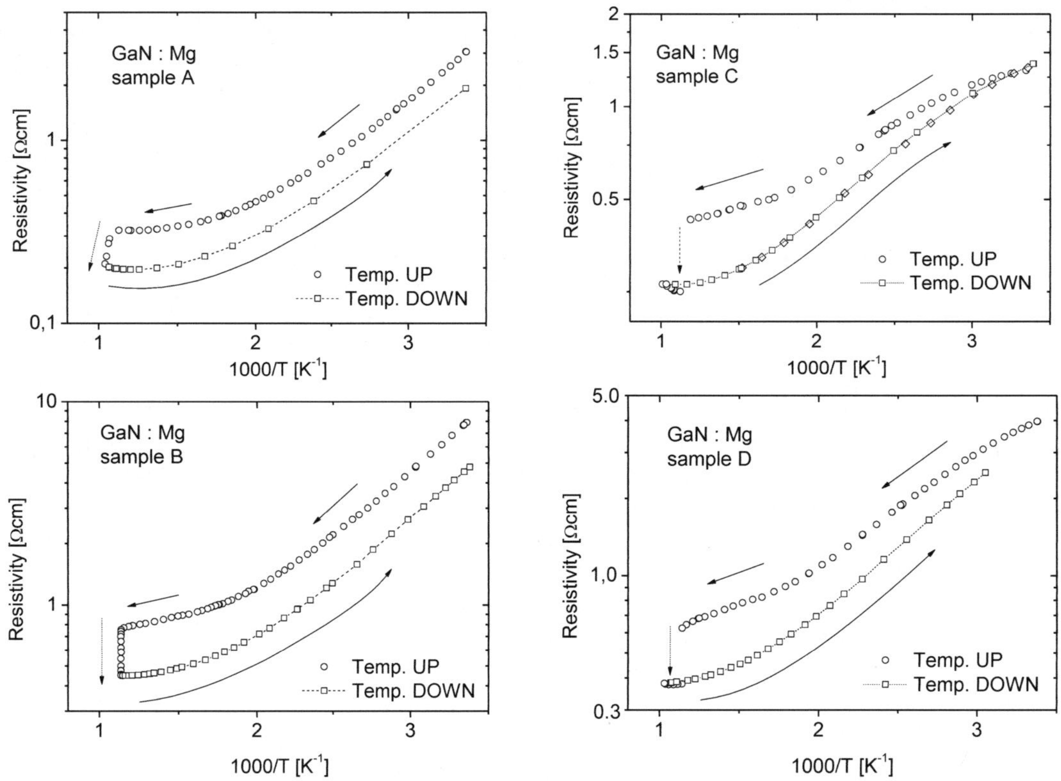

Figure 1. Variation of the resistivity ρ(1/T) of the Mg-doped GaN samples A, B, C and D *vs* reciprocal of temperature as well as the resistivity changes during isothermal annealing. The circles correspond to the ρ(1/T) variation during the heating procedure and the squares to the cooling after the annealing process.

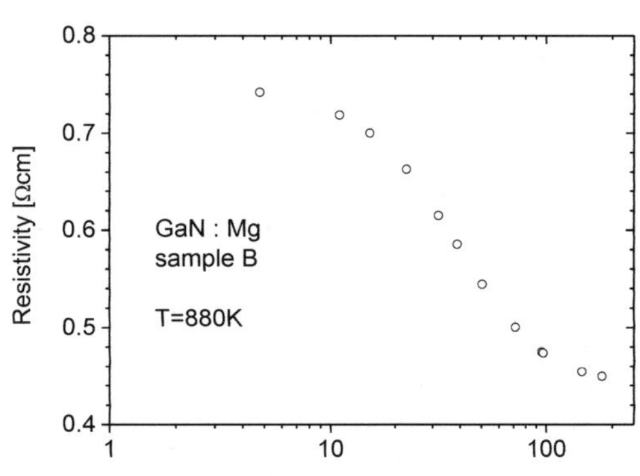

Figure 2. Resistivity *vs* time during an isothermal annealing at a temperature of 880 K (sample B).

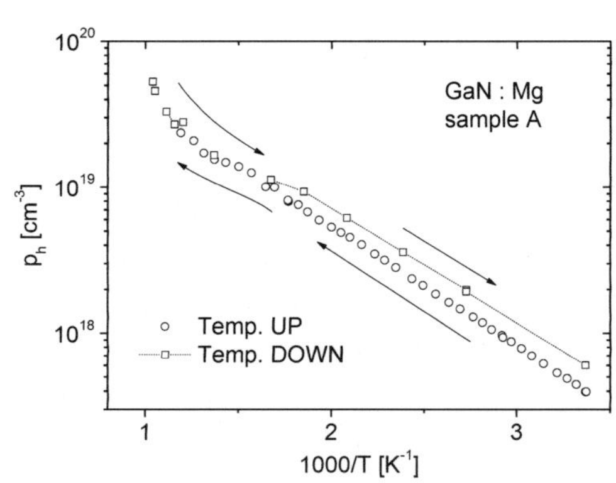

Figure 3. Hall concentration *vs* reciprocal of temperature before (circles) and after (squares) annealing (sample A).

This variation of sample resistivity is caused by the changes in free carrier concentration. This can be seen on Figure 3 where the temperature dependence of the hole concentration in sample A before and after annealing is presented. The annealing process increases undoubtedly the hole concentration in the sample but the character of the $p_H(T)$ variation remains unchanged. It seems that this conclusion is valid also for mobility changes but in this case, the effect is less pronounced. The effect of the annealing process is irreversible. It can be seen on Figure 4 where two successive ρ(1/T) runs after t annealing are presented (sample D). Both sets of experimental results overlap perfectly up to 660 K.

3. Discussion and conclusions

The Mg-doped GaN epitaxial layers under study were grown by MBE. In this growth technique the Mg acceptor is electrically active in the as-grown material [2] and effectively our samples had high hole concentrations without any activation procedure. However from resistivity and Hall Effect measurement carried out in vacuum it was shown that at T > 870 K, an annealing process takes place leading to an increase of free holes concentration in the samples. As a results, the significant decrease of resistivity and weak increase of mobility take place. According to the case of MOVPE one can guess that this effect is related to the some residual dissociation of Mg-H complexes. However this process should occur already at temperatures significantly lower, of the order of 650 K [4]. The other explanation might be due to the fact that secondary ion mass spectrometry indicates a high concentration of oxygen, comparable to that of magnesium, in the samples (this oxygen, most likely, is coming from ammonia [5]). Indeed, it was reported that the acceptor (Mg) – donor (O) complexes in bulk GaN material dissociate at temperatures above 600°C [6]. Moreover, preliminary low temperature photoluminescence study of sample A, presented on Figure 5, shows some evolution of the donor-acceptor pair (DAP) band after the annealing process.

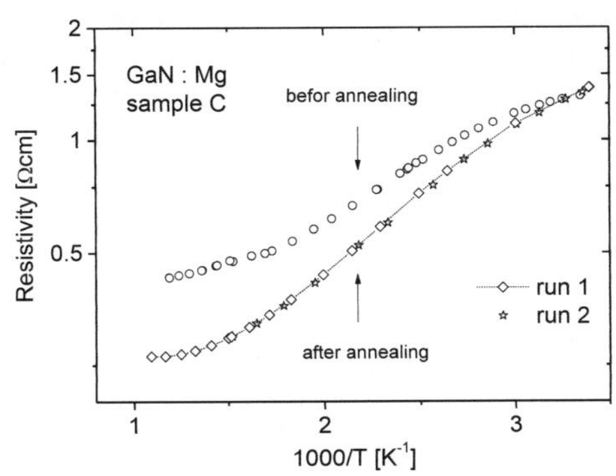

Figure 4. Resistivity vs reciprocal temperature for sample C: circles–before annealing; squares–after annealing; stars–second run after annealing.

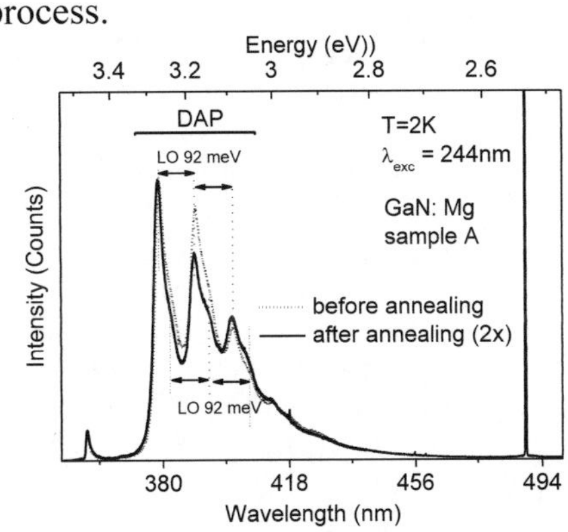

Figure 5. Low temperature PL spectra for sample A: dots line – results before annealing ; solid line – results after annealing

Acknowledgments
This work is supported by ANR Project: <ANR-14-CE26-0025-01> "NANOGANUV"

References

[1] Nakamura S, Mukai T, Senoh M and Iwasa N 1992, *Jpn. J. Appl. Phys.*, 31, L139
[2] Wook K, Salvador A, Botchkarev A E, Aktas O, Mohammad S N and Morcoç H 1996, *Appl. Phys. Lett.* 69 (4), pp.559-561
[3] Dussaigne A, Damilano B, Brault J, Massies J, Feltin E and Grandjean N, 2008, *J. Appl. Phys.* 103, 013110
[4] Litwin-Staszewska E, Piotrzkowski R, Dmowski L, Prystawko P, Czernecki R, Konczewicz L 2006, *J. Appl. Phys.*99, 033703
[5] Wook K, Botchkarev A E, Salvador A, Popovici G, Tang H and Morkoç H 1997, *J. Appl. Phys.* 82, 219
[6] Litwin-Staszewska E, Suski T, Piotrzkowski R, Grzegory I, Bockowski M, Robert J-L, Konczewicz L, Wasik D, Kamińska E, Cote D and Clerjaud B 2001, *J. Appl. Phys.* 89, 12, p.7960

Improved Ohmic-contact to AlGaN/GaN using Ohmic region recesses by self-terminating thermal oxidation assisted wet etching technique

J Liu, J Wang, H Wang, L Zhu and W Wu

Institute of Microelectronics, Peking University, Beijing 100871, China

E-mail: wangjinyan@pku.edu.cn

Abstract. Lower Ti/Al/Ni/Au Ohmic contact resistance on AlGaN/GaN with wider rapid thermal annealing (RTA) temperature window was achieved using recessed Ohmic contact structure based on self-terminating thermal oxidation assisted wet etching technique (STOAWET), in comparison with conventional Ohmic contacts. Even at lower temperature such as 650°C, recessed structure by STOAWET could still obtain Ohmic contact with contact resistance of 1.97Ω·mm, while conventional Ohmic structure mainly featured as Schottky contact. Actually, both Ohmic contact recess and mesa isolation processes could be accomplished by STOAWET in one process step and the process window of STOAWET is wide, simplifying AlGaN/GaN HEMT device process. Our experiment shows that the isolation leakage current by STOAWET is about one order of magnitude lower than that by inductivity coupled plasma (ICP) performed on the same wafer.

1. Introduction

Improvement of Ohmic contacts is essential to fully utilize the advantages of GaN heterostructure electron devices. Etching the Ohmic regions by plasma before deposition of metals is one of the most widely used ways to improve the performance of Ohmic contacts on AlGaN/GaN [1-5], which may result in serious damages, and the power of plasma and the etching time of Ohmic region recesses should be accurately controlled [2, 4, 5].

In this letter, self-terminating thermal oxidation assisted wet etching technique (STOAWET) with thermal oxidation of AlGaN layer followed by etching in KOH solution, which has been applied to gate recess etching [6], is proposed to be used in the fabrication of Ohmic region recess and mesa isolation at one process step, simplifying the device process. The experimental results show that lower Ti/Al/Ni/Au Ohmic contact resistance (Rc) and wider rapid thermal annealing (RTA) temperature window could be obtained by STOAWET in comparison with conventional Ohmic contacts structure, where all Ohmic contacts were fabricated on the same wafer. The isolation leakage current by STOAWET is about one order of magnitude lower than that by inductivity coupled plasma (ICP) performed on the same wafer, implying the practicability of STOAWET in the isolation on AlGaN/GaN heterostructure.

2. Experiments

AlGaN/GaN epi-wafer used in this work was grown on (0001) sapphire substrate by metal organic chemical vapour deposition (MOCVD). The wafer structure consists of a 2μm unintentionally doped

GaN buffer layer, a 22nm unintentionally doped $Al_{0.23}Ga_{0.77}N$ barrier layer and a GaN cap layer with thickness of 2.8nm. Transfer length method (TLM) structures were fabricated. Firstly, 200nm plasma-enhanced chemical vapour deposition SiO_2 was deposited on top of the sample as mask layer. Then, Ohmic recess and isolation regions were defined by optical lithography followed by the removal of SiO_2 using SF_6-based plasma. The sample was subsequently treated by thermal oxidation for 45 min at 650°C followed by 45 min etching in KOH solution at 70°C. After SiO_2 was then removed by buffered oxide etching, Ti/Al/Ni/Au (20nm/160nm/50nm/100nm) metals for Ohmic contact were formed by electron beam evaporation followed by lift-off process. For comparison, TLM structures without Ohmic region recesses (conventional structures) were also fabricated on the same wafer with SiO_2 protecting the Ohmic contact regions during oxidation and wet etching. Then the wafer was diced into several samples, annealed in N_2 ambience at different temperatures to investigate the effect of RTA temperatures. Channel width of the TLM structures is 95μm and the contact resistance was derived by TLM patterns with separations of 5μm, 10μm, 15μm, 20μm, 25μm and 30μm. Partial cross-section of the recessed TLM structure is shown in Figure 1. As depicted in [6], the whole AlGaN layer in etched regions is etched away by STOAWET without affecting the underlying GaN layer. To compare the isolation efficiency of STOAWET and ICP, Cl_2/BCl_3 based plasma was used to etch the isolation regions with conventional Ohmic contact structure on another sample from the same wafer.

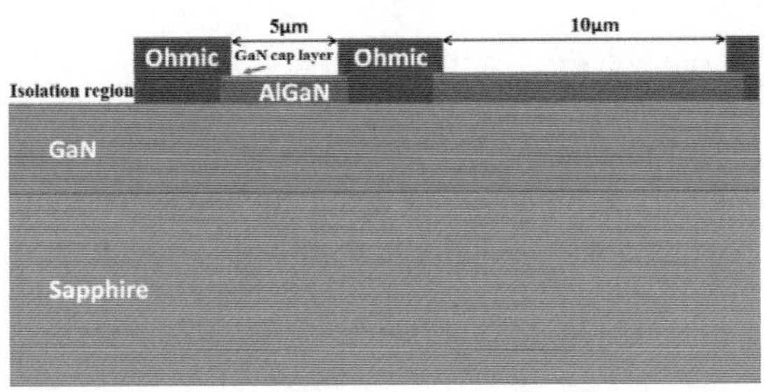

Figure 1. Partial cross-section of recessed TLM structure

3. Results and discussion

Figure 2(a) shows the lowest Rc of recessed Ohmic contact achieved at various RTA temperatures. It could be seen that Rc from 0.76Ω·mm to 0.94Ω·mm could be achieved annealed at the temperature from around 780°C to 900°C for recessed Ohmic contact. However, conventional Ohmic contact could only achieve the lowest Rc of about 1.15Ω·mm at 830°C for 35s, larger than those of recessed structure annealed from 750°C to 920°C. Conventional Ohmic contact became worse (not straight in linear regions of I-V curves) annealed at 810°C as shown in Figure 2(b) compared with recessed structure as shown in Figure 2(c) and after annealing at 850°C for 32s average Rc as large as 2.62Ω·mm was obtained, indicating a narrow RTA temperature window (<40°C, defined by Rc of ~1Ω·mm). Therefore, etching the Ohmic contact regions by STOAWET before metals deposition could effectively decrease Rc and expand RTA temperature window (>120°C, defined by Rc<1Ω·mm), making this technique easier to achieve lower Rc with a wider Ohmic annealing temperature window improving the tolerance of temperature fluctuation in practical production. It is mentioned that the best Ohmic performance is obtained when the two-dimensional electron gas (2DEG) channel is completely removed [2] and minimum contact resistance is obtained just before etching through the barrier [4], therefore recessed Ohmic structure by STOAWET just meets the condition for improving Ohmic contacts to GaN heterostructures.

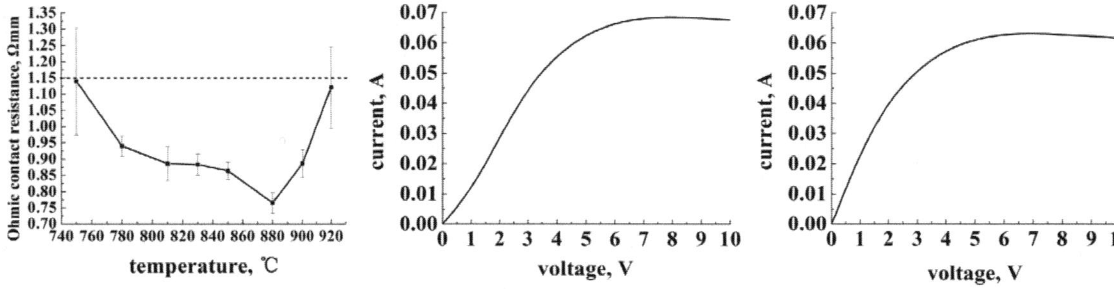

(a) (b) (c)

Figure 2. (a) Lowest recessed Ohmic contact resistance achieved at various RTA temperatures derived by TLM patterns with separations of 5μm, 10μm, 15μm, 20μm, 25μm and 30μm. RTA time differs from 20s to 50s. Dotted line marks the lowest Rc of conventional Ohmic contact, which was achieved at 830°C for 35s. (b) I-V curve of conventional Ohmic structure and (c) I-V curve of recessed Ohmic structure with separation of 5μm and width of 95μm annealed at 810°C for 40s.

Some samples were also annealed at 650°C for different time and achieved the best Rc of 1.97Ω·mm for recessed structure after 5min RTA. I-V curves of recessed and conventional Ohmic structures with separation of 10μm and width of 95μm annealed at 650°C are shown in Figure 3, indicating that conventional Ohmic structure could not achieve Ohmic contact at such low temperature at all.

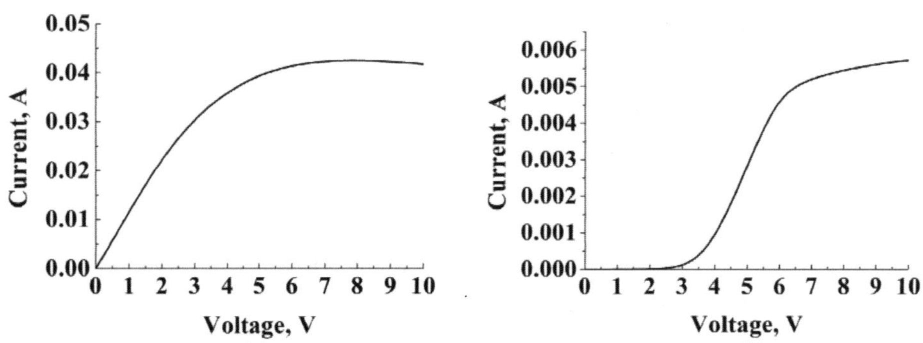

(a) (b)

Figure 3. I-V curves of (a) recessed and (b) conventional Ohmic structures with separation of 10μm and width of 95μm annealed at 650°C for 5min

Samples by STOAWET and ICP isolation with conventional Ohmic contact structure were annealed at 830°C for 35s. The leakage current at a bias of 10V between two pads with separation of 40μm and width of 95μm isolated by STOAWET is measured as 4.8pA, which is about one order of magnitude lower than that between two mesas isolated by ICP (24pA). It is reported that low surface roughness of ~0.27nm could be achieved by STOAWET [6] and damages could be caused by ICP [7]. Therefore, it is speculated that smoother GaN surface by STOAWET with fewer surface damages results in lower isolation leakage current.

It should be noted that by using STOAWET recessed Ohmic structure and device isolation could be fabricated in one process step. Moreover, as reported in [6], the window of oxidation temperature, oxidation time and wet etching time of STOAWET is wide enough, making the condition of Ohmic region recessing easy to control using STOAWET, improving the convenience and consistence of practical AlGaN/GaN device process, in comparison with dry etching method.

4. Conclusion

It is demonstrated that by using STOAWET Ohmic region recesses and mesa isolation could be fabricated in one process step. Lower Ohmic contact resistance and wider RTA temperature window of recessed Ohmic contact could be achieved by STOAWET in comparison with conventional Ohmic contact. According to our experimental results, the recessed contact resistance achieved by STOAWET in this work is obviously improved compared with the contact resistance of conventional Ohmic structure. At lower temperature such as 650°C where conventional Ohmic structure cannot form Ohmic contact, recessed structure by STOAWET could still obtain Ohmic performance with Rc of 1.97Ω·mm. About one order of magnitude lower leakage current between two mesas is achieved by STOAWET than by ICP, implying the practicability of STOAWET in AlGaN/GaN isolation. Due to the wider process widows of Ohmic region recessing and RTA and the integration of Ohmic region recess and device isolation processes, STOAWET could improve the convenience and consistence of practical AlGaN/GaN device process.

Acknowledgments

The authors would like to thank Nanofabrication facility in Suzhou Institute of Nano-Tech and Nano-Bionics （SINANO） for the help in sample fabrication. This work was supported by the National Natural Science Foundation of China (No. 60406004), Beijing Natural Science Foundation (No.4112025), and National Key Micrometer/Nanometer Processing Lab.

References

[1] Lau W, Tan J and Singh B 2009 ‘Formation of Ohmic contacts in AlGaN/GaN HEMT structures at 500°C by Ohmic contact recess etching’ *Microelectron. Reliab.* **49** 5 pp 558-561

[2] Wang L, Kim D and Adesida I 2009 ‘Direct contact mechanism of Ohmic metallization to AlGaN/GaN heterostructures via Ohmic area recess etching’ *Appl. Phys. Lett.* **95** 17 p 172107

[3] Wang C, He Y, Zheng X, Zhao M, Mi M, Li X, Mao W, Ma X and Hao Y 2015 ‘Low ohmic-contact resistance in AlGaN/GaN high electron mobility transistors with holes etching in ohmic region’ *Electron. Lett.* **51** 25 pp 2145–47

[4] Fagerlind M and Rorsman N 2011 ‘Optimization of recessed ohmic contacts for AlGaN/AlN/GaN heterostructures using C(V) characterization of MSHM structures’ *Phys. Status Solidi C* **8** 7-8 pp 2204–06

[5] Buttari D, Chini A, Meneghesso G, Zanoni E, Moran B, Heikman S, Zhang N, Shen L, Coffie R, DenBaars S and Mishra U 2002 ‘Systematic characterization of Cl_2 reactive ion etching for improved ohmics in AlGaN/GaN HEMTs’ *IEEE Electron Device Lett.* **23** 2 pp 76-78

[6] Xu Z, Wang J, Liu Y, Cai J, Liu J, Wang M, Yu M, Xie B, Wu W, Ma X, and Zhang J 2013 ‘Fabrication of normally off AlGaN/GaN MOSFET using a self-terminating gate recess etching technique’ *IEEE Electron Device Lett.* **34** 7 pp 855-857

[7] Cao X, Pearton S, Dang G, Zhang A, Ren F and Hove J 2000 ‘GaN n- and p-type Schottky diodes: effect of dry etch damage’ *IEEE Trans. Electron. Devices* **47** 7 pp 1320-24

Influence of sample design on the strong light-matter coupling in ZnSe-based periodic structures

K. Sebald[1], SK. S. Rahman[1], T. Kaya[1], J. Gutowski[1], T. Klein[2,3], S. Klembt[2,4], A. Gust[2], D. Hommel[2,5]

[1] Semiconductor Optics, [2] Semiconductor Epitaxy, Institute of Solid State Physics, University of Bremen, Germany

E-mail: ksebald@ifp.uni-bremen.de

Abstract. Cavity-exciton polaritons have attracted much interest because these light-matter quasiparticles are very promising for various optoelectronic applications. Bragg-polaritons have been discussed as new tools for tailoring light-matter interactions. These structures can be created by incorporating quantum wells (QWs) periodically into a distributed Bragg reflector (DBR). The advantage of this sample type is the high number of QWs which can be embedded into the sample in order to increase the Rabi-splitting energy. Calculations show, that the coverage of the sample with a thin metal layer results in an increase of the temperature stability of the strong coupling regime. In addition, this concept enables a specific spectral variation of the cavity resonance allowing for the manipulation of the light-matter interaction.

1. Introduction

The realization and manipulation of the strong coupling regime in microcavities (MCs) are one of the current research objectives, due to the unique properties of the generated exciton-polaritons for novel optical devices as well as for fundamental physics [1, 2]. Strong coupling in classical microcavities has been demonstrated for different material systems. In particular, the use of wide-bandgap semiconductor enabled polaritonic lasing at room temperature [3–5]. In recent years, Bragg-polariton samples, where the QWs are directly inserted into the DBR, have been discussed as a new tool for tailoring light-matter coupling [6, 7]. The advantage of this type of an unfolded MC sample is the simplification of the sample design and the large number of QWs which can be inserted into the structure without increasing the mode volume. In this case, the light-matter interaction occurs between the QW excitons and the Bragg-mirror modes forming a Bragg polariton. The first report on the realization of strong coupling for an unfolded MC concerned single InGaAs QWs embedded in a 30-periods distributed Bragg reflector (DBR) stack with a Rabi-splitting energy of 9.3 meV [6]. In order to control and manipulate the polariton eigenstate, as it is necessary for fundamental physics as well as for application, metal films can be utilized. This results in a local energy blueshift of the cavity mode which leads to a lower energy position of the polariton. In such a way potential barriers of about 200 μeV [8] can be created. An additional positive effect is the increase of the internal reflectivity of the device

[3] Present address:BIAS, Klagenfurter Str. 2, 28359 Bremen, Germany
[4] Technische Physik, University of Würzburg, Am Hubland, Germany
[5] Institute of Experimental Physics, University of Wroclaw and Wroclaw Research Center EIT, Poland

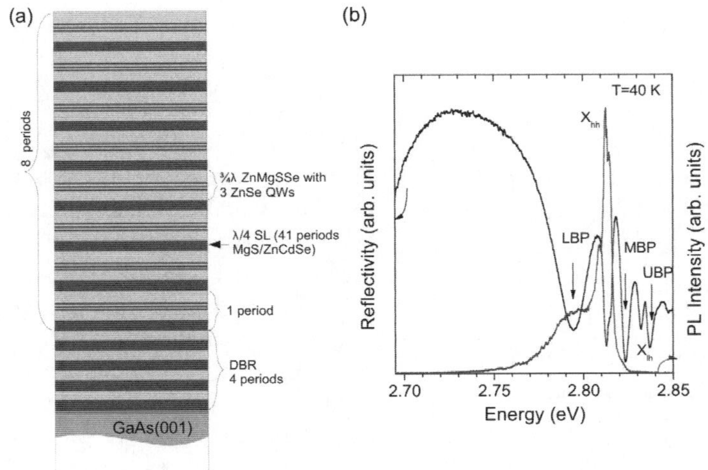

Figure 1. (a) Principle drawing of the sample setup of the unfolded cavity structure. (b) Reflectivity and photoluminescence spectra of the Bragg-polariton sample measured at T = 40 K (excitation energy 3.03 eV). The spectral position of the lower (LBP), middle (MBP) and upper Bragg-polariton (UBP) branch are indicated together with those of the X_{hh} and X_{lh} state.

improving its optical quality. In this contribution we will present investigations of the optical properties of a Bragg-polariton sample and discuss the possibility to achieve strong coupling at room temperature by the usage of metal films.

2. Experimental setup and results

The Bragg-polariton sample was grown by molecular beam epitaxy. For the Bragg-layers $Zn_{0.79}Mg_{0.21}S_{0.23}Se_{0.77}$ is chosen as the high-index material, while the low-index material consists of a superlattice of MgS and $Zn_{0.63}Cd_{0.37}Se$. Thus, a refractive index contrast in the order of $\Delta n = 0.4$ is obtained. Each three 8 nm thick ZnSe QWs are embedded in the center of any of the high-index material with a total thickness of $3\lambda/4$ (144 nm). The low-index material possesses a thickness of $\lambda/4$ (44.7 nm) (see Fig. 1 (a)). Eight of these Bragg-polariton layer combinations were grown on four plain DBR pairs, where the latter ensure a precise *in-situ* control of the respective $\lambda/4$-thickness. Micro-reflectivity and -photoluminescence (excitation energy 3.03 eV) measurements were performed and the experimental findings are compared to calculations using a transfer matrix (TM) method (CAMFR program [9]). Fig. 1 (b) shows the reflectivity of the sample in comparison to its photoluminescence at T = 40 K. The reflectivity exhibits five distinct minima. The second minimum centered at 2.812 eV originates from the ZnSe QW heavy-hole (X_{hh}) exciton, while the fourth minimum at 2.832 eV can be attributed to the absorption of light-hole exciton (X_{lh}). This classification is underscored by the PL spectrum showing a dominant contribution of the ZnSe QW X_{hh} together with a signature probably due to trionic recombination on the low-energy side. We have already shown earlier that the implementation of the QWs into the DBR structure results in the interaction of the first Bragg mode with the X_{hh} and X_{lh} states leading to the formation of three polariton states, the lower (LBP), middle (MBP) and upper Bragg-polariton (UBP) branch [10]. The strong indication for the existence of strong coupling is traceable up to a temperature of 200 K with Rabi-splitting energies of 24 meV and 13 meV between the Bragg-polaritons. In the reflectivity spectrum in Fig. 1 (b) the LBP, MBP and UBP can be identified at 2.7945 eV, 2.823 eV and 2.837 eV (first, third and fifth minimum). In Fig. 2 (a) the photoluminescence spectra are depicted in the temperature region between 40 K and 160 K. The emission of the X_{hh} and X_{lh} as well as the one at the position of LBP can be identified. The relative detuning of the LBP band with regard to the respective energy position of X_{hh} at any of the given temperatures can be easily derived if the appropriate energy position of X_{hh} is taken as zero. Besides the increasing spectral width of the excitonic components with growing temperature the spectral shift of the emission attributed to the LBP shows a decrement

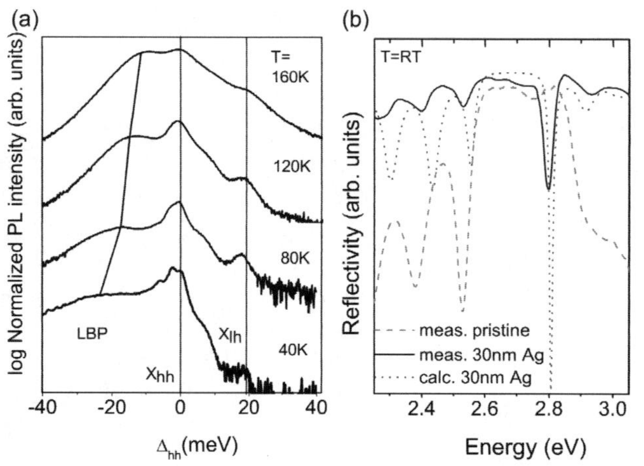

Figure 2. (a) PL spectra of the Bragg-polariton sample taken for temperatures between 40 and 160 K and vertically shifted for clarity. The relative energy position of the X_{hh} is taken as zero for the different temperatures. The spectral positions of the X_{hh} and X_{lh} are marked by a vertical lines while the relative shift of the emission at the position of the LBP is labeled by a line as guide to the eye. (b) Measured reflectivity of a pristine and a metal covered (30 nm Ag) microcavity (dashed and solid line) together with the calculated spectrum (dotted line).

for increasing temperatures. This is due to the anticrossing of the Bragg mode with the X_{hh}, which we have already reported to occur at T = 130 K [10]. For temperatures exceeding 200 K the strong-coupling regime is broken due to the spectral broadening of the excitonic components. These results have shown that a remarkably large Rabi-splitting energy can even be realized with eight Bragg-polariton stacks. The observed interaction strength is comparable to a complete MC structure consisting of an 18-fold bottom DBR, a λ cavity, and a 15-pair top DBR, for which a Rabi-splitting energy of 19 meV was reported at 70 K [11]. In order to fully exploit the promising properties of a Bragg-polariton sample one has to achieve the strong coupling regime near room temperature without scaling up the fabrication effort. One possibility is the usage of a thin metal layer which can drastically increase the optical quality of the structure, as already reported [12]. To emphasize this idea, we present a micro-reflectivity measurement at RT of a planar cavity sample which consists of a 12-pair bottom DBR, a $\lambda/2$ cavity, and a 1.5-pair top DBR, as shown in Figure 2 (b). The spectrum for the pristine sample (dashed line) possesses a cavity dip at $E_C = 2.742$ eV, and the reflectivity is close to 94 %. This relatively low value is due to the small number of top DBR layer pairs. When the MC is covered by a 30 nm thick Ag layer the reflectivity minimum shifts to higher energies what goes along with an enhancement of the transmission. The reflectivity minimum of that sample possesses a full width at half maximum of 33 meV. The measured spectrum is displayed along with the calculated one (dotted line) in which the band possesses a spectral width of 13 meV. The calculated blueshift of the reflectivity minimum amounts to 65 meV, being about a factor of 1.7 times larger than the experimental value. The maximal reflectivity is increased to 99.5%. These results indicate the large improvement of the optical properties of a microcavity achieved by applying a metal deposition. The deviation of the measurement relative to the calculation is due to an oxidation of the thin layer which is not completely reproduced by the calculation, although an oxidation degree of 15% was assumed. We have applied this concept theoretically to the Bragg-polariton sample by performing transfer matrix calculations. We have assumed that the sample is covered by a 30 nm thick Ag layer and as calculation parameters we have inserted the spectral width and position of the QWs excitonic components at room temperature. The spectral position of the Bragg mode can be tuned relative to X_{hh} by changing the thickness of the top layer by an appropriate factor as given in Fig. 3 (a). The resulting map of the gray-coded reflectivity shows

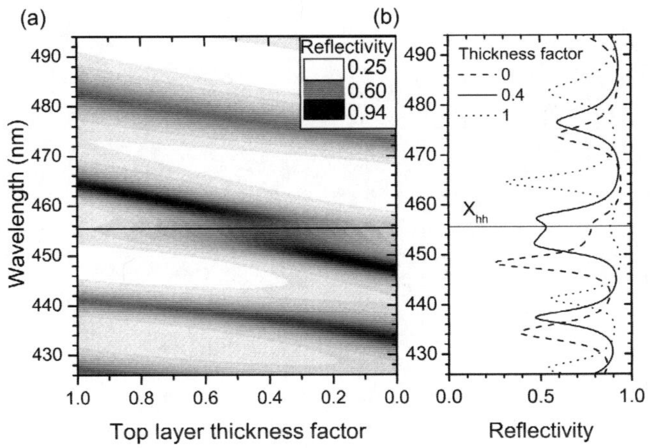

Figure 3. Calculated reflectivity for different top layer thicknesses of a Bragg-polariton sample covered with 30 nm of Ag. The parameter for the calculation matches the excitonic spectral position and spectral width of a ZnSe QW reference sample at room temperature. An anticrossing at the position of the X_{hh} (horizontal line) is observed. (b) shows the corresponding reflectivity spectra.

an anticrossing for a thickness factor of 0.4 at the position of the X_{hh}. The energy splitting can be deduced to 30 meV from the reflectivity spectra shown in Fig. 3 (b). This calculation indicates that a device operating at room temperature in the strong coupling regime can be achieved with a rather simple Bragg-polariton sample design by adding a thin metal layer.

3. Conclusion

The realization of strong coupling in an unfolded microcavity was verified by analyzing the photoluminescence for different temperatures. Calculations indicate that the application of a thin metal layer may help to achieve strong coupling at room temperature due to the improved reflectivity.

Acknowledgment: We gratefully acknowledge financial support from the Deutsche Forschungsgemeinschaft (Se 1846/1-2, Ho 1388/33-2).

References

[1] Kasprzak J, Richard M, Kundermann S, Baas A, Jeambrun P, Keeling J M J, Marchetti F M, Szymańska M H, André R, Staehli J L, Savona V, Littlewood P B, Deveaud B and Dang L S 2006 *Nature* **443** 409
[2] Amo A, Lefrére J, Pigeon S, Adrados C, Ciuti C, Carusotto I, Houdré R, Giacobino E and Bramati A 2009 *Nature Physics* **5** 805–810
[3] Christmann G, Butté R, Feltin E, Carlin J F and Grandjean N 2008 *Applied Physics Letters* **93** 051102
[4] Li F, Orosz L, Kamoun O, Bouchoule S, Brimont C, Disseix P, Guillet T, Lafosse X, Leroux M, Leymarie J, Malpuech G, Mexis M, Mihailovic M, Patriarche G, Réveret F, Solnyshkov D and Zuniga-Perez J 2013 *Applied Physics Letters* **102** 191118
[5] Klein T, Klembt S, Durupt E, Kruse C, Hommel D and Richard M 2015 *Applied Physics Letters* **107** 071101
[6] Askitopoulos A, Mouchliadis L, Iorsh I, Christmann G, Baumberg J J, Kaliteevski M A, Hatzopoulos Z and Savvidis P G 2011 *Phys. Rev. Lett.* **106** 076401
[7] Sedov E S, Iorsh I V, Arakelian S M, Alodjants A P and Kavokin A 2015 *Phys. Rev. Lett.* **114** 237402
[8] Lai C W, Kim N Y, Utsunomiya S, Roumpos G, Deng H, Fraser M D, Byrnes T, Recher P, Kumada N, Fujisawa T and Yamamoto Y 2007 *Nature* **450** 529
[9] Bienstmann P 2001 *Rigorous and efficient modeling of wavelength scale photonic components* (Universiteit Gent, Belgien)
[10] Sebald K, Rahman S, Cornelius M, Kaya T, Gutowski J, Klein T, Gust A, Hommel D and Klembt S 2016 *Applied Physics Letters* **108** 121105

[11] Sebald K, Seyfried M, Klembt S, Bley S, Rosenauer A, Hommel D and Kruse C 2012 *Applied Physics Letters* **100**
[12] Sebald K, Rahman S, Cornelius M, Gutowski J, Klein T, Klembt S, Kruse C and Hommel D 2015 *Applied Physics Letters* **107** 062101

IOP Conf. Series: Journal of Physics: Conf. Series **864** (2017) 012021 doi:10.1088/1742-6596/864/1/012021

Optical investigations and strain effect in AlGaN/GaN epitaxial layers

M. Jayasakthi[1,2], S. Juillaguet[1*], H. Peyre[1], L. Konczewicz[1], M. Moret[1], O. Briot[1], K. Baskar[2] and S. Contreras[1]

[1] Laboratoire Charles Coulomb (L2C), UMR 5221 CNRS-Université de Montpellier, Montpellier, FR-34095, France
[2] Crystal Growth Centre, Anna University, Chennai – 600 025, India

*sandrine.juillaguet@umontpellier.fr

Abstract. $Al_xGa_{1-x}N$ epilayers with x ranging from 0.20 to 0.50 have been grown on c-plane sapphire substrate by metal-organic chemical vapor deposition. The thickness, composition, strain and stress values of the AlGaN were determined by high resolution X-ray diffraction. The optical properties of the epilayers were studied using photoluminescence (PL) and reflectivity measurements. The effect of the stress on the bandgap can be explained by room temperature PL. The temperature dependent PL result shows the well-known "S-shape" behavior.

1. Introduction

AlGaN-based deep-ultraviolet (DUV) light emitting diodes (LEDs) have received tremendous attention due to an increased demand for their high performance and thermal stability for numerous applications for: backside lighting in liquid crystal displays (LCDs); bio-chemical identification; air- and water-purification; high-density optical data storage; solid state lighting; short distance communication links [1,2]. Epitaxial layers of group-III nitrides are most commonly grown on c-plane sapphire as substrate material due to their low cost and stability at high temperatures. The commonly used sapphire substrate presents a large lattice mismatch with the group-III nitrides resulting highly defective in AlGaN films when they are grown directly on it [3]. Different methods such as a deposition of a low temperature buffer layer of GaN or AlN have been adopted to reduce defect density in the metal-organic chemical vapor deposition (MOCVD) growth of AlGaN [3, 4]. Another method to reduce defect density in the AlGaN layers is the use of a high temperature GaN template [5, 6]. The growth of high quality AlGaN-based alloys is still challenging [7]. The AlGaN structural, electronic, and optical properties are not completely understood, limiting the realization of the devices with required performance. Therefore, measuring and characterizing the lattice strain and optical properties of AlGaN layer is very important for optimizing the design and fabrication of the AlGaN/GaN heterostructures for device applications.

2. Experimental Details

The AlGaN samples were grown on (0001) sapphire substrates by MOCVD. Trimethylgallium (TMGa), trimethylaluminum (TMAl), and ammonia (NH_3) were used as gallium, aluminum (Al), and nitrogen precursors, respectively. A thin GaN nucleation layer was deposited onto the substrate, at 500 °C. Then, the temperature was increased to 1050 °C to grow a thick GaN template and AlGaN epilayer was grown at 1080 °C. During the growth TMAl flow was varied and all growth conditions were kept constant. Five samples of AlGaN with different Al mole percent and thickness were grown.

The composition and thickness of AlGaN films on GaN were measured by using a high-resolution X-ray diffraction (HRXRD) (PANalytical X'Pert Pro MRD, Netherland). Optical properties of AlGaN layers were studied by Photoluminescence (PL) and reflectivity. PL was performed by exciting the samples with a continuous-wave frequency-doubled Ar^+ laser (λ= 244 nm). For low temperature (LT) PL a Triax spectrometer from Jobin-Yvon Horiba fitted with a 600 gr/mm grating and a cooled CCD camera completed the set-up. Samples were mounted in a He-flow cryostat offering the possibility to adjust the temperature from low temperature (LT) at 5 K, to room temperature (RT). The RTPL spectra were measured using a UV-enhanced photomultiplier tube. Reflectivity spectrum of AlGaN layer has been measured at 10 K using a Halo light as the source and collected by a photomultiplier tube.

3. Results and discussion

The compositions and thicknesses of AlGaN layers were determined from HRXRD ω-2θ scans of the (0002) reflection and calculated values are given in Table 1. Figure 1 shows the ω-2θ scans of (0002) diffraction plane of sample B.

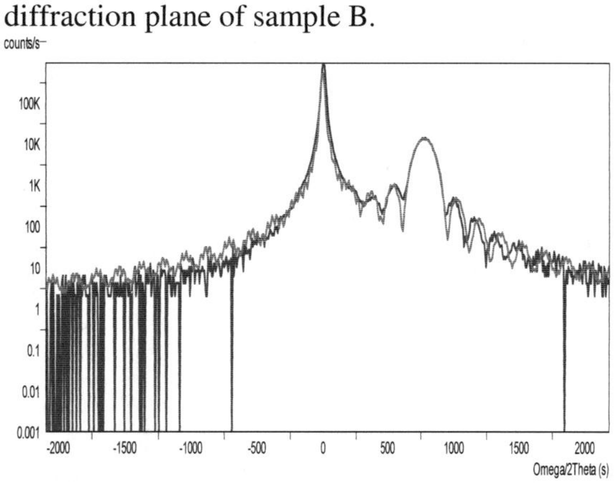

Figure 1. HRXRD (0002) refection plan of sample B (blue) with simulation (red).

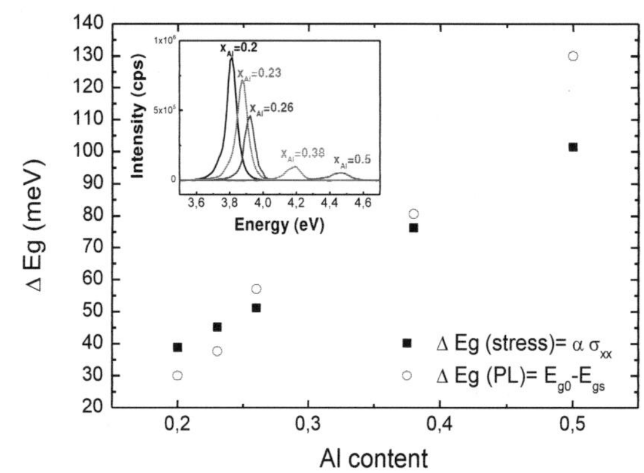

Figure 2. Effect of stress on the bandgap as a function of Al content. Inset RTPL spectra

The out of plane (ε_{zz}) and in plane strains (ε_{xx}) were calculated by the following equation [8]:

$$\varepsilon_{zz} = \frac{c - c_0}{c_0} \qquad \varepsilon_{xx} = \frac{a - a_0}{a_0}, \tag{1}$$

where c and a are the lattice parameters of the strained AlGaN films calculated from X-ray diffraction data and c_0 and a_0 are the unstrained lattice parameters of AlGaN. For AlGaN alloys, a_0 and c_0 were determined according to Vegard's law. The (0004) reflection and the (10-15) reflection of AlGaN have been chosen to determine the lattice parameters c and a respectively. The in plane stress can be expressed in terms of elastic modules, σ_{xx} shown below:

$$\sigma_{xx} = \varepsilon_{xx}(C_{11} + C_{12} - 2\frac{C_{13}^{\ 2}}{C_{33}}) \tag{2}$$

The value of stresses for AlGaN layers estimated using Eq. (2) for AlGaN samples are presented in the Table 1. The results show that the in plane stress in AlGaN layer for all the samples is of tensile type and the out of plane strain reveals compressive type. Low Al content samples provide minimum in plane tensile stress compared to higher Al content samples. Considering the effect of stress on the band gap, we can use the following equation [9]:

$$E_{g0}(x) - E_{gs}(x) = \alpha\sigma_{xx} \tag{3}$$

where σ_{xx} is the calculated in plane stress (see Table 1), E_{gs}, E_{g0} and α are the RT measured optical band gap, relaxed band gap, and the pressure coefficient (meV/GPa), respectively. The relaxed band gap can be calculated by the following equation using a bowing value b equal to 0.9 eV [9]

$$E_{g0_{AlGaN}}(x) = x\,E_{oAlN} + (1-x)E_{oGaN} - b\,x(1-x) \tag{4}$$

Note that from ref [9] the pressure coefficient is assumed to be not dependent of Al content (for $x < 0.4$) and equal to 25 meV/GPa. In this case, we can plot (Figure 2) the effect of stress on the bandgap as a function of Al content. The square symbols correspond to the effect of the biaxial stress (Eq. 3) indeed the open symbols correspond to the difference energy between the relaxed band gap (Eq. 4) and the measured optical bandgap at RT (see inset RTPL spectra). The results confirm the effect of the stress on the bandgap at room temperature.

Table 1. The composition, thickness, out of plane strain (ε_{zz}), in plane strain (ε_{xx}) and in plane stress (σ_{xx}) values of AlGaN layers are calculated from HRXRD

samples	XRD Al content (%)	AlGaN thickness (nm)	ε_{zz}	ε_{xx}	σ_{xx} (GPa)
A	20	110	-0.00173	0.0033	1.546
B	23	100	-0.00202	0.0038	1.793
C	26	90	-0.00222	0.0043	2.008
D	38	46	-0.00307	0.00648	3.051
E	50	23	-0.00385	0.00862	4.061

The low temperature PL spectra of the samples A, B and C have been measured at low temperature (5 K) (Figure 3). For sample D and E, the emission intensity of this peak cannot be observed, because our detection limit at LT is out of range. The high intensity peak energy is found to represent near band-edge emission of AlGaN. As expected, this peak exhibits a blue shift with increasing Al concentration. At Low energy, two additional peaks of the main emission peak can be clearly observed due to longitudinal optical (LO) phonon replicas (1LO and 2LO). We found a LO phonon energy ~95 meV, in good agreement the LO phonon energy of AlGaN [10].

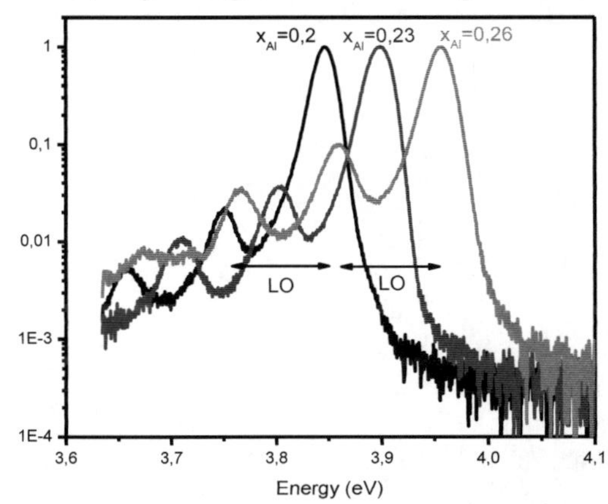

Figure 3. Normalized LTPL spectra of the samples A, B and C.

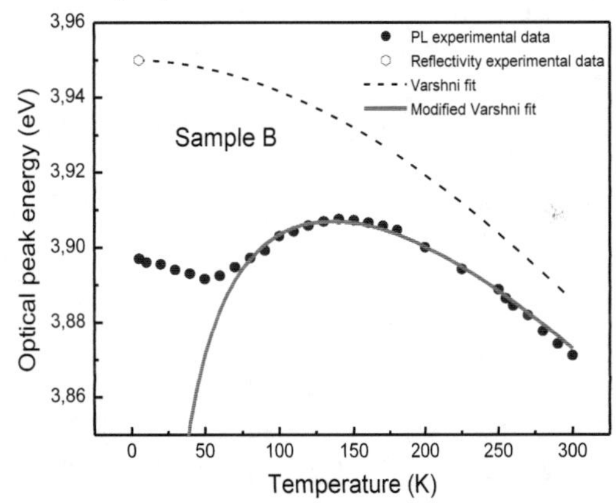

Figure 4. PL peak energy position as a function of temperature for sample B.

Figure 4 shows the PL peak energy position for sample B as a function of temperature. The temperature varied from 5 K up to room temperature. The temperature dependent PL result shows an "S-shape" behavior observed in the AlGaN epilayers. The shift in the peak position does not follow the expected trend due to strain. The tensile strain is always smaller than at growth temperature and the S-shape behaviour cannot be explained by plastic deformation [11]. This "S-shape" behavior is attributed to exciton localization due to compositional fluctuations in the AlGaN epilayers. The temperature dependence of the band gap can approximated using the Varshni model (Eq. 5):

$$E(T) = E_g(0) - \frac{\alpha T^2}{T+\beta} \tag{5}$$

The Varshni equation is valid only for binary and is no longer valid for ternary alloys when accounting for composition fluctuations. The modified expression can be expressed [12]:

$$E(T) = E_g(0) - \frac{\alpha T^2}{T+\beta} - \frac{\sigma^2}{K_B T} \quad (6)$$

Here $E_g(0)$ is the band gap of AlGaN at T = 0 K, α and β are Varshni's coefficients for band gap reduction with increasing temperature, and σ is the standard deviation of the Gaussian distribution of the band gap fluctuations due to the random fluctuations. The low temperature reflectivity spectrum of AlGaN has been measured at 10 K and is used to determine the band gap of this alloy. We obtained an energy value equal to 3.95 eV (open symbol in Fig. 4). Fitting the experimental data of sample B using the Eq. (6), we obtained a good agreement (line in Fig. 4) leading to Varshni constants $\alpha = 10^{-3}$ eV/K, $\beta = 1100$ K and the $\sigma = 18$ meV. In the same figure, we have plotted Varshni fit (Eq. 5) using the same values of α and β to compare the results (dashed line in Fig. 4). Our result is in good agreement with the results reported by Bell *et al.* [11].

4. Conclusion

AlGaN epilayers have been grown on GaN/sapphire using MOCVD. The composition, thickness and strain of AlGaN films were measured using HRXRD. The in plane strain and in plane stress in AlGaN layers are tensile type and the out of plane strain reveals compressive type. At room temperature in AlGaN samples an effect of the stress is observed. Two low intensity additional peaks on the lower energy side of the main emission peak can be clearly observed from our AlGaN epilayers in LTPL, which are due to LO phonon replicas. The temperature dependent PL result shows an "S-shape" behavior observed in AlGaN epilayers. We deduced the standard deviation of the potential fluctuations σ (18 meV) in AlGaN from modified Varshni equation.

Acknowledgement: One of the authors (M. Jayasakthi) would like to thank European commission for the Erasmus Mundus Scholarship (Svagata).

References

[1] Vinod A, Jianping Z, Ashay C, Wu S, Jie S, Radhika P, Maxim S and Muhammad A K 2002 *Jpn. J. Appl. Phys.* **41** L435
[2] Chang S J, Lai W C, Su Y K, Chen J F, Liu C H, and Liaw U H, 2002 *IEEE J. Sel. Topics Quantum Electron.* **8** 278
[3] Satoshi K, Motoaki I, Nobuaki H, Tetsuya T, Hiroshi A, Isamu A, Satoshi W, Yawara K, Norihide Y 2001 *J. Cryst. Growth* **223** 83
[4] Shih C F, Chen N C, Lin S Y, Liu K S 2005 *Appl. Phys. Lett* **86** 211103
[5] Kenji I, Kazumasa H, Hiroshi A, Isamu A 1990 *J. Cryst. Growth* **104** 533
[6] Bethoux J M, Vennéguès P, Natali F, Feltin E, Totterau O 2003 *J. Appl. Phys.* **94** 6499
[7] Kuzuhara M, Miyamoto H, Ando Y, Inoue T, Okamoto Y and Nakayama T 2003 *Phys. Stat. Sol. A* **200** 161
[8] Ozturk M K Altuntas H, Corekci S, Hongbo Y, Ozcelik S and Ozbay E 2011 *Strain* **47** 19
[9] Goyal A, Kapoor A K, Raman R, Dalal S, Mohan P and Muralidharan R 2015 *J. Appl. Phys.* **117** 225702
[10] Cros A, Angerer H, Ambacher O, Stutzmann M, Hopler R, and Metzger T 1997 *Solid State Commun.* **104** 35
[11] Bell A, Srinivasan S, Plumlee C, Omiya H, Ponce F A, Christen J, Tanaka S, Fujioka A and Nakagawa Y 2004 *J. Appl. Phys* **95** 4670
[12] Wang T, Raviprakash G, Ranalli F, Harrison C N, Bail J, David J P R, Parbrook P J, Ao J P and Ohno Y 2004 *J. Appl. Phys.* **97** 083104

IOP Conf. Series: Journal of Physics: Conf. Series **864** (2017) 012022 doi:10.1088/1742-6596/864/1/012022

Theoretical analysis of AlGaN/GaN resonant tunnelling diodes with step heterojunctions spacer and sub-quantum well

Y Liu, B Gao and M Gong

Key Laboratory for Microelectronics, College of Physical Science and Technology, Sichuan University, Chengdu Sichuan, 610064, China

Correspondence author e-mail: gaobo@scu.edu.cn

Abstract. In this paper, we proposed to use step heterojunctions emitter spacer (SHES) and InGaN sub-quantum well in AlGaN/GaN/AlGaN double barrier resonant tunnelling diodes (RTDs). Theoretical analysis of RTD with SHES and InGaN sub-quantum well was presented, which indicated that the negative differential resistance (NDR) characteristic was improved. And the simulation results, peak current density J_P=82.67 mA/μm^2, the peak-to-valley current ratio PVCR=3.38, and intrinsic negative differential resistance R_N=-0.147Ω at room temperature, verified the improvement of NDR characteristic brought about by SHES and InGaN sub-quantum well. Both the theoretical analysis and simulation results showed that the device performance, especially the average oscillator output power presented great improvement and reached 2.77mW/μm^2 magnitude. And the resistive cut-off frequency would benefit a lot from the relatively small R_N as well. Our works provide an important alternative to the current approaches in designing new structure GaN based RTD for practical high frequency and high power applications.

1. Introduction

Resonant tunneling diodes (RTDs) characterized by negative differential resistance (NDR) have attracted researchers' attention because of its potential wide range application prospects, among which terahertz oscillator is an important branch and has been studied in many papers[1-4]. The lower junction capacitance of RTD comparing with other NDR devices, such as Esaki tunnel and transferred electron devices [5], makes it an excellent candidate in the generation and detection of terahertz waves. And enormous amounts of previous researches focused on AlAs/InGaAs/AlAs quantum well structural RTDs. However, it showed a tremendous challenge to design a stable and reliable terahertz oscillator when it comes to high frequency and high power. Apparently, the output power and temperature stability have become the bottleneck for conventional RTDs (such as GaAs based RTD). Naturally, gallium nitride (GaN) based RTDs gained researchers' attention gradually these years due to its large electron effective mass[5] and longitudinal optical phonon energy[6] enable it very high speed devices, especially the advantage of wide bandgap (~3.43eV), large conduction band discontinuity, high carrier mobility, and the thermal stability[8] comparing with second generation semiconductors paved the way toward higher frequency, higher power applications at room temperature for GaN based RTDs.

Despite of the great promises for GaN based RTDs, the measurements of devices, theoretical analysis, and simulations all have indicated that there are still plenty of works, for example, decreasing device series resistance (R_S), realizing reproducible NDR characteristic and higher output power etc., need to do before practical applications. The typical figures of merit in NDR devices are the intrinsic negative differential resistance R_N ($R_N = \Delta V \times \Delta I^{-1}$), peak-to-valley current ratio PVCR ($PVCR = I_P \times I_V^{-1}$), and average oscillator output power P_{MAX} ($P_{MAX} = 3\Delta I \times \Delta V \times 16^{-1}$ [9]). The R_N is particularly important for increasing the resistive cutoff frequency f_R ($f_R \propto (R_N C)^{-1}$) for RTD. In recent researches, GaN/AlGaN (GaN/AlN) and GaN/InAlN (GaN/InN) are the hotspots, and many groups have spent a lot of attention on them these years [5, 8, 10-12]. The works of Bayram, C. et al. verified that the reproducibility of NDR benefit a lot from the low Al content AlGaN barrier [5, 8, 10]. Subhra C. et al. combined RTD and HEMT together [13]. Yang et al. have simulated the RTD oscillator instantaneous characteristics [14]. And for AlGaN/GaN/AlGaN double barrier RTDs, its peak current density (J_P) and PVCR reached around 1.23~11.2mA/μm^2 and 1.03~9.03[6, 15], respectively. However, papers reported about GaN based RTDs rarely focus on f_R and P_{MAX} at present, which means the problems about increasing peak current (I_P) and PVCR and decreasing valley current (I_V), R_N and R_S are still great challenges. So for the sake of promoting RTD device performance, in this paper, we proposed to use SHES and InGaN sub-well for AlGaN/GaN/AlGaN double barrier RTDs to increase J_P, and decrease R_N. Meanwhile, P_{MAX} was promoted a lot as well.

2. Device structure and energy band analysis

The device structure of GaN based RTD with SHES and sub-quantum well is shown in figure 1 (a). First, the double barrier (DB) region consists of $Al_{0.2}Ga_{0.8}N$/GaN/$Al_{0.2}Ga_{0.8}N$ (1.5nm/1.5nm/1.5nm), thus, an inherent lattice mismatch between materials is inevitable [10]. However, the critical thickness above which dislocations are generated is inversely proportional to Al content for GaN homo-epitaxial growth [10]. And in our simulations, the NDR phenomenon weakened with the increasing of the DB region thickness, and when the thickness was greater than 1.8nm, the NDR phenomenon disappeared nearly. This NDR degradation verified the critical thickness (~1.8nm for $Al_{0.2}Ga_{0.8}N$) and the negative effect of the dislocations on the NDR phenomenon. So in our works, low aluminum content (20%) and thin barrier (1.5nm) was employed to form DB region which could minimize active layer related dislocations [10]. Next, AlGaN (5nm) layer with its Al content reducing in ladder type is the emitter spacer, namely the step heterojunctions emitter spacer (SHES) region in figure 1 (a). Then $In_yGa_{1-y}N$ (2nm) layer was inserted between DB region and SHES region, which formed a sub-quantum well before the DB region. The collector spacer consists of 5nm GaN layer. Finally, emitter region and collector region is composed of heavily doped, 1×10^{19}cm^{-3}, $Al_{0.4}Ga_{0.6}N$ layer on the top and GaN layer on the bottom, respectively. The heavily doped emitter and collector region are aimed to ensure the Ohmic contact between the device and electrodes.

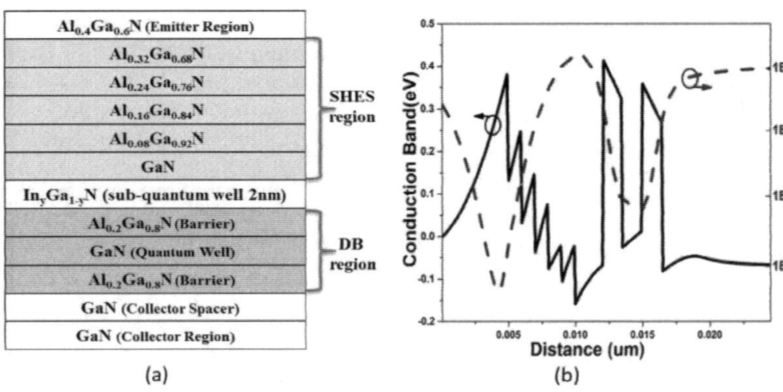

Figure 1. (a) The device structure and (b) energy band diagram and electron density before resonant of GaN based RTD with SHES and sub-quantum well.

The energy band diagram of GaN based RTD with SHES and sub-quantum well is shown in figure 1 (b). As you can see, the InGaN sub-quantum well turned the tunneling mode for electrons from 3D-2D-3D mechanism to 2D-2D mechanism which means higher tunneling efficiency and consequently higher peak current. Then, the SHES formed continuous heterojunctions, which means continuous electron triangular quantum wells are formed, thus 2-dimentional electron gas (2DEG) was produced in these triangular quantum wells. The 2DEG accumulated before DB region which you can see from the electron density curve in figure 1 (b), with much higher mobility than common electron carriers accelerated electrons tunneling process from emitter region to the DB region, which could boost the peak current immensely. It is indicated that the SHES and InGaN sub-well promoted the emitter region's electron carrier mobility, and transmission coefficient which could reduce the carrier's transportation time in the device, consequently enabling RTD higher frequency. The heavily doped $Al_{0.4}Ga_{0.6}N$ layer acts as the emitter region offering the electron source for the tunneling process.

3. Simulation

In our simulation, the mesa was set to be 30μm^2. Figure 2 (a) shows the IV characteristic comparison of different structure RTDs. Apparently, the IV characteristic of RTD with SHES and InGaN sub-well is the best, having the I_P=2.48A, R_N=-0.147Ω and P_{MAX}=83mW. And the device performance was elevated a lot by the SHES and sub-well comparing with previous works [6, 15]. As you can see, the peak current increase from 0.282A to 2.48A and P_{MAX} increased from 13.03mW to 83mW, both of the two typical figures of merit presented around one order of magnitude promotion while the PVCR maintained at 3.38. It is also found that the peak voltage (V_P) increased with the appearing of SHES. V_P is related with series resistance (R_S) and the difference between the bound state energies ($\Delta E_{(n+1,\,n)}$) in the quantum well. And we calculated R_S using equation (1). Because the second positive resistance region of the IV curves in figure 2 (a) are approximately parallel but I_D increased a little with the SHES appearing, R_S of RTD with SHES and sub-quantum well is a little greater than RTD without them, which contributed to the increasing of V_P partly.

$$R_S = \frac{dV}{dI} - (\frac{kT}{q})\frac{1}{I_D} \tag{1}$$

Then, V_P is also proportional to $\Delta E_{(n+1,\,n)}$, but equation (2) indicates that the $\Delta E_{(n+1,\,n)}$ should be the same because of the same DB active regions of different structure RTDs. So the increasing of V_P may be due to the increasing of the difference between the tunnelling energy levels. And the SHES and sub-quantum well do have lowered the energy level before the DB active region which you can see from the energy band diagram in figure 1 (b).

$$\Delta E_{(n+1,n)} = E_{n+1} - E_n = \frac{\hbar^2}{2m^*}(\frac{\pi}{L_W})^2\left[(n+1)^2 - n^2\right] \tag{2}$$

After calculating the R_N, it was found that the absolute value of R_N decreased from 1.023 to 0.147, so for the resistive cut-off frequency f_R, it could be increased about one order of magnitude. Terahertz oscillator application of RTD will benefit a lot from the relatively small R_N. And after doing more optimizing works to decrease R_S, the maximum oscillating frequency (f_{MAX}) is expected to reach 2.07THz when the R_S=1/2R_N.

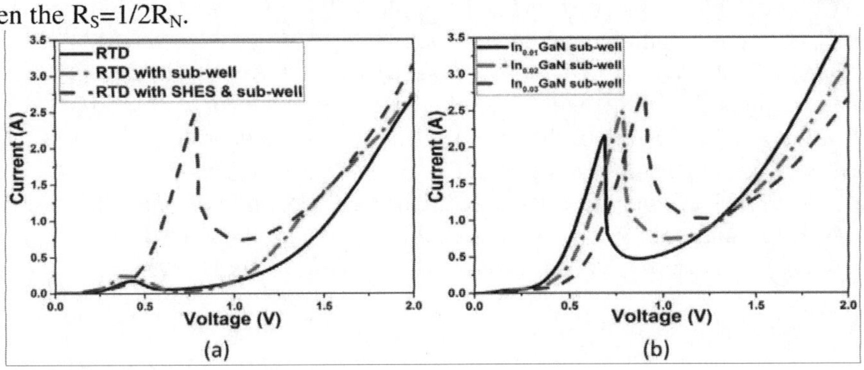

Figure 2. (a) IV characteristic comparison of different structure RTD (b) I-V characteristic of DB RTD with different In content $In_yGa_{1-y}N$ sub-quantum well.

Then, we studied the In content of $In_yGa_{1-y}N$ sub-quantum well (y range from 0.01 to 0.03), its influence on the performance of RTD is shown in figure 2 (b). Both the I_P and I_V are increasing with the increasing of the In content from 0.01 to 0.03, consequently, the P_{MAX} increased from 51.54 to 99.4 mW. However the PVCR decreased from 4.61 to 2.68 and absolute value of R_N increased from 0.098 to 0.181 which is not good for increasing the resistive cut-off frequency. However, after applying equation (1) to the IV curve of figure 3 (b), it was found that R_S was decreasing with the increasing of the In content. This phenomenon may indicate a feasible method for decreasing R_S, thus realizing R_S=1/2R_N, even if sacrificing some of frequency characteristic. In our previous works, to make a trade-off between the frequency and output power characteristic, we chose the 0.02 In content for $In_yGa_{1-y}N$ sub-quantum well.

4. Conclusion

In this paper, we presented a comparative analysis of the electrical characteristics of GaN based RTDs with and without step heterojunctions emitter spacer (SHES) and sub-quantum well. Through theoretical analysing on the energy band and simulating the electrical characteristics at room temperature, it was verified that the SHES and sub-quantum well could elevate the RTD performance greatly. In our works, the J_P and P_{MAX} reached 82.67mA/μm^2 and 2.77mW/μm^2, respectively, and the PVCR maintains at 3.38. The results indicated that most of the typical figures of merits in RTD realized great magnitude promotion; especially our peak current density and output power are superior to the papers reported to date. And due to the pretty small intrinsic negative differential resistance R_N, while the series resistance R_S meets the equation of R_S=1/2R_N, the maximum oscillating frequency is expected to be 2.07THz magnitude. Meanwhile, we found a feasible method to decreasing the series resistance which is favourable for realizing maximum oscillating frequency. The results of GaN based RTD with SHES and sun-quantum well in our works make RTD more applicable for high frequency and high power applications. The works we did also activate further researches toward deceasing valley current I_V for increasing PVCR and decreasing R_S for realizing the expected maximum oscillating frequency.

References

[1] Kitagawa S, Suzuki S, and Asada M 2016 *Electronic Letters.* **52** 479-81
[2] Kitagawa S, Suzuki S, and Asada M 2014 *IEEE Electron Device Letters.* **35** 1215-17
[3] Lee J, Kim M, and Yang K 2016 *IEEE Transactions on Terahertz Science and Technology.* **6** 336-40
[4] Jacobs K J P, Stevens B J, and Wada O 2015 *IEEE Electron Device Letters.* **36** 1295-98
[5] Bayram C, Vashaei Z, and Razeghi M 2010 *Applied Physics Letters.* **97** 181109
[6] Li D, Tang L, and Edmunds C 2012 *Applied Physics Letters.* **100** 252105
[7] Bayram C, Vashaei Z, and Razeghi M 2010 *Applied Physics Letters.* **97** 181109
[8] Vashaei Z, Bayram C, and Razeghi M 2010 *Journal of Applied Physics.* **107** 83505
[9] Kim C S and Brandli A 1961 *IRE Transaction on circuit theory.* **8** 416-25
[10] Bayram C, Vashaei Z, and Razeghi M 2010 *Applied Physics Letters.* **97** 92104
[11] Golka S, Pflu¨gl C, and Schrenk W 2006 *Applied Physics Letters.* **88** 172106
[12] Boucherit M, Soltani A, and Monroy E 2011 *Applied Physics Letters.* **99** 182109
[13] Subhra C, Swarnabha C and Dheubes B 2015 *Journal of Semiconductor.* **36** 044001
[14] Yang L A, He H, and Mao W 2011 *Applied Physics Letters.* **99** 153501
[15] Chen H, Yang L, and Hao Y 2014 *Journal of Applied Physics.* **116** 74510

IOP Conf. Series: Journal of Physics: Conf. Series **864** (2017) 012023 doi:10.1088/1742-6596/864/1/012023

Source-Drain Punch-Through Analysis of High Voltage Off-State AlGaN/GaN HEMT Breakdown

H Jiang[1], X Li[2], J Wang[1], L Zhu[1], H Wang[1], J Liu[1], M Wang[1], M Yu[1], W Wu[1], Y Zhou[3], and G Dai[3]

[1] Institute of Microelectronics, Peking University, Beijing 100871, China
[2] Global Energy Interconnection Research Institute, Beijing 102211, China
[3] Microsystem and Terahertz Research Center, China Academy of Engineering Physics, Chengdu 610200, China

E-mail: wangjinyan@pku.edu.cn

Abstract. AlGaN/GaN high-electron mobility transistor's (HEMT's) off-state breakdown is investigated using conventional three-terminal off-state breakdown I–V measurement. Competition between gate leakage and source-injection buffer leakage (SIBL) is discussed in detail. It is found that the breakdown is dominated by source-injection which is sensitive to gate voltage and gate length at large gate-to-drain spacing (Lgd > 7μm), where a threshold drain voltage of the occurrence of the SIBL current in GaN buffer exists, and after this threshold voltage the SIBL current continually increased till the buffer breakdown. Our analysis showed that due to the punch-through effect in the buffer, a potential barrier between 2DEG and GaN buffer at the source side mainly controlled by the drain voltage determines the buffer leakage current and the occurrence of the following buffer breakdown, which could explain the experimentally observed breakdown phenomenon.

1. Introduction

The AlGaN/GaN high electron mobility transistors (HEMTs) are considered to be promising candidates for the next generation high power and high frequency devices, due to their outstanding fundamental physical properties, such as large breakdown fields, low on-state resistance and high electron mobility [1]-[3]. One of the major issues that limit the performance of AlGaN/GaN HEMTs is the off-state breakdown voltage (V_{BRoff}), because it determines the maximum working voltage and output power. It is well known that GaN buffer layer has a large density of deep-level acceptor traps originated from dislocations, defects unintentionally or intentionally introduced by the epi-growth, which can result in high-resistance buffer to enhance device performance, for example, prevents punch-through and buffer leakage current [4][5].

In this work, AlGaN/GaN HEMT's off-state breakdown is investigated by conventional three-terminal off-state breakdown I–V measurement. Competition between gate leakage and the source-injection buffer leakage [6] (SIBL) breakdown mechanisms is discussed.

2. Device fabrication

The AlGaN/GaN HEMT structure in this work was grown on a [0001] polished sapphire substrate by MOVCD. As shown in Fig. 1, a 2 μm unintentionally doped GaN channel layer, a 2 nm AlN nucleation layer and a 22 nm unintentionally doped $Al_{0.3}Ga_{0.7}N$ were grown on the substrate from bottom to up one by one. From room temperature Hall measurements, it could be seen that at the AlGaN/GaN interface, the 2DEG density is 1.2×10^{13} cm^{-2} and the electron mobility is 1159 cm^2/Vs.

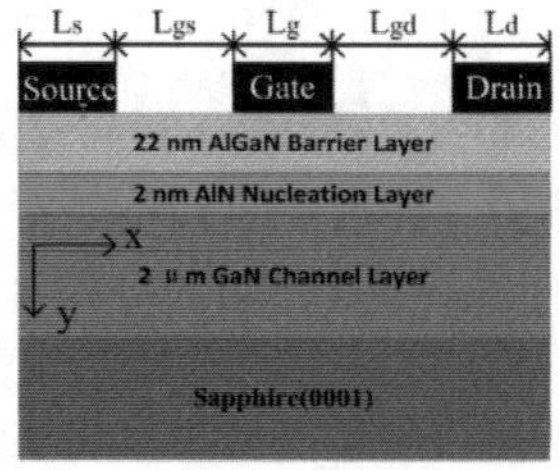

Figure 1. Cross-section of AlGaN/GaN HEMT

A complete AlGaN/GaN HEMT process flow in this paper contains source/drain ohmic contact formation, mesa isolation etching, gate Schottky contact, passivation and contact via etching. The source/drain ohmic contact was formed by evaporating Ti/Al/Ni/Au and then annealing at 830 °C for 30 seconds under N_2 atmosphere. The mesa isolation was performed by RIE. The Schottky contact came from e-beam evaporated Ni/Au layer. And a 200 nm SiN layer was deposited as passivation material.

3. Experiment results

The output and transfer characteristics of a fabricated AlGaN/GaN HEMT with Lg=1 μ m, Lgs=1 μ m and Lgd=1 μ m showed that the saturation drain current density could reach 0.75A/mm@Vg=2V, the gate threshold voltage, V_{th}, is -3.5V and the maximum transconductance is 136mS/mm, which were used for verification of the following simulation of AlGaN/GaN HEMT.

The conventional three-terminal off-state breakdown I-V measurement was adopted to characterize high-voltage performance of the AlGaN/GaN HEMT with different Lgd, where the breakdown voltage, V_{BR}, is defined as the drain voltage at which the drain current reaches 1mA/mm with $V_g < V_{th}$. The comparison of gate current, Ig, and source current, Is, for devices with different Lgd at Vgs=-7V and Vs=0V is shown in Fig. 2(a). It is obviously shown that there are different breakdown mechanisms for devices with different Lgd. For gate-to-drain spacing Lgd ≤ 7μm, gate current, Ig, is the major part of the drain current, Id, and the device breakdown originates from Ig. However, for larger Lgd, Ig is very small and Is dominates Id, the leakage current path is inside GaN buffer instead of barrier layer between the gate and drain. Due to interests in high voltage (>300V) breakdown, our main analysis focussed on the buffer off-state breakdown.

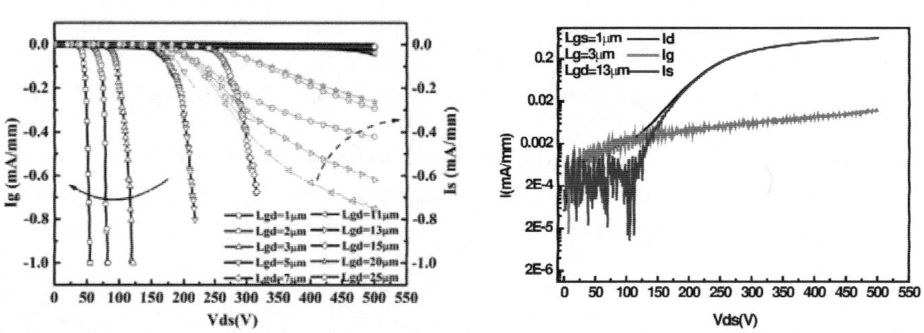

Figure 2. Gate and source currents by three-terminal off-state breakdown measurements at Vg = -7V and Vs = 0V for HEMT with Lg = 3μm, Lgs = 1μm, W = 50μm and different Lgd in (a) *linear and* (b) *logarithmic coordinate.*

IOP Conf. Series: Journal of Physics: Conf. Series **864** (2017) 012023 doi:10.1088/1742-6596/864/1/012023

For device with larger Lgd (>7μm), Figure 2(a) shows that the source current should determine the device breakdown when Vd > 150V. Figure 2(b) shows that for device with Lgd=13 μ m, Ig dominate Id when Vds<150V and Is becomes the major part of Id when Vds>150V till the occurrence of device breakdown. The above experimental results suggest that there exists a buffer current driven device breakdown.

4. Simulation and analysis

Simulation models for the same device structures as the samples of AlGaN/GaN HEMT were established based on Sentaurus TCAD, where all model parameters were extracted and verified by the experimental data. Since the source-drain buffer current is the focus of this work, the gate leakage is excluded from the model. The simulated 2D electron density distribution at different Vds is given in Fig. 3, which shows the existence of a SIBL current path in GaN buffer layer, induced by the large positive drain voltage. With the increasing drain voltage, the SIBL path is gradually approaching 2DEG channel at the source side through GaN buffer from 2DEG channel at the drain side, and then the source-drain punch-through occurs. When Vds=160V, the current path could be obviously observed.

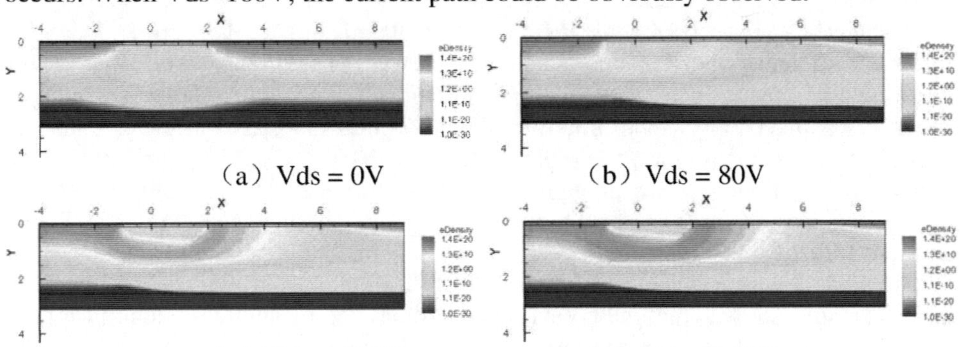
(a) Vds = 0V (b) Vds = 80V

(c) Vds = 160V (d) Vds = 240V

Figure 3. Simulated electron density distribution at different Vds (Lg = 3 μ m, Lgd = 5 μ m, Vg = -7V, Vs = 0V).

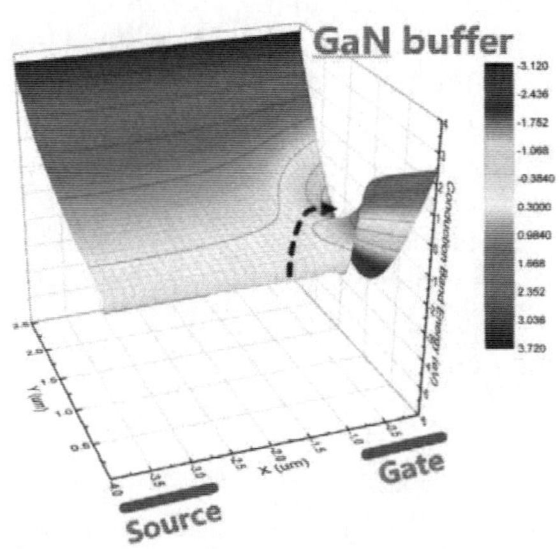

Figure 4. Simulated distribution of bottom of conduction band at Vds = 240V

Moreover, the simulated distribution of the bottom of conduction band is given in Figure 4, which shows the significant decrease of the potential barrier occurs for source-injected electrons from 2DEG channel at source side after large drain voltage is applied. It was found that as the drain voltage increasing, the minimum barrier height along the SIBL path is lowered, while the barrier position moved close to the source terminal and the AlGaN/GaN interface.

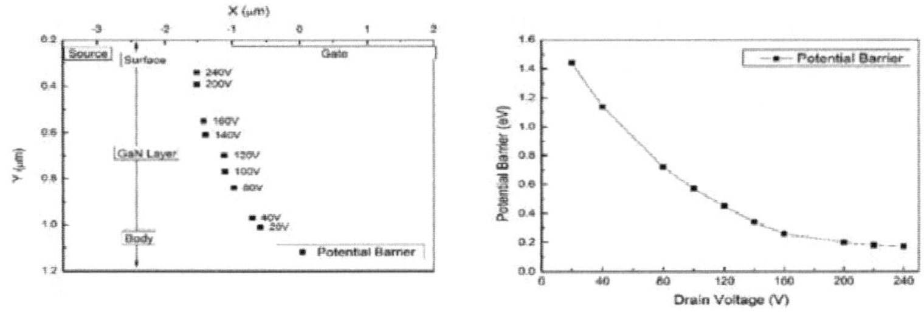

(a) (b)

Figure 5. Simulated dependence of position (a) and height (b) of the minimum potential barrier between 2DEG channel and GaN buffer on Vds.

In order to accurately characterize the electron barrier in GaN buffer, the heights and positions of minimum electron barrier are quantitatively extracted under different drain voltages, shown in Fig. 5(a) and (b). As the drain voltage increasing, the barrier position is moving close to the source terminal and the AlGaN/GaN interface (Fig. 5(a)), while the barrier height is lowering (Fig. 5(b)). It should be noted that, the minimum potential barrier approaches a saturated value, 0.2eV, while the corresponding positions of minimum potential barrier move close to below 0.2μm from AlGaN/GaN interface, which indicates the electron in 2DEG channel could easily inject into the GaN buffer. The above physical description could explain that Is starts to dominate Id when Vds = 150V in Figure 2(b), at which the minimum potential barrier become low enough for electrons to inject into GaN buffer.

5. Conclusion

The off-state breakdown mechanisms of AlGaN/GaN HEMTs with different Lgd are investigated. Our experimental results showed that the breakdown is dominated by source-injection for devices with large gate-to-drain spacing (Lgd > 7μm), where a threshold drain voltage (about 150V) exists, above which the SIBL current continually increases till the occurrence of breakdown. The following theoretical analysis based on Sentaurus TCAD simulation showed the existence of a current path in GaN buffer layer, mainly induced by the large positive drain voltage. With the increasing drain voltage, the SIBL path gradually approaches 2DEG channel at the source side from 2DEG channel at the drain side, while the minimum potential barrier between the 2DEG and GaN buffer at source side decreases until electrons in the source-side 2DEG channel could inject into the buffer, which finally results in the subsequent buffer breakdown.

Acknowledgments

The authors would like to thank Nanofabrication facility in Xidian University for the help to accomplish device fabrication. This work was supported by the National Natural Science Foundation of China (No. 60406004), Beijing Natural Science Foundation (No.4112025), and National Key Micrometer/Nanometer Processing Lab.

References

[1] N. Ikeda, Y. Niiyama, H. Kambayashi, Y. Sato, T. Nomura and S. Kato, Proc IEEE, 98, p.1151 (2010).
[2] U. K. Mishra, L. Shen, T. E. Kazior and Y. F. Wu, Proc IEEE, 96, p.287 (2008).
[3] W. Saito, Y. Takada, M. Kuraguchi, K. Tsuda, I. Omura, T. Ogura and H. Ohashi, IEEE Trans. Electron Devices, 50, p.2528 (2003).
[4] W. Saito, T. Noda, M. Kuraguchi and Y Takada, IEEE Trans. Electron Devices, 56, p.1371 (2009).
[5] M.Silvestri, M. J. Uren and M, Kuball, Appl. Phys. Lett., 102, p.073501 (2013).
[6] M. J. Wang and K. J. Chen, IEDM Tech. Dig., p.149–152 (2008).

IOP Conf. Series: Journal of Physics: Conf. Series **864** (2017) 012024 doi:10.1088/1742-6596/864/1/012024

Physical origin of the 1/4-cycle phase shift in the radiation-induced oscillatory magnetoresistance in 2DES.

Jesús Iñarrea[1,2]

[1]Escuela Politécnica Superior,Universidad Carlos III,Leganes,Madrid,28911 ,Spain
[2]Unidad Asociada al Instituto de Ciencia de Materiales, CSIC, Cantoblanco,Madrid,28049,Spain.

Abstract. We explain the 1/4-cycle phase shift of minima in the microwave-induced magnetoresistance oscillations in high mobility two dimensional electron systems. We calculate the minima positions obtaining an exact coincidence with the experimental ones. We find that the physical origin of the 1/4-cycle phase shift is a delay of $\frac{\pi}{2}$ of the radiation-driven Landau guiding center with respect to radiation, demonstrating the oscillating nature of the Landau states under irradiation.

1. Introduction

Microwave-induced magnetoresistance (R_{xx}) oscillations (MIRO)[1, 2], show up in high mobility two-dimensional electron systems (2DES) when they are irradiated with microwaves (MW) at low temperature ($T \sim 1K$) and under low magnetic fields (B) perpendicular to the 2DES. At high enough MW power (P) maxima and minima oscillations increase but the latter evolve into zero resistance states (ZRS)[1, 2]. Despite the fact of the important experimental[3, 4, 5, 6, 7, 8, 9, 10, 11, 12, 13, 14, 15, 16, 17] and theoretical efforts [18, 19, 20, 21, 22, 23, 24, 25, 26, 27, 28, 29, 30, 31] that have been made on MIRO and ZRS, their physical origin still remains controversial and far from reaching a definite consensus among the people devoted to this field. Among the different features defining MIRO we can underline one of them that can be consider as fundamental: MIRO present a 1/4-cycle phase shift in the minima position[3, 6]. In this paper we present, based on the radiation-driven electron orbits model[18, 19], a theoretical analysis on MIRO where we explain the 1/4-cycle shift of minima. Thus, the reason for the shift is that the Landau orbit guiding center displacement lags behind the driving force (radiation) by a definite phase constant of $\frac{\pi}{2}$.

2. Theoretical Model

The total electronic hamiltonian H of a MIRO problem can be exactly solved and the solution for the total wave function of H[18, 27, 32, 33, 34]reads: $\Psi_n(x,t) \propto \phi_n(x - X_0 - x_{cl}(t), t)$, where ϕ_n is the solution for the Schrödinger equation of the unforced quantum harmonic oscillator. Thus, the obtained wave function (Landau state or Landau orbit) is the same as the one of the standard quantum harmonic oscillator where the guiding center of the Landau state, X_0 without radiation, is displaced by $x_{cl}(t)$. $x_{cl}(t)$ is the classical solution of a negatively charged, forced

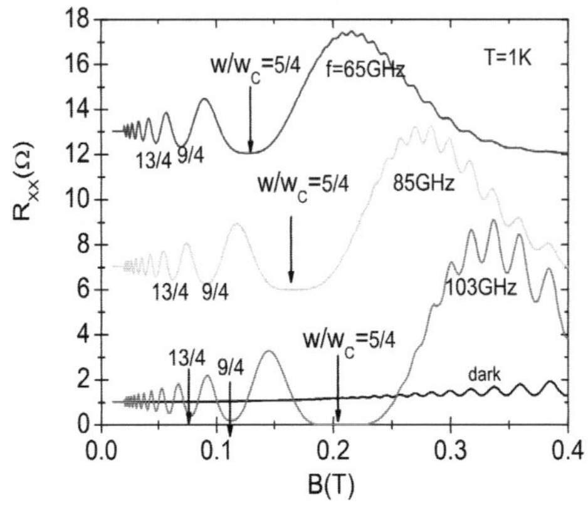

Figure 1. Dependence of MIRO minima with low radiation frequencies. Calculated R_{xx} versus B under radiation for several frequencies (65, 85 and 103 GHz). The curves are shifted for clarity. The dark curve is also presented.

and damped, harmonic oscillator[35, 36]: $x_{cl}(t) = \frac{-eE_o}{m^*\sqrt{(w_c^2-w^2)^2+\gamma^4}}\cos(wt-\beta) = -A\cos(wt-\beta)$ where E_0 is the amplitude of the MW electric field and β is a phase constant. β is the phase difference between the radiation-driven guiding center and the driving radiation; its expression reads, $\tan\beta = \frac{\gamma^2}{w_c^2-w^2}$. Thus, the guiding center lags behind the time-dependent driving force, (radiation), a phase constant of β. In the above calculations radiation has been expressed as $E(t) = E_0\cos wt$. γ is a phenomenologically-introduced damping factor for the interaction of electrons with the lattice ions giving rise to the emission of acoustic phonons. When the damping parameter γ is important, $(\gamma > w \Rightarrow \gamma^2 >> w^2)$, then $\tan\beta \to \infty$ and $\beta \to \frac{\pi}{2}$. Now, the time-dependent guiding center is, $X(t) = X_0 + x_{cl}(t) = X_0 - A\sin wt$.

When we turn on radiation, the guiding centers begin to harmonically oscillate according to the expression of $X(t)$. Later, in a time t_i, scattering starts and the electron jumps from $\Psi_n(t)$ at $X_n(t)$ to $\Psi_m(t)$ at $X_m(t)$ in a total time $t_f = t_i + \tau$. In the latter expression τ is the time it takes the electron to "fly" from the initial to the final orbits. We describe this time as "flight time" and is smaller than the total scattring time. The average advanced distance will change accordingly: $\Delta X(t) = X_m(t_f) - X_n(t_i) = \Delta X_0 - A\sin w(t_i+\tau) + A\sin wt_i$. On average the scattering will start when the electron is at its mid position in the MW-driven oscillation of the initial Landau state. Then, $wt_i = 2\pi n$ n being an integer and therefore we obtain the final expression: $\Delta X(t) = \Delta X_0 - A\sin w\tau$. The longitudinal conductivity σ_{xx} is given by[37] $\sigma_{xx} \propto \int dE\frac{(\Delta X(t))^2}{\tau_q}$ being E the energy and $\frac{1}{\tau_q}$ is the remote charged impurity scattering rate[18]. To obtain R_{xx} we use the common tensorial relation $R_{xx} = \frac{\sigma_{xx}}{\sigma_{xx}^2+\sigma_{xy}^2} \simeq \frac{\sigma_{xx}}{\sigma_{xy}^2}$, where $\sigma_{xy} \simeq \frac{n_i e}{B}$, n_i being the electrons density, and $\sigma_{xx} \ll \sigma_{xy}$. Thus, the final expression for the longitudinal resistance reads, $R_{xx} \propto -A\sin w\tau$. Importantly, according to the last expression, the MIRO minima positions are given by: $w\tau = \frac{\pi}{2} + 2\pi j \Rightarrow w = \frac{2\pi}{\tau}\left(\frac{1}{4}+j\right)$, j being a positive integer. If we compare the calculated expression with the ones previously obtained in experiments by Mani et al[3, 6]:

$$\begin{array}{ccc} theory & \Longleftrightarrow & experiment \\ min \to w = \frac{2\pi}{\tau}\left(\frac{1}{4}+j\right) & \Leftrightarrow & \frac{w}{w_c} = \left(\frac{1}{4}+j\right) \end{array} \quad (1)$$

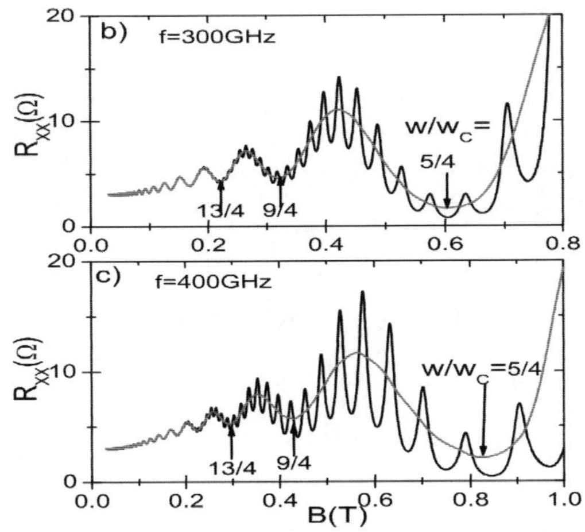

Figure 2. Calculated irradiated magnetoresistance, R_{xx}, vs the magnetic field B for frequencies of the terahertz regime, 300 GHz for the upper panel and 400 GHz for the lower panel. Both panels exhibit total R_{xx}, with radiation-induced resistance oscillations and Shubnikov-de Haas (SdHO) oscillations, (black curves online), and R_{xx} without Shubnikov-de Haas (SdHO) oscillations, (red curves online).

and thus, $\tau = \frac{2\pi}{w_c} = T_c$, i.e., the flight time equals the cyclotron period. Now, the 1/4-cycle phase shift of the MIRO minima can be traced back to a phase constant of $\frac{\pi}{2}$ that is the delay that presents the driven guiding center with respect to radiation in the whole range of B. The final expression of the irradiated magnetoresistance turns out to be: $R_{xx} \propto -\frac{eE_o}{m^* \sqrt{(w_c^2-w^2)^2+\gamma^4}} \sin\left(2\pi \frac{w}{w_c}\right)$.

3. Results

In Fig.1 we exhibit the dependence of MIRO minima with low radiation frequencies. We show calculated irradiated R_{xx} versus B and the dark case too. First minima positions show up for all frequencies labelled according to equation (1). We observe, as expected, that minima displace to higher B as the frequency increases. Minima positions and their behavior with B are in total agreement with experiments. Fig. 2 exhibits calculated irradiated R_{xx} vs B for a much higher radiation frequency corresponding to the terahertz regime: 300 GHz in the upper panel and 400 GHz in the lower panel. For both panels we represent the total R_{xx} (black curves online) and the averaged out R_{xx} (without SdHO, red curves online). We can see intense radiation-induced oscillations in the TH regime that clearly fulfill for both frequencies the 1/4-cycle phase shift of the oscillations minima, $(w/w_c = 5/4, 9/4, 13/4...)$. These calculated results demonstrate and explain that the corresponding shift is mainly dependent on the phase difference between radiation and the harmonic displacement of the Landau orbit guiding center.

4. Conclusions

In summary, we have presented a theoretical analysis of one of the universal features of MIRO as is the 1/4-cycle phase shift of minima. We have used the radiation driven electron orbits model. We have obtained an exact coincidence between the calculated values for minima and the experimental ones. The key outcome of our analysis is that this shift is an evidence of a delay of $\frac{\pi}{2}$ between the oscillating Landau state guiding center and radiation.

5. Acknowledgments

This work is supported by the MINECO (Spain) under grant MAT2014-58241-P and ITN Grant 234970 (EU). GRUPO DE MATEMATICAS APLICADAS A LA MATERIA CONDENSADA, (UC3M), Unidad Asociada al CSIC.

6. References

[1] R. G. Mani, J. H. Smet, K. von Klitzing, V. Narayanamurti, W. B. Johnson, and V. Umansky, Nature(London) **420**, 646 (2002)
[2] M. A. Zudov, R. R. Du, L. N. Pfeiffer, and K. W. West, Phys.Rev. Lett. **90**, 046807 (2003)
[3] R. G. Mani, J. H. Smet, K. von Klitzing, V. Narayanamurti, W. B. Johnson, and V. Umansky , Phys. Rev. Lett. **92**, 146801 (2004).
[4] R. G. Mani, J. H. Smet, K. von Klitzing, V. Narayanamurti, W. B. Johnson, and V. Umansky , Phys. Rev. B**69**, 193304 (2004).
[5] R. L. Willett, L. N. Pfeiffer, and K. W. West, Phys. Rev. Lett. **93**, 026604 (2004).
[6] R. G. Mani, Physica E (Amsterdam) **22**, 1 (2004);
[7] J. H. Smet, B. Gorshunov, C. Jiang, L. Pfeiffer, K. West, V. Umansky, M. Dressel, R. Meisels, F. Kuchar, and K. von Klitzing , Phys. Rev. Lett. 95, 118604 (2005).
[8] Z. Q. Yuan,C.L. Yang, R.R. Du, L.N.Pfeiffer and K.W. West , Phys. Rev. B**74**, 075313 (2006).
[9] R. G. Mani, W.B. Johnson, V.Umansky, V. Narayanamurti and K. Ploog, Phys. Rev. B**79**, 205320 (2009).
[10] S. Wiedmann, G.M. Gusev, O.E. Raichev, A.K. Bakarov, and J.C. Portal, Phys. Rev. Lett., **105**, 026804, (2010)
[11] S. Wiedmann, G.M. Gusev, O.E. Raichev, A.K. Bakarov, and J.C. Portal, Phys. Rev. B, **81**, 085311, (2010)
[12] D. Konstantinov and K. Kono, Phys. Rev. Lett. **103**, 266808 (2009)
[13] S. I. Dorozhkin, L. Pfeiffer, K. West K, K. von Klitzing, J.H. Smet JH, NATURE PHYSICS, **7**, 336-341, (2011)
[14] R. G. Mani, C. Gerl, S.Schmult, W. Wegscheider and V. Umansky, Phys. Rev. B**81**, 125320, (2010)
[15] R.G. Mani, A.N. Ramanayaka and W. Wegscheider, Phys. Rev. B., **84**, 085308, (2011)
[16] Jesus Inarrea, R.G. Mani and W. Wegscheider, Phys. Rev. ,**82** 205321 (2010)
[17] R. G. Mani, Int. J. Mod. Phys. B, **18**, 3473, (2004); Physica E, **25**, 189 (2004)
[18] J. Iñarrea and G. Platero, Phys. Rev. Lett. **94** 016806, (2005)
[19] J. Iñarrea and G. Platero, Appl. Phys. Lett., **89**, 052109, (2006)
[20] J. Iñarrea and G. Platero, Phys. Rev. B, **76**, 073311, (2007)
[21] J. Iñarrea, Appl. Phys. Lett. **90**, 172118, (2007)
[22] A.C. Durst, S. Sachdev, N. Read, S.M. Girvin, Phys. Rev. Lett.**91** 086803 (2003) (2005)
[23] X.L. Lei, S.Y. Liu, Phys. Rev. Lett.**91**, 226805 (2003)
[24] Ryzhii et al, Sov. Phys. Semicond. 20, 1299, (1986)
[25] P.H. Rivera and P.A. Schulz, Phys. Rev. B **70** 075314 (2004)
[26] M.G. Vavilov et. al., Phys. Rev. B, **70**, 161306(2004)
[27] J. Iñarrea and G. Platero, Appl. Phys Lett. **93**, 062104, (2008)
[28] J. Iñarrea and G. Platero, Phys. Rev. B, **78**, 193310,(2008)
[29] J. Iñarrea, Appl. Phys Lett. **92**, 192113,(2008)
[30] Jesus Inarrea and Gloria Platero, Appl. Phys. Lett. **95**, 162106, (2009)
[31] J. Iñarrea, Appl. Phys Lett. **90**, 262101,(2007)
[32] J. Inarrea and G. Platero, Appl. Physl Lett. **89**, 172114, (2006)
[33] E.H. Kerner, Can. J. Phys. **36**, 371 (1958).
[34] K. Park, Phys. Rev. B **69** 201301(R) (2004).
[35] A.P. French, *Vibrations and waves.*, W.W. Norton and Company, New York, 1971.
[36] I.G. Main, *Vibrations and waves in Physics.*, Cambridge Unversity Press, 1993.
[37] B.K. Ridley. *Quantum Processes in Semiconductors*, 4th ed. Oxford University Press, (1993).

IOP Conf. Series: Journal of Physics: Conf. Series **864** (2017) 012025 doi:10.1088/1742-6596/864/1/012025

Influence of environment and geometry on measured quantum coherence in InGaAs mesoscopic wires and stadia

Jean J Heremans[1], **Yuantao Xie**[1] **and Clément Le Priol**[2]

[1]Virginia Tech, Department of Physics, Blacksburg, Virginia 24061, USA

[2]Ecole Polytechnique, Department of Physics, 91128 Palaiseau, France

E-mail: `heremans@vt.edu`

Abstract. Lithographically fabricated mesoscopic wires and stadia, coupled to a decohering classical environment by two side-wires, were studied using low-temperature quantum transport to obtain quantum phase coherence lengths. Longer phase coherence lengths were observed in longer wires, an effect due to environmental coupling decoherence being mitigated by averaging over longer wire lengths. In stadia the dominant decoherence mechanism appears as device-device, here stadium-wire, coupling decoherence rather than environmental coupling decoherence. Consistency between the experimental results and a decoherence mechanism due to stadium-wire coupling is found in the relation between lower decoherence rate and relative geometrical similarity between the stadia and their connecting wires.

1. Introduction

Investigations in quantum coherence promise insight in applications of quantum technologies, and help elucidate the boundary between classical and quantum systems. Quantum coherence can be characterized by a quantum phase coherence length L_ϕ, the length scale over which carriers' wave functions maintain their phase coherence. L_ϕ can be reduced by decoherence mechanisms, such as electron-phonon and electron-electron scattering [1, 2], and by decoherence mechanisms introduced during measurements, such as level broadening, and by decoherence via interaction with a wider classical environment [3–5]. The latter will be studied in this work. Another decoherence effect, *device-device* coupling decoherence due to connections between different geometrical parts of devices, is barely investigated, and will be studied here as well. Devices were fabricated on an $In_{0.53}Ga_{0.47}As/In_{0.52}Al_{0.48}As$ heterostructure by electron-beam lithography. Sample dimensions and geometries are listed in Table 1, with examples shown in Fig. 1. The quantum well in the heterostructure contains a two-dimensional electron system (2DES), with mobility μ_{2D} = 1.49 $m^2/(Vs)$ and areal carrier density $N_s = 2.02 \times 10^{16}$ m^{-2}, at a temperature $T = 0.38$ K. The electron Fermi wavelength $\lambda_F = 17.6$ nm. The electron mean-free-path ℓ_e = 0.77 μm was derived from N_s and μ_{2D} using a Γ-point effective mass $0.0353m_e$(m_e, free electron mass) and a low T band gap 813 meV. Samples were measured in a ^{3}He cryostat, with T varying from 0.38 K to 5.0 K. The transport properties of the heterostructure do not appreciably change over this temperature range. Quantum interference effects visible in quantum transport, weak-antilocalization (WAL) [6–11] and universal conductance fluctuations (UCFs) [12–14], are are used to extract values for L_ϕ from measured device conductance G as function of magnetic field B applied normally to the 2DES.

Table 1: Type and characteristic dimensions of the wire samples (wire length L) and of stadia samples (width W and length L of the side-wires), referring to geometries in Fig. 1. Values listed for L_ϕ are measured at $T = 0.38$ K.

Sample	A	B	C	D	E	F	G	H
Geometry	wire	wire	wire	wire	stadium	stadium	stadium	stadium
W (µm)	0.75	0.75	0.75	0.75	1.4	1.0	1.0	0.6
L (µm)	11	6	4	2	1.0	1.0	3.0	1.0
W/L (µm)					1.4	1.0	0.33	0.6
L_ϕ (µm)	5.10	4.87	4.65	4.13	2.17	2.06	1.91	1.81

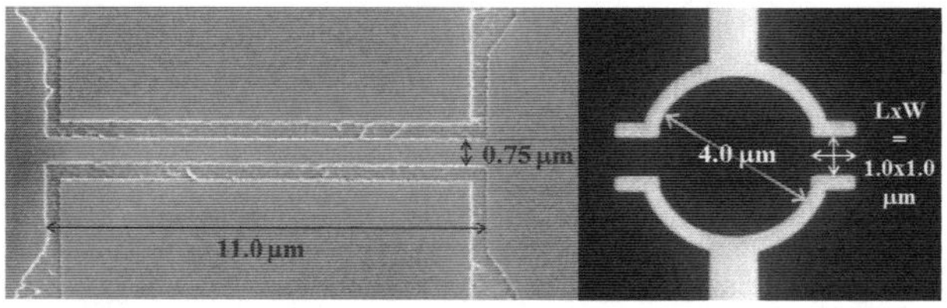

Figure 1: Micrographs of two example sample geometries. Left panel: two U-shape trenches delineate a wire, in this example with length L=11 µm (the lithographic width of all wires is 0.75 µm). Right panel: an example stadium sample with side-wires with W = 1 µm and L = 1 µm. The inner diameter of 4 µm is constant for all stadia. The lighter regions denote etched trenches, which form barriers for the electrons.

2. Models of WAL and UCFs

Quantum corrections to the magnetoconductance $G(B)$ of the wires and stadia contain both WAL and UCF signatures. An example of measured $G(B)$ *vs* B is depicted in Fig. 2 for stadium sample F. Data analysis choices to extract values for L_ϕ were made dependent on the relative amplitudes of WAL and UCFs. In the wires, UCFs are relatively weaker, while in the stadia analysis of WAL is hindered by strong UCFs. Thus L_ϕ in wires and stadia were extracted by analysis of WAL and UCFs respectively. Concerning the 1D WAL analysis, under spin-orbit interaction (SOI), singlet and triplet states are formed in pairing of time-reversed trajectories. The quantum correction to conductance $\delta G(B)$ due to each state is proportional to a corresponding effective coherence length. The sum of corrections for all four states is the WAL correction expressed as [6, 7, 15, 16]

$$\delta G(B) = -\frac{1}{2}\frac{e^2}{\pi\hbar}\frac{1}{L}(L_{1,1} + L_{1,-1} + L_{1,0} - L_{0,0}) \tag{1}$$

in which $L_{0,0} = \left(L_\phi^{-2} + L_B^{-2}\right)^{-\frac{1}{2}}$ is the effective length for the singlet state, and $L_{1,\pm1} = \left(L_\phi^{-2} + L_{so}^{-2} + L_B^{-2}\right)^{-\frac{1}{2}}$ and $L_{1,0} = \left(L_\phi^{-2} + 2L_{so}^{-2} + L_B^{-2}\right)^{-\frac{1}{2}}$ are effective lengths for the triplet states, with L_{so} the characteristic length of spin decoherence due to SOI [15]. For narrow wires with $\ell_e \gtrsim 0.6\,W$, the magnetic length L_B is expressed as $L_B = l_m\sqrt{C_1 l_m^2 \ell_e / W^3}$ at low B [17], with $l_m = \sqrt{\hbar/eB}$, and $C_1 = 4.75$ for specular boundary scattering [6, 7, 18], valid in this work due to the existence of side depletion layers.

Concerning UCF analysis, the correlation function shows discernible quantum coherence effects, and can be used to extract L_ϕ. The correlation function is defined as,

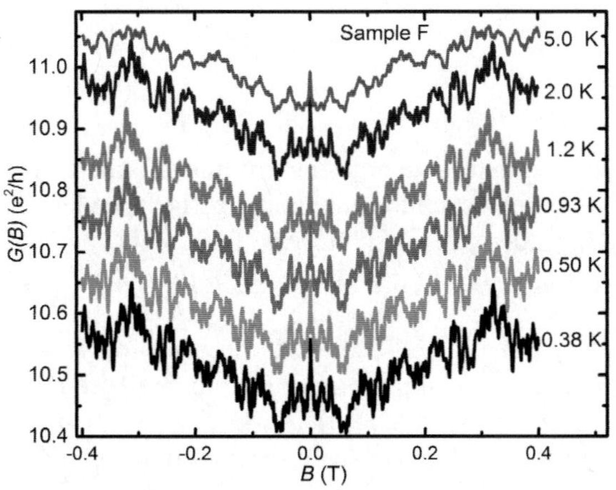

Figure 2: Magnetoconductance data $G(B)$ *vs* B for stadium sample F, for various T. The amplitude of the UCFs decreases with increasing T, as expected. While the WAL signature is visible over a very small range of B, UCFs were used in the stadia samples to obtain L_ϕ.

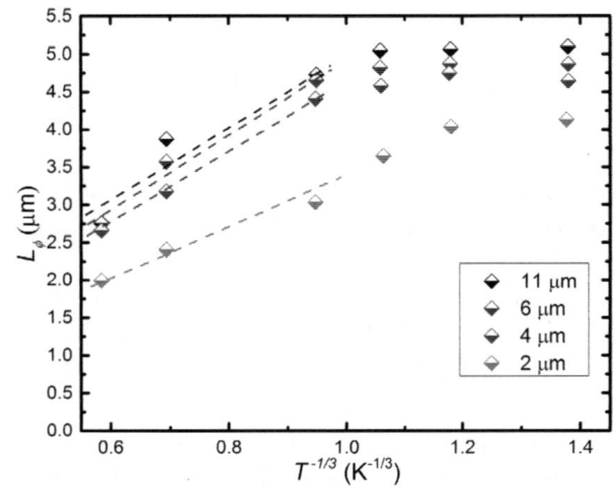

Figure 3: L_ϕ plotted *vs* $T^{-1/3}$ for the wire samples. Dashed lines are guidelines rather than fitting curves, illustrating the presence of 1D Nyquist scattering limiting L_ϕ for $T > 1$ K. For $T > 1$ K, $L_\phi \sim T^{-1/3}$, while L_ϕ approaches to saturation values at lower T. L_ϕ increases with increasing wire length L.

$$\overline{\delta G(B)\delta G(B+\Delta B)} = \langle [G(B) - \langle G(B)\rangle][G(B+\Delta B) - \langle G(B)\rangle]\rangle \tag{2}$$

in which the angled brackets denote an average over a range of B. We define $B_\phi = \hbar/(4eL_\phi^2)$, the B under which one flux quantum h/e threads the area $8\pi L_\phi^2$. If $B >> B_\phi$, we have an expression allowing determination of L_ϕ:

$$\overline{\delta G(B)\delta G(B+\Delta B)} = \overline{\delta G^2(B)}\frac{1}{b}\Psi'(\frac{1}{2}+\frac{1}{b}) \tag{3}$$

in which $b = \Delta B/2B_\phi$, and $\Psi(x)$ is the digamma function.

3. Experimental Results & Discussion

Extracted values for L_ϕ in wires are plotted in Fig. 3. The dependence of L_ϕ on T is clear. L_ϕ decreases with increasing T, in agreement with previous work [2, 6, 12, 14, 19]. L_ϕ for T>1 K follows $L_\phi \sim T^{-p}$, with $p \approx 0.34 \pm 0.02$. This result is consistent with quasi-elastic Nyquist electron-electron scattering in a 1D wire [14, 20]. For the stadia however the T-dependence of L_ϕ is weak ($p \approx 0.07$, not pictured), implying that a dominant decoherence mechanism for the stadia can be found in T-independent processes.

The results in Fig. 3 show that for wires L_ϕ increases with increasing of L in the entire range of T of the measurements. The reason lies in decoherence via interaction at the wire endpoints with the wider classical environment. Since this effect is averaged out in long wires, mitigating its effect on measured L_ϕ, longer wires show longer L_ϕ as observed. L_ϕ measured in stadia samples at $T = 0.38$ K are listed in Table 1, where the dimension dependence cannot be explained solely by environmental coupling decoherence (which would predict that the wider the side wire is, the shorter L_ϕ is). Device-device, here stadium-wire, coupling decoherence is invoked to explain the results. Device-device coupling decoherence is induced by low wavefunction hybridization efficiency between different geometries. With the stadium-wire coupling decoherence rate noted as γ_α, the total decoherence rate can be expressed as

$$\frac{1}{\tau_\phi} = \frac{1}{\tau_{\phi 0}} + \frac{1}{\tau_d} + 2\gamma_\alpha \tag{4}$$

where the factor 2 is due to 2 connection points, and the second term denotes environmental interaction decoherence via the dwell time τ_d [1,21]. The decoherence rate of stadium-wire coupling is dependent on geometrical similarity between inner circular stadia and the side-wires. High levels of internal coherence were observed between identical dots in quantum dot arrays [4, 22], implying a low *dot-dot* (identical in geometry) coupling decoherence rate γ_α. In the present work γ_α can be large and even dominate over $1/\tau_d$ due to the difference in geometry between stadia and side-wires. W and L of side-wires are comparable to the 2 μm radius of the stadia and lower W/L of the side-wires imply higher geometrical similarity with the stadia. In Table 1, for stadia E, F, G, H, L_ϕ decreases with decreasing W/L, except in the comparison between G and H (*cfr* below). A decrease in W/L, here caused either by narrower W or longer L, lowers wavefunction hybridization efficiency and thus leads to shorter L_ϕ. For stadium G, the side-wire with long $L = 3$ μm impacts the results, and environmental interaction decoherence obscures the effect of stadium-wire coupling decoherence.

4. Conclusions

The dependence of L_ϕ on mesoscopic device geometry is studied in this work. Decoherence by interaction with a classical environment is detected in wires, leading to a decreasing L_ϕ in shorter wires. For stadia, *stadium-wire* coupling decoherence dominates over environmental interaction decoherence effects, except for samples with longer side-wires. The work shows that values for L_ϕ in mesoscopic devices should be interpreted in the light of device geometry. The work was supported by the U.S. Department of Energy, Office of Basic Energy Sciences, Division of Materials Sciences and Engineering under award DOE DE-FG02-08ER46532.

References

[1] Hackens B, Faniel S, Gustin C, Wallart X, Bollaert S, Cappy A and Bayot V 2005 *Phys. Rev. Lett.* **94** 146802
[2] Lin J J and Bird J P 2002 *J. Phys.: Condens. Matter* **14** R501
[3] Bird J P, Micolich A P, Linke H, Ferry D K, Akis R, Ochiai Y, Aoyagi Y and Sugano T 1988 *J. Phys.: Condens. Matter* **10** L55
[4] Elhassan M, Bird J P, Akis R, Ferry D K, Ida T and Ishibashi K 2005 *J. Phys.: Condens. Matter* **17** L351
[5] Bird J P, Micolich A P, Ferry D K, Akis R, Ochiai Y, Aoyagi Y and Sugano T 1998 *Solid-State Electron.* **42** 1281
[6] Kallaher R L, Heremans J J, Goel N, Chung S J and Santos M B 2010 *Phys. Rev. B* **81** 035335
[7] Kallaher R L, Heremans J J, Roy W V and Borghs G 2013 *Phys. Rev. B* **88** 205407
[8] Iordanskii S V, Lyanda-Geller Y B and Pikus G E 1994 [JETP Lett 60, 206(1994)] *Pis'ma Zh. Eksp. Teor. Fiz* **60** 199
[9] Bergmann G 2010 *Int. J. Mod. Phys. B* **24** 2015
[10] McPhail S, Yasin C E, Hamilton A R, Simmons M Y, Linfield E H, Pepper M and Ritchie D A 2004 *Phys. Rev. B* **70** 245311
[11] Deo V, Zhang Y, Soghomonian V and Heremans J J 2015 *Sci. Rep.* **5** 9487
[12] Hackens B, Delfosse F, Faniel S, Gustin C, Boutry H, Wallart X, Bollaert S, Cappy A and Bayot V 2002 *Phys. Rev. B* **66** 241305(R)
[13] Ferry D K, Akis R A, Pivin D P, Bird J P and Holmberg N 1998 *Phys. E* **3** 137
[14] Rudolph M and Heremans J J 2011 *Phys. Rev. B* **83** 205410
[15] Heremans J J, Kallaher R L, Rudolph M, Santos M, Roy W V and Borghs G 2014 *Proc. of SPIE* **9167** 91670D
[16] Zduniak A, Dyakonov M I and Knap W 1997 *Phys. Rev. B* **56** 1996
[17] Ren S L, Heremans J J, Vijeyaragunathan S, Mishima T D and Santos M B 2015 *J. Phys.: Condens. Matter* **27** 185801
[18] Beenakker C W J and van Houten H 1988 *Phys. Rev. B* **38** 3232
[19] Bird J P, Ishibashi K, Ferry D K, Ochiai Y, Aoyagi Y and Sugano T 1995 *Phys. Rev. B* **51** 18037
[20] Natelson D, Willett R L, West K W and Pfeiffer L N 2001 *Phys. Rev. Lett.* **86** 1821
[21] Hackens B, Faniel S, Gustin C, Wallart X, Bollaert S, Cappy A and Bayot V 2006 *Physica E* **34** 511
[22] Elhassan M, Bird J P, Shailos A, Prasad C, Akis R, Ferry D K, Takagaki Y, Lin L H, Aoki N, Ochiai Y, Ishibashi K and Aoyagi Y 2001 *Phys. Rev. B* **64** 085325

Maximum entropy mobility spectrum analysis of LPE-grown and anodic oxidated $Hg_{1-x}Cd_xTe(x=0.237)$

Z Y Song[1], L Y Shang[1, *], T Lin[2], Y F Wei[2], J H Chu[1,2]
[1]Key Laboratory of Polar Materials and Devices, Ministry of Education, East China Normal University, Shanghai 200062, China
[2]National Laboratory for Infrared Physics, Shanghai Institute of Technical Physics, Chinese Academy of Science, Shanghai 200083, China

[*]The corresponding author: lyshang@ee.ecnu.cn

Abstract. In this paper, magneto-transport properties of the LPE-grown and anodic oxidated p-type $Hg_{1-x}Cd_xTe(x=0.237)$ films have been studied by using maximum entropy mobility spectrum analysis (ME-MSA) technique. It can be found that the high-mobility electron($\mu_e \sim 2\times 10^4 cm^2/Vs$) has considerable contributions to the conduction of anodic oxidated $Hg_{1-x}Cd_xTe(x=0.237)$ film, but not in LPE-grown $Hg_{1-x}Cd_xTe(x=0.237)$ film. The high-mobility electron maintains dominant contributions from 11k to 150k, which can be attributed to two-dimensional electron gas in the inversion layer of anodic oxidated p-type $Hg_{1-x}Cd_xTe(x=0.237)$ film. In addition, we also observe the nonphysical contributions of low mobility electrons ($\mu_e \sim 0.08\times 10^4 cm^2/Vs$) in mobility spectrum of both LPE-grown and anodic oxidated p-type HgCdTe films. The low-mobility electrons, so-called mirror peaks, can be interpreted as a consequence of magnetic freeze-out of holes in vacancy-doped HgCdTe, which disappeared at T=150k.

1. Introduction

HgCdTe(MCT) is the most important modern semiconductor alloy system for infrared application. Nevertheless, owning to the relative instability of Hg atoms in HgCdTe, the surface treatments play a crucial role in determining the performance of the devices. The most common method to protect the surface of HgCdTe is anodic oxidation[1,2,9], which still introduce some positive oxidation defects, such as oxygen vacancies, in anodic oxidated p-type MCT films[2]. These positive oxidation defects, limiting the detector's performance, lead to inverted surface of p-type HgCdTe films providing us a two dimensional electron system which could carry on many basic physics research. In addition, due to the narrow band of HgCdTe, thermal excited electrons, light holes tend to make significant contributions to the conductions of MCT samples[3,6,7]. Multicarrier transport studies[7] were performed on a set of MCT samples by using mobility spectrum with changing temperature[6,7]. But, little quantitative research has been performed on transport properties of anodic oxidated MCT system.

Standard measurements of the resistivity and Hall coefficient at a single magnetic field provide only averaged values of the carrier densities and mobilities[5], which can not give the actual individual species and their accurate mobility. To overcome this shortcoming, Beck and Anderson[3] proposed in 1987 an approach known as the mobility-spectrum analysis(MSA).

Furthermore, Maximum entropy mobility spectrum analysis(ME-MSA)[5] approach minimizes the fit deviation from measured data with the principle of entropy maximization. It doesn't allow nonphysical negative conductivities and every probable distribution is concerned, with the help of maximum entropy principle. ME-MSA estimates the mobility spectrum by assigning "an entropy" to each probability distribution and choosing the one with the highest entropy.

In this paper, we investigated magnetotransport properties of LPE-grown and anodic-oxidated HgCdTe materials. First, we will introduce the experimental details of sample processing; Then the mobility spectrums of LPE-grown and anodic oxidated p-type $Hg_{1-x}Cd_xTe$(x=0.237) films will be given and we find that the high-mobility electron maintains the dominant contribution the conduction of anodic oxidated MCT until T=150k, which can be attributed to two dimensional electron gas in the inversion layers of anodic oxidated p-type $Hg_{1-x}Cd_xTe$(x=0.237).

2. Experiments

A set of p-type HgCdTe wafers were grown by standard liquid-phase epitaxy (LPE) on a lattice matched CdZnTe substrate. The surface was passivated by anodic grown native oxidase. The samples were cut into $7 \times 7\mu m$ squares and then indium was used to make Ohmic contacts. The magneto-transport measurements were performed by using Vander Pauw configuration with an applied magnetic field between 0 T and 9 T and from 11k to 300 K in a liquid-helium-cooled continuous-flow cryostat.

3. MS of LPE-grown and anodic oxidated $Hg_{1-x}Cd_xTe$ (x=0.237)

Table1. Selected transport parameters extracted from MS of LPE-grown HgCdTe and anodic oxidated HgCdTe at 20 k.

	Name	μ (m^2/Vs)	σ $(10^{-4}sv^{-1})$	N $(10^{11}cm^2)$
LPE-grown HgCdTe	E3	0.09	0.98	6.73
	T1	0.08	1.54	13.76
	T1b	0.29	0.21	0.46
Anodic oxidated HgCdTe	E1	2.76	35.2	7.97
	E2	0.036	1.51	3.06
	T2	0.064	2.68	37.19
	T2b	0.051	2.13	6.06

Figure.1 shows the mobility spectrums for electron(Fig.a) and hole(Fig.b) at 20 k of as-grown HgCdTe(black circles) and the anodic oxidated one(red triangles). It can be identified that three kinds of carriers in MS of electrons(Fig.1a) at 20K: the low-mobility electron (E2, E3) and high-mobility electron E1. Notably, the high-mobility electron was only found in the anodic oxidated HgCdTe, and the percentage of E1 electrons to the conductivity is 84% . The contributions of E2 and E3 electrons are not expected to the conductions of two samples at 20k. And the E2 and E3 peaks disappear at 150k, as is shown in figure. 2a. But the high-mobility electron(E1) still maintain the dominant contribution to the conductivity in anodic oxidated HgCdTe at 150k (figure.2a), Furthermore the concentration of

E1 electron was almost constant with the increasing temperature (figure.4), which could be the typical characteristics of two dimensional electron gas. We deduced that a two dimensional electron gas appears in the surface layer of MCT since the process of anodic oxidation would introduces some optimistic oxygen vacancies which have positive charges[1,2].

There were four kinds of holes which contribute to the conduction of two samples, as is shown in figure.1b. The low-mobility holes(T1 and T2) are the expected heavy holes. The light holes were easily excited due to the narrow bandgap($E_g = 0.118eV$) of $Hg_{1-x}Cd_xTe$(x=0.237) and ME-MSA also gave the mobility of light holes(T1b and T2b) which consisted with the results of past reports[6,7,8,9].

We could also find that the low-mobility electron(E2 and E3) disappeared at 150K. The mobilities of E2 and E3 electrons are very low in HgCdTe. Similar features have been observed in other studies of HgCdTe[10]. A possible origin of the low-mobility electrons is the magnetic freezeout of the carriers. The magnetic freezeout[4,10] was to be induced by an increase of the activation energy through a localization of the impurity states, which causes a reduction of the number of carriers as the magnetic field increases. Figure.2b shows that heavy holes(T2) appear in the conducting system with the increasing temperature.

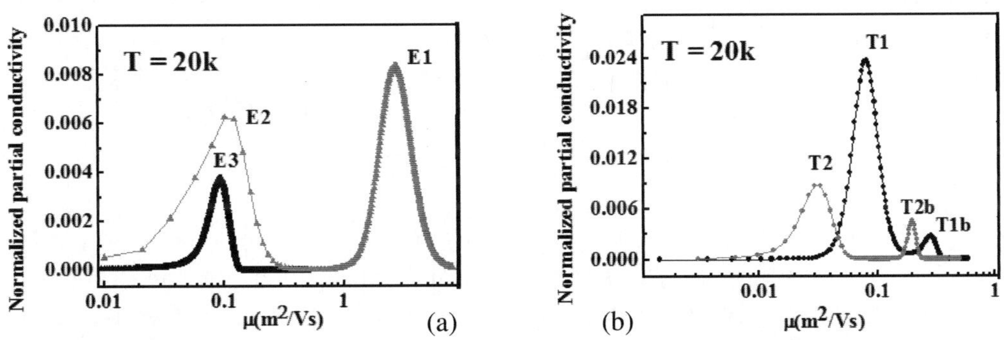

Figure.1. MS of electrons(a) and holes(b) for LPE-grown(black circles) and anodic oxidated(red triangles) HgCdTe at T=20k.

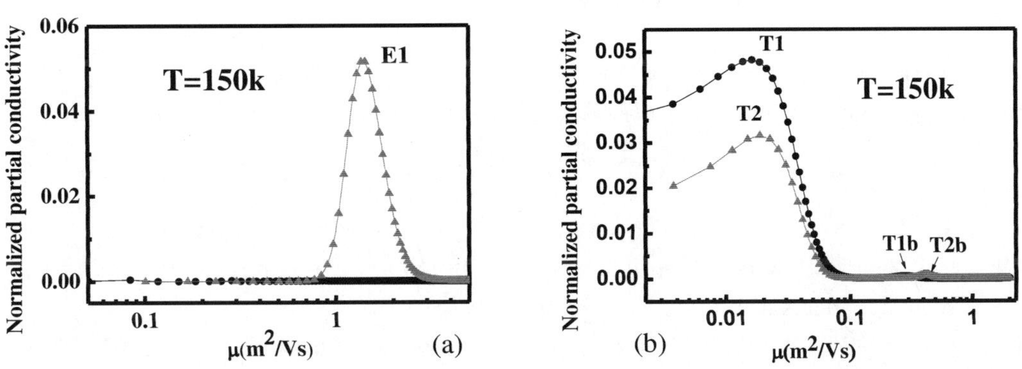

Figure.2. MS of electrons(a) and holes(b) for LPE-grown(black circles) and anodic oxidated(red triangles) HgCdTe at T=150k.

Figure.3 shows the temperature dependence of mobility of heavy holes (T1) of as-grown HgCdTe. It can be shown that the mobility of T1 holes increases until 77k and then decreases as the temperature is increased from 77k to 200k. The iconized-impurities scattering is the dominated mechanism below 77k since the mobilities and temperatures satisfies the relationship of $\mu \sim T^{3/2}$. As temperature is increasing, phonon scattering becomes the predominated scattering mechanism, which gives a decrease of the mobility of T1 holes. The dominated phonon scattering also gives a slight decrease of the mobility of E1 electron, as is indicated in figure.4.

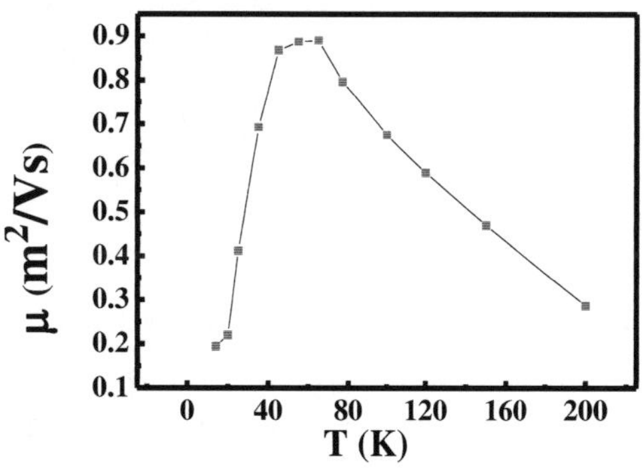

Figure.3.Temperature-dependent of hole mobility for LPE-grown HgCdTe.

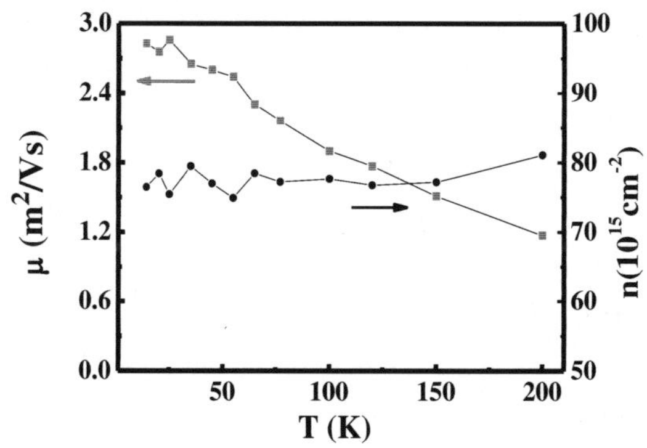

Figure.4.Temperature-dependent of electron concentration(black circles) and mobility(red squares) characteristics for anodic oxidated HgCdTe.

4. Conclusions

Mobility spectrum analysis has been employed to study transport properties of as-grown and anodic oxidated HgCdTe films. It shows that electrons are the main contribution to the conduction of anodic oxidated p-type HgCdTe films. We think those electrons were 2DEG located in the invert surface of the anodic oxidated p-type HgCdTe films. The low-mobility electrons found in both two kinds samples can be assigned to magneto freeze-out carriers. MSA based studies of electronic transport properties of HgCdTe films can be employed to determine carrier transport parameters, enabling identification of the main defects limiting the properties of infra HgCdTe detectors.

Acknowledgments

The work was sponsored by the STCSM(13ZR1453700) and the NSFC(61306119) of China.

Reference :

[1] Virt I, Bilyk M, Stefaniuk I and Kuzma M 2001 *Solid-State Electronics* **45** 1743
[2] Yang M J, Yang C H, Kinch M A and Beck J D 1989 *Appl. Phys. Lett.***54** 265
[3] Beck W A and Anderson J R 1987 *J. Appl. Phys.* **62** 541
[4] Rothman J, Meilhan J, Perrail G, Belle J,and Greavrand O *J. Electron. Mater.* 2006. **35** 1174.
[5] Kiatgamolchai S et al 2002 *Phys.Rev.B.* **66** 036705.
[6] Antoszewski J, Faraone L, Vurgaftman I, MEYER J R, and Hoffman C A 2004 *J. Electron. Mater.* **33** 673.
[7] Umana-Membreno G A et al *J. Electron. Mater.* 2010. **39** 1023.
[8] Umana-Membreno G A, Antoszewski J, and Faraone. 2013 *J, Microelectron. Eng* **109** 232.
[9] Nemirovsky Y, Burstein L and Kidron 1985 J. Appl. Phys. **58** 366.
[10] Gold M C and Nelson D A 1986 *J. Vac. Sci. Technol. A* **4** 2040.

IOP Conf. Series: Journal of Physics: Conf. Series **864** (2017) 012027 doi:10.1088/1742-6596/864/1/012027

Charge carrier mobility in semiconductor solid solutions and percolation phenomena

E Rogacheva, O Nashchekina and E Martynova
National Technical University "Kharkov Polytechnic Institute", Kharkov, UKRAINE

E-mail: rogachova.olena@gmail.com

Abstract. The existence of a range of an anomalous growth in charge carrier mobility under the transition to heavy doping is established for Bi_2Te_3 -Sb_2Te_3 solid solutions. This confirms our suggestion about the universal character of critical phenomena accompanying the transition from impurity discontinuum to impurity continuum. The experimental results are analyzed in terms of percolation theory taking into account alloy scattering and spatial correlations of impurity centers.

1. Introduction

Materials based on heavily doped semiconductors are widely used in various semiconductor devices [1]. That is why studying the transition from weak to heavy doping when the interaction between impurity atoms starts to make a substantial contribution to the formation of crystal properties, is of special interest. In the available literature on heavy doping, the behavior of the electron subsystem under changing impurity concentration is usually considered (e.g. the Mott transition). At the same time, it is the lattice subsystem that determines such important parameters as charge carrier mobility, phonon thermal conductivity, lattice specific heat, and so on. This stimulates detailed studies of the dependences of properties on the composition in the range of small impurity concentrations.

Earlier, we revealed anomalies in the behavior of different physical properties in IV-VI – based semiconductor solid solutions in the range of small impurity concentrations (0.5-1.0 at.%) and suggested that those anomalies have a universal character and are connected with the transition to heavy doping (see, for example [2-7]). Such transition is expected to be accompanied by critical phenomena, which, in turn, manifest themselves through anomalies in the concentration dependences of different properties. To confirm the suggestion about the universal character of the concentration-dependent anomalies of properties in solid solutions, it is necessary to expand the scope of objects and properties to be studied.

The objects studied in this work are Bi_2Te_3- and Sb_2Te_3-based semiconducting solid solutions in the Bi_2Te_3 -Sb_2Te_3 system. Bi_2Te_3 and Sb_2Te_3 have a rhombohedral crystal lattice of a tetradymite type and form between themselves a continuous series of solid solutions [8]. The stoichiometric Bi_2Te_3 and Sb_2Te_3 exhibit *p*-type conductivity due to the presence of antisite defects. Bi_2Te_3-Sb_2Te_3 solid solutions belong to the best low temperature thermoelectric materials, broadly applied for manufacturing various types of refrigerating devices, and the most efficient materials for use near room temperature [8-10]. One of the most important tasks of the development of efficient thermoelectric materials is increasing the ratio of charge carrier mobility μ to lattice thermal conductivity, which is usually attained by forming solid solutions. That is why understanding the behavior of properties of those materials under the introduction of new components is very important from not only the scientific but also the practical point of view.

Recently, interest in investigating the V_2VI_3 semiconducting compounds has grown sharply due to the prediction and subsequent experimental evidence of their special properties characteristic of 3D-topological insulators. It is pointed out that one of the ways of using these unique surface properties in practice, in particular for thermoelectric applications, is the introduction of impurities with a view to reducing charge carrier concentration in the volume of a crystal [11-13].

The goal of the present work is to study the behavior of the Hall charge carrier mobility in the range of compositions corresponding to the transition from weak to heavy doping in Bi_2Te_3 -Sb_2Te_3 solid solutions.

2. Experimental

Two series of solid solutions in the Bi_2Te_3 -Sb_2Te_3 system were synthesized: Bi_2Te_3-based (0 – 7 mol.% Sb_2Te_3) solid solutions and Sb_2Te_3-based (0 – 7 mol.% Bi_2Te_3) solid solutions. Polycrystalline samples were prepared by direct melting of high-purity elements in evacuated quartz ampoules at 1070 K with subsequent annealing at ~ 650 K for 300 h. X-ray spectroscopy and selective chemical analysis confirmed the homogeneity of the samples and showed that deviation of the actual chemical composition of the prepared samples from the intended one did not exceed ± 0.02 at. %. The electrical conductivity σ and the Hall coefficient R_H were measured at room temperature using the van der Pauw method with an error not exceeding 5 %. The Hall carrier mobility μ_H was calculated as $\mu_H = \sigma \cdot R_H$. All the alloys exhibited *p*-type conductivity.

3. Results and discussions

In figure 1, the room-temperature dependences of Hall charge carrier mobility on the Sb_2Te_3 (for the Bi_2Te_3-based solid solutions) or Bi_2Te_3 (for the Sb_2Te_3-based solid solutions) content are presented. It is seen that for both the Bi_2Te_3-based and Sb_2Te_3-based solid solutions, after the initial drop in μ_H which occurs under increasing concentration of Sb_2Te_3 (or Bi_2Te_3) up to ~ 0.5 mol.%, μ_H grows in the interval between ~ 0.5 and ~ 1 mol.% and then decreases.

It is natural to attribute the drop in μ_H under the introduction of the first portions of the impurity to an increase in scattering of holes by impurity defects, but the subsequent sharp growth in hole mobility cannot be interpreted in terms of the traditional model of alloy scattering which is usually applied to explain a decrease in μ_H in solid solutions when an impurity is introduced.

The anomalous increase in charge carrier mobility that we observed in the two studied systems within a narrow concentration interval in the range of small impurity concentrations indicates the existence of a phase transition. Within the framework of ideas we have developed earlier [14,15], this is a percolation-type transition from an impurity discontinuum to an impurity continuum.

According to modern notions [16-18], there is a direct analogy between percolation theory and fluctuation theory of second-order phase transitions. In both cases, in the vicinity of a transition, properties of the system are determined by strongly developed interacting fluctuations, peculiarities of thermodynamic quantities obey a power law, and their exponents are called critical exponents.

The effects observed in this work provide additional support for our proposition about the universal character of critical phenomena accompanying the transition from weak to heavy doping. For comparison, we also show previously obtained data for the PbTe - PbSe system [2]. It is seen that in that system, in a relatively narrow concentration range of a second component (~0.5-2.0 mol.%), the anomalous increase in μ_H is also observed.

Applying the short-range interaction approximation, assuming a random distribution of impurity atoms and using the ideas of percolation theory, one can estimate the radius of the impurity atom's "action sphere" R_o, knowing the concentration of impurity atoms corresponding to the percolation threshold x_C at which a continuous chain of overlapping deformation fields passing through the entire crystal (infinite cluster) is formed. In accordance with one of the problems of percolation theory – the problem of spheres – one can write:

$$\frac{4}{3}\pi N_c (2R_0)^3 = 2.7 \quad (1)$$

where N_c is the average number of sphere centers per volume unit, corresponding to the percolation threshold. Determining the percolation threshold as the concentration of the impurity component (Bi_2Te_3 or Sb_2Te_3) corresponding to the beginning of the anomalous section of the curve (0.5 mol. %), we obtain that for the Bi_2Te_3–based (Sb_2Te_3-based) solid solutions $R_0 \approx 1.5\ a_0$ (a_0 is the Bi_2Te_3 (Sb_2Te_3) pseudocubic unit cell parameter calculated as $a_0 = V^{1/3}$, where V is the unit cell volume of the Bi_2Te_3 (Sb_2Te_3) crystal).

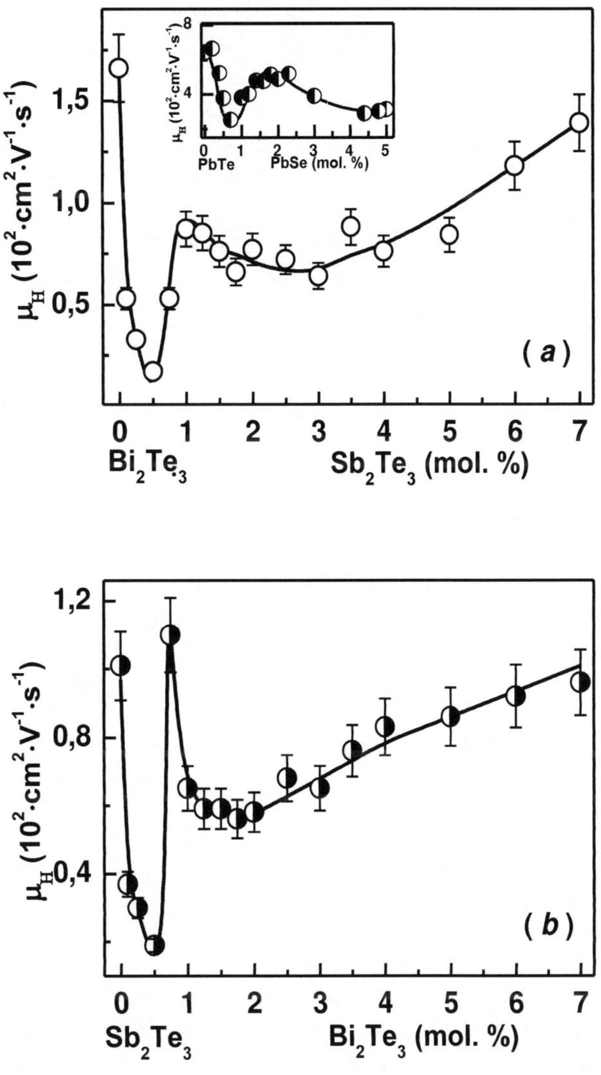

Figure 1. Room-temperature dependences of Hall charge carrier mobility μ_H on Sb_2Te_3 (a) or Bi_2Te_3 (b) concentration in Bi_2Te_3-Sb_2Te_3 solid solutions. Inset: the analogous dependences for the PbTe - PbSe system [2]

The formation of continuous chains of interacting impurity atoms can stimulate a redistribution of impurity atoms leading to the realization of their configuration corresponding to a minimum of the thermodynamic potential. Possible self-organization processes may include a long-range ordering of impurity atoms. The sharp change in the properties in the vicinity of the percolation threshold suggests that the formation of an infinite cluster may be a trigger mechanism for the ordering of atoms, for example as a result of deformational interactions.

On the other hand, the Coulomb attraction between charged defects of opposite signs stimulates processes of chemical interaction leading to the formation of neutral molecular complexes. In this case new structural elements appear, and the formation of percolation channels through these elements becomes possible. Such processes are more probable in heterovalent solid solutions than in isovalent ones, and we observed them, for example in PbTe-Bi_2Te_3 solid solutions [19]. We introduced the concept of "the complex formation threshold" [20], which is somewhat similar to the percolation threshold but in this case percolation occurs not via separate atoms but via complexes. In the isovalent Bi_2Te_3-Sb_2Te_3 solid solution such processes are less probable, than in heterovalent ones, and, judging from figure 1, no additional anomalies in the μ_H dependence on composition are observed.

4. Conclusions

It is shown that in the Bi_2Te_3-Sb_2Te_3 solid solutions, at impurity concentrations in the range 0.5-1.0 mol % in the dependences of Hall charge carrier mobility on composition, an anomalous growth in μ_H is observed. We attribute these peculiarities to the transition, occurring under increasing impurity content, to the impurity continuum when the interaction of impurity atoms becomes cooperative. The results obtained in this work for the isovalent solid solutions formed by cation substitution represent another evidence for the proposition about the universal character of critical phenomena accompanying the transition from weak doping to the formation of impurity continuum.

Acknowledgement
This work was supported by the Ukrainian Ministry of Education and Science (Project # M3925).

References

[1] Bonch-Bruevich V L and Kalashnikov S G 1978 *Semiconductor Physics* (Moscow: Nauka)
[2] Rogacheva E I and Vodorez O S 2013 *J. Thermoelectricity* **2** 61
[3] Rogacheva E I, Sinelnik N A and Nashchekina O N 1993 *Acta Phys. Pol.* **84** 729
[4] Rogacheva E and Krivulkin I 2002 *Semiconductors* **36** 966
[5] Rogacheva E 2003 *J. Phys. Chem. Solids* **64** 1579
[6] Rogacheva E, Vodorez O and Nashchekina O 2013 *J. Phys. Chem. Solids* **74** 35
[7] Rogacheva E *et al* 2014 *Phys. St. Sol.* B. **251** 1231
[8] Rowe D M 2012 CRC *Handbook of Thermoelectrics* (Boca Raton, FL: CRC Press, Taylor & Francis Group)
[9] Goldsmid H J 2014 *Materials* **7** 2577
[10] Bell L E 2008 *Science* **321** 1457
[11] Zhang H *et al* 2009 *Nat. Phys.* **5** 438
[12] Chen Y L *et al* 2009 *Science* **325** 178
[13] Xia Y *et al* 2009 *Nat. Phys.* **5** 398
[14] Rogacheva E I 1993 *Jpn. J. Appl. Phys.* **32** 775
[15] Rogacheva E I, Nashchekina O N and Dresselhaus M S 2005 *J. Thermoelectricity* **4** 82
[16] Stauffer D and Aharony A 1992 *Introduction to percolation theory* (London; Washington, DC: Taylor & Francis Group)
[17] Shklovskii B I and Efros A L 1984 *Electronic Properties of Doped Semiconductors* (New York: Springer-Verlag).
[18] Patashinskiy A Z, Pokrovskiy V L 1982 *The fluctuation theory of phase transitions* (Moscow Nauka).
[19] Rogacheva E I, Vodorez O S, and Nashchekina O N 2013 *J. Phys. Chem. Solids* **74** 35
[20] Rogacheva E I, 2009 *Phys Status Sol C.* **6** 1307

Longitudinal magnetoconductivity and magnetodielectric effect in bilayer graphene

U Zülicke[1,2] and R Winkler[3,4]

[1] School of Chemical and Physical Sciences and MacDiarmid Institute for Advanced Materials and Nanotechnology, Victoria University of Wellington, PO Box 600, Wellington 6140, New Zealand
[2] Kavli Institute for Theoretical Physics, University of California, Santa Barbara, CA 93106, USA
[3] Department of Physics, Northern Illinois University, DeKalb, Illinois 60115, USA
[4] Materials Science Division, Argonne National Laboratory, Argonne, Illinois 60439, USA

E-mail: `uli.zuelicke@vuw.ac.nz`

Abstract. It was recently shown that a finite imbalance between electron densities in the $\mathbf{K}$ and $\mathbf{K}'$ valleys of bilayer graphene induces a magnetoelectric coupling. Here we explore ramifications of this electronically tunable magnetoelectric effect for the optical conductivity and dielectric permittivity of this material. Our results augment current understanding of longitudinal magnetoresistance and magnetocapacitance in unconventional materials.

1. Introduction

The coupling of electric and magnetic degrees of freedom in materials continues to be studied intensely, because of both its interesting physical origin and its potential for useful applications [1–5]. The recent discovery [6, 7] of magnetoelectric couplings in bilayer graphene (BLG) has established an intriguing link between magnetoelectricity and atomically thin materials [8, 9]. Here we investigate ramifications of this effect for electric transport, pointing out intriguing similarities, but also crucial differences, with transport characteristics [10–15] arising from the chiral anomaly [16] in Dirac [17] and Weyl [18, 19] semimetals.

As a relevant context, we summarize the basic constitutive relations that represent the chiral magnetic effect [10–16] in Dirac and Weyl semimetals with two valleys characterized by opposite chirality. The most important feature is the generation of a finite imbalance ϱ_{v} of electron densities between the valleys when both an electric field $\boldsymbol{\mathcal{E}}$ and a magnetic field $\boldsymbol{\mathcal{B}}$ are applied, which also depends on the relaxation time τ_{v} for inter-valley scattering of charge carriers:

$$\varrho_{\mathrm{v}} = -\frac{e^3}{2\pi^2\hbar^2}\,\tau_{\mathrm{v}}\,\boldsymbol{\mathcal{B}}\cdot\boldsymbol{\mathcal{E}} \quad . \tag{1}$$

Combined with the fact that ϱ_{v} in the presence of a magnetic field causes an electric current $\delta\mathbf{j}$ to flow,

$$\delta\mathbf{j} = -\frac{e}{2\pi^2\hbar^2\,D(E_{\mathrm{F}})}\,\boldsymbol{\mathcal{B}}\,\varrho_{\mathrm{v}} \quad , \tag{2}$$

a correction to the magnetoconductivity tensor emerges that is of the form $\delta\underline{\sigma}(\boldsymbol{\mathcal{B}}) = \delta\sigma(\mathcal{B})\,\hat{\boldsymbol{\mathcal{B}}}\hat{\boldsymbol{\mathcal{B}}}$, with

$$\delta\sigma(\mathcal{B}) = \frac{e^4\,\tau_\mathrm{v}}{4\pi^4\hbar^4\,D(E_\mathrm{F})}\,\mathcal{B}^2 \quad . \tag{3}$$

Here $D(E_\mathrm{F})$ denotes the density of states at the Fermi energy E_F. See, e.g., Refs. [10,12,14] for relevant derivations and Ref. [20] for a recent experimental observation.

As discussed in greater detail further below, the existence of a magnetoelectric coupling in BLG whose magnitude depends on ϱ_v [6] gives rise to a set of constitutive relations that are similar to those given above for Dirac and Weyl semimetals.

2. Magnetoelectric coupling in BLG for in-plane electric & magnetic fields

For the purpose of the present discussion, we only consider electric fields $\boldsymbol{\mathcal{E}}_\parallel$ and magnetic fields $\boldsymbol{\mathcal{B}}_\parallel$ *parallel to the plane* of the BLG sheet. (Effects of field couplings in the general case are presented in Refs. [6,7].) The single-particle Hamiltonian describing the charge-carrier dynamics is given by

$$H = H_0(\mathbf{k} + e\boldsymbol{\mathcal{A}}) + e\,\xi_\parallel\,\boldsymbol{\mathcal{E}}_\parallel \cdot \boldsymbol{\mathcal{B}}_\parallel\,\tau_z \quad , \tag{4}$$

where $H_0(\mathbf{k})$ describes the low-energy band structure for the field-free situation in the two valleys associated with the high-symmetry $\mathbf{K}$ and $\mathbf{K}'$ points in the BLG Brillouin zone, and $\boldsymbol{\mathcal{A}}$ is the electromagnetic vector potential. The diagonal Pauli matrix τ_z operates in valley (isospin) space, and $\xi_\parallel$ is a material-dependent parameter quantifying the strength of magnetoelectric coupling.

Several unusual physical effects arise from the presence of the term proportional to $\boldsymbol{\mathcal{E}}_\parallel \cdot \boldsymbol{\mathcal{B}}_\parallel\,\tau_z$. Due to its asymmetric coupling to the $\mathbf{K}$ and $\mathbf{K}'$ valleys, simultaneous application of electric and magnetic fields gives rise to a valley-charge imbalance

$$\varrho_\mathrm{v} \equiv \varrho^{\mathbf{K}} - \varrho^{\mathbf{K}'} = 2e^2\,\xi_\parallel\,D(E_\mathrm{F})\,\boldsymbol{\mathcal{B}}_\parallel \cdot \boldsymbol{\mathcal{E}}_\parallel \quad . \tag{5}$$

This represents an intriguing analogy with the situation in the Dirac or Weyl semimetals, see Eq. (1) above. Furthermore, mirroring phenomena associated with axion electrodynamics [21,22] or, more generally, with the magnetoelectric effect [23–25], the in-plane magnetic field in conjunction with a finite ϱ_v induces an electric current [6]

$$\delta\mathbf{j} \equiv \mathbf{j}^{\mathbf{K}} + \mathbf{j}^{\mathbf{K}'} = \xi_\parallel\,\boldsymbol{\mathcal{B}}_\parallel\,\partial_t\varrho_\mathrm{v} \quad . \tag{6}$$

While there is some similarity between (6) and the constitutive relation (2) for the Weyl semimetals, there is the crucial difference that the current arising from magnetoelectric coupling in BLG is not proportional to ϱ_v itself but to its time derivative.

3. Magnetotransport from magnetoelectric coupling in BLG

The main focus of the present work is to elucidate how electric-transport characteristics of BLG are modified by the magnetoelectric coupling. Hence, we assume $\boldsymbol{\mathcal{B}}_\parallel$ to be static, but $\boldsymbol{\mathcal{E}}_\parallel$ to be time dependent, $\boldsymbol{\mathcal{E}}_\parallel(t)$. Assuming furthermore spatial uniformity, the continuity equation for ϱ_v becomes

$$\partial_t\varrho_\mathrm{v} + \frac{\varrho_\mathrm{v}}{\tau_\mathrm{v}} = 2e^2\,\xi_\parallel\,D(E_\mathrm{F})\,\boldsymbol{\mathcal{B}}_\parallel \cdot \partial_t\boldsymbol{\mathcal{E}}_\parallel \quad . \tag{7}$$

Straightforward solution using Fourier transformation yields

$$\varrho_\mathrm{v}(\omega) = 2e^2\,\xi_\parallel\,D(E_\mathrm{F})\,\boldsymbol{\mathcal{B}}_\parallel \cdot \boldsymbol{\mathcal{E}}_\parallel(\omega)\,\frac{1}{1 - i/(\omega\tau_\mathrm{v})} \quad . \tag{8}$$

Inserting (8) into the Fourier-transformed Eq. (6), we obtain a contribution to the BLG-sheet conductivity tensor that is due to the magnetoelectric coupling, $\delta\underline{\sigma}(\omega, \boldsymbol{\mathcal{B}}_\parallel) = \delta\sigma(\omega, \mathcal{B}_\parallel)\, \hat{\boldsymbol{\mathcal{B}}}_\parallel \hat{\boldsymbol{\mathcal{B}}}_\parallel$, with

$$\delta\sigma(\omega, \mathcal{B}_\parallel) = -2e^2\, \frac{D(E_\mathrm{F})\, \xi_\parallel^2}{\tau_\mathrm{v}}\, \mathcal{B}_\parallel^2\, \frac{(\omega\tau_\mathrm{v})^2}{\sqrt{1+(\omega\tau_\mathrm{v})^2}}\, \mathrm{e}^{-i\arctan(\omega\tau_\mathrm{v})} \quad . \tag{9}$$

Thus the valley-density-dependent magnetoelectric coupling causes a reduced conductivity in the direction parallel to an applied in-plane magnetic field. The magnitude of the associated response is set by the scale

$$\sigma_\mathrm{ME}(\mathcal{B}_\parallel) = 2e^2\, \frac{D(E_\mathrm{F})\, \xi_\parallel^2}{\tau_\mathrm{v}}\, \mathcal{B}_\parallel^2 \quad , \tag{10}$$

which depends quadratically on $\mathcal{B}_\parallel$ but inversely on τ_v. In the static limit $\omega\tau_\mathrm{v} \ll 1$, the dissipative part of the conductivity dominates but is also suppressed by a small factor $(\omega\tau_\mathrm{v})^2$. The opposite (high-frequency) limit $\omega\tau_\mathrm{v} \gg 1$ is characterized by the conductivity correction being dominantly capacitive with magnitude proportional to the large factor $\omega\tau_\mathrm{v}$, with the dissipative part saturating at $\sigma_\mathrm{ME}(\mathcal{B}_\parallel)$.

4. Magnetodielectric effect in BLG

Using the familiar identity relating the 3D-bulk conductivity and the dielectric permittivity [26], we find a magnetoelectric-coupling-induced change to the dielectric tensor,

$$\frac{\delta\underline{\varepsilon}(\omega, \boldsymbol{\mathcal{B}}_\parallel)}{\varepsilon_0} \equiv i\frac{\delta\underline{\sigma}(\omega, \boldsymbol{\mathcal{B}}_\parallel)}{\varepsilon_0\, \omega\, d} = \frac{\delta\varepsilon(\omega, \mathcal{B}_\parallel)}{\varepsilon_0}\, \hat{\boldsymbol{\mathcal{B}}}_\parallel \hat{\boldsymbol{\mathcal{B}}}_\parallel \quad , \tag{11}$$

$$\frac{\delta\varepsilon(\omega, \mathcal{B}_\parallel)}{\varepsilon_0} = -\frac{2e^2}{\varepsilon_0}\, \frac{D(E_\mathrm{F})\, \xi_\parallel^2}{d}\, \mathcal{B}_\parallel^2\, \frac{\omega\tau_\mathrm{v}}{\sqrt{1+(\omega\tau_\mathrm{v})^2}}\, \mathrm{e}^{i\,\mathrm{arccot}(\omega\tau_\mathrm{v})} \quad . \tag{12}$$

Here d is the electronic width of BLG. Thus this material exhibits a negative in-plane magnetodielectric effect [27, 28] with a magnitude that saturates at the susceptibility scale

$$\chi_\mathrm{ME}(\mathcal{B}_\parallel) = \frac{2e^2}{\varepsilon_0}\, \frac{D(E_\mathrm{F})\, \xi_\parallel^2}{d}\, \mathcal{B}_\parallel^2 \tag{13}$$

in the high-frequency limit $\omega\tau_\mathrm{v} \gg 1$ but which acquires a prefactor $\omega\tau_\mathrm{v}$, and therefore vanishes, in the static limit $\omega\tau_\mathrm{v} \ll 1$. The imaginary part of the dielectric function exhibits the lineshape of the derivative of a Lorentzian, thus its measurement would allow convenient extraction of the inter-valley relaxation time τ_v. See Fig. 1 for an illustration of the frequency dependencies exhibited by the real and imaginary parts of the contributions to the conductivity and the dielectric constant due to the magnetoelectric effect.

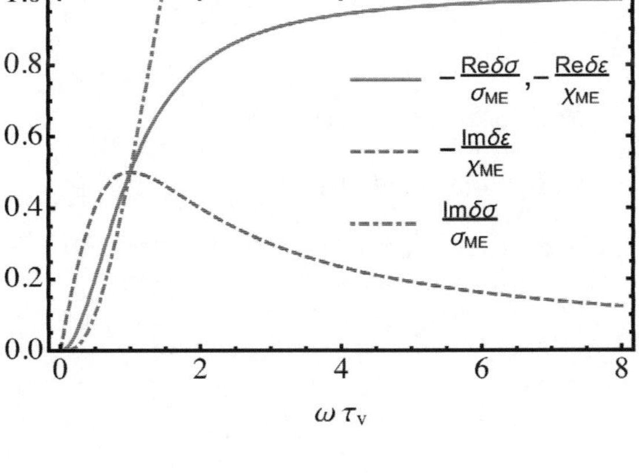

Figure 1. Frequency dependence of the real and imaginary parts of the contribution $\delta\sigma$ to the longitudinal sheet conductivity in BLG and the contribution $\delta\varepsilon$ to the dielectric constant arising due to the valley-density-dependent magnetoelectric effect. τ_v is the inter-valley relaxation time.

5. Discussion and conclusions

Our study has revealed the existence of a longitudinal magnetoconductivity and an associated magnetocapacitance in BLG arising from its valley-density-dependent magnetoelectric effect. More broadly, this behavior will be exhibited also by other materials whose symmetries are similar to BLG. The ability to generate a valley-density imbalance ϱ_{v} by simultaneously applied non-orthogonal electric and magnetic fields constitutes an intriguing analogy between BLG and the Dirac and Weyl semimetals. Furthermore, in both material classes, a magnetic field causes electric current to flow when the valley densities are unequal, but this imbalance must also be time-varying in BLG whereas a current is generated already by a static ϱ_{v} in Dirac and Weyl semimetals. Thus, while the unusual magnetotransport properties emerging in BLG and those associated with the chiral anomaly in Dirac and Weyl semimetals turn out to share certain features, they have a fundamentally different origin and also exhibit different dependencies on physical parameters such as the inter-valley relaxation time.

Acknowledgments

UZ thanks M. S. Fuhrer, I. Martin, and J. E. Moore for useful discussions and acknowledges support from NSF Grant No. PHY11-25915 while at the Kavli Institute for Theoretical Physics. RW was supported by the NSF under grant No. DMR-1310199. Work at Argonne was supported by DOE BES under Contract No. DE-AC02-06CH11357.

References

[1] Fiebig M 2005 *J. Phys. D* **38** R123–52
[2] Ramesh R and Spaldin N A 2007 *Nat. Mater.* **6** 21–9
[3] Khomskii D 2009 *Physics* **2** 20
[4] Kimura T 2012 *Annu. Rev. Condens. Matter Phys.* **3** 93–110
[5] Dong S, Liu J, Cheong S and Ren Z 2015 *Adv. Phys.* **64** 519–626
[6] Zülicke U and Winkler R 2014 *Phys. Rev. B* **90** 125412
[7] Winkler R and Zülicke U 2015 *Phys. Rev. B* **91** 205312
[8] Novoselov K S, Jiang D, Schedin F, Booth T J, Khotkevich V V, Morozov S V and Geim A K 2005 *Proc. Natl. Acad. Sci. USA* **102** 10451–53
[9] Xu M, Liang T, Shi M and Chen H 2013 *Chem. Rev.* **113** 3766–98
[10] Zyuzin A A and Burkov A A 2012 *Phys. Rev. B* **86** 115133
[11] Grushin A G 2012 *Phys. Rev. D* **86** 045001
[12] Son D T and Spivak B Z 2013 *Phys. Rev. B* **88** 104412
[13] Vazifeh M M and Franz M 2013 *Phys. Rev. Lett.* **111** 027201
[14] Burkov A A 2015 *Phys. Rev. B* **91** 245157
[15] Goswami P, Pixley J H and Das Sarma S 2015 *Phys. Rev. B* **92** 075205
[16] Nielsen H and Ninomiya M 1983 *Phys. Lett. B* **130** 389–96
[17] Yang B J and Nagaosa N 2014 *Nat. Commun.* **5** 4898
[18] Vafek O and Vishwanath A 2014 *Annu. Rev. Condens. Matter Phys.* **5** 83–112
[19] Hasan M Z, Xu S Y and Bian G 2015 *Phys. Scr.* **T164** 014001
[20] Xiong J, Kushwaha S K, Liang T, Krizan J W, Hirschberger M, Wang W, Cava R J and Ong N P 2015 *Science* **350** 413–16
[21] Wilczek F 1987 *Phys. Rev. Lett.* **58** 1799–802
[22] Franz M 2008 *Physics* **1** 36
[23] Hehl F W, Obukhov Y N, Rivera J P and Schmid H 2008 *Phys. Lett. A* **372** 1141–46
[24] Qi X L, Hughes T L and Zhang S C 2008 *Phys. Rev. B* **78** 195424
[25] Essin A M, Turner A M, Moore J E and Vanderbilt D 2010 *Phys. Rev. B* **81** 205104
[26] Jackson J D 1999 *Classical Electrodynamics* 3rd ed (Hoboken, NJ: Wiley) Section 7.5 C
[27] Kimura T, Kawamoto S, Yamada I, Azuma M, Takano M and Tokura Y 2003 *Phys. Rev. B* **67** 180401
[28] Singh M P, Prellier W, Simon C and Raveau B 2005 *Appl. Phys. Lett.* **87** 022505

Structural and electronic inhomogeneity of graphene revealed by Nano-ARPES

Chaoyu Chen[1], José Avila[1], Hakim Arezki[2], Fei Yao[3], Van Luan Nguyen[3], Young Hee Lee[3,4], Mohamed Boutchich[2] & Maria C. Asensio[1*]

[1] ANTARES Beamline, Synchrotron SOLEIL, Université Paris-Saclay, L'Orme des Merisiers, Saint Aubin-BP 48, 91192 Gif sur Yvette CEDEX, France.
[2] Group of electrical engineering-Paris, UMR CNRS 8507, CentraleSupélec, Univ. Paris-Sud, Université Paris-Saclay,
Sorbonne Universités, UPMC Univ Paris 06
3 & 11 rue Joliot-Curie, Plateau de Moulon 91192 Gif-sur-Yvette CEDEX, France
[3] IBS Center for Integrated Nanostructure Physics (CINAP), Institute for Basic Science, Sungkyunkwan University, Suwon 440-746, Korea
[4] Department of Energy Science, Sungkyunkwan University, Suwon 440-746, Korea

E-mail: maria-carmen.asensio@synchrotron-soleil.fr

Abstract. Electronic structure describes the distribution of electronic states in reciprocal space, being one of the most fundamental concepts in condensed matter physics, since it determines the electrical, optical and magnetic behaviours of materials. Due to its two-dimensional honeycomb lattice with covalent bonding, pristine graphene exhibits unsurpassed in-plane stiffness and stable structural properties. Here by employing angle-resolved photoemission spectroscopy with spatial resolution $\sim$ 100 nm (Nano-ARPES), we discuss in detail the structural and electronic properties of graphene grown on cooper by chemical vapour deposition (CVD). Our results reveal the spatial inhomogeneity of graphene film, demonstrating the power of Nano-ARPES to detect the microscopic inhomogeneity of electronic structure for different materials.

1. Introduction

Despite its rapid exploration and application, graphene overlayers on metal surfaces still present various phenomena which are far away from the being completely understood[1, 2, 3]. Recently, commercially available cooper foil has become an efficient and inexpensive substrate for scalable growth of large-area graphene films[4, 5]. However, the dramatic variation of graphene quality from case to case implies that the interaction between graphene and Cu can not be simply considered as "weak". This limits the ultimate application of massively produced graphene and raises the inevitable requirement of highly precise electronic structure description.

Although high resolution angle-resolved photoemission spectroscopy (ARPES) has been well established as a powerful technique to probe distinctly the electronic structure of complex materials[6, 7], conventional ARPES has its limitation to describe the impact of the polycrystalline nature on the low energy quasiparticle behaviour of graphene films on copper foils[8, 9]. Here we employ the first ARPES facility with spatial resolution down to sub-micrometer scale (Nano-ARPES or *k-microscope*)[10, 11] to study the graphene grown on Cu foil at both macroscopic and microscopic scales. We will show the phase complexity of graphene overlayer on Cu substrate. This conclusion may stimulate further interest to explore more novel graphene phases based on transition metal substrates.

2. Experimental details

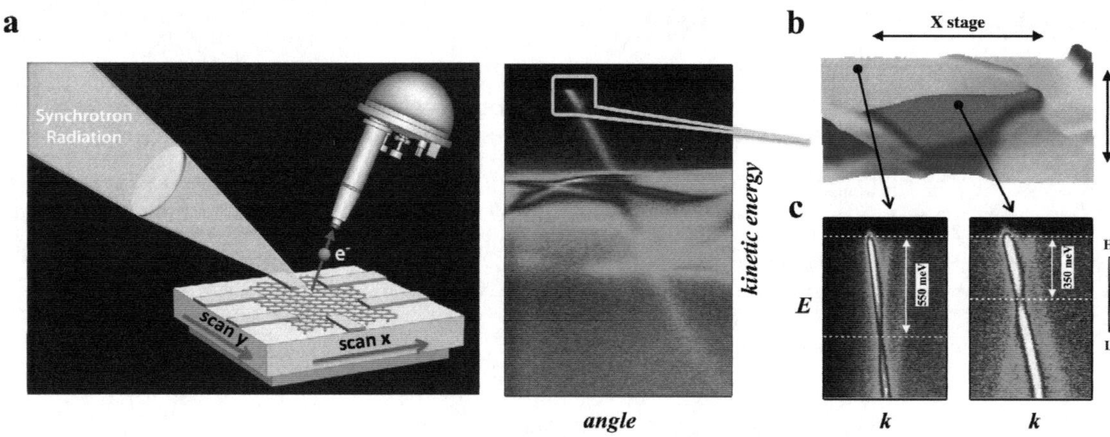

Figure 1. **a,** Working principle of Nano-ARPES. **b,** Graphene Dirac cone spectra intensity distribution image and spatially resolved Dirac cone detailed structure.

To begin with, the working principle of Nano-ARPES is illustrated in Fig. 1a. Fresnel zone plate is used to focus the synchrotron radiation beam into nanometer size. The sample is mounted on a high-precision scanning stage with five degrees of freedom. Piezoscanners allow zero backlash and nanometric accuracy in the linear motion. Interferometers are used to control the simultaneous movement of sample and zone plate. Images of the sample chemical composition and angle-resolved valence-band states can be obtained by synchronized scanning of the sample and collecting the photoelectron by energy analyzer.

As shown in Fig .1b, the spectra intensity distribution image of graphene upper Dirac cone was generated by integrating the corresponding energy-angle window and scan the sample along two in-plane axises. The variation of spectra intensity clearly illustrates the electronic inhomogeneity of graphene on Cu. From any position within the scanning range we can extract or remeasure the detailed electronic structure. As shown the Fig. 1c, the inhomogeneity shown in the image indeed comes from the electronic phase separation. The Dirac cone shown in the left panel, with weak upper Dirac cone intensity actually is highly doped, with the Dirac level lying 550 meV below the Fermi level. For the Dirac cone shown in the right panel with strong intensity, the doping level is 350 meV. This clear contrast not only demonstrates the ability of Nano-ARPES to detect the microscopic electronic inhomogeneity, but also reveal the complexity of seeming simple structure of CVD graphene on Cu.

3. Macroscopic electronic structure of graphene on cooper

The optical image shown in Fig. 2a highlights one particular Cu grain with different contrast (top middle) compared with the surrounding ones. The spatial resolved-capability of Nano-ARPES allow us to locate this grains from the intensity distribution image of Dirac cone (Fig. 2b). The Fermi surface, mapped in normal ARPES mode with spatial resolution $\sim 50\ \mu$m, is shown in Fig. 2c. It is clear that graphene in this area has multiple orientations while one orientation indicated by the red hexagon is dominating. From the Fermi surface we can also see

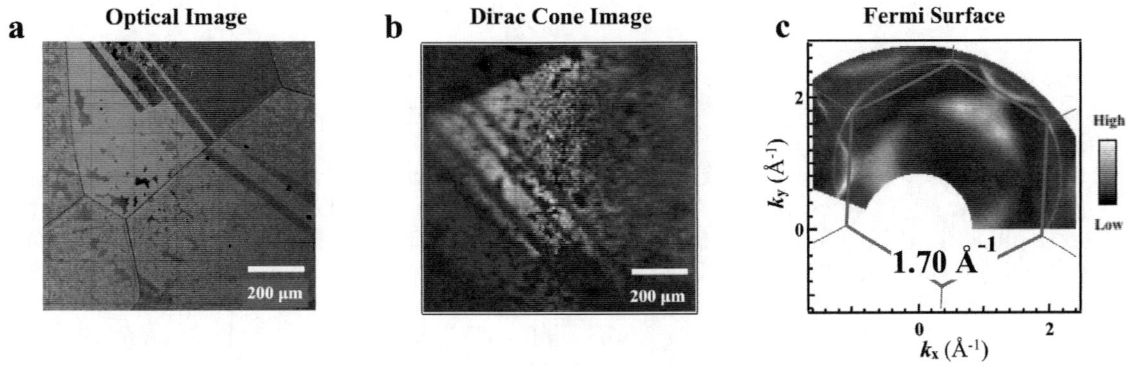

Figure 2. **a,** Optical image of Gr/Cu. **b,** Nano-ARPES real space image of the Dirac cone intensity distribution. Scale bar 200 μm. **c,** Fermi surface of graphene grown on Cu (111) from the same area as shown in **a** and **b**.

the part from Cu *sp* bands, of which the hexagonal symmetry helps us to define that, for this grain, graphene is covering Cu (111) surface. These information gives us the general picture of

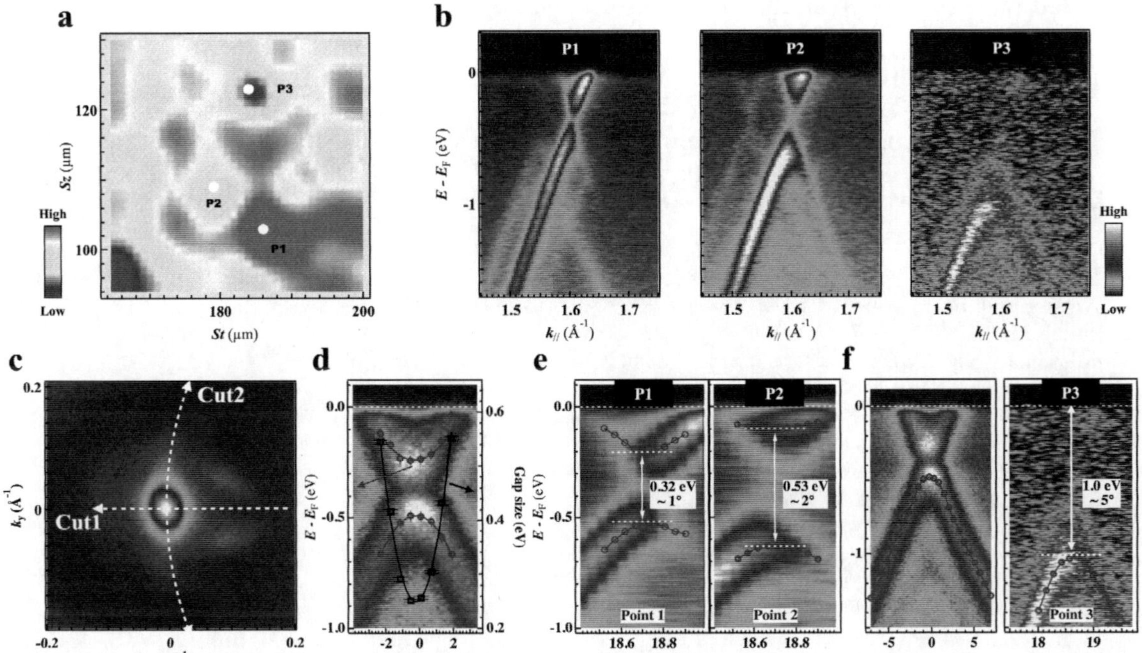

Figure 3. **a,** Dirac cone intensity distribution image. **b,** Detailed electronic structure of graphene Dirac cone corresponding to different points shown in **a**. **c,** Constant energy (Dirac energy) contour for Dirac cone of graphene. **d,** Dispersions extracted from Cut2 shown in **c**. **e,** Detailed analysis of the dispersions shown in **b**. **f,** Left: dispersion extracted from Cut2 shown in **c**. Right: detailed analysis for dispersion from **P3**. Cut1 represent the direction where dispersions shown in **b, e** and right panle of **f** are along.

the sample being studied, allowing further detailed analysis as will be shown in the following.

4. Microscopic electronic structure of graphene on cooper

we now turn to discuss microscopic rotational domains in this area. Fig. 3a shows the upper Dirac cone intensity image in microscopic scale. From the Dirac cone spectra from different positions shown in Fig. 3b, it is clear that the contrast shown in intensity image comes from the orientation change of the graphene. That is to say, while for **P1**, the dispersion is taken across the K point within $\pm 1\ ^o$ accuracy, for **P2** and **P3**, the dispersion cut is slightly away from K point. As the measurement plane is fixed for photoemission process, this angle difference means that graphene in **P2** and **P3** is rotated respect to **P1**.

As analysed in Fig. 3d, by fitting the upper and lower Dirac cone bands along *KK'* direction (Cut2), we obtain the relation between the gap size and rotation angle, indicated by the black curve. From this relation, we can determine the relative rotation angle to be $\sim$ 2 and 5 ^{o} for **P2** and **P3** respectively. In reference[12], we prove that the graphene lattice structure changes dramatically with respect to the rotational angle, due to the interaction with the underlying Cu. Here, by microscopically show the existence of several rotational domains , we may expect very different structural behaviour for graphene.

By exploring the electronic structure of graphene on copper foil from both macroscopic and microscopic scale, we demonstrated the phase complexity of CVD graphene on Cu (111). Detailed Nano-ARPES analysis shows the existence of graphene domains with different rotational angle and possible structural change, deserving further detailed exploration by advanced tools such as Nano-ARPES, microRaman, STM and so on.

5. Acknowledgments

The Synchrotron SOLEIL is supported by the Centre National de la Recherche Scientifique (CNRS) and the Commissariat à l'Energie Atomique et aux Energies Alternatives (CEA), France. We thank Dr D. Alamarguy and Stephane Lorcy for the technical support, Dr A. Ouerghi, Dr Q. S. Wu, Dr Yue Chen for helpful discussions.

6. References

[1] Wintterlin J and Bocquet M L 2009 *Surface Science* **603** 1841-1852
[2] Batzill M 2012 *Surface Science Reports* **67** 83-115
[3] Voloshina E and Dedkov Y *Phys. Chem. Chem. Phys.* **14** 13502-13514
[4] Li X et al 2009 *Science* **324** 1312-1314
[5] Yu Q et al 2011 *Nature Materials* **10** 443-449
[6] Damascelli A, Hussain Z and Shen Z X 2003 *Rev Mod Phys* **75** 473-541
[7] Lu D et al 2012 *Annual Review of Condensed Matter Physics* **3** 129-167
[8] Avila J et al 2013 *Scientific Reports* **3** 8
[9] Walter A L et al 2011 *Physical Review B* **84** 195443
[10] Avila J and Asensio M C 2014 *Synchrotron Radiation News* **27** 24-30.
[11] Avila J et al 2013 *J. Phys.: Conf. Series* **425** 192023
[12] Chen C Y et al 2016 *submitted*

Charge mobility modification of semiconducting carbon nanotubes by intrinsic defects

Hongcun Bai* [1, 2], **Yujia Ma** [1, 2], **Jinsuo Ma** [1, 2], **Jingnan Mei** [1, 2], **Yan Tong** [1, 2], **Yongqiang Ji** [1, 2]

1) State Key Laboratory Cultivation Base of Natural Gas Conversion, Ningxia University, Yinchuan, Ningxia 750021, China

2) School of Chemistry Science and Engineering, Ningxia University, Yinchuan, Ningxia 750021, China

E-mail: hongcunbai@Gmail.com, hongcunbai@nxu.edu.cn

Abstract. Charge carrier mobility is a central transport property in nanoscale electronics. Carbon nanotubes (CNTs) are supposed to have high carrier mobility. The preparation methods of CNTs have been greatly improved, but the defects always exist. This work presented first-principle investigations on the charge carrier mobility of carbon nanotubes containing several intrinsic defects. The charge carrier mobilities of zigzag (10, 0) tubes with Stone–Wales, mono vacant and 5/8/5 defects were studied as an example to explore the role of defects. Most carrier mobilities were decreased, but several values of mobility are unexpectedly increased upon the appearance of the defects. This interesting result is discussed based on the changes of the stretching modulus, the effective mass of the carrier and deformation potential constant induced by the defects.

Keywords: CNTs, Intrinsic Defects, Carrier Mobility, Charge Transport, DFT

1. Introduction

Charge carrier mobility is a central transport property in nano-scaled electronics. As one of the exiting carbon forms, carbon nanotubes (CNTs) are good candidates for the next generation electronics due to high carrier mobility [1-2].

CNTs commonly are fabricated by using various methods, such as arc discharge, chemical vapors deposition and laser vaporization [3]. However, the accuracy of material processing at the atomic scale is still a great challenge currently. The preparation methods are improved a lot, but different kinds of defects always exist in any materials. So defect is also unavoidably in the obtained samples of CNTs [4-6].

It is known that the electronic properties of the materials can be modulated by the geometrical structures. This is also true for CNTs. For instance, the tube diameter, chirality, substitutions, physical or chemical adsorptions and encapsulations are important for tailoring the electronic structures of

CNTs [7-11]. Previous studies also investigated the CNTs containing intrinsic defects, and revealed the impact of the defects on the relative stabilities, electronic properties and young's modulus [12-14]. However, up to date, few understanding is available on the carrier mobility of CNTs induced by defects according to our best knowledge. Thus this paper detailed studies on the carrier mobility of semiconducting zigzag CNTs induced by intrinsic defects are investigated from first-principle point view.

2. Models and Computational Details

The charge carrier mobility of the CNTs with and without defects is studied under the framework of deformation potential (DP) theory and effective mass approach [15-18]. The carrier mobility μ of 1D crystal can be expressed as:

$$\mu = \frac{e\hbar^2 C}{(2\pi k_B T)^{1/2} \left|m^*\right|^{3/2} E_1^2} \tag{1}$$

where C is the stretching modulus of 1D crystal, m^* is the effective mass of charge carriers and E_1 is DP constants. The detailed descriptions can be seen from our previous studies [17, 18]. It should be aware that only longitudinal acoustic (LA) phonon scattering is considered in the models adopted here. Other scattering mechanisms, such as optical phonon scattering, inter-subband scattering and radial-breathing phonon scattering that may also limit the charge mobility, are not addressed in this work [19].

First-principle studies are conducted based on the Kohn-Sham density functional theory (DFT) combined with self-consistent field crystal orbital (SCF-CO) method. We have chosen hybrid B3LYP method because it is more accurate to explore various carbon-based systems and the methods used here could give rather good results compared with those obtained by various different functional and basis sets according to the previous calculations [20]. All calculations are performed by using the linear combination of atomic orbital (LCAO) technique with CRYSTAL2014 code. A double-ζ plus polarization basis set 6-21G (*d*, *p*) implemented in the CRYSTAL program for solid-state calculations and 41 k-points sampling in the first Brillouin zone were adopted in the SCF-CO calculations.

3. Results and Discussion

It is known that the structure and physical property of SWCNT mainly depend on the diameter as well as chirality. We limit to semiconducting CNTs with zigzag edges in this work, following previous studies [2]. The armchair tubes are not considered since they all exhibit metallic property. The semiconducting zigzag CNT (10, 0) was selected as a case study to explore the carrier mobility induced by defects. A super-cell was adopted to avoid interaction between adjacent defects; thus, each defect is presented in three unit cells of CNT (10, 0). The obtained structures with fully geometrical optimizations of pristine CNT (10, 0) and four defective structures are shown in **Figure 1**.

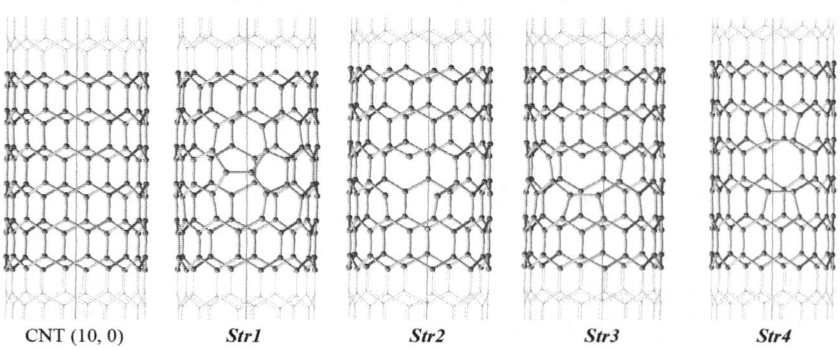

Figure 1. The structures of pristine CNT (10, 0) and four defective tubes.

Among them, ***Str1*** is the model containing Stone–Wales (SW) defect. A SW defect occurs when one C–C bond rotates 90° and forms two 5-membered rings and two 7-membered rings. It is also the most commonly observed defects in carbon-based materials with *sp2* hybridization, such as fullerenes, CNTs and graphene. ***Str2*** is the original mono vacant defected structure. It can be obtained by removing one carbon atom from the hexagonal network directly, with three unsaturated bonds left. ***Str3*** can be seen as the configurations derived from ***Str2*** by geometrical reconstructions. ***Str4*** is the 5/8/5 defect, which can be seen as the structure derived from double vacant defect by geometrical reconstructions. To confirm the stability of these defective structures, molecular dynamics (MD) simulations are conducted by using the density functional based tight binding (DFTB) method. We performed MD simulations at 300 K with each for 3 ps under the constant-temperature constant-volume condition with the time step size as 1 fs. It is found that the geometries of these tubes with mono-vacant or related defects are well kept according to the snapshot images of the equilibrium structure at the end of each MD simulation as well as the potential energies with time evolution. Thus there is no sign of these structures studied here according to the MD simulations.

The calculated electronic band structures are plotted in **Figure 2**. We can see that pristine (10, 0) tube is a semiconductor with a direct band gap of 1.064 eV, which agrees quite well with previous DFT calculations [7]. As the defect is introduced into the tube, it is observed that the semiconducting property is well maintained for all the defective tubes. Nevertheless, the obtained band gaps are all obviously reduced compared with that of pristine tube.

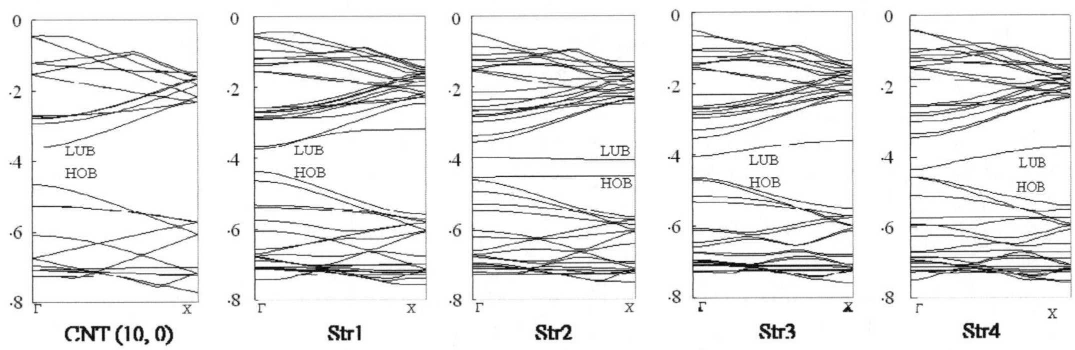

Figure 2. Band structures of pristine CNT (10, 0) and four defective tubes.

The calculated electron and hole mobilities for the semiconducting tubes at room temperature are listed in **Table 1**. The electron and hole mobilities (μ_e and μ_h) of the pristine (10, 0) tube were calculated to be 1.096×10^4 and 1.390×10^3 $cm^2V^{-1}s^{-1}$, respectively, which agree well with previous DFT calculations by Wang and co-workers [7]. Furthermore, the obtained electron mobility is one order larger than the hole. Thus, electron is the majoring carrier, for which the transport is more favorable in pristine (10, 0) tube than hole carrier.

Table 1. Stretching modulus (C, in eV/Å), deformation potential constant (E_1, in eV), effective mass of carrier (m^*, in m_e^0) and mobility (μ, in $cm^2V^{-1}s^{-1}$) for defective and pristine tubes.

tubes	C	E_{1V}	E_{1C}	m_h^*	m_e^*	μ_h	μ_e
CNT(10, 0)	573	11.907	4.242	0.0812	0.0812	1.390×10^3	1.096×10^4
Str1	560	11.592	3.429	0.0653	0.0663	1.986×10^3	2.219×10^4
Str2	558	3.641	1.465	2.604	1.252	7.971×10^1	1.477×10^3
Str3	563	11.374	1.469	0.0882	0.1265	1.322×10^3	4.614×10^4
Str4	547	14.150	13.116	0.3026	0.1149	1.300×10^2	6.490×10^2

Now we turn to the tubes containing defects. It is noticed that carrier mobility is mainly determined by three parameters according to equation, the stretching modulus (C), the effective mass of the carrier (m^*) and DP constant. Firstly we focus on the stretching modulus of the tubes upon the incorporation of the vacancy or related defects. From **Table 1**, it is found that the obtained the stretching modules of the defective tubes are in the range of 547-563 eV/Å, and a bit smaller than that of the pristine CNT (10, 0). Thus the charge mobility of the CNTs influenced by the mechanical property of CNTs induced by defects is not significant, despite the structure of the different defects. Secondly, we turn to the deformation potential constant. E_{1V} and E_{1C} of pristine CNT (10, 0) are 11.907 and 4.424 eV, respectively. For E_{1V}, it is found that E_{1V} of ***Str2*** is greatly reduced, while others are almost unchanged. For E_{1C}, the values of ***Str2*** and ***Str3*** are reduced, while that of ***Str4*** is increased obviously. Thus these variations induced by different defects would affect the charge carrier transport of the CNTs according to equation (1). For instance, the increased E_{1C} of ***Str4*** is unfavoured to the electron mobility. Thirdly, we pay attention to the effective mass of charge carriers. m_h^* and m_e^* of pristine CNT (10, 0) are both 0.0812 m_e^0. As the defect is introduced, it is found that m_h^* and m_e^* of ***Str1*** are both decreased to about 0.066 m_e^0. As to other defective structures, the effective mass of charge carriers are all increased. In particular, m_h^* and m_e^* of ***Str2*** are 2.604 and 1.252 m_e^0, which are enhanced by two orders of magnitude. It is noticed that the increased effective mass would reduce the charge carrier mobility according to equation (1). Thus the variation of effective mass induced by different defects would also affect the charge carrier transport of the CNTs.

The calculated electron and hole mobilities for the defective tubes at room temperature are listed in **Table 1**. The hole mobilities (μ_h) of ***Str2*** and ***Str4*** are calculated to be 7.971×10^1 and 1.300×10^2 $cm^2V^{-1}s^{-1}$, respectively, while electron mobilities (μ_e) of are 1.477×10^3 and 6.490×10^2 $cm^2V^{-1}s^{-1}$, respectively. Thus both electron and hole mobilities of ***Str2*** and ***Str4*** are decreased by one or two orders upon the defects incorporation. Obviously these two defects would limit the carrier transport of CNT (10, 0). As is expected, this fact comes without much surprise. Furthermore, it is aware that that hole mobility of ***Str3*** is 1.322×10^3 $cm^2V^{-1}s^{-1}$, which is very close to that of pure (10, 0) tube. As a matter of fact, the change on the hole mobility induced by ***Str3*** defect is less than 5%. However, it is interesting that μ_h and μ_e of the tube with SW defects are calculated to be 1.986×10^3 and 2.219×10^4 $cm^2V^{-1}s^{-1}$, respectively. μ_e of ***Str3*** is 4.614×10^4 $cm^2V^{-1}s^{-1}$. These values are as high as about 1.5-4.2 times compared with that of the pristine (10, 0) tube. Thus unexpectedly some defects would increase the mobility of the tube. This is a very interesting result. On the basis of our common sense, defects usually are supposed to block the movement of carriers, leading to a decreased mobility. However, theoretical predictions based on DFT studies in this work reveal that the carrier mobilities of zigzag CNTs with some intrinsic defects can be unexpectedly increased compared with those of the pristine tubes. Further understanding with more insights about the roles of the vacancy or related defects on the charge carrier mobility of CNTs is still on going in our research group.

4. Conclusion

The carrier mobilities of zigzag CNT (10, 0) tube with SW, single vacant and 5/8/5 defects are investigated and compared with those of the pristine tube by using first-principle calculations in this work. It is found that the semiconducting properties of the tubes can be well preserved upon the appearance of the different intrinsic defects. It is found that both electron and hole mobilities of ***Str2*** and ***Str4*** are decreased by one or two orders due to the defects. The hole mobility of ***Str3*** is very close to that of pure (10, 0) tube. However, μ_h and μ_e of the tube with SW defects, as well as μ_e of ***Str3*** are obviously increased compared with that of the pristine (10, 0) tube.

Acknowledgements

This work is mainly supported by National Natural Science Foundation of China (21363017). The authors J. Ma, J. Mei and Y. Tong appreciate the financial support from Training Program of Innovation and Entrepreneurship for College Students of Ningxia (Q201610749038).

References

[1] T. Dürkop, B. M. Kim, M. S. Fuhrer. 2004 *J. Phys.: Condensed Matter* **16** R553–R580.
[2] B. Xu, Y. D. Xia, J. Yin, X. G. Wan, K. Jiang, A. D. Li, D. Wu, Z. G. Liu. 2010 *Appl. Phys. Lett.* **96** 183108.
[3] C. Wang, K. Takei, T. Takahashi, A. Javey. 2013 *Chem. Soc. Rev.* **42** 2592–2609.
[4] Chen, C., Hayazawa, N. & Kawata, S. 2014 *Nature Commun.* **5**, 3312.
[5] Hashimoto, A., Suenaga, K., Gloter, A., Urita, K.; Iijima, S. 2004 *Science* **430** 870–873.
[6] Lin, W. & Moon, K. 2010 *ACS Nano* **4** 1716.
[7] G. Wang, Y. Huang. 2008 *J. Phys. Chem. Solids* **69** 2531–2534.
[8] H. Bai, W. Qiao, Ying Zhu and Y. Huang. 2015 *Curr. Appl. Phys.* **15** 342–351.
[9] W. Qiao, H. Bai, Y. Zhu and Y. Huang. 2012 *J. Phys.: Condens. Matter* **24** 185302.
[10] L. Muzi, F. Tardani, C. L. Mesa, A. Bonincontro, A. Bianco and G. Risuleo. 2016 *Nanotechnol.* **27**, 155704.
[11] P. Lou. 2015 *RSC Adv.* **5** 55458–55467.
[12] F. Ding. 2005 *Phys. Rev. B* **72** 245409.
[13] S. Sharma, R. Chandra, P. Kumar, N. Kumar. 2014 *Comput. Mater. Sci.* **86** 1–8.
[14] J. Kotakoski, A. V. Krasheninnikov, and K. Nordlund. 2006 *Phys. Rev. B* **74** 245420.
[15] Y. Huang, R. Liu. 1991 *Chem. Res. Chi. Univ.* **7** 107–113.
[16] F. B. Beleznay, F. Bogár, J. Ladik. 2003 *J. Chem. Phys.* **119** 5690.
[17] H. Bai, Y. Ai and Y. Huang. 2011 *Phys. Status Solidi B* **248** 969–973.
[18] Y. Zhu, H. Bai, Y. Huang. 2016 *J. Phys.: Condens. Matter* **28** 045303.
[19] Jinyang Xi, Dong Wang, Zhigang Shuai. 2015 *WIREs: Comput. Mol. Sci.* **5** 215–227.
[20] H.-W. Wang, B.-C. Wang, W.-H. Chen, and M. Hayashi. 2008 *J. Phys. Chem. A* **112** 1783.

IOP Conf. Series: Journal of Physics: Conf. Series **864** (2017) 012031 doi:10.1088/1742-6596/864/1/012031

Imaging Electron Motion in a Few Layer MoS_2 Device

S Bhandari[1], K Wang[2], K Watanabe[3], T Taniguchi[3], P Kim[1,2] and R M Westervelt[1,2]

[1]School of Engineering and Applied Sciences, Harvard University, Cambridge, MA 02138, U.S.A.
[2]Department of Physics, Harvard University, Cambridge, MA 02138, U.S.A
[3]National Institute for Materials Science, 1-1 Naniki, Tsukuba, 305-0044, Japan

Email: sbhandar@fas.harvard.edu

Abstract. Ultrathin sheets of MoS_2 are a newly discovered 2D semiconductor that holds great promise for nanoelectronics. Understanding the pattern of current flow will be crucial for developing devices. In this talk, we present images of current flow in MoS_2 obtained with a Scanned Probe Microscope (SPM) cooled to 4 K. We previously used this technique to image electron trajectories in GaAs/AlGaAs heterostructures and graphene. The charged SPM tip is held just above the sample surface, creating an image charge inside the device that scatters electrons. By measuring the change in resistance ΔR while the tip is raster scanned above the sample, an image of electron flow is obtained. We present images of electron flow in an MoS_2 device patterned into a hall bar geometry. A three-layer MoS_2 sheet is encased by two hBN layers, top and bottom, and patterned into a hall-bar with multilayer graphene contacts. An SPM image shows the current flow pattern from the wide contact at the end of the device for a Hall density $n = 1.3\text{x}10^{12}$ cm $^{-2}$. The SPM tip tends to block flow, increasing the resistance R. The pattern of flow was also imaged for a narrow side contact on the sample. At density $n = 5.4\text{x}10^{11}$ cm^{-2}; the pattern seen in the SPM image is similar to the wide contact. The ability to image electron flow promises to be very useful for the development of ultrathin devices from new 2D materials.

1. Introduction

With the discovery of two-dimensional (2D) materials, Scanning Probe Microscope (SPM) imaging has become an important tool in shedding light on the physics of devices made from these devices at the nanoscale. SPM imaging technique has been used to image electron trajectories in GaAs/AlGaAs heterostructures [1-9] and graphene [10,11]. A recently discovered material is an ultrathin sheet of MoS_2 which forms a 2D semiconductor. MoS_2 displays unique electronic properties such as thickness-dependent band structure [13-15], spin-valley physics, and valley Hall effects [16-18]. It also shows a good carrier mobility, compared with other 2D semiconductors [15,18-22]. At liquid-He temperatures, the mobility in MoS_2 is limited, as a result of scattering by surface impurities and the substrate [21]. Little is known about the sources of scattering and disorder in MoS_2. It is speculated that intrinsic factors such as lattice defects and extrinsic factors such as charged impurities and substrate adsorbates contribute to the scattering [20, 23]. Atomically flat hexagonal boron nitride (hBN) substrate and top layers greatly improve the mobility in graphene [24] and MoS_2 [20]. However, the mobility in hBN-encapsulated MoS_2 is still lower than the theoretically predicted value [20]. Recent reports have suggested that interfacing MoS_2 with multilayer graphene greatly lowers the contact resistance [20,25].

IOP Conf. Series: Journal of Physics: Conf. Series **864** (2017) 012031 doi:10.1088/1742-6596/864/1/012031

In this paper, we use a cooled SPM to image electron flow in a three-layer MoS_2 sheet. The MoS_2 device is encapsulated by hBN layers on the top and bottom and patterned into a hall-bar with multilayer graphene contacts. Using the SPM, we observe images of current flow in MoS_2 cooled to 4.2 K. A charged SPM tip held above the sample surface, creates an image charge inside the MoS_2 device that scatters electrons. By measuring the change in resistance ΔR while the tip is raster scanned above the sample, an image of electron flow is obtained. The SPM images show an electron flow pattern that peaks at the center of a contact and decays exponentially into the device. The SPM images were taken at two different contacts – a wide contact at the end of the device and a narrow contact along the side. The electron density n inside the device was measured using the Hall effect. For the wide contact at $n = 1.3 \times 10^{12}$ cm^{-2}, the SPM image shows a partially blocked region in the middle of the contact that exponentially decays with characteristic length is $L = 250\,nm$. Similarly, for the narrow contact at density $n = 5.4 \times 10^{11}$ cm^{-2}, a partially blocked region occurs with exponential decay with $L = 200\,nm$. The decay lengths are compared with the mean free path from a simple 2D electron gas model of transport.

2. Method

Figure 1(a) shows a schematic diagram of the set up used to image electron flow in the MoS_2 device with a cooled scanning probe microscope (SPM). The Hall bar is patterned from a hBN/MoS_2/hBN sandwich. The dimensions are 11.0x5.0 μm^2, with two narrow (width 1 μm) contacts along each side separated by 3.0 μm center-to-center, and wide contacts (width 3 μm) at either end. To improve the contact resistance, each of the contacts is interfaced with multilayer graphene which is then contacted by metal electrodes. The heavily doped Si substrate acts as a back-gate, covered by a 285 nm insulating layer of SiO_2. The back-gate capacitance is $C_G = 11.5$ nF. The density n can be tuned to be in the conductive or insulating regime by applying an appropriate gate voltage V_G between the backgate and the MoS_2. The density n is obtained from measurements of the Hall resistance $R_H = 1/ne$ where e is the electron charge.

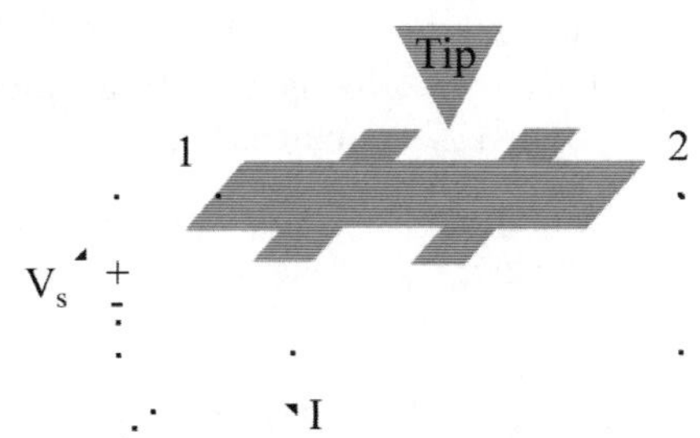

Figure 1: Experimental setup. Schematic diagram illustrating the scanning gate microscope tip above the MoS_2 device to image electron flow.

In our measurements, a 20 nm diameter SPM tip is scanned at a constant height above the MoS_2 device while the conductance of the sample is measured. As shown in Fig. 1, a dc voltage V is applied between the two wide contacts while current, while the resulting current I between them is measured. For this paper the electron density of the device is tuned by the backgate voltage to be in the conductive regime. An image of electron flow is obtained by displaying the change in resistance ΔR of the device as the tip is raster scanned above the sample.

3. Results and Discussion

SPM images of the MoS_2 device are shown in Fig 2(b) a.nd 2(c). Figure 1(a) shows an optical image of the hall bar device. The white outlines in Fig. 2(a) – a square and an ellipse - indicate the region of image scans. Figure 2(b) shows the resistance change ΔR as the tip is scanned over the elliptical region

IOP Conf. Series: Journal of Physics: Conf. Series **864** (2017) 012031 doi:10.1088/1742-6596/864/1/012031

shown in Fig. 2(a) in the wide contact with electron density n = 1.3 x 10^{12} cm^{-2}. The black solid lines denote the edges of the contact. Similarly, Fig. 2(c) shows the resistance change ΔR as the tip is scanned over the square region for the narrow side contact shown in Fig. 2(a). The resistance change ΔR pattern is similar to the wide contact with resistance change peaking at the center and decaying outward. The electron density was tuned to n = 5.4 x 10^{11} cm^{-2}.

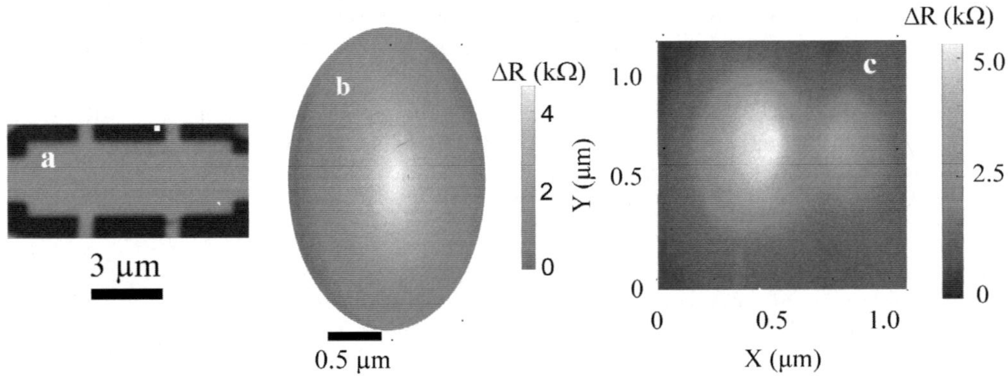

Figure 2: (a) Optical image of MoS_2 sample. (b) Cooled SPM image at 4.2K showing the current flow pattern through the wide contact at n = 1.3x10^{12} cm^{-2}, (c) Cooled SPM images showing current flow through the narrow contact at n = 5.4x10^{11} cm^{-2}.

A line cut of the resistance change ΔR through the middle of the partially blocked region of Fig. 2(b) is plotted in Fig. 3(a). The data is a semi-log plot of ΔR *vs*. the depth X into the device from the contact. As shown in Fig. 3(a), ΔR drops exponentially from the end of the contact into the device with a decay length $L_1 \approx 250\,nm$. Similarly, Fig. 3(b) is a semi-log plot of ΔR *vs*. depth X into the device for the narrow contact for Fig. 2(c). The resistance change also drops exponentially, with decay length $L_2 \approx 200\,nm$.

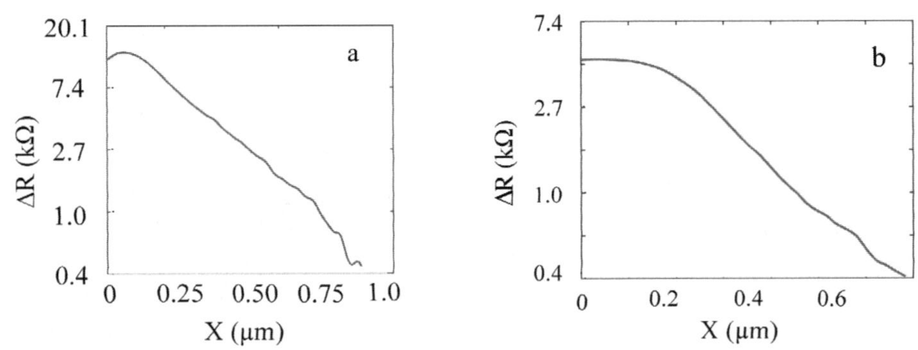

Figure 3: Semi-log plots of resistance change ΔR vs. depth X into the device from the end of the contact. The plots are line cuts through the middle of the partially blocked region in Fig. 2 for (a) wide contact and (b) narrow contact. ΔR decays exponentially into the device.

IOP Conf. Series: Journal of Physics: Conf. Series **864** (2017) 012031 doi:10.1088/1742-6596/864/1/012031

To interpret these images, we compare the exponential fall off observed in the MoS_2 device with the mean free path estimated using a simple model. The mean free path for electrons L_m is

$$L_m = v_F \tau \quad (1)$$

For a 2D free electron gas, the Fermi wavenumber k_F is

$$k_F = (m^*/\hbar) v_F = (2\pi n)^{1/2} \quad (2)$$

Using Eq. 1 and 2, we find the 2D sheet resistance ρ

$$1/\rho = (e^2/h) k_F L_m \quad (3)$$

and the mean free path

$$L_m = \left(\frac{h}{e^2}\right) \frac{1}{\rho (2\pi n)^{1/2}} \quad (4)$$

For the wide contact the density and sheet resistance in Hall measurements are $n = 1.3\text{x}10^{12}$ cm^{-2} and $\rho = 2{,}250\ \Omega$. The estimated mean free path $L_m = 40$ nm is well below the measured decay length $L_l = 250$ nm. Similarly, for the narrow contact with $n = 5.4\text{x}10^{11}$ cm^{-2} and $\rho = 5{,}130\ \Omega$ the estimated mean free path $L_m = 27$ nm is below the measured decay length $L_l = 200$ nm.

The estimated mean free paths are based on n and R measurements averaged over the whole area of the device. The discrepancy between the measured decay length and the mean free path may be due to changes in the local density n and density of states near the MoS_2 layer edge and inside the contact. The band structure of ultrathin MoS_2 sheets changes with thickness [13-15] and additional states are expected at the edge. The energy bands are very sensitive to strain [26,27] which may be present in the contact regions due to the structure of the encapsulated hBN/MoS_2/hBN device near the contacts [28, 29]. In addition, the surface states at the edge of MoS_2 could change the charge density along the edge [30]. These effects could cause the mean free path to vary across the sample and differ from their average value near the contacts.

4. Conclusion

A cooled scanning probe microscope allows us to locally investigate the electronic motion in nanoscale structures. Recently discovered 2D materials such as MoS_2 offer new approaches to electronics and photonics, as well as an opportunity to probe physics in low-dimensional systems. Several unique electronic properties such as thickness/strain dependent band structure [13-15], spin-valley physics and valley Hall effects have been predicted in MoS_2. However, the carrier mobility has been limited to moderate values. Using our cooled SPM, we are able to locally probe electron motion at the nanoscale, inside an ultrathin MoS_2 device. We imaged electron motion close to two contacts, discovering a region where the contact joins the channel where the tip partially blocks electron flow. The resistance change ΔR drops exponentially with tip position as the tip moves into the sample, with a decay lengths 250 nm and 200 nm for the wide and narrow contacts. A simple model to estimate the mean free path L_m using values of n and R for the device finds smaller values. The electronic density and band structure of MoS_2 are expected to vary near the edge and near the contacts, offering a possible resolution.

Acknowledgements

The SPM imaging research and the ray-tracing simulations were supported by the U.S. DOE Office of Basic Energy Sciences, Materials Sciences and Engineering Division, under grant DE-FG02-07ER46422. MoS_2 sample fabrication was supported by Air Force Office of Scientific Research contract FA9550-14-1-0268 and Army Research Office contract W911NF-14-1-0247. Growth of hexagonal boron nitride crystals was supported by the Elemental Strategy Initiative conducted by the MEXT, Japan and a Grant-in-Aid for Scientific Research on Innovative Areas No. 2506 "Science of Atomic Layers" from JSPS. Nanofabrication was performed in the Center for Nanoscale Systems (CNS) at Harvard University, a member of the National Nanotechnology Coordinated Infrastructure Network (NNCI), which is supported by the National Science Foundation under NSF award ECCS-1541959.

References

[1] Topinka M A et al 2001 *Nature* **410** 183
[2] Tessmer S H et al 1998 *Nature* **392** 51
[3] Yacoby A et al 1999 *Sol. State Comm.* **111** 1-13
[4] Steele G A et al *Phys. Rev. Lett.* **95** 136804
[5] Topinka M A et al 2000 *Science* **289** 5488
[6] LeRoy B J et al 2002 *Appl. Phys. Lett.* **80** 23
[8] Aidala K E et al 2007 *Nat. Phys.* **3** 464
[9] Fallahi P et al 2005 *Nano Lett.* **5** 223
[10] Martin J et al 2008 *Nat. Phys.* **4** 144-148
[11] Bhandari S et al 2016 *Nano Lett.* **16** 3
[12] Fallahi P et al 2005 *Nano Lett.* **5** 223
[13] Lee C H et al 2014 *Nat. Nanotech.* **9** 676-681
[14] Mak K F et al 2010 *Phys. Rev. Lett.* **105** 136805
[15] Kim S et al 2012 *Nat. Comm.* **3** 1011
[16] Mak K et al 2014 *Science* **344** 1489-1492
[17] Mak K F et al 2012 *Nat. Nanotech.* **7** 494-498
[18] Xiao D et al 2012 *Phys. Rev. Lett.* **108** 196802
[19] Baugher B et al 2013 *Nano Lett.* **13** 4212-4216
[20] Cui X et al 2015 *Nat. Nanotechnol.* **10** 6
[21] Qiu H et al 2012 *Appl. Phys. Lett.* **100** 123104
[22] Ayari A et al 2007 *J. Appl. Phys.* **101** 014507
[23] Radisavljevic B 2013 *Nat. Mat.* **12** 815-820
[24] Dean C R et al 2010 *Nat. Nano.* **5** 722-726
[25] Yu L et al 2014 *Nano Lett.* **14** 3055-3063
[26] Feng J et al 2012 *Nat. Phot.* **6** 866-872
[27] Ghorbani M et al 2013 *Phys. Rev. B.* **87** 235434
[28] Singh S K et al 2013 *Phys. Rev B.* **87** 184106
[29] Murray R et al 1979 *J. Appl. Cryst.* **12** 312-315
[30] Zhang C et al 2014 *Nano Lett.* **14** 2443-2447

IOP Conf. Series: Journal of Physics: Conf. Series **864** (2017) 012032 doi:10.1088/1742-6596/864/1/012032

Synthesis of bilayer MoS_2 and corresponding field effect characteristics

Mingxu Fang, Yulin Feng, Fang Wang, Zhengchun Yang, Kailiang Zhang*

School of Electronic Information Engineering, Tianjin Key Laboratory of Thin Film Electronics and Communication Devices, Tianjin University of Technology, 391 West Binshui Road,Tianjin,China,300384.

Mingxu Fang and Yulin Feng are co-first authors; they contributed equally to the work.

Corresponding Author 's Email: kailiang_zhang@163.com

Abstract: Two-dimensional transition-metal dichalcogenides such as MoS_2 are promising materials for next-generation nano-electronic devices. The physical properties of MoS_2 are determined by layer number according to the variation of band-gap. Here, we synthesize large-size bilayer-MoS_2 with triangle and hexagonal nanosheets in one step by chemical vapor deposition, Monolayer and bilayer-MoS_2 back-gate field effect transistors are also fabricated and the performance including mobility and on/off ratios are compared. The bilayer-MoS_2 back-gate field effect transistor shows superior performance with field effect mobility of ~$21.27 cm^2V^{-1}s^{-1}$, and I_{on}/I_{off} ratio of ~3.9×10^7.

1. Introduction

With the miniaturization of device critical size in integrated circuit, traditional silicon-based semiconducts have ran into their bottleneck, Thus it is urgent to find novel materials to replace and continuing improve the performance of the device. The emergence of two-demensional(2D) material including grapheme, boron nitride and transition-metal dichalcogenides have attracted extensive attention in electronic and optoelectronic devices duo to the ultrathin body and special physical property. MoS_2,as one of most potential material in TMDs, has been widely studied due to the suitable band-gap range from 1.2eV-1.9eV from monolayer to bulk form[1,2]. At this condition, preparation of high quality MoS_2 with different layer numbers become particularly crucial for its applications. Thus far, chemical vapor deposition is one of the most effective strategy to synthesize high quality, uniform and large-scale monolayer MoS_2 [3-5]. However, there has few study on controllable growth of bi-layer or few-layer MoS_2 nanosheets, which limits their application in demanded fields.

In this work, We obtained large-size(~40μm), uniform and high quality bilayer-MoS_2 by chemical vapor deposition in one step. The electrical performance of monolayer and bilayer MoS_2 based back-gate field effect transistors(FETs) are compared. The results indicate that bilayer-MoS_2 FET shows superior performance with field effect mobility of ~21.27$cm^2V^{-1}s^{-1}$ and I_{on}/I_{off} of ~3.9×10^7.

2. Experiment

2.1 Synthesis of bilayer MoS_2 and fabrication of MoS_2-FETs

In this work, MoS_2 bilayer was synthesized on 275nm SiO_2/Si with MoO_3 and S as precursors by using a two zones tube furnace. The substrates were treated by piranha solution (H_2SO_4:H_2O_2=3:1) for 2h to remove the impurity, followed by sonication in acetone and isopropanol for 10 minutes. 10mg MoO_3 was placed at the center zone in high temperature with substrate upper side, and 20mg S was located upstream in a low temperature zone. Pump the furnace to 30mTorr,then fill with Ar and H_2 (500:50sccm) to ambient pressure, After reducing the total flow to 55sccm(Ar:H_2=50:5),the center zone gradually increases to 750℃ with a heating rate of 15℃/min, and S increases to 150℃ when the center zone reached 600℃,the growth process is conducted at 750℃ for 3 minutes.

The as-synthesized MoS_2 is transferred onto a 300nm SiO_2/Si(p^{++}) substrate by etching the SiO_2 with HF solution.

MoS_2 back-gate field effect transistors are fabricated by electron beam lithography (EBL, Raith e-LINE plus). Au(100nm)/Ti(20nm) are deposited as the source and drain electrodes. And then annealed in Ar atmosphere at 300℃ for 2h.

2.2 Characterization of MoS_2

The surface topography of monolayer and bilayer MoS_2 are observed by optical microscope (Nikon LV150N) and AFM (Agilent 5600LS), Raman spectra is obtained by confocal Raman microscopic (ThermoFisher DXR). The electrical characteristics are measured by semiconductor parameter analyzer (Agilent B1500A) at the pressure of 10^{-3} Pa.

3. Results and Discussion

Figure 1 shows the optical images of MoS_2 in different growth conditions. The results indicate that growth in horizontal direction is much easier at high temperature, whereas the H_2 could promote the vertical growth of MoS_2.Low temperature and small flow of H_2 resulted in monolayer and small-size of MoS_2 triangle or hexagonal (Figure1a-c), and high temperature and larger flow of H_2 form MoS_2 particles (Figure1g-i). After optimization, large-size (~40μm) bilayer MoS_2 triangle and hexagonal nanosheets(Figure1e) are acquired at 750℃ with 5sccm H_2.The Raman curves in Figure 1j show two typical Raman active modes centered at 403.93cm^{-1}(404.01cm^{-1}) and 383.63cm^{-1}(382.86cm^{-1}) for monolayer (and bilayer),respectively. The two Raman peaks are associated with the out-of plane vibration of S (A_{1g}) atoms and the in-plane vibration of Mo and S atoms (E^1_{2g}). The frequency difference (Δ) value between the two peaks was ~20.3cm^{-1} and (21.15cm^{-1}, indicating the MoS_2 are monolayer and bilayer). The AFM image (Figure1k) shows the thickness of monolayer and bilayer-MoS_2.

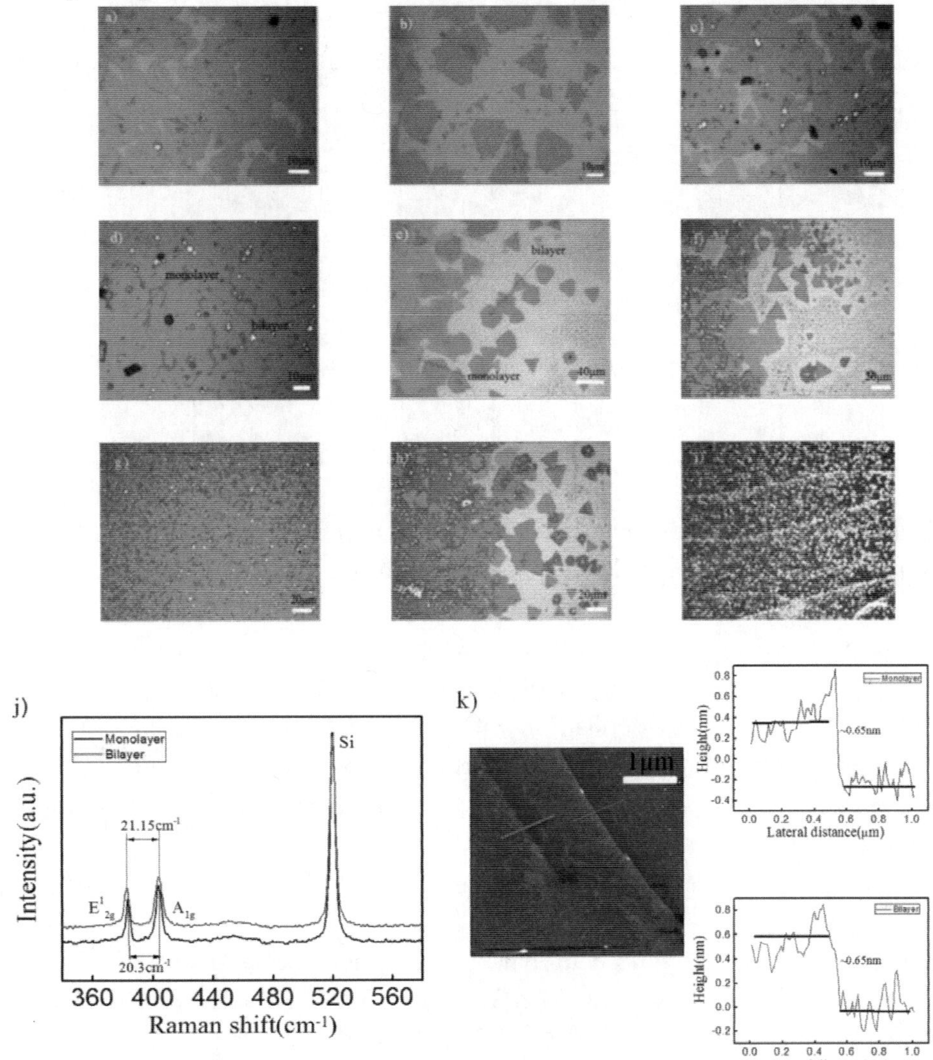

Figure 1 Optical images of (a-c) 700℃ with 1,5,10sccm H_2, (d-f) 750℃ with 1,5,10sccm H_2,and (g-i) 800℃ with 1,5 10sccm H_2; (j) Raman spectrum of monolayer and bilayer-MoS_2; (k) AFM image of monolayer and bilayer-MoS_2

Figure 2(a-d) shows the field effect properties of bilayer MoS_2. From Figure 2a, the field-effect mobility of ~21.27cm^{2}V^{-1}s^{-1} could be extracted by using the equation $\mu=[dI_{ds}/dV_{ds}]\times[L/(WC_iV_{ds})]$,where L and W is the channel length(1μm) and width(1.5μm),respectively. C_i is the capacitance between the channel and the back-gate per unit area. The on-current could reach ~9.5μA/μm and the I_{on}/I_{off} of ~3.9×10^7 is also be deduced by using a semi-logarithmic coordinate in Figure 2b.The field-effect mobility, I_{on}/I_{off} and on-current of monolayer-MoS_2 field effect transistor are 5.36cm^{2}V^{-1}s^{-1}, 9.6×10^6 and

1.1μA/μm (Figure 2e-h),respectively. The output characteristics (I_{ds}-V_{ds}) of bilayer-MoS_2 and monolayer-MoS_2 field effect transistors in Figure2c and 2g exhibited excellent ohmic contacts.

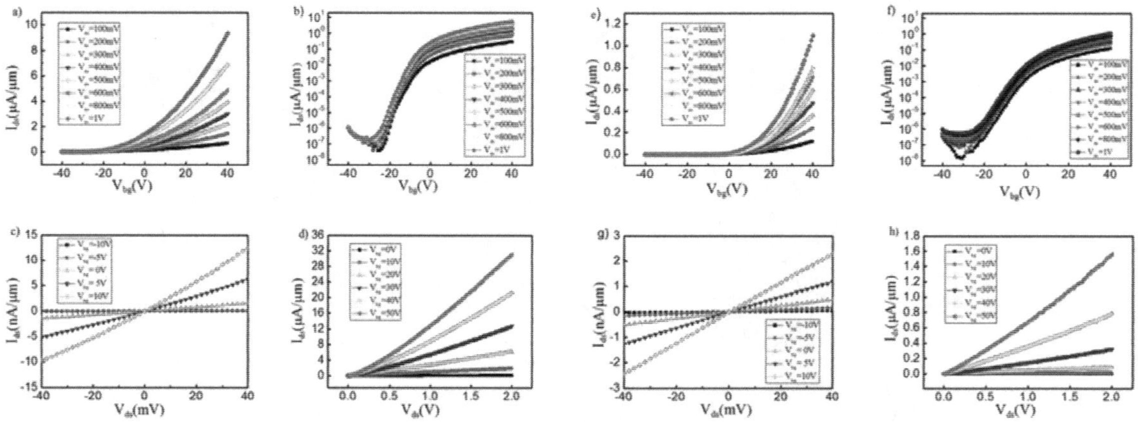

Figure 2 Typical transfer characteristics (I_{ds}-V_{bg}) and output characteristics (I_{ds}-V_{ds}) of bilayer-FETs (a-d) and monolayer MoS_2-FETs (e-h), respectively.

4. Conclusion

In this paper, lager-size (~40μm) of bilayer-MoS_2 triangle and hexagonal have been synthesized by chemical vapor deposition in one step. Bilayer-MoS_2 back-gate field effect transistors show the high performance with field effect mobility of ~$21.27cm^2V^{-1}s^{-1}$ and I_{on}/I_{off} ratio of ~3.9×10^7.The growth of high quality bilayer-MoS_2 expands their applications in nano-electronic and optoelectronic devices.

Acknowledgments

This work is supported by the National Natural Science Foundation of China(Grant Nos 61274113, 61404091, 61505144, 51502203, and 51502204) and Tianjin Natural Science Foundation(Grant Nos 14JCZDJC31500, 14JCQNJC00800).

Reference

[1] Radisavljevic B et al 2011 Single-Layer MoS_2 Transistors *[J].Nat. Nanotechnol* **6** 147-150

[2] Wang Q H et al 2012 Electronics and optoelectronics of two-dimensional transition metal dichalcogenides *[J]. Nat. Nanotechnol* **7** 699-712

[3] Ling X et al 2014 Role of the Seeding Promoter in MoS_2 Growth by Chemical Vapor Deposition *[J]. Nano Lett* **14** 464-472

[4] Liu K K et al 2012 Growth of large-area and highly crystalline MoS_2 thin layers on insulating substrates *[J]. Nano Lett* **12** 1538-1544

[5] Feng Y L et al 2015 Synthesis of Large Area Highly Crystalline MonolayerMolybdenum Disulfide with Tunable Grain Size in H_2 Atmosphere *[J]. ACS App. Mater. Interfaces* **7** 22587–22593

IOP Conf. Series: Journal of Physics: Conf. Series **864** (2017) 012033 doi:10.1088/1742-6596/864/1/012033

Exciton spectrum in two-dimensional transition metal dichalcogenides: The role of Diracness

Maxim Trushin[1], Mark Oliver Goerbig[2] and Wolfgang Belzig[1]
[1]University of Konstanz, Fachbereich Physik, D-78457 Konstanz, Germany
[2]Laboratoire de Physique des Solides, Univ. Paris-Sud, Université Paris-Saclay, CNRS UMR 8502, F-91405 Orsay, France
[1]University of Konstanz, Fachbereich Physik, D-78457 Konstanz, Germany

E-mail: maxim.trushin@uni-konstanz.de

Abstract. The physics of excitons, electron-hole pairs that are bound together by their mutual Coulomb attraction, can to great extent be understood in the framework of the quantum-mechanical hydrogen model. This model has recently been challenged by spectroscopic measurements on two-dimensional transition-metal dichalchogenides that unveil strong deviations from a hydrogenic spectrum. Here, we show that this deviation is due to the particular relativistic character of electrons in this class of materials. Indeed, their electrons are no longer described in terms of a Schrödinger but a massive Dirac equation that intimately links electrons to holes. Dirac excitons therefore inherit a relativistic quantum spin-1/2 that contributes to the angular momentum and thus the exciton spectrum. Most saliently, the level spacing is strongly reduced as compared to the hydrogen model, in agreement with spectroscopic measurements and *ab-initio* calculations.

1. Introduction

In this paper, we focus on the problem of excitonic spectra in two dimensional (2d) semiconducting transition metal dichalcogenides (2dTMDs). An exciton is a bound state of an electron and a hole which are attracted to each other by the Coulomb force [1]. The conventional 2d hydrogen-like exciton spectrum is given by

$$E_{nm} = \Delta - \frac{e^4\mu}{2\epsilon^{2-}\hbar^{2}} \frac{1}{(n + |m| + 1/2)^2}, \qquad (1)$$

where e is the elementary charge, ϵ is the dielectric constant, $\hbar$ is the Planck constant, Δ is the bandgap, and $\mu^{-1} = m_e^{-1} + m_h^{-1}$ is the reduced mass with $m_{e,h}$ being the electron/hole mass, and $n = 0, 1, 2...$, $m = 0, \pm 1, \pm 2$ are the radial and magnetic quantum numbers, respectively. However, the exciton spectrum recently observed in 2d TMDs [2, 3, 4] does not resemble the conventional Rydberg series (1). A few very recent Letters [5, 6, 7] propose different explanations of the nonhydrogenic exciton spectra based on the Berry's phase and non-local screening. Multiple ab-initio [8, 9, 10, 11] and other numerical calculations [12] have managed to reproduce the non-hydrogenic spectrum, but understanding is still missing from our point of view. We claim that the main mechanism responsible for the nonhydrogenic Rydberg series in 2dTMDs was overlooked until now. Here, we develop a new model which takes into account additional angular momentum (aka Diracness) coming from the pseudospin degree of freedom unavoidable in 2d semiconducting materials with honeycomb structure. The pseudospin changes the magnetic quantum number m in Eq. (1) to the total angular momentum $j = m + 1/2$ and the spectrum is then given by Eq. (6). The model allows for a transparent interpretation of existing experimental and theoretical data in simple terms, hence, providing understanding that we are striving for.

2. Model

The one-particle Hamiltonian for carriers in 2dTMDs is given by [13],

$$H_1 = \begin{pmatrix} \Delta/2 & \hbar v k e^{-i\theta} \\ -\hbar v k e^{i\theta} & -\Delta/2 \end{pmatrix}, \tag{2}$$

where $\tan\theta = k_y/k_x$, and v is the velocity parameter which can be either measured [14] or calculated [13]. Due to the spin-orbit coupling in real 2dTMDs H_1 splits into two versions with the bandgaps usually denoted by Δ_A and Δ_B. The pseudospin degree of freedom encoded in the matrix structure of Eq. (2) suggests the possibility that atypical quantum effects can play a role in bound states (1). The eigenvalues of (2) are $\pm\sqrt{(\hbar v k)^2 + \Delta^2/4}$, which in parabolic approximation suggest the same effective mass $m_{e,h} = \Delta/(2v^2)$ for electrons and holes.

We assume that the center of mass does not move for optically excited e-h pair [1], and the electron and hole momenta have the same absolute values but opposite directions. The two-particle Hamiltonian without Coulomb interactions is therefore given by the tensor product [15] $H_2 = H_1 \otimes I_2 - I_2 \otimes (T H_1 T^{-1})$ (here I_2 is the 2×2 unit matrix, and $T H_1 T^{-1}$ is the time reversal of H_1), and reads

$$H_2 = \begin{pmatrix} 0 & \hbar k v e^{i\theta} & -\hbar k v e^{-i\theta} & 0 \\ -\hbar k v e^{-i\theta} & \Delta & 0 & \hbar k v e^{-i\theta} \\ -\hbar k v e^{i\theta} & 0 & -\Delta & \hbar k v e^{i\theta} \\ 0 & \hbar k v e^{i\theta} & -\hbar k v e^{-i\theta} & 0 \end{pmatrix}. \tag{3}$$

Eq. (3) can be block-diagonalized [16] into a matrix $H_2 = H_2^+ \oplus H_2^-$ with

$$H_2^\pm = \begin{pmatrix} -\frac{\Delta}{2} \pm \sqrt{\hbar^2 v^2 k^2 + \frac{\Delta^2}{4}} & -\hbar k v e^{i\theta} \\ -\hbar k v e^{-i\theta} & \frac{\Delta}{2} \pm \sqrt{\hbar^2 v^2 k^2 + \frac{\Delta^2}{4}} \end{pmatrix}, \tag{4}$$

where H_2^+ has the eigenvalues $E_0 = 0$ and $E_k = 2\sqrt{\hbar^2 v^2 k^2 + \Delta^2/4}$ describing the excitonic states with vanishingly weak interaction. The diagonal terms in (4) can be written within the effective mass approximation, but the matrix remains in the peculiar mixed "Dirac-Schrödinger" form: The off-diagonal "Dirac" terms $\hbar v k e^{\pm i\theta}$ couple the "Schrödinger" states. Our goal is to write an effective-mass Hamiltonian which mimics this feature, but remains tractable at the analytical level. To do that we expand the diagonal terms in H_2^+ up to k^2-terms, switch on the Coulomb interaction $V(r) = -e^2/\epsilon r$, and change the momenta to the corresponding operators. The resulting Hamiltonian reads

$$\hat{H} = \begin{pmatrix} \frac{2\hbar^2 v^2}{\Delta}(\hat{k}_x^2 + \hat{k}_y^2) + V(r) & \sqrt{2}\hbar v(\hat{k}_x - i\hat{k}_y) \\ \sqrt{2}\hbar v(\hat{k}_x + i\hat{k}_y) & \Delta + V(r) \end{pmatrix}, \tag{5}$$

which contains no pseudo-differential operators, such as $\sqrt{\Delta^2/4 + \hbar^2 v^2 \hat{k}^2}$ that we would have to deal with starting directly from Eq. (4). Note, that our model (5) has not been obtained directly from the original 4×4 model (3) of coupled 2d Dirac fermions — indeed, the block-diagonalizing transformation depends on the lattice momentum k and therefore does not commute with the potential $V(r)$. However, our hypothesis is that the quantum mechanical nature of pseudospin is more important than that of momentum. That is why this approximation has been employed even though the quantum-mechanical non-commutativity between r and k will lead to corrective terms that could e.g. be treated perturbatively at a later stage. The major merit of our model (5) is to reproduce the relevant excitonic bands while retaining the off-diagonal terms whose manifestation we are investigating here. In order to find the eigen values of (5) we parametrize the interaction by $\alpha = e^2/\sqrt{2}\epsilon^{-}hv$and assume that $\alpha \ll 1$. This limit corresponds to the experimentally relevant regime of shallow bound states near the edge of continuous spectral branch E_k. The spectrum has been derived in [16] and reads

$$E_{nj} = \Delta - \frac{e^4\mu}{2\epsilon^{2-}\hbar} \frac{1}{(n + |j| + 1/2)^2}, \tag{6}$$

where $j = m + 1/2$ is the *total* (i.e. orbital and pseudospin [13]) angular momentum. Here, the binding energy is $E_b = e^4\mu/(2\epsilon^{2-}\hbar)$ is four times smaller than in the conventional model (1). As compared with Eq. (1), this spectrum shows much better agreement with the measurements [2, 3, 4] and numerical calculations [12, 8] and, along with the effective Hamiltonian (5), represents our main finding.

3. Discussion and conclusion

First of all, we would like to emphasize that our model Hamiltonian (5) cannot be continuously deformed towards the standard hydrogen-like one because of the matrix structure of the former in which the pseudospin angular momentum is encoded. Formally, one can obtain the conventional series (1) out of our model at $\alpha \to \infty$ since the diagonal (i.e. "Schrödinger") part in the Hamiltonian (5) dominates in this limit. However, the state with $E_0 = 0$ still remains and can overlap with the deep bound states making the whole model invalid.

Table 1. Optical transition energies for absorption peaks A (ground state of the exciton A), A' (its 1st excited state), B (ground state of the exciton B), and B' (its 1st excited state) obtained from different studies. Our model employs parameters from [14]: $\hbar v = 1.01\text{eV} \times 3.193\text{Å}$, $\Delta_{SO} = \Delta_B - \Delta_A = 0.16$ eV. The environment is SiO_2 with the dielectric constant $\epsilon = 3.9$.

Reference	A	A'	B	B'	Comments
Ref. [4]	1.87 eV	not resolved	2.03 eV	2.24 eV	photoluminescence spectroscopy
Ref. [12]	1.93 eV	2.13 eV	2.05 eV	2.27 eV	tight-binding calculations
Ref. [8]	1.88 eV	2.20 eV	2.02 eV	2.32 eV	first-principles calculations
Eq. (6)	1.87 eV	2.14 eV	2.01 eV	2.30 eV	$\Delta_A = 2.24$ eV, $\Delta_B = 2.4$ eV

In Table 1, we compare the exciton spectrum (6) with the previous numerical models [12, 8], which also include Diracness along with many other effects. Our model, however, does not take into account non-Coulomb interactions due to the non-local screening in thin semiconductor films [17, 2, 5], as we consider them less important than Diracness. The non-local screening makes the dielectric constant ϵ dependent on the exciton radius which increases with n [2, 5], whereas Diracness modifies the very backbone of the exciton model — the fundamental $1/(n + 1/2)^2$ spectral series. In our model, the dielectric constant ϵ is solely determined by the SiO_2 substrate, and the Coulomb interaction strength is simply parametrized by the parameter α. Nevertheless, our model qualitatively agrees with the numerical outcomes [12, 8], as shown in Table 1. The

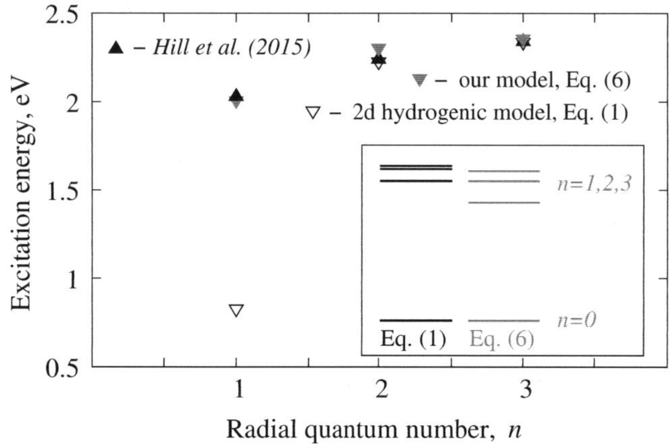

Figure 1. B-exciton spectrum for 2d MoS_2: theory and measurements. The measurements for MoS_2 are taken from Ref. [4]. The bandgap size Δ_B is 2.4eV and the band parameter $\hbar v = 1.01\text{eV} \times 3.193\text{Å}$ [14] result in $\alpha = 0.81$. The comparison of our model with measurements for WS_2 and WSe_2 has been done in [16]. The inset demonstrates the relative level spacing in our and conventional models.

quantitative version of our model should involve the exact diagonalization of (5) with the Keldysh potential [17] as $V(r)$.

We also compare the theoretical data with the exciton transition energies measured in [4] for MoS_2. As one can see, all the models overestimate the measured excitation energy of B' peak, but more or less agree with each other. The latter confirms that our model captures the main mechanism hidden in the numerical calculations. Note, however, that the measurements [4] allow for at least two different assignments of the exciton spectral lines to the theoretical Rydberg series. Nevertheless, we compare the B-exciton spectrum measured there with the outcomes of our model. One can see that the Rydberg spectrum (1) overestimates the binding energy, whereas Eq. (6) demonstrates much better fit.

To conclude, the pseudospin angular momentum (Diracness) does not lead just to a correction of the hydrogen-like exciton spectra but qualitatively modifies the excitonic Hamiltonian which suggests lower binding energy and reduced level spacing.

References

[1] Elliott R J 1957 *Phys. Rev.* **108** 1384
[2] Chernikov A, Berkelbach T C, Hill H M, Rigosi A, Li Y, Aslan O B, Reichman D R, Hybertsen M S and Heinz T F 2014 *Phys. Rev. Lett.* **113** 076802
[3] He K, Kumar N, Zhao L, Wang Z, Mak K F, Zhao H and Shan J 2014 *Phys. Rev. Lett.* **113** 026803
[4] Hill H M, Rigosi A F, Roquelet C, Chernikov A, Berkelbach T C, Reichman D R, Hybertsen M S, Brus L E and Heinz T F 2015 *Nano Letters* **15** 2992
[5] Olsen T, Latini S, Rasmussen F and Thygesen K S 2016 *Phys. Rev. Lett.* **116** 056401
[6] Zhou J, Shan W Y, Yao W and Xiao D 2015 *Phys. Rev. Lett.* **115** 166803
[7] Srivastava A and Imamoğlu A 2015 *Phys. Rev. Lett.* **115** 166802
[8] Qiu D Y, da Jornada F H and Louie S G 2013 *Phys. Rev. Lett.* **111** 216805
[9] Komsa H P and Krasheninnikov A V 2012 *Phys. Rev. B* **86** 241201
[10] Shi H, Pan H, Zhang Y W and Yakobson B I 2013 *Phys. Rev. B* **87** 155304
[11] Cheiwchanchamnangij T and Lambrecht W R L 2012 *Phys. Rev. B* **85** 205302
[12] Wu F, Qu F and MacDonald A H 2015 *Phys. Rev. B* **91** 075310
[13] Xiao D, Liu G B, Feng W, Xu X and Yao W 2012 *Phys. Rev. Lett.* **108** 196802
[14] Kim B S, Rhim J W, Kim B, Kim C and Park S R 2016 Experimental determination of the massive Dirac-fermion parameters in MoS_2, $MoSe_2$, WS_2, and WSe_2 *APS Meeting Abstracts, preprint arXiv:1601.01418*
[15] Rodin A S and Castro Neto A H 2013 *Phys. Rev. B* **88** 195437
[16] Trushin M, Goerbig M O and Belzig W 2016 *Phys. Rev. B* **94** 041301(R)
[17] Keldysh L V 1979 *Pis'ma Zh. Eksp. Teor. Fiz.* **29** 716

Enhanced current fluctuation in Coulomb blockade regime of multilevel quantum dot

Atsushi Iwasaki and Mikio Eto

Faculty of Science and Technology, Keio University, 3-14-1 Hiyoshi, Kohoku-ku, Yokohama 223-8522, Japan

E-mail: eto@rk.phys.keio.ac.jp

Abstract. Nonequilibrium current fluctuation is theoretically studied through a multilevel quantum dot in the Coulomb blockade regime, where the higher-order tunneling processes, "cotunneling," are dominant in the transport. The current fluctuation shows a Poisson (super-Poisson) noise in the elastic (inelastic) cotunneling regime when the bias voltage eV between the leads is smaller (larger) than the excitation energy Δ from the ground level in the quantum dot. The super-Poisson noise is remarkably enhanced by the degeneracy of the excited states around the beginning of inelastic cotunneling at $eV \approx \Delta$, qualitatively in good agreement with a recent experimental result. An analytical expression is given for the current fluctuation as a function of degeneracy.

1. Introduction

Nonequilibrium current fluctuation through a quantum dot yields information on the dynamic motion of single electron transport [1]. In the present paper, we theoretically examine the current fluctuation in the Coulomb blockade regime, where the transport takes place by the higher-order tunnel processes, cotunneling. Although the nonequilibrium fluctuation of the cotunneling current was reported by several groups [2, 3, 4], we pay special attention to the roles of excited levels in the quantum dot.

Let us consider a quantum dot with N energy levels ($E_1, E_2, \cdots, E_N$) attached to two leads L and R, as shown in Fig. 1. The bias voltage V is applied, $eV = \mu_L - \mu_R$ with μ_L and μ_R being the electrochemical potentials in the leads ($\mu_R < \mu_L$). At low temperatures, only the elastic cotunneling through the ground level E_1 takes place when eV does not exceed $\Delta_{12} = E_2 - E_1$, whereas inelastic processes are also possible when $eV > \Delta_{12}$. In the latter regime, Okazaki *et al.* found a large current fluctuation when an excited level is within the bias window; $E_1 < \mu_R < E_2 < \mu_L$ [5]. This is attributable to the sequential transport through level E_2, besides the cotunneling through level E_1. A large difference in the time scale between sequential and cotunneling transports results in a super-Poisson noise (Fano factor ≈ 3) for the current fluctuation. Recently, an enhancement of the current fluctuation (Fano factor ≈ 1.5) was observed when the excited levels are out of the bias window ($E_1, \cdots, E_N < \mu_R$) in the transport through a carbon-nanotube (CNT) quantum dot [6]. The degeneracy of the energy levels should play a role in this current fluctuation, which stems from the spin and valley degrees of freedom in the CNT.

To understand the mechanism of enhanced current fluctuation in CNT, we examine a simple model of a quantum dot with N energy levels for spinless electrons. We show that a super-Poisson noise in the inelastic regime can be significantly enhanced by the degeneracy of excited levels.

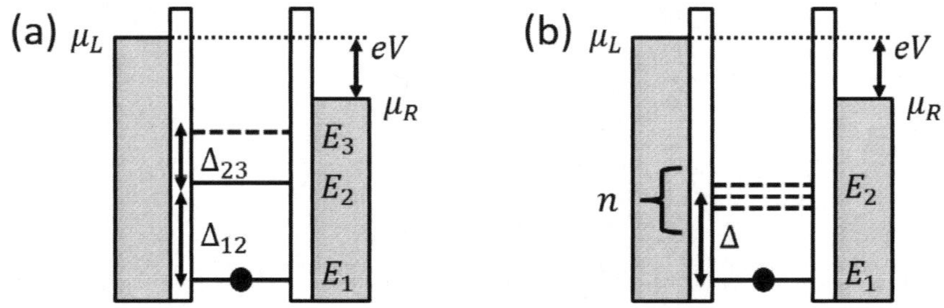

Figure 1. Two models for a quantum dot in the Coulomb blockade regime with one electron. Electrons are spinless. The quantum dot is attached to two leads with electro-chemical potentials, μ_L and μ_R ($\mu_L > \mu_R$). The bias voltage V is applied between the leads in such a way as $\mu_L = eV/2$ and $\mu_R = -eV/2$. (a) There are two or three energy levels in a quantum dot. The level spacings are denoted by $\Delta_{i,i+1} = E_{i+1} - E_i$. (b) A quantum dot possesses a non-degenerate ground level E_1 and an excited level E_2 with n-fold degeneracy. $\Delta = E_2 - E_1$.

2. Model and Calculation Method

In our model, spinless electrons are described by the Hamiltonian, $H = H_0 + H_R + H_T$:

$$H_0 + H_R = \sum_{i=1}^{N} E_i n_i + U \sum_{i>j} n_i n_j + \sum_{\alpha=L,R} \sum_{i=1}^{N} \sum_{k} \epsilon_k \, c_{\alpha k i}^{\dagger} c_{\alpha k i}, \tag{1}$$

$$H_T = \sum_{\alpha=L,R} \sum_{i=1}^{N} V_{\alpha i} d_i^{\dagger} c_{\alpha k i} + \text{h.c.}, \tag{2}$$

with $n_i = d_i^{\dagger} d_i$, where d_i and $d_i^{\dagger}$ are annihilation and creation operators of an electron in the quantum dot with energy level E_i ($i = 1, \cdots, N$), respectively. $c_{\alpha k i}$ and $c_{\alpha k i}^{\dagger}$ are those in lead α with wavenumber k and channel i. (We assume that level i in the quantum dot is coupled to channel i in lead α by $V_{\alpha i}$ in the tunnel Hamiltonian H_T, which is the situation in CNT with spin and valley degrees of freedom.) The Coulomb interaction is denoted by U in the quantum dot.

We examine the Coulomb blockade regime with one electron in the quantum dot, $E_1 < \mu_L, \mu_R < E_1 + U$. The bias voltage $eV = \mu_L - \mu_R$ is so small that all the excited levels are out of the bias window, $E_i < \mu_R$ ($i = 1, \cdots, N$). First, we study a model for the quantum dot with two or three levels [Fig. 1(a)]. The level spacings are denoted by $\Delta_{i,i+1} = E_{i+1} - E_i$. Second, the degeneracy of the excited levels is studied by another model in Fig. 1(b).

The current $\langle I \rangle$ and current fluctuation $S(\omega = 0)$, which is the zero-frequency component of Fourier transformation of the current-current correlation function, $\langle I(t)I(0)\rangle - \langle I \rangle^2$, are evaluated using the full counting statistics to solve the master equation for the stationary state [4] . The transition rate $\Gamma_{i\to j}^{\alpha\to\beta}$ from level E_i to level E_j in the quantum dot, accompanied by the single-electron transport from lead α to β is given by the second-order perturbation with respect to H_T. For $i = j$, it is

$$\Gamma_{i\to i}^{\alpha\to\beta} = \frac{2\pi}{\hbar} \nu^2 \int d\epsilon_k \left[\frac{|V_{\alpha i}|^2 |V_{\beta i}|^2}{(E_i - \epsilon_k)^2} + \sum_{l\neq i} \frac{|V_{\alpha l}|^2 |V_{\beta l}|^2}{(\epsilon_k - E_l - U)^2} \right] f_\alpha(\epsilon_k)[1 - f_\beta(\epsilon_k)], \tag{3}$$

where ν is the density of states in the leads and $f_\alpha(\epsilon)$ is the Fermi distribution function in lead α. For $i \neq j$

$$\Gamma_{i\to j}^{\alpha\to\beta} = \frac{2\pi}{\hbar} \nu^2 |V_{\alpha j}|^2 |V_{\beta i}|^2 \int d\epsilon_k \left(\frac{1}{E_i - \epsilon_k} + \frac{1}{\epsilon_k - E_j - U} \right)^2 f_\alpha(\epsilon_k)[1 - f_\beta(\epsilon_k - E_j + E_i)]. \tag{4}$$

IOP Conf. Series: Journal of Physics: Conf. Series **864** (2017) 012034 doi:10.1088/1742-6596/864/1/012034

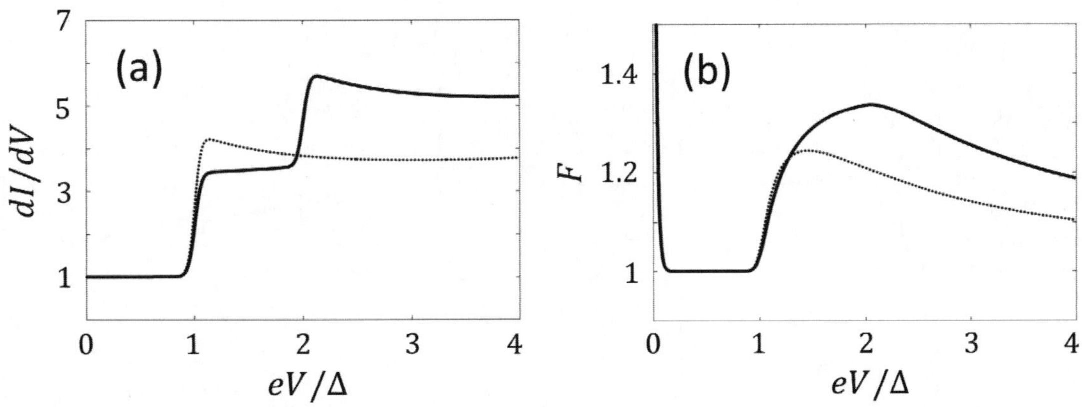

Figure 2. (a) dI/dV and (b) Fano factor F as a function of the bias voltage eV for the model in Fig. 1(a). The case of two levels ($\Delta_{12} = \Delta$) is denoted by dotted lines, whereas that of three levels ($\Delta_{12} = \Delta_{23} = \Delta$) is by solid lines. dI/dV is normalized by its value at $eV/\Delta = 0$. $E_1 = -10\Delta$, $U = 20\Delta$, and temperature $k_BT = 0.02\Delta$. All the tunnel couplings $V_{\alpha i}$ are identical to each other.

The Fano factor, $F = S(\omega = 0)/2e\langle I\rangle$, is useful to characterize the nonequilibrium current fluctuation at $eV > k_BT$, so-called shot noise. $F = 1$ for the Poisson processes, whereas $F > 1$ ($F < 1$) for the super- (sub-) Poisson processes.

3. Calculated Results

Figure 2 presents (a) dI/dV and (b) Fano factor F as a function of bias voltage eV for the model in Fig. 1(a). In the case of two levels ($\Delta_{12} = \Delta$; dotted line), dI/dV is almost constant at $eV < \Delta$, reflecting the elastic cotunneling through the lower level E_1 only, whereas dI/dV suddenly increases around $eV = \Delta$ owing to the possible inelastic processes at $eV > \Delta$. A small peak in the dI/dV curve is seen at the beginning of inelastic cotunneling, which is caused by the increase in the probability P_2 that an electron occupies level 2 in the quantum dot with increasing eV ($dP_2/dV > 0$) [7]. The Fano factor is almost unity at $k_BT < eV < \Delta$ (large F at $eV < k_BT$ is due to the thermal noise). $F = 1$ in the elastic cotunneling regime means that the cotunneling takes place as a Poisson process from lead L to R. F is enhanced to 1.25 at $eV > \Delta$. This super-Poisson noise is caused by more than one time-scale for both elastic and inelastic processes.

In the case of three equi-distant levels ($\Delta_{12} = \Delta_{23} = \Delta$; solid line), dI/dV shows two rapid increases around $eV = \Delta$ and 2Δ. Note that the Fano factor is larger than that in the two-level system at $\Delta < eV < 2\Delta$, which is ascribable to the additional transport accompanied by the transition from level E_2 to E_3. Since the probability P_2 to occupy level E_2 is finite, an electron can be excited to level E_3 from E_2, which enlarges the current fluctuation further.

Figure 3 is devoted to the results for the model in Fig. 1(b), where the higher level is n-fold degenerate in the quantum dot ($n = 1$, 2, and 4). The peak in the dI/dV curve at the beginning of inelastic cotunneling becomes larger for larger n. This is because the probability to occupy the higher level, P_2, increases with V more rapidly (dP_2/dV is larger) for larger n.

We find a larger peak in the Fano factor at $\Delta < eV < 2\Delta$ for larger n. This is due to the larger difference in the time scales for the transport through level E_1 and E_2 for larger n. In contrast, the Fano factor becomes smaller for larger n at $eV > 3\Delta$, where the role of the lower level in the transport is less for larger n.

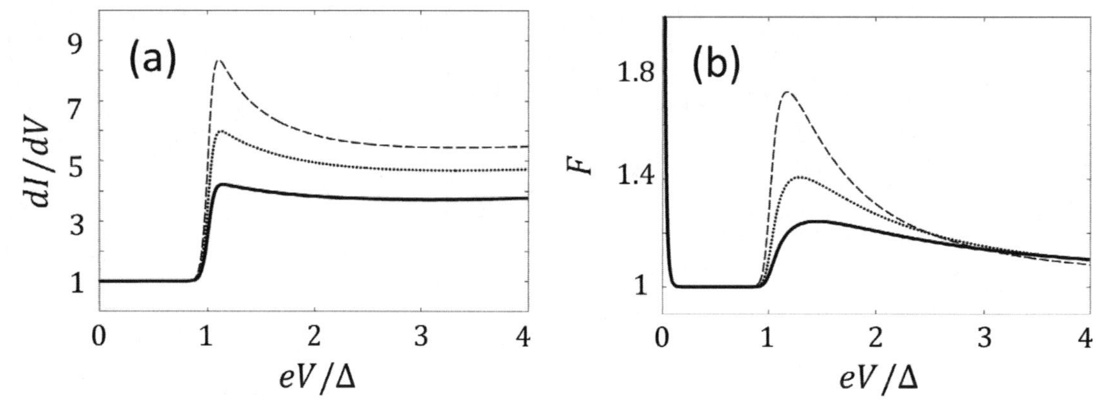

Figure 3. (a) dI/dV and (b) Fano factor F as a function of the bias voltage eV for the model in Fig. 1(b). The degeneracy of the higher level is $n = 1$ (solid line), 2 (dotted line), and 4 (broken line). dI/dV is normalized by its value at $eV/\Delta = 0$. $E_1 = -10\Delta$, $U = 20\Delta$, and temperature $k_B T = 0.02\Delta$. All the tunnel couplings $V_{\alpha i}$ are identical to each other.

We obtain an analytical expression for the Fano factor at $eV > \Delta$:

$$F = 1 + \frac{8n(n+1)(x-1)[3(n+1)x+n+9]}{[(n+1)x+3-n]^2[(5n+1)(n+1)x^2-(5n+1)(n-3)x-16n]}, \tag{5}$$

where $x = eV/\Delta$. This expression is justified when $-E_1$, $E_1 + U \gg \Delta$, eV. The curves in Fig. 3(b) are well reproduced by this expression for $eV > \Delta$.

4. Conclusion

We have studied the nonequilibrium current and its fluctuation through a multilevel quantum dot in the Coulomb blockade regime, where the cotunneling processes are dominant in the transport. The current fluctuation shows a Poisson noise in the elastic regime and a super-Poisson noise in the inelastic regime. The super-Poisson noise is remarkably enhanced by the degeneracy of the excited states around the beginning of inelastic cotunneling, which is qualitatively in good agreement with a recent experimental result using a CNT quantum dot. The analytical expression for this enhanced noise has been given as a function of degeneracy.

Acknowledgments

This work is partly supported by a Grant-in-Aid for Scientific Research (S) (No. 26220711) from Japan Society for the Promotion of Science and MEXT Grant-in-Aid for Scientific Research on Innovative Areas "Science of hybrid quantum systems" (No. 15H05870).

Reference

[1] Blanter Y M and Büttiker M 2000 *Phys. Rep.* **336** 1
[2] Sukhorukov E V, Burkard G and Loss D 2001 *Phys. Rev. B* **63** 125315
[3] Thielman A, Hettler M H, König J and Schön G 2005 *Phys. Rev. Lett* **95** 146806
[4] Kaasbjerg K and Belzig W 2015 *Phys. Rev. B* **91** 235413
[5] Okazaki Y *et al.* 2013 *Phys. Rev. B* **87** 041302(R)
[6] Fujiwara R, Arakawa T and Kobayashi K private communications.
[7] Eto M 2001 *Jpn. J. Appl. Phys.* **40** 1929

IOP Conf. Series: Journal of Physics: Conf. Series **864** (2017) 012035 doi:10.1088/1742-6596/864/1/012035

The role of electron-phonon interactions on the coherence lifetime of monolayer transition metal dichalcogenides

C. E. Stevens[1], P. Dey[1], J. Paul[1], Z. Wang[2], H. Zhang[1], A. H. Romero[3], J. Shan[2], D. J. Hilton[4], D. Karaiskaj[1]

[1]Department of Physics, University of South Florida, 4202 East Fowler Ave., Tampa, Florida 33620
[2]Department of Physics, Pennsylvania State University, University Park, PA 16802, USA
[3]Physics Department, West Virginia University, Morgantown, WV, 26506-6315, USA
[4]Department of Physics, University of Alabama at Birmingham, Birmingham, Alabama 35294, USA

E-mail: `karaiskaj@usf.edu`

Abstract. We investigate the excitonic dephasing of transition metal dichalcogenides, namely MoS_2, $MoSe_2$ and WSe_2 atomic monolayer thick and bulk crystals, in order to understand the factors that determine the optical coherence in these materials. Coherent nonlinear optical spectroscopy, temperature dependent absorption combined with theoretical calculations of the phonon spectra, reveal the important role electron-phonon interactions plat in dephasing process. The temperature dependence of the electronic band gap and the excitonic linewidth combined with '*ab initio*' calculations of the phonon energies and the phonon density of state reveal strong interaction with the E' and E" phonon modes.

1. Introduction

Transition metal dichalcogenides (TMDs) have recently received a lot of attention and have been suggested for numerous electronic and optoelectronic applications. Therefore, understanding the fundamental many-body interactions and how these interactions influence the optical properties of these materials is crucially important [1]. As bulk materials TMDs are indirect band semiconductors with rather poor optical properties. However, when reduced to atomic monolayer they transition to direct band gap materials leading to greatly enhanced emission [2]. When photon absorption occurs electrons are excited from the valence band to the conduction band. The Coulomb interaction between the positively charged hole and the electron leads to a new quasi-particle, namely the exciton. Excitons are of importance for the optical properties of semiconductors, therefore understanding their properties, such as coherence loss or dephasing can be key for many optoelectronic devices. Many-body effects play a crucial role in determining the optical coherence and lifetime of excitons through electron-phonon interactions and exciton-exciton scattering [3]. Recent experimental studies have measured the radiative recombination lifetime of excitons in WSe_2 to be exceedingly short and therefore limit the coherence time [4]. Strong exciton-exciton scattering or excitation induced dephasing was thought to be a limiting factor in monolayer $MoSe_2$ as well [5]. Finally, theoretical studies have attributed the dephasing mechanism to phonon assisted scattering of excitons to a lower lower lying dark state [6].

2. Experiment

The optical excitation in the four-wave mixing (FWM) experiments was provided by s laser oscillator, within the tuning range of the Ti:Sapphire crystal, whereas outside the oscillator spectral range, an optical parametric oscillator was used. In the three pulse FWM, three pulses are incident on the sample with wavevectors $-\vec{k}_a$, $\vec{k}_b$, and $\vec{k}_c$. The nonlinear interaction gives rise to a signal in the direction $-\vec{k}_a + \vec{k}_b + \vec{k}_c$. The phase conjugate pulse $-\vec{k}_a$ and the second pulse $\vec{k}_b$ are separated by the time delay τ whereas pulse k_b and the third pulse k_c are separated by the population time T. By varying the time delay τ and monitoring the time-integrated FWM intensity, the dephasing time of excitons can be measured.

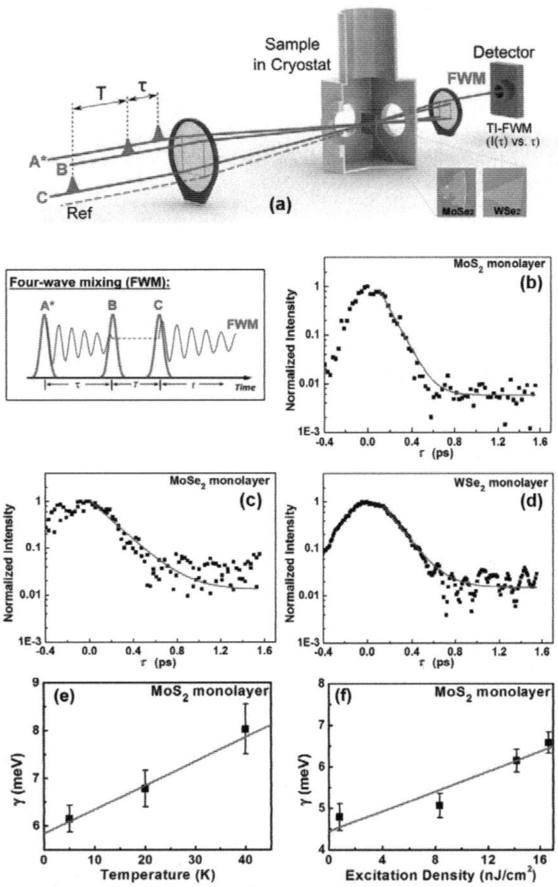

Figure 1. (a) The four phase-stabilized linearly polarized beams obtained from the multi-dimensional optical nonlinear spectrometer (MONSTR) instrument are focused on the sample, which is held in the cryostat at 5 K. The FWM signal can be measured as a function of the time delays τ and T using a detector, which integrates over the time delay t. The sequence of the laser pulses used in the experiments, where A* corresponds to the phase conjugated pulse. The time delay τ corresponds to the time between the pulses A* and B, T is the time delay between the pulses B and C, and t is the evolution of the echo in 'real time'.(b-d) Time-integrated FWM of atomic monolayers (a) MoS_2 (b) $MoSe_2$ and (c) WSe_2. The black symbols are the experimental data whereas the red line is the single exponential fit.(e) Temperature dependence of the homogeneous linewidth for atomic monolayer MoS_2. Squares are the measured values whereas the red line is the linear fit. (f) Excitation density dependence of the homogeneous linewidth for atomic monolayer MoS_2. Squares are the measured values whereas the red line is the linear fit.

3. Results and discussion

The strength of electron-phonon interactions often determines the optical properties of materials and can be a limiting factor for many applications. While the inhomogeneous linewidth of excitons is determined by material imperfections and can be reduced by improving the material quality, the homogeneous linewidth is often limited by the fundamental interactions taking place, which place limitations on the material system. The excitonic coherence is a direct measure of the homogeneous linewidth, therefore processes leading to dephasing can be the limiting factor [3].

We start by discussing the FWM measurements shown in Fig. 1. The dephasing time T_2 provides the homogeneous linewidth and was measured for atomic monolayers of MoS_2, WSe_2 and $MoSe_2$ to be 200 fs (~6.58 meV), 279 fs (~4.72 meV), and 394 fs (~3.34 meV) respectively, showing a rapid decay for all the materials. The temperature dependence of the homogeneous linewidth for atomic monolayer MoS_2 is shown in Fig. 1 (e) and the homogeneous linewidth γ increases very rapidly with temperature. At temperatures higher than 40 K it exceeds our ability to measure it. The scattering rate of excitons with acoustic phonons at low temperature is described by a linear relationship $\gamma = \gamma^* + aT$, where γ^* is the temperature independent broadening originating from other sources. The linear fit leads to an acoustic phonon scattering coefficient a ~ 45 μeV/K, which is large and indicates strong electron-phonon interactions. Furthermore, the excitation density dependence of the homogeneous linewidth γ is shown in Fig. 1 (f) and exhibits some excitation induced dephasing [3]. However, at the limit of 'zero' excitation power the residual homogeneous linewidth independent of excitation density is ~4.5 meV.

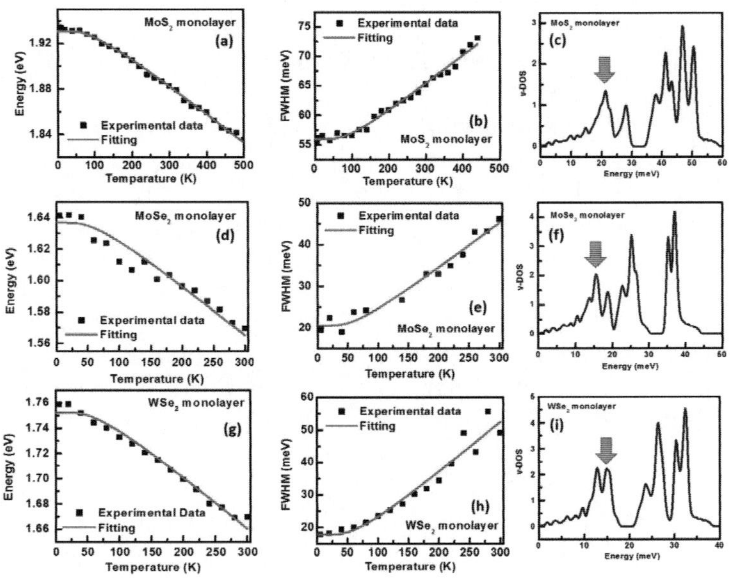

Figure 2. (a) Electronic band gap shift, temperature broadening of the direct excitons, and calculated phonon density of states for atomic monolayer MoS_2 (a-c), $MoSe_2$ (d-f), and WSe_2 (g-i), respectively. The blue arrows indicate the phonon energies used in the fittings.

In order to obtain a quantitative understanding of the electron-phonon interactions in TMDs we measure the energy shift of the electronic band gap with temperature for both bulk and atomic monolayer samples. Temperature dependent absorption spectra of the excitons provide the energy shift of the band gap with temperature and also the temperature broadening of the excitonic resonance. These effects are in large part a result of electron-phonon interactions [7]. The energy shift of the exciton resonance with temperature for bulk MoS_2, $MoSe_2$, and WSe_2 have been discussed elsewhere [1]. The fitting procedure was performed using,

$$E_g = E_0 - E_1 \left[2(\exp\left(\frac{\Theta}{kT}\right) - 1)^{-1} + 1 \right] \quad (1)$$

where E_0 and E_1 are fitting parameters, whereas Θ is either a dominant phonon or an average phonon energy. The absorption spectra of the excitons in TMDs at low temperature are predominantly inhomogeneously broadened. However, as the temperature rises, the phonon scattering leads to observable broadening. The linewidths of the exciton absorption spectra as a function of temperature for bulk $MoSe_2$ and WSe_2 were also discussed elsewhere [1]. The theoretical fit was performed according to

$$\gamma = \gamma_I + \frac{b}{\exp(\Theta/kT) - 1} \tag{2}$$

where γ_I is the temperature independent inhomogeneous broadening. Both the energy shift and broadening of the excitonic resonance can be fitted simultaneously by using a phonon energy $\Theta = 22.1$ meV, 16.3 meV and 15.5 meV for MoS_2, WSe_2, and $MoSe_2$, respectively.

Electronic band gap shift and temperature broadening of the excitons for atomic monolayer MoS_2 (a-b), $MoSe_2$ (d-e), and WSe_2 (g-h) are shown in Fig. 2. In order to be able to identify the phonon energies and gain a deeper understanding of the dephasing mechanism we performed *'ab initio'* calculations of the phonon spectra and phonon DOS for bulk and atomic monolayer TMDs [1]. The phonon DOS for monolayer MoS_2, $MoSe_2$ and WSe_2, are shown in Fig. 2 (c), (f), and (i), respectively, where the blue arrows marks the energies of the Θ phonons obtain from the fitting of the experimental data. The calculations also reveal two phonon modes at these energies corresponding to the E' and E" representation. The calculated phonon energies are 23.3 (E') meV and 23.5 (E") meV for MoS_2, 12.6 (E') and 15.8 (E") for WSe_2, and 15.8 meV (E') and 17.7 meV (E") for $MoSe_2$. These energies are very close to the observed phonon energies. However, the two modes are energetically very close which makes it difficult to distinguish them experimentally.

In conclusion, we provide a detailed description of the electron-phonon interactions and their possible role on the dephasing mechanism of direct excitons in atomic monolayer TMDs. The temperature dependence and excitation density dependence of the dephasing in atomic monolayers indicate that the electron-phonon interactions to play an important role in the rapid dephasing, and as a result, the large residual homogeneous linewidths in this material system. Calculations of the phonon energies and phonon DOS determined these phonon energies as the E' and E" phonon modes, which could facilitate the rapid dephasing observed in TMDs. The large residual homogeneous linewidth at low excitation densities and low temperature observed experimentally for MoS_2 is well in agreement with the efficient exciton relaxation in WS_2 through phonon emission at all temperatures, reported in the theoretical studies [6].

3.1. Acknowledgments
The research at USF, Penn. State, and UAB is supported by the U.S. Department of Energy, Office of Basic Energy Sciences, Division of Materials Sciences and Engineering under Award DE-SC0012635. AHR acknowledges the support from the Extreme Science and Engineering Discovery Environment (XSEDE), which is supported by National Science Foundation Grant No. ACI-1053575, the support of NSF under project 1434897 and the Donors of the American Chemical Society Petroleum Research Fund for partial support of this research under contract 54075-ND10.

References
[1] Dey P, Paul J, Wang Z, Stevens C E, Liu C, Romero A H, Shan J, Hilton D J and Karaiskaj D 2016 *Phys. Rev. Lett.* **116** 127402
[2] Mak K F, Lee C, Hone J, Shan J and Heinz T F 2010 *Phys. Rev. Lett.* **105** 136805
[3] Shah J 1999 *Ultrafast Spectroscopy of Semiconductors and Semiconductor Nanostructures* (Springer-Verlag)
[4] Poellmann C, Steinleitner P, Leierseder U, Nagler P, Plechinger G, Porer M, Bratschitsch R, Schller C, Korn T and Huber R 2015 *Nature Mat.* **14** 889
[5] Moody G, Dass C K, Hao K, Chen C H, Li L J, Singh A, Tran K, Clark G, Xu X, Berghuser G, Malic E, Knorr A and Li X 2015 *Nat. Commun.* **6** 1038
[6] Selig M, Berghäuser G, Raja A, Nagler P, Schüller C, Heinz T F, Korn T, Chernikov A, Malic E and Knorr A *https://arxiv.org/abs/1605.03359*
[7] Viña L, Logothetidis S and Cardona M 1984 *Phys. Rev. B* **30** 1979

High-resolution Electronic and Chemical imaging of wonder nanomaterials beyond graphene

José Avila, Chaoyu Chen, Stephane Lorcy & Maria C. Asensio*

ANTARES Beamline, Synchrotron SOLEIL& Université Paris-Saclay, L'Orme des Merisiers, Saint Aubin-BP 48, 91192 Gif sur Yvette Cedex, France.

E-mail: `asensio@synchrotron-soleil.fr`

Abstract. Despite the great progress made recently in spectroscopic imagery and even the remarkable success achieved, the challenge still remain concerning the precise determination of the chemical and electronic imagery of advances materials, which usually are available as heterogeneous large crystals or tiny homogeneous monocrystals. Here we report, a recently developed novel X-ray microscope, labelled, *k-microscope* or NanoARPES (Nano Angle Resolved Photoelectron Spectroscopy) particularly well suited to provide both high resolved chemical and electronic information in the real and reciprocal space of complex materials with nano-scale resolution.

1. Introduction

In today's state-of-the-art material science, length scales from the nano- to the mesoscopic scales control and characterise chemical and electronic properties of matter. In recent decades we have witnessed exponential advances in the development of different areas of new nanotechnologies. These advances, seen particularly important in nanoelectronics, nanomagnetism and nanochemistry, among others, affect almost every aspect of our lives and are major players in our evolution towards the 'information and automation age".

New materials are often initially isolated or synthesized as small crystallites, which makes difficult their precise exploration. Also, it can be difficult to produce materials without fluctuations in their chemical composition, bringing out heterogeneities at variable scales. Polycrystalline materials, with chemically homogeneous and well-structured grains are also frequent, where the crystalline orientation is challenging to unravel as well as its consequences on the transport and magnetic properties are unknown. In the field of nano-objects also, even if these smart "building blocks" may show remarkable properties, they would have remained unexploited if , we had not developed new tools capable of viewing and scrutinizing objects on a wide range of scales, from a few microns to a few tens of nanometers.

Recently, great progress has been made as a result of the rapid expansion in the range of modern microscopies. However, if they have achieved nanometer spatial resolution, the challenge still remains to provide powerful high-energy-resolution spectroscopic tools for probed nano and micro-areas, as illustrated in Fig. 1. The challenge is to quantify and analyze the electronic properties of advanced materials on a nano and micrometer scale. For such a result, analysis of the electronic structure must be comprehensive, not only with regard to detection of core

Figure 1. Different heterogeneous samples. Chemical and electronic determination at varied scales is needed to characterize systems presenting phase separation, low dimensional films, confinement in nano-objects among others inhomogeneous samples.

levels, but especially to study the structure of delocalized valence bands, directly responsible for chemical bonds, electrical transport and the thermal and mechanical properties.

Traditionally, ARPES is the only technique capable of making sufficiently precise measurements of the dispersion of the band structure of materials in reciprocal space. The state of the art ARPES equipments installed at synchrotron radiation sources is such that it can offer energy and angular resolution of better than 5 meV and 0.1°, respectively. Yet, until now, no instrument has been capable of performing spatially resolved ARPES experiments on the nanometer scale.

In this paper, we present the first results of the NanoARPES microscope recently installed on the ANTARES beamline at the French Synchrotron SOLEIL[1]. With a spatial resolution of several tens of nanometers, this sophisticated instrument is able to carrying out the direct imaging of core levels and their chemical shifts, band electronic structures in reciprocal space and constant energy surfaces in reciprocal space, especially the Fermi surfaces.

2. Principle and performance of *k-microscope*

The *k-microscope* is a scanning photoemission microscopy, which combines linear and angle sweeps of the sample to perform precise electronic band structure determination by using ARPES and chemical imaging by core level detection, respectively. As shown in Fig. 2, the photon source has been combined with an advanced microscope, which has precise sample handling abilities. Moreover, it is fully compatible with a high angular and energy resolution hemispherical analyzer and a set of Fresnel Zone Plates (FZP) able to focalise the beam spot up to a few tenths of nanometers, depending on the spatial resolution of the selected FZP. More detailed instrumental and conceptual design of the instrument have been described in several previous publications[2, 3, 4, 5]. The main difference between the ANTARES microscope and other conventional ARPES instruments is that for the *k-microscope*, the specimens are mounted on the high-precision plate that ensure their nano-positioning in the x and y directions. The microscope has two operating modes, discontinuous punctual NanoARPES detection and sequential spectroscopic imaging. The first mode is like conventional but with a focalised nano-spot size, while for the second mode, the nano-spot of X-ray is raster scanned relative to the sample to produce an image with one pixel at a time. The complete set-up consists of fourteen independent axis piezo stages, which focus the beam and perform the sample scanning with nanometric precision. Moreover, the sample position relative to the focal distance of the FZP for experiments at constant and variable photon energy is controlled by an automatic feedback interferometric system.

IOP Conf. Series: Journal of Physics: Conf. Series **864** (2017) 012036 doi:10.1088/1742-6596/864/1/012036

3. Chemical and electronic imaging of graphene polycrystalline films on copper foils grown by chemical vapour deposition

To demonstrate the capabilities of the *k-microscope*, we present the first results obtained from a specimen of single layer graphene film grown by chemical vapour deposition (CVD) on polycrystalline copper foils[6]. As it is shown in Fig. 2, by collecting exclusively the "sp" copper states (those confined in the red box of Fig. 2b), the location, shape and orientation of Cu grains and their grain boundaries could be precisely determined. Fig. 2c shows one of the resulting NanoARPES images, in which the grain structure of the copper substrate can be clearly identified. If instead of copper states, we record the π^* states (green box of Fig. 2b), the resulting nano-ARPES image provides the size, distribution and orientation of the graphene grains covering the copper grains, as can be observed in Fig. 2d. As the graphene π and π^* bands are only close to the Fermi level at the K points of the graphene grains investigated, the NanoARPES image detects not only the presence of the pristine grains, but also their orientation.

In Figure 3, ARPES results show, that both multi- and single-grain ARPES spectra present a well-defined conical Dirac dispersion, with n-type extrinsic doping. Due to the continuous orientation distribution of the graphene grains, corroborated by LEED and the Fermi surface results, the multi-grain ARPES data, taken along any ΓK direction display "filled Dirac cones", (see Fig. 3e), which is a consequence of the Dirac cone overlapping from grains adjacently oriented. Comparison between panel (b) and (c) of figure 3 shows that individual priscine graphene grains should be measured in order to determine correctly the existence and size of the Dirac gap. Multigrain Dirac cone shows apparent gaps up to 80% bigger than the real gap,

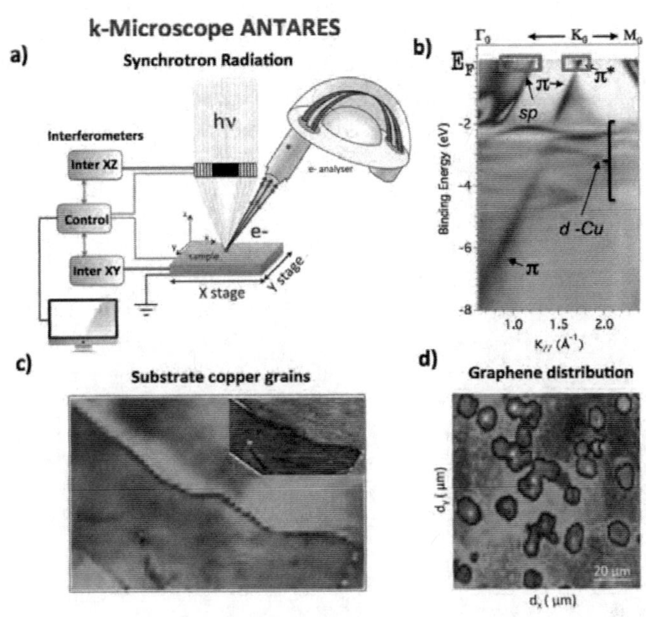

Figure 2. **(a)** Schematic of the different components of the ANTARES *k-microscope*. **(b)** ARPES data showing the d and sp bands of the copper substrate together with the massless band π of graphene. **(c)** real-space image of the copper states intensity (red box states of panel(b)) obtained by NanoARPES mapping presented on a linear scale as a false-color image. The inset of panel **(c)** shows the sample optical image where the copper border grains are visible. Panel**(d)** shows the real-space image of graphene grains distribution, obtained by monitoring the graphene states intensity (green box states of panel (b)).

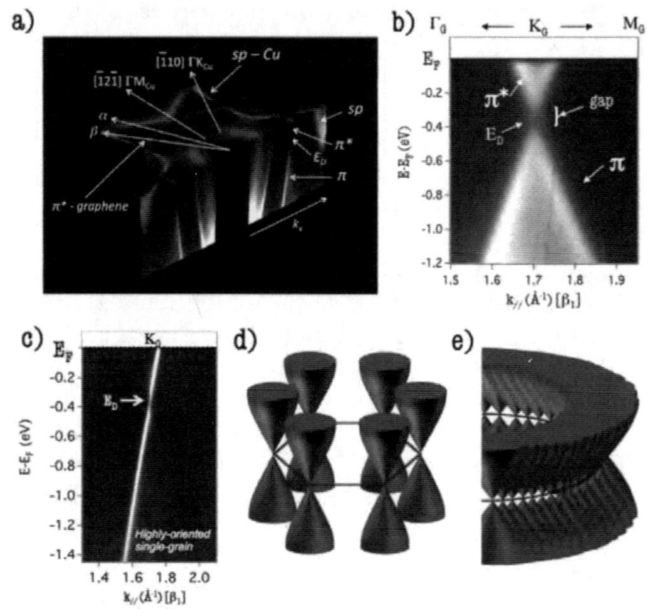

Figure 3. Panel(**a**) shows the Fermi surface map of the graphene multigrain film recorded with a spot size of $\sim$ 100 um and a photon energy of 100 eV, covering several individual graphene grains randomly oriented. (**b**) depicts multi-grain energy-momentum dispersion relations of π and π^* bands near E_F, using circularly polarized light of 30 eV. Panel (**c**) shows the $E(k)$ dispersions measured by NanoARPES, along the high symmetry direction ΓK, in a single pristine graphene. Finally, panels (**d**) and (**e**) show the reciprocal space of single- and multi- graphene grain, respectively, where the superposition of Dirac cones of graphene grains randomly oriented has been schematized.

due to the random orientation of the graphene grains.

The results presented here indicate that by employing NanoARPES, we are able to measure the complete electronic structure, of polycrystalline graphene films, which is impossible to tackle by conventional ARPES. This allows to use NanoARPES for future studies in a wide domain of heterogeneous materials.

4. Acknowledgments

The Synchrotron SOLEIL is supported by the Centre National de la Recherche Scientifique (CNRS) and the Commissariat à l'Energie Atomique et aux Energies Alternatives (CEA), France.

5. References

[1] Avila J. and Asensio M C. 2014 *Synchrotron Radiation News* **27** 24-30
[2] Bostwick A. et.al 2014 *Synchrotron Radiation News* **25** 19-25
[3] Avila J. et al. 2013 *Journal of Physics Conference Series (SRI2012)* **425** 192023
[4] Avila J. et al. 2013 *Journal of Physics Conference Series (SRI2012)* **425** 132013
[5] Chen C Y et al. *J. Phys.: Condens. Matter* **29** 183001
[6] Avila J. et al. 2013 *Scientific Reports* **3** , 2439

MoS_2 materials synthesized on SiO_2/Si substrates via MBE

Linjie Zhan[‡], Wen Wan[‡], Zhenwei Zhu[‡], Tien-Mo Shih[‡], and Weiwei Cai[‡*]

[‡] Department of Physics, State Key Laboratory of Physical Chemistry of Solid Surfaces, Xiamen University, 361005, China

*Email: wwcai@xmu.edu.cn

Abstract. Two-dimensional (2D) MoS_2 materials possess indirect-to-direct bandgap tunability, and have enjoyed wide applications in electronics and optoelectronics. Most of investigators have ubiquitously synthesized these materials by using the chemical vaporization deposition (CVD) method. Here we have adopted MoO_3 source materials to synthesize MoS_2 on 280-nm SiO_2/Si substrates via molecular beam epitaxy (MBE). We have obtained triangular nucleation, tens-of-micron domain, and monolayer MoS_2. This MBE technique can be applied to synthesizing other members of semiconducting layered transition metal dichalcogenides (TMDCs), such as WS_2, $MoSe_2$, and WSe_2 materials.

1. Introduction

Layered TMDCs [1-3] have recently attracted great interest because of their specific optical and electronic properties. Within this class of 2D materials, MoS_2 possesses high in-plane mobility [4], large Seebeck coefficient [5], compatible nanopores [6], having enjoyed wide applications in electronics and optoelectronicsis [7-11]. Similarly as graphene films [12], MoS_2 films can be physically exfoliated from the bulk. However, generally the process requires much labor, yields low repeatability, and cannot conveniently produce entirely uniform monolayers. Additionally, CVD method has been widely used, and is capable of conveniently synthesizing deposited materials based on the nucleation mechanism. Sources of bulky quantities must be injected into the vacuumed tube, and are deposited on the substrate under relatively high temperatures. During this method, the sulfurization of MoO_3 jointly with CVD method is being commonly used. An approach for synthesizing MoS_2 flakes in rhomboid shapes with controllable number of layers by the layer-by-layer sulfurization of MoO_2 microcrystals [13] has also been developed with high-crystallinity domains occupying tens of microns. However, how to controllably prepare high-quality uniform monolayer MoS_2 films for the practical application of MoS_2 in electronics and optoelectronics has so far remained challenging.

In comparison, first, MBE demands higher-degreed vacuum, leading to less contamination. Second, the substrate tends to be heated to low temperatures such that numbers of thermal-stress-induced wrinkles and cracks in the crystal structure decrease. Third, materials tend to be slowly synthesized, facilitating the control of crystal-structure accuracies. Finally, the MBE is capable of maneuvering intermittent syntheses, which allow impromptu changes of source materials in order to regulate epitaxial compositions and impurities. Consequently, the MBE technology widens the low-dimension-material research horizon. Here we have adopted MoO_3 source materials to synthesize MoS_2 on 280-nm SiO_2/Si substrates via MBE. Addition, we have obtained triangular nucleation, tens-of-micron domain and monolayer MoS_2. This MBE technique can be aptly applied to synthesizing other members of TMDCs and their heterojunctions.

2. MoS_2 materials synthesized on SiO_2/Si substrates

2.1. The nucleation and further growing of MoS_2

Owing to the reflection-enhancing effect on specific wavelengths of visible light between MoS_2 and 280-nm thick SiO_2 oxide layer on Si, MoS_2 layers can be observed by an optical microscope. Figure 1a shows the optical image of MoS_2 domains, and can be enlarged by the scanning electron microscopy (Figure 1b). Some domains look like triangles, indicating that MoS_2 crystallites possess a good crystallinity; some exhibit rhomboidal shapes [14], implying that crystals have not been fully sulfurized. When the grown time is extended, MoS_2 crystallites can be enlarged to tens of micron, which can be characterized by Raman spectrum, as shown in Figure 1c. The Raman spectrum of larger MoS_2 domain, shown in Figure 1d, is characterized by the Raman shift of E_{2g}^1 and A_{1g} modes lying in 382.7 and 403.1 cm^{-1}. The frequency difference between the E_{2g}^1 and A_{1g} modes is 20.4 cm^{-1}, confirming that the MoS_2 domain is monolayer.

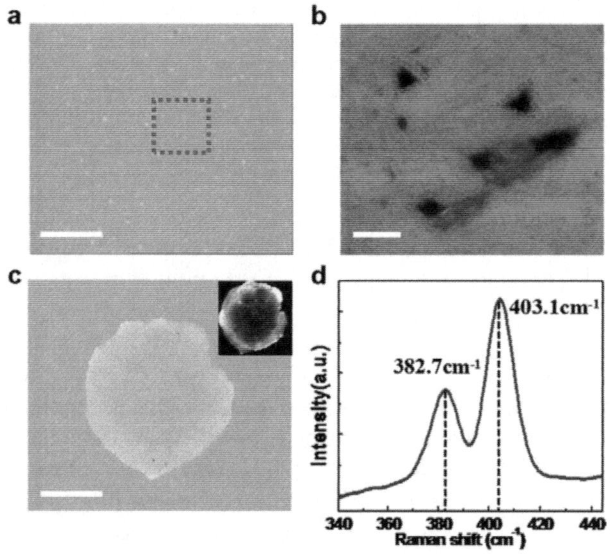

Figure 1. The nucleation and further growing of MoS_2. (a) Optical image showing the nucleation of MoS_2 on SiO_2/Si. Scale bar, 20 μm. (b) Enlarged SEM image of the area indicated in the red dashed line in (a). Scale bar, 2 μm. (c) Optical image of MoS_2 in large domain. The inset shows Raman mapping of characteristic E_{2g}^1 peak (d) The Raman spectrum of MoS_2 in (c), showing the E_{2g}^1 and the A_{1g} peaks.

A conspicuous spiral-like MoS_2 domain is depicted in Figure 2a, of which the monolayer, bilayer, and bulk are represented by black, green, and read circles, respectively. In Figure 2b, Raman spectra of these circled MoS_2 are shown. Their E_{2g}^1 Raman-shift peaks are located almost at the same position (384.6 cm^{-1}), whereas A_{1g} counterparts are correspondingly situated at 404.6, 406.4, and 410.1 cm^{-1}. Frequency differences between E_{2g}^1 and A_{1g} peaks approximately equal 20.0, 21.8, and 25.5 cm^{-1} for monolayer, bilayer, and bulk MoS_2, respectively.

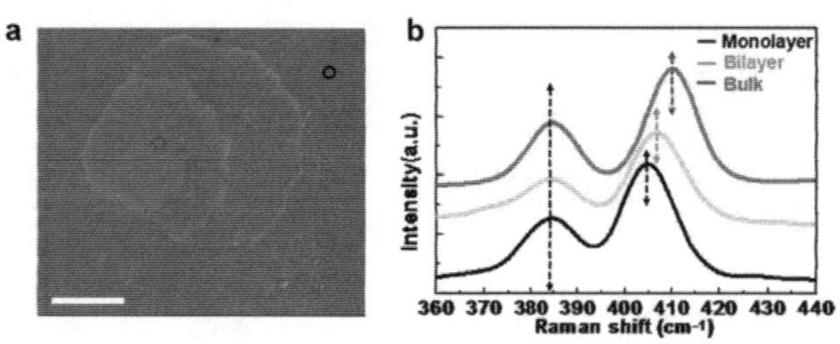

Figure 2. Spiral-like MoS_2 domain. (a) Optical image of spiral-like MoS_2 domain. Scale bar, 20 μm. (b) The Raman spectrum of corresponding area in (a).

2.2. Monolayer MoS2 in large scale

Large-scale monolayer MoS_2 has been readily prepared by our own lab-built MBE instrument, simply by extending the growing time to three hours or even longer. Figure 3a shows the optical image of SiO_2/Si substrate that is covered with monolayer MoS_2. Raman mapping (the intensity integration of A_{1g} peak) of edges, marked with the red dashed line, is shown in the inset, indicting a high uniformity of monolayer MoS_2. For the purpose of confirming the thickness, monolayer MoS_2 on SiO_2/Si is further investigated with photoluminescence spectroscopy (PL) measurements. The PL is a useful technique for readily distinguishing the thickness of MoS_2 material. At room temperature, the PL emission of monolayer MoS_2 lies typically between 1.82 and 1.89 eV [15]. The PL spectrum measured at room temperature (Figure 3b) shows two strong energy emission peaks at 1.84 eV (672 nm) and 2.01 eV (616 nm) which are regarded as the signature of monolayer MoS_2.

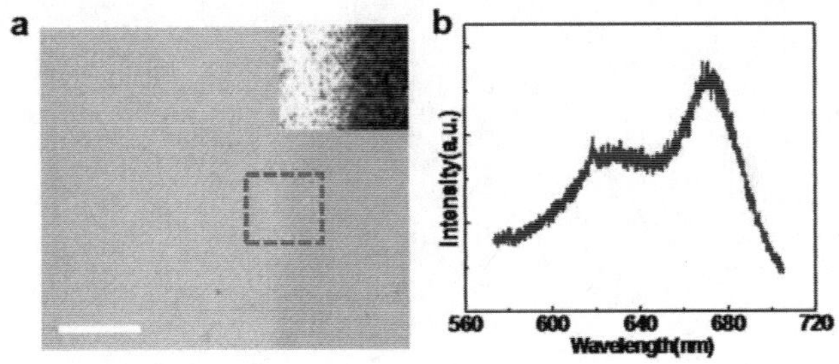

Figure 3. Monolayer MoS_2 in large scale. (a) Optical image of monolayer MoS_2. Inset: Raman mapping of the area indicated in the red dashed line. Scale bar, 60 μm. (b) Photoluminescence spectra of monolayer MoS_2.

3. Conclusion

In summary, we have synthesized triangular nucleation, tens-of-micron domain, and monolayer MoS_2 on 280-nm SiO_2/Si in the present study. Characterizations of the atomically-thin film using optical microscopes, Raman spectroscopy, and PL spectroscopy provide the evidence that morphology, layer thickness, optical properties, and electrical properties of the synthesized MoS_2 material bear the quality similar to those of the exfoliated monolayer MoS_2. As a controllable approach, the MBE technique enables us to study the nucleation and to further synthesize MoS_2 materials in large-scaled areas. It can be suitably applied to studying syntheses of other members of TMDCs and their heterojunctions.

Acknowledgments

This work is supported by the National Natural Science Foundation of China (Grant Nos. 10975115) and the Natural Science Foundation of Fujian Province of China (Grant No. 2012J06002).

References

[1] E. del Corro et al., Nano Lett. 2016, **16,** 2363.
[2] Y. W. Zhang et al., Appl. Phys. Lett. 2016, **108,** 153114.
[3] H. Liu et al., ACS Nano 2015, **9,** 6619.
[4] W. Z. Bao, X. H. Cai, D. Kim, K. Sridhara, and M. S. Fuhrer, Appl. Phys. Lett. 2013, **102,** 042104.
[5] M. Buscema et al., Nano Lett. 2013, **13,** 358.
[6] K. Liu, J. D. Feng, A. Kis, and A. Radenovic, ACS Nano 2014, **8,** 2504.
[7] Q. Ji et al., Nano Lett. 2013, **13,** 3870.
[8] Y. Gong et al., Nature Mater. 2014, **13,** 1135.
[9] B. Radisavljevic, A. Radenovic, J. Brivio, V. Giacometti, and A. Kis, Nature Nanotechnol. 2011,**6,** 147.
[10] A. Rezk et al., Nano Lett. 2016, **16,** 849.
[11] W. Wan et al., *RSC Adv*. 2016, **6,** 323.
[12] A. K. Geim, and K. S. Novoselov, Nature Mater. 2007, **6,** 183.
[13] X. S. Wang, H. B. Feng, Y. M. Wu, and L. Y. Jiao, J. Am. Chem. Soc. 2013, **135,** 5304.

[14] Y. M. Shi et al., Nano Lett. 2012, **12,** 2784.
[15] K. M. McCreary et al., Adv. Funct. Mater. 2014,**41,** 6449.

IOP Conf. Series: Journal of Physics: Conf. Series **864** (2017) 012038 doi:10.1088/1742-6596/864/1/012038

Magneto-optical evidence of the topological phase transition in (111)-$Pb_{1-x}Sn_xTe$

T Phuphachong[1], B A Assaf[2], V V Volobuev[3], G Bauer[3], G Springholz[3], L A de Vaulchier[1] and Y Guldner[1]

[1] Laboratoire Pierre Aigrain, Ecole Normale Supérieure, CNRS, PSL Research University, Université Pierre et Marie Curie, Université Denis Diderot, 24 rue Lhomond, 75005 Paris, France
[2] Département de Physique, Ecole Normale Supérieure, CNRS, PSL Research University, 24 rue Lhomond, 75005 Paris, France
[3] Institut für Halbleiter und Festkörperphysik, Johannes Kepler Universität, Altenberger Strasse 69, 4040 Linz, Austria

E-mail: thanyanan.phuphachong@ens.fr

Abstract. Recently, the study of topological crystalline insulators has become of great interest in condensed matter physics. In this work, we present infrared magneto-optical investigation of such novel quantum states in narrow gap rocksalt IV-VI semiconductor $Pb_{1-x}Sn_xTe$ which exhibits a topological phase transition at a critical value $x_c \approx 0.40$. High-mobility Bi-doped trivial and non-trivial (111)-$Pb_{1-x}Sn_xTe$ ($0 \leq x \leq 0.56$) films grown on BaF_2 substrates by MBE were examined at 4.5K and magnetic fields B=0-15T. Massive and massless Dirac fermion models are used to analyze transmission spectra. We are able to determine the band parameters of the bulk and the topological surface states of such material.

1. Introduction

The search for topological states of matter beyond the existing Z_2 topological insulators (TIs) protected by time-reversal symmetry has recently become one of the major topics in physics of semiconductors. A novel topological state called topological crystalline insulator (TCI) protected by mirror symmetry of the crystal has been discovered in narrow gap rocksalt IV-VI semiconductors such as $Pb_{1-x}Sn_xTe$ and $Pb_{1-x}Sn_xSe$ alloys [1]. Such materials were shown to exhibit semiconducting bulk states accompanied by gapless Dirac surface states. In $Pb_{1-x}Sn_xTe$ a band inversion occurs at four equivalent L-points in the three-dimensional Brillouin zone (3DBZ) when the Sn content is above the critical value $x_c \approx 0.40$, leading to a trivial to non-trivial topological phase transition for $x \geq x_c$ [2]. Since the L-points are mirror symmetric with respect to six equivalent {110} diagonal crystallographic planes, this band inversion results in an even number of surface Dirac cones at points of (001), (110) or (111) surfaces of the crystal where the L-points are projected [3].

In this work, (111)-oriented $Pb_{1-x}Sn_xTe$ films were studied. $Pb_{1-x}Sn_xTe$ crystallizes in a face-centered-cubic structure as NaCl. The small direct band gap of the compound is located at the L-points of the 3DBZ shown in figure 1a. (111)-$Pb_{1-x}Sn_xTe$ possesses four ellipsoidal bulk carrier pockets at the L-points: one bulk longitudinal valley (black) of which the major axis is parallel to the [111] direction and three bulk oblique valleys (red) of which the major axes are tilted by ±70.5° with respect to the [111] direction. A band inversion occurs at four equivalent L-points when going from PbTe to SnTe as illustrated in figure 1b. In the topological trivial phase, the lowest conduction and the highest valence bands are labelled respectively by L_6^- and L_6^+. The energy gap (E_g>0) initially decreases with increasing Sn content, closes at x=x_c, and then re-opens when x>x_c (E_g<0). In the topological non-trivial phase (inverted regime), L_6^+ and L_6^- states form respectively the conduction and valence bands. This band inversion results in the emergence of the topological surface states (TSS) at four different points on the corresponding two-dimensional Brillouin zone (2DBZ). As can be seen in figure 1a, the projections of the L-points are located at the $\bar{\Gamma}$ and $\bar{M}$ points of the (111) 2DBZ. The bulk longitudinal valley results in the $\bar{\Gamma}$-point Dirac cone (blue), while the three oblique valleys give rise to the $\bar{M}$-point Dirac cones (green).

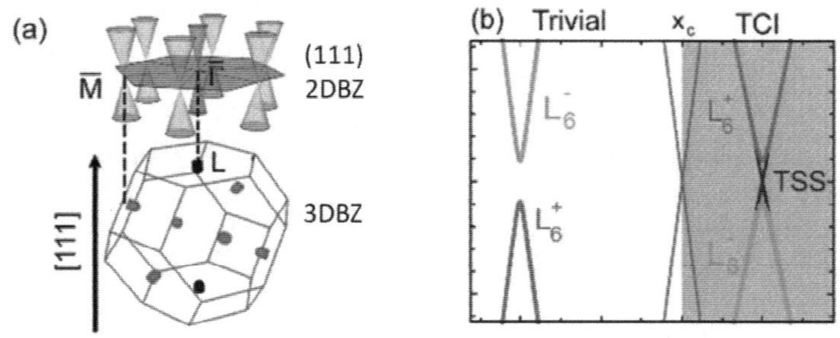

Figure 1. (a) The first 3D Brillouin zone of $Pb_{1-x}Sn_xTe$ accompanied by a corresponding (111) 2D Brillouin zone. (b) Schematic illustration of topological phase transition in $Pb_{1-x}Sn_xTe$ system.

2. Methods

Bi-doped (111)-$Pb_{1-x}Sn_xTe$ ($0 \leq x \leq 0.56$) films were grown by molecular beam epitaxy (MBE) on BaF_2 substrates. PbTe and SnTe effusion cells were employed to control the layer composition [4]. Light Bi-doping is necessary for high Sn content $Pb_{1-x}Sn_xTe$ films for compensating the intrinsic p-type carrier concentration (p>10^{19} cm^{-3}) originating from native (Pb,Sn) vacancies [5]. Bulk carrier densities ~1×10^{18} cm^{-3} and high mobilities ~10000 cm^2/Vs are obtained from transport measurements at 77K. Far- and mid-infrared (FIR and MIR) magneto-optical absorption measurements in the Faraday geometry were performed on $Pb_{1-x}Sn_xTe$ samples at 4.5K and magnetic fields B=0-15T. The spectral range is 4-450 meV. The sample pieces were placed in a superconducting coil. B is applied parallel to the [111] direction. Transmission spectra were acquired and analyzed by a Bruker Fourier transform infrared spectrometer. The relative transmission at a fixed B is defined as the normalization of the sample transmission T(B) by the zero-field transmission T(0). Further information about the sample growth, characterization and magneto-optical absorption experiment is given in [6].

3. Experimental results and analyses

3.1. Bulk states

Figure 2(a,b) show typical IR transmission spectra obtained for x=0.11 and x=0.56 at 4.5K. Strong transmission minima dispersing as a function of B can be clearly seen. Two series of transitions are associated with different bulk valleys, known to have a massive Dirac dispersion. The strongest series denoted by black arrows is attributed to the bulk longitudinal valley. Other transitions marked by red arrows result from the bulk oblique valleys.

The transmission minima are plotted as a function of B in figure 2(c,d) in order to construct transition fan diagrams. Full black circles and empty red circles denote respectively the transitions in the longitudinal and oblique valleys. The green rectangle represents the BaF_2 substrate reststrahlen band (22-55 meV) where no absorption can be observed. A massive Dirac model is used to fit the data for both types of bulk valleys. As the joint density of states is optimal for k_z=0 (**B** and **z**‖[111]), the Landau levels (LLs) at k_z=0 are sufficient to describe the observed magneto-optical interband transitions. The interband transition ($N^v - (N-1)^c$) energies, for p-type materials, in the massive Dirac model are given by [6] and [7]:

IOP Conf. Series: Journal of Physics: Conf. Series **864** (2017) 012038 doi:10.1088/1742-6596/864/1/012038

$$E^{\mathrm{c}}_{N-1} - E^{\mathrm{v}}_{N} = \sqrt{\Delta^2 + 2v_{\mathrm{F}}^2\hbar eB(N-1)} + \sqrt{\Delta^2 + 2v_{\mathrm{F}}^2\hbar eBN} \quad (1)$$

The conduction and valence energies of a LL index N are denoted by E^{c}_{N} and E^{v}_{N}, respectively. Δ is defined to be half of the energy gap E_{g}, $\Delta = E_{\mathrm{g}}/2$. v_{F} is the Fermi velocity. The band edge mass of Dirac fermions is given by $m^* = |\Delta|/v_{\mathrm{F}}^2$. e and $\hbar$ have their usual meaning.
The ground cyclotron resonance (CR) can be written as:

$$E^{\mathrm{v}}_{0} - E^{\mathrm{v}}_{1} = -|\Delta| + \sqrt{\Delta^2 + 2v_{\mathrm{F}}^2\hbar eB} \quad (2)$$

The black and red solid lines in figure 2(c,d) are the calculated transition energies. The CR-L and CR-O refer to the CR originating from the bulk longitudinal and oblique valleys, respectively. An excellent agreement with the massive Dirac fermion model is obtained in all investigated samples.

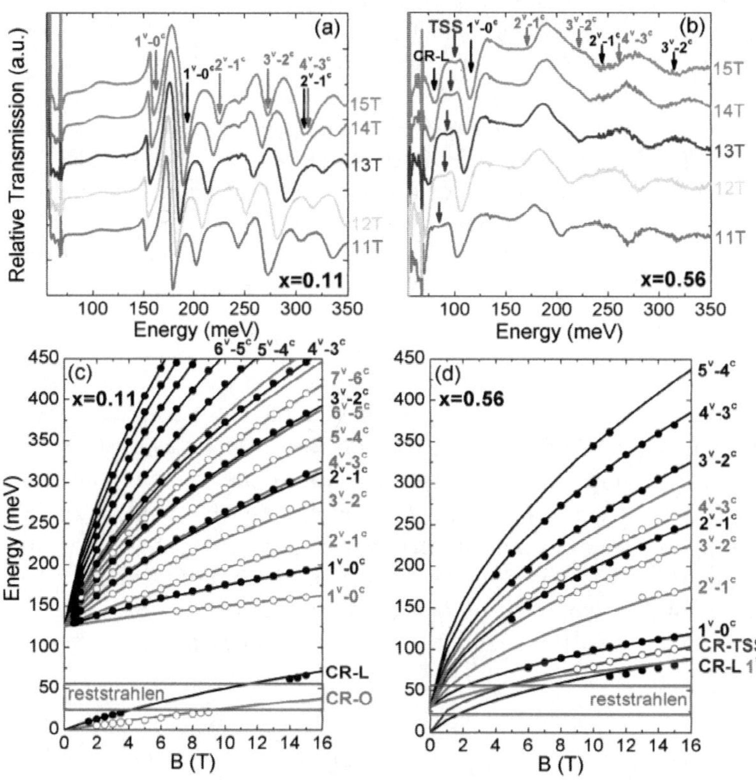

Figure 2. Magneto-optical transmission spectra of (111)-$Pb_{1-x}Sn_xTe$ measured at 4.5K and B=11-15T for x=0.11 (a) and x=0.56 (b). The black and red arrows mark transmission minima resulting from the longitudinal and oblique valleys, respectively. All spectra are shifted vertically for clarity. Transition fan charts are constructed for x=0.11 (c) and x=0.56 (d). Black and red circles denote respectively the transitions in the longitudinal and oblique valleys. Black and red fits are calculated from a massive Dirac model. The transmission minima shown by blue arrows in (b) and blue circles in (d) are interpreted as the CR-TSS using a massless Dirac model. No absorption can be observed in the range 22-55 meV due to the BaF_2 reststrahlen band.

The massive Dirac model allows us to precisely determine band parameters such as E_{g}, v_{F} and m^*. Their values for the two samples x=0.11 and x=0.56 are shown in table 1.

Table 1. Magneto-optical band parameters obtained from the massive Dirac model analysis in $Pb_{1-x}Sn_xTe$ for x=0.11 and x=0.56 measured at 4.5K. The letters l and o denote respectively longitudinal and oblique valleys. m_0 is the electron rest mass.

Composition	E_{g} (meV)	$v_{\mathrm{F}}^{\mathrm{l}}$ ($\times10^5$ m/s)	$v_{\mathrm{F}}^{\mathrm{o}}$ ($\times10^5$ m/s)	m_{l}^*	m_{o}^*
$Pb_{0.89}Sn_{0.11}Te$	125±5	8.15±0.10	5.4±0.1	$(0.017\pm0.001)m_0$	$(0.038\pm0.003)m_0$
$Pb_{0.44}Sn_{0.56}Te$	-30±10	7.1±0.1	4.9±0.1	$(0.005\pm0.002)m_0$	$(0.011\pm0.004)m_0$

Moreover, one can estimate the band edge Fermi energy E_{F} by calculating the Landau level energy at a fixed B where the transmission disappears. In $Pb_{0.89}Sn_{0.11}Te$ the black transition $1^{\mathrm{v}} - 0^{\mathrm{c}}$ is measured down to 0.6T, thus E_F~7.5 meV below the valence band edge of the bulk longitudinal valley. In $Pb_{0.44}Sn_{0.56}Te$ the bulk longitudinal valley transition $1^v - 0^c$ vanishes at about 4-5T, one can thus get E_F~40 meV.

3.2. Topological surface states

Besides the CR-L and the first interband transition $1^v - 0^c$ associated with the bulk longitudinal valley, additional weaker transmission minima indicated by blue arrows in figure 2b are observed in $Pb_{0.44}Sn_{0.56}Te$. Such minima can be interpreted as the CR of the TSS (CR-TSS) and are seen to unambiguously satisfy a massless Dirac model. The surface LL transition energies can be simply obtained by replacing Δ=0 in equation 2. The energy of the CR-TSS is then given by:

$$E_{CR-TSS} = \sqrt{2v_F^2\hbar eB} \qquad (3)$$

The blue solid line in figure 2d is the calculated CR-TSS transition using the same Fermi velocity as the bulk longitudinal valley of $Pb_{0.44}Sn_{0.56}Te$, v_F=(7.1±0.1)×10^5 m/s. Since the Fermi velocities of the longitudinal bulk states and the TSS are found to be equal, this indicates that the observed TSS occur at the $\bar{\Gamma}$-point of the (111) 2DBZ. These observations are in good agreement with recent theoretical calculations [3]. Moreover, since no CR-TSS transition below 8T can be observed, one can deduce a Fermi level in the $\bar{\Gamma}$-Dirac cone E_F~75 meV measured from the Dirac point. Such a CR-TSS is systematically measured in the non-trivial samples and is not observed in the trivial regime. Note that the TSS interband transitions are nearly located at the energy of the strong interband bulk transitions. This can be easily seen by considering the square root terms in equation 1. For $N \geq 1$ and $B>1$T, $(2v_F^2\hbar eBN)^{1/2} \approx (\Delta^2 + 2v_F^2\hbar eBN)^{1/2}$, so that the TSS interband transitions cannot be experimentally resolved.

4. Conclusion

FIR and MIR magneto-optical absorption spectroscopy allows us to map out the LL spectrum in high-mobility Bi-doped (111)-oriented $Pb_{1-x}Sn_xTe$ ($0 \leq x \leq 0.56$) films grown on BaF_2 substrates using MBE. The bulk band parameters (E_g, v_F, m^*) of all samples can be accurately determined using the massive Dirac model. Topologically non-trivial samples ($x>0.40$) are experimentally demonstrated to exhibit massless Dirac TSS occurring at the $\bar{\Gamma}$-point. This is one of the first observations of TSS-related features by magneto-optics. In topologically trivial samples ($x<0.40$), only bulk band transitions can be observed. This study paves the way for further work on the optical properties of topological Dirac fermions that are predicted to hold great promise for quantum electronic devices and data storage applications.

References

[1] Ando Y and Fu L 2015 *Annu. Rev. Condens. Matter Phys.* **6** 361-81
[2] Bauer G 1980 Magnetooptical Properties of IV-VI Compounds *Narrow Gap Semiconductors Physics and Applications* vol 133, ed Zawadzki W (Berlin Heidelberg: Springer) pp 427-46
[3] Safaei S, Kacman P and Buczko R 2013 *Phys. Rev.* B **88** 045305
[4] Springholz G and Bauer G 2013 Landolt-Börnstein – Group III Condensed Matter *Growth and Structuring* vol 34A, ed Klingshirn C (Verlag Berlin Heidelberg: Springer) pp 415-561
[5] Ueta A Y, Springholz G, Schinagl F, Marschner G and Bauer G 1997 *Thin Solid Films* **306** 320
[6] Assaf B A, Phuphachong T, Volobuev V V, Inhofer A, Bauer G, Springholz G, de Vaulchier L A and Guldner Y 2016 *Sci. Rep.* **6** 20323
[7] Mitchell D L and Wallis R F 1966 *Phy. Rev.* **151** 581

Acknowledgements

We acknowledge support from the LabEx ENS-ICFP: ANR-10-LABX-0010/ANR-10-IDEX-0001-02 PSL and support from Austrian Science Fund IRON SFB F2504-N17.

IOP Conf. Series: Journal of Physics: Conf. Series **864** (2017) 012039 doi:10.1088/1742-6596/864/1/012039

Cr doped topological insulator Bi_2Se_3 under external electric field: A first-principle study

Ruqian Lian[1,2], Jian-Min Zhang[1,2,*], Yanmin Yang[1,2], Guigui Xu[1,2], Kehua Zhong[1,2], Zhigao Huang[1,2,*]

[1]Fujian Provincial Key Laboratory of Quantum Manipulation and New Energy Materials, College of Physics and Energy, Fujian Normal University, Fuzhou 350117, China

[2]Fujian Provincial Collaborative Innovation Center for Optoelectronic Semiconductors and Efficient Devices, Xiamen 361005, China

Email: jmzhang@fjnu.edu.cn (J.-M. Zhang), zghuang@fjnu.edu.cn (Z. G. Huang).

Abstract. In this paper, we investigated the magnetic topological insulator (MTI) Cr-doped Bi_2Se_3 film using first principles calculations based on the density functional theory (DFT). The band structure of Cr doped 3QL-Bi_2Se_3 film was calculated comparing with pure Bi_2Se_3 film. Our results demonstrate that the doping of Cr atom changes the degenerate surface state of pure Bi_2Se_3, inducing the ferromagnetism. Under the external electric field, the band gap of pure Bi_2Se_3 films is determined by the charge transfer and the effect of spin-orbital coupling (SOC). For the MTI, the electric field will redistribute the electrons and enhance the magnetism. Our results will further promote the development of the electronic and spintronic applications of topological insulator.

1. Introduction

Topological insulators (TIs) have become a research hotspot due to its gapless state in the surface and energy gap in the bulk [1-6]. Specifically, the prediction and implementation of three dimensions TIs make it is of great interesting [7-8]. The intricate interplay between the topological order and the ferromagnetic order is promising to induce a series of unconventional spintronic effects which will drive the new device paradigms growth [9-12]. As the magnetism introduced, TI is expected to realize the long-sought quantized anomalous Hall effect (QAHE) [13,14]. It has been revealed that Cr atom can be doped into Bi_2Se_3 with both the ferromagnetic and insulating properties [15]. In Cr doped Bi_2Se_3, the Curie temperature has an estimated value of 76K under the doping level of 7.4% [16]. Therefore, the approach to control the band gap of TIs, such as inducing the external electric field, is significant for the application of TIs. Recent study revealed that the pure Bi_2Se_3 film becomes gapless under the external electric field due to the redistribution of the electron [17]. However, the mechanism for MTIs under the electric field, such as Cr doped Bi_2Se_3 films, is still not clear.

In this paper, we study the property of 3QL-Bi_2Se_3 film with one Cr atom doped under the external electric field comparing with the pure case. The total magnetic moment of the system and the band structure are studied using first-principles calculations based on the density functional theory. The result demonstrates that the doped case can be spontaneous stable. In particular, the electric field will slightly enhance the magnetic moment, which has been revealed to break the time reversal symmetry (TRS), and hence open a band gap. Furthermore, the electron transfer induced by the electric field can also increase the band gap.

2. Method and Model

We perform the first-principle density functional theory calculation using Vienna *ab initio* simulation package (VASP) [18,19]. We use the projected augmented wave (PAW) potentials [20] describing the interrction of ion-electron with the Perdew-Burke-Ernzerhof generalized gradient approximation (GGA-PBE) [21,22]. GGA+U calculations that U=3 eV and J=0.87 eV are performed to describe the strong electron-electron correlation. Spin-orbital coupling (SOC) is considered. Figure 1 shows the crystal structure for Cr doped 3QL-Bi_2Se_3 film in our VASP calculation. A Cr atom is replaced for the Bi atom in the upper surface. The experimental lattice constants a=4.138 Å and c=28.64 Å is used for Bi_2Se_3. We set the cutoff energy at 300eV. The Brillouin zone for the 3QL-Bi_2Se_3 is sampled by a gamma-centered 11×11×1 mesh. We use a vacuum layer of 20Å and define the positive direction of the electric field from the doping surface to the un-doping surface.

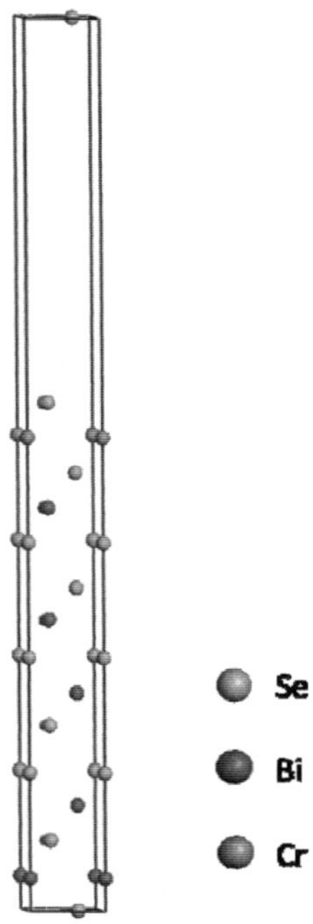

Figure 1. Crystal structure illustration of one Cr atom doped in the 3QL-Bi_2Se_3 surface.

3. Results and Discussions

3.1. Electronic property without the external electric field

We calculated band structure for pure Bi_2Se_3 and Cr doped Bi_2Se_3 respectively, as show in Fig. 2. We find that the pure Bi_2Se_3 has a band gap with the value of 0.111 eV. The gap is induced by the coupling between two surfaces. But for the Cr doped Bi_2Se_3 film, the gap only 0.099 eV can be obtained because of the nondegenerate band after doping. The result indicates that the magnetic atoms of Cr can be effectively doped in the Bi_2Se_3 film with the ferromagnetic order and the insulativity, rendering it is promising for realizing quantized anomalous Hall effect.

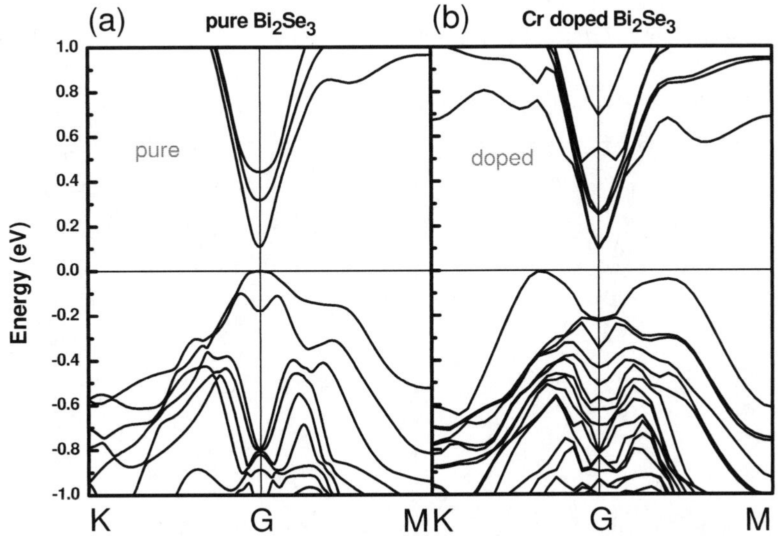

Figure 2. Band structures for (a) pure Bi_2Se_3 and (b) Cr doped Bi_2Se_3 3QL film.

3.2. Electronic property of the pure Bi_2Se_3 film under the electric field

Under the electric field, there are two major reasons can change the energy band gap: the magnetic moment of system and the redistribution of electrons. In the former case, it has been revealed that the gap in surface states of a TI can be opened up by the magnetic perturbation [23]. On the other hand, it is also been demonstrated that external electric field can control the magnetism, but also redistribute the electrons to move in the opposite direction of it that induces an internal electric field and make the charge redistribute again.

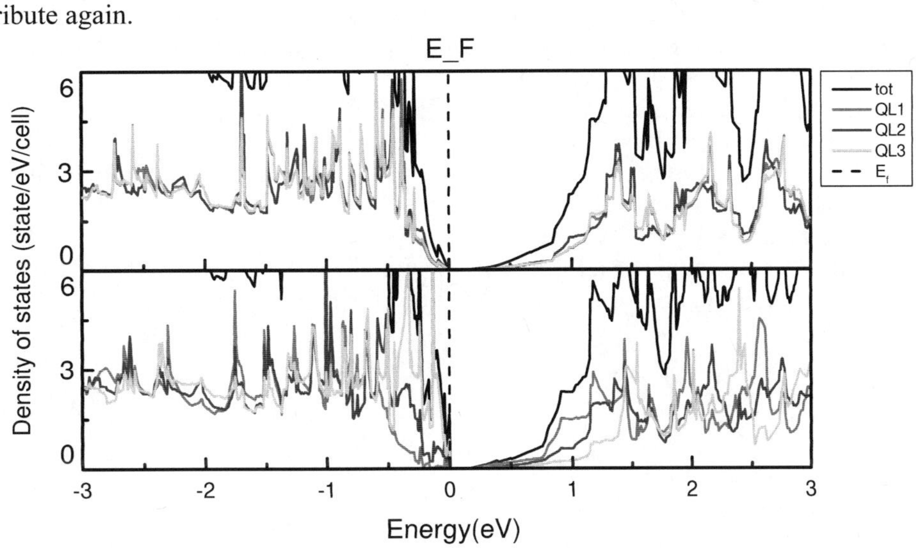

Figure 3. Density of states of 3QLs-Bi_2Se_3 film and the partial densities of states calculated for QL1 (red), QL2 (blue) and QL3 (green) under the electrical field of 0 and 0.02 V/Å

Figure 3 shows the density of states of 3QLs-Bi_2Se_3 film under the electrical field of 0 and 0.02 V/Å. We find the partial densities of states of the upper (red) and the lower (green) QLs are almost overlapping, because of its symmetrical structure. With the increase of electric field, the degenerate state is removed. And the gap of pure Bi_2Se_3 film decrease gradually, as shown in Fig. 4. It suggests that electric field causes the electron transfer and hence to change the gap.

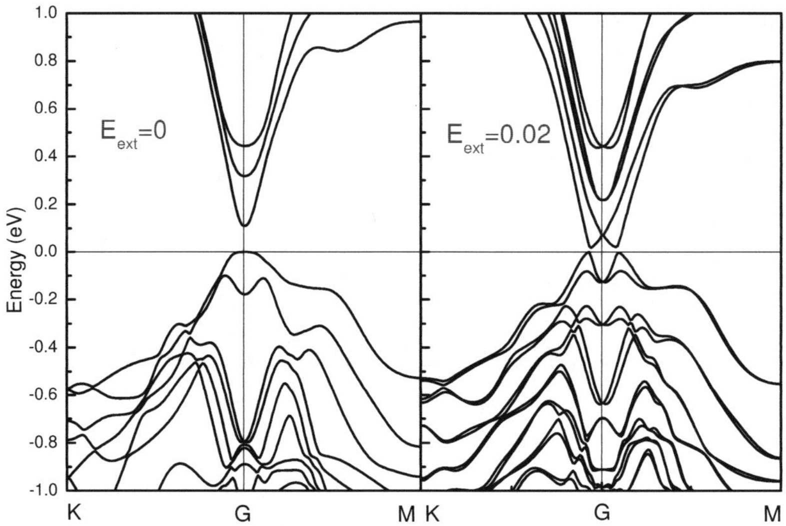

Figure 4. The band structure of 3QLs-Bi_2Se_3 film under the electrical field of 0 and 0.02 V/Å

3.3. Electronic property of the Cr doped Bi_2Se_3 film under the electric field

Comparing with the pure Bi_2Se_3 film, we study the Cr doped Bi_2Se_3 film under the electrical field. For the Cr doped Bi_2Se_3 film, the band gap gradually enlarges as the external electric field increases. While, when E_{ext} = 0.02 V/Å, the band gap is slightly increase by 0.002 eV. From Fig. 6, we can see the gap rises to 0.144 eV for E_{ext} = 0.1 V/Å, from the value of 0.099 eV under the zero fields. This result is different with the pure case. The inevitable reason is that the two surface states are nondegenerate, and the contribution at the top of valence band belongs to the upper surface, which is containing Cr atom, as shown in Fig. 5.

To discuss the magnetic properties of the Cr doped Bi_2Se_3 film, we calculate the ferromagnetic moment under increasing electric fields. With the field increasing from 0 to 0.1 V/Å, the magnetic moment rise from 2.979 μ_B to 2.985 μ_B. It suggests the increasing of electric field enhances the magnetic moment, enlarging the band gap. This result indicates that the external electric field plays an important role on the magnetic and electric properties of magnetic topological insulators (MTIs).

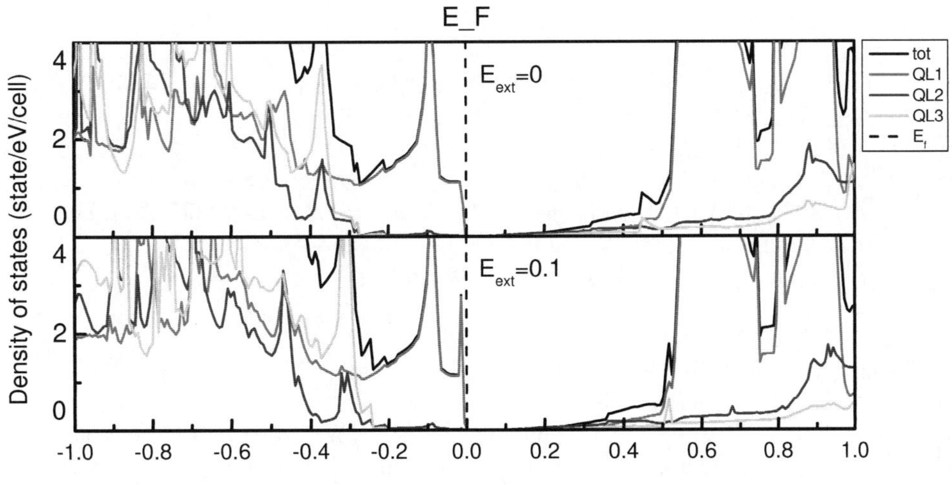

Figure 5. Total densities of states (black) for Cr doped Bi_2Se_3 3QL film and the partial densities of states calculated for QL1 (red), QL2 (blue) and QL3 (green) under the electric field of 0 and 0.1 V/Å.

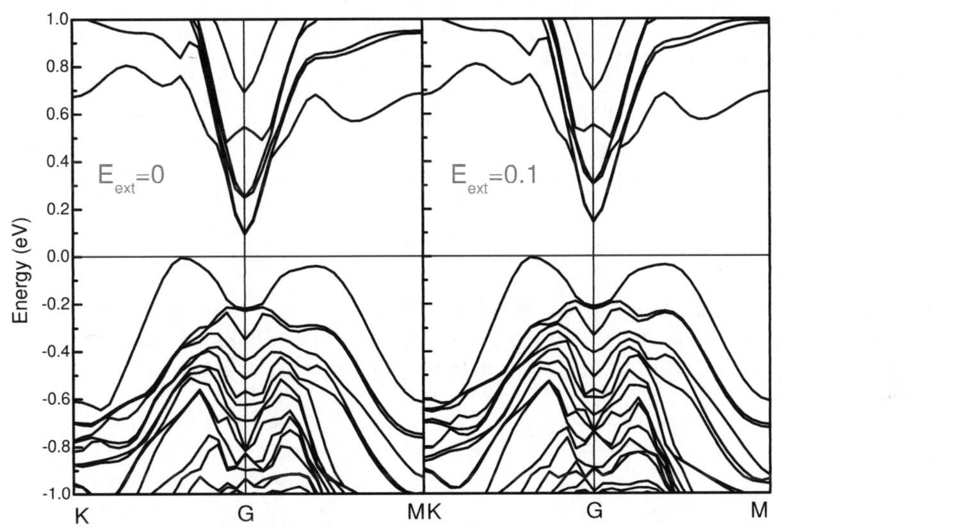

Figure 6. The band structure of Cr doped Bi_2Se_3 3QL film under the electrical field of 0 and 0.1 V/Å

4. Conclusion

In summary, we have studied for the magnetic atom Cr doped Bi_2Se_3 films under external electric field using first-principles calculations based on DFT. We found that the pure Bi_2Se_3 film is more sensitive to the electric field than the Cr doped Bi_2Se_3 film. Our calculated band structure indicates that substitution of Cr makes system become nondegenerate, resulting in a smaller gap. With the increase of electric fields, the charge transfer and magnetism of system will enhance, resulting the band gap of Cr doped Bi_2Se_3 enlarged. Our result indicates that Cr doped Bi_2Se_3 film is electrically tunable, providing significant guidelines towards the study of MTIs.

Reference

[1] Hasan M Z and Kane C L 2010 *Rev. Mod. Phys.* **82** 3045–3067; Qi X L and Zhang S C 2011 *Rev. Mod. Phys.* **83** 1057
[2] Kane C L, Mele E J 2005 *Phys. Rev. Lett.* **95** 146802

[3] Bernevig B A, Hughes T L, Zhang S C 2006 *Science* **314** 1757–1761
[4] Moore J E, Balents L 2007 *Phys. Rev. B: Condens. Matter* **75** 121306
[5] Liu C X, Qi X L, Dai X, Fang Z, Zhang S C 2008 *Phys. Rev. Lett.* **101** 146802
[6] Tse W K, MacDonald A H 2010 *Phys. Rev. Lett.* **105** 057401
[7] Zhang H, Liu C X, Qi X L, Dai X, Fang Z and Zhang S C 2009 *Nat. Phys.* **5** 438
[8] Xia Y, Qian D, Hsieh D, Wray L, Pal A, Lin H 2009 *Nat. Phys.* **5** 398
[9] Liu Q, Liu C-X, Xu C, Qi X-L and Zhang S-C 2009 *Phys. Rev. Lett.* **102** 156603
[10] Qi X-L, Li R, Zang J and Zhang S-C 2009 *Science* **323** 1184
[11] Garate I and Franz M 2010 *Phys. Rev. Lett.* **104** 146802
[12] Wray L A, Xu S-Y, Xia Y, Hsieh D, Fedorov A V, Hor Y S, Cava R J, Bansil A, Lin H and Hasan M Z 2011 *Nat. Phys.* **7** 32
[13] Yu R, Zhang W, Zhang H-J, Zhang S-C, Dai X and Fang Z 2010 *Science* **329** 61
[14] Chang C-Z et al 2013 *Science* **340** 167
[15] Zhang J-M, Zhu W, Zhang Y, Xiao D and Yao Y 2012 *Phys. Rev. Lett.* **109** 266405
[16] Zhang J-M, Ming W, Huang Z, Liu G-B, Kou X, Fan Y, Wang K L and Yao Y 2013 *Phys. Rev. B* **88** 235131
[17] Yang H, Peng X Y, Liu W L, Wei X L, Hao G L, He C Y, Li J, Malcol Stocks G and Zhong J X 2014 *J. Phys.: Condens. Matter* **26** 395005
[18] Kresse G and Hafner J 1993 *Phys. Rev. B* **48** 13115-13118
[19] Kresse G and Furthmüller J 1996 *Phys. Rev. B* **54** 11169
[20] Blöchl P E 1994 *Phys. Rev. B* **50** 17953
[21] Kresse G and Furthmüller J 1996 *Comput. Mater. Sci.* **6** 15
[22] Perdew J P, Burke K and Ernzerhof M 1996 *Phys. Rev. Lett.* **77** 3865
[23] Chen YL, Chu J-H, Analytis J G, Liu Z K, Igarashi K, Kuo H-H, Qi X L, Mo S K, Moore R G, Lu D H, Hashimoto M, Sasagawa T, Zhang S C, Fisher I R, Hussain Z and Shen Z X 2010 *Science* **329** 659

Acknowledgments

We acknowledge the financial support by the National Natural Science Foundations of China (Nos. 61574037, 21203025, 11344008, 11404058, 61404029) and the Natural Science Foundations of Fujian Province of China (Nos. 2015J05005, 2016J01011, 2016J05151)

IOP Conf. Series: Journal of Physics: Conf. Series **864** (2017) 012040 doi:10.1088/1742-6596/864/1/012040

Magnetotransport response in the 3D topological insulator Bi_2Te_3 with indium superconducting electrodes

Zhuo Wang, Tianyu Ye, and R. G. Mani

Department of Physics and Astronomy, Georgia State University, Atlanta, GA 30303, USA

E-mail: zwang16@student.gsu.edu

Abstract. This report studied the magneto-transport properties in the hybrid structure which combines thin flakes of 3D topological insulator (TI) Bi_2Te_3 with topside indium superconducting electrodes. The observed anomalous magnetoresistance suggests two critical transitions. The first transition, obtained approximately at $T = 3.4$ K, is attributed to superconductivity in the indium electrodes. While the second transition, observed at lower temperature, is attributed to the proximity effect generated from the interface between the TI and the superconductor.

1. Introduction

Over the past decade, a new class of quantum spin Hall insulators, called Topological Insulators (TI), have been examined both theoretically and experimentally.[1, 2, 3, 4, 5, 6, 7, 8, 9, 10, 11] Topological insulators (TIs) are electronic materials with a bulk insulating band gap including metallic states on their edges or surfaces in the 2- and 3- dimensional cases, respectively. These exotic gapless surface states of TI display a linear dispersion relation within a single "Dirac cone" and these surface states are topologically protected by the time-reversal symmetry from backscattering by impurities [1, 2, 3, 4, 5, 6, 7]. Among investigations of TI, transport studies of TI-superconductor hybrid structures are thought to be particularly interesting because electrical experimental approaches via the proximity effect have been suggested to find the theoretically predicted Majorance fermions at the TI-superconductor interface [8, 9, 10]. Beyond the conventional proximity effect, which leads to a leakage of Cooper pairs from the superconductor side to the normal metal side, the 3D topological insulators in the presence of an s-wave superconductor are expected to host an exotic p-wave superconductivity capable of hosting Majorana fermions. Thus, motivated by such possibilities, this study examines the magnetotransport response in flakes of 3D topological insulator Bi_2Te_3,[11] contacted with indium superconductor electrodes.[12, 13, 11] We have observed two anomalous transitions on the magnetoresistance below $T = 4.2$ K, which are attributed here to the superconductivity effect in the contacts and generalized proximity effect at the interface, respectively.[11]

2. Experiments and Results

Thin Bi_2Te_3 flakes (~ 25 μm) were mechanically exfoliated from high-quality single crystals and transferred onto Si/SiO_2 substrates. TI/superconductor junctions were realized by directly pressing indium on the surface of Bi_2Te_3 in a Hall bar configuration, see the schematic of measurement in Fig.1(a). The magnetic field was applied perpendicular to the Bi_2Te_3 surface in the range of $-0.3\text{T} \leq B \leq 0.3\text{T}$. The four-terminal lock-in technique was utilized to obtain the magnetoresistance response within a measured section with the length-to-width ratio, $L/W \approx 2$. Various temperatures in range of 1.7 K $\leq T \leq$ 3.8 K were obtained in a ^{4}He cryostat by controlling the vapor pressure of liquid helium. Here, indium is a s-wave superconductor with the critical temperature $T_c = 3.41$ K. Hall effect measurements indicated the carrier concentration is 10^{19} cm^{-3} in the measured samples.

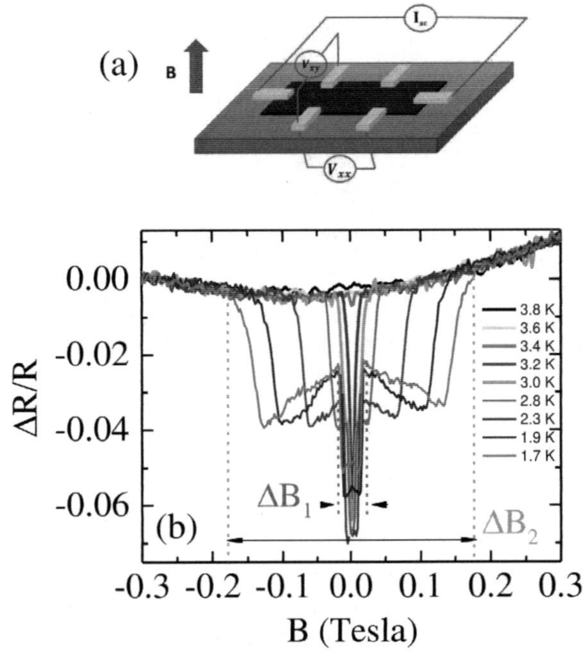

Figure 1. (*a*) The schematic of device measurement. Four-terminal measurement was carried out on the cleaved Bi_2Te_3 flake (black) through indium superconducting contacts (gray) in a Hall bar configuration.(*b*) Normalized magnetoresistance, $\Delta R/R$, vs. the magnetic field, B, at various temperatures in range of 1.7 K $\leq T \leq$ 3.8 K

The normalized magnetoresistance, $\Delta R/R$, which is defined as $\Delta R/R = [R(B) - R(B_N)]/R(B_N)$, where $B_N = -0.3\text{T}$ is shown in Fig.1(b) for temperatures $1.7 \leq T \leq 3.8K$. Above $T = 3.4$ K, $\Delta R/R$ exhibits similar featureless weak positive magnetoresistance of $B = 0.3\text{T}$. Remarkably, at $T = 3.4$ K, which is slightly below the critical temperature of the indium superconductor, there appears a sudden and narrow drop in $\Delta R/R$ near zero magnetic field. This dip mainly grows in depth rapidly with the decreasing temperature until $T = 3.0$ K, while the width, denoted here as ΔB_1 increases. We found that down to $T = 1.7$ K, the critical field for this narrow dip, defined as $B_{c1} = \Delta B_1/2$, is smaller than the critical field of indium superconductor, i.e., $B_c^{In}(0K) = 0.0281\text{T}$. As temperature is reduced below $T = 3.0\text{K}$, there occurs a second, broad drop in the vicinity of $B = 0$. In contrast to the evolution of the first drop, this second dip becomes much broader in width, denoted as ΔB_2, but not depth, at lower temperatures. The critical field of this broad drop is defined as $B_{c2} = \Delta B_2/2$, exceeds the critical field of the indium superconductor at $T = 1.7$ K.

Fig. 2 (a) compares the magnitude of $\Delta R/R$ vs. T at selected constant magnetic field values, which are $B = 0$ mT, $B = 60$ mT, and $B = 120$ mT, respectively. The result demonstrates that the $\Delta R/R$ shows a sudden reduction with decreasing temperatures, and then plateaus. Moreover, the temperature at which the reduction appears is strongly dependent on B, with lower transition temperatures for higher magnetic fields. Next, we plot the critical field for the two terms, i.e., B_{c1} and B_{c2}, with temperature. Fig.2(b) and (c) exhibit the B_{c1} and B_{c2} which

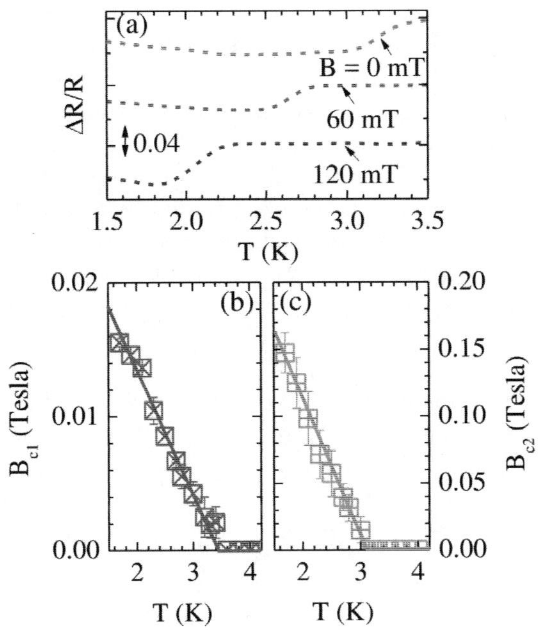

Figure 2. (a) Temperature dependence of $\Delta R/R$ at fixed magnetic fields of $B = 0$ mT, $B = 60$ mT, and $B = 120$ mT, respectively. (b) The critical field associated with the narrow drop, B_{c1}, vs. T. (c) The critical field associated with the broad term, B_{c2}, vs. T.

are extracted from the first derivative of the measured magnetoresistance data by identifying the point of substantial change in the slope. The plot suggests that both critical fields exhibit an approximately linear increase with the decreasing temperatures. Apparently, the rate of increase of B_{c2} with decreasing T is nearly 10 times than that of B_{c1}.

3. Discussion and Summary

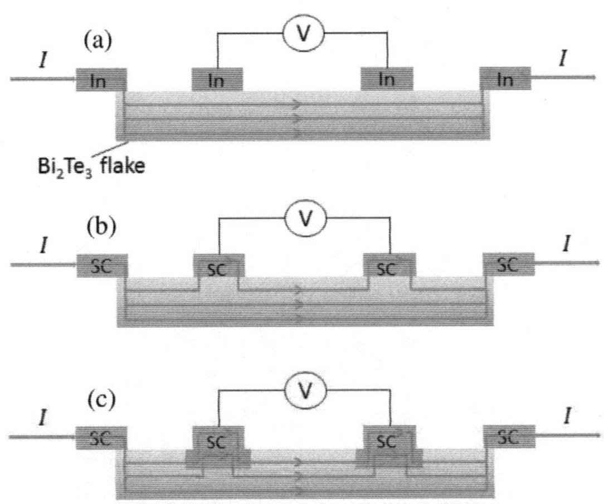

Figure 3. The schematic illustration for explaining the observed magnetoresistance behavior. (a) The image shows the current I flowing through Bi_2Te_3 flake when the temperature is above indium critical temperature, T_{c1}. (b) This illustration for temperature below T_{c1} shows that a fraction of the current I is shunted via indium contacts due to contact superconductivity. (c) The scenario as the temperature further decreases below T_{c2}. The Cooper pairs injected into the Bi_2Te_3 flake induce an additional shunted current in the vicinity of indium/TI interface due to the proximity effect.

The sketch in Fig.4 provides one possible scenario for the observed anomalous magnetoresistance behavior in TI/superconductor hybrid structure, based on previous studies of current flow anomalies.[12, 13, 14, 15] In these experiments, the current was applied through source and drain indium contacts, and voltage was measured through a pair of indium contacts on one side as in the typical four-terminal transport measurement. When $T \geq T_{c1}$, see Fig.4(a), the applied current is carried via both the bulk and the surface states of Bi_2Te_3. Note that in such a high temperature condition, the indium probes act as a normal metal. As the temperature is further reduced below T_{c1}, the indium probes become superconductors, which induces an additional fraction of current I to be shunted through the superconducting probes, as sketched in Fig.4(b). This scenario serves to explain the narrow drop in $\Delta R/R$ observed at $T \approx 3.4$K as mentioned above. Fig.4(c) depicts the situation that occurs with a further decrease in temperature below T_{c2}, where the broad drop in $\Delta R/R$ shows up. In this low temperature condition, proximity effect is induced at the interface in the vicinity of indium superconducting probes, as Cooper pairs are injected into Bi_2Te_3 side. The proximity effect regime, (see the gray areas in Fig.4(c)), results in another shunted current, which gives rise to the second correction in the magnetoresistance occurred at lower temperatures. The only unresolved issue in the experiments is the observed $B_{c2} > 0.028$T in experiment.

In summary, we examined the magnetoresistance in a 3D topological insulator Bi_2Te_3 flake with indium superconducting electrodes. The results indicate two critical transitions, which are associated with two anomalous drops observed in the magnetoresistane. The first transition, with the critical field smaller than the indium critical field, is attributed to the superconductivity effect. The second transition, with the critical field much exceeding the indium critical field is considered as a generalized proximity effect in the TI-superconductor system.

4. Acknowledgments

ZW, TY, and magnetotransport studies Georgia State University have been supported by the U.S. Department of Energy, Office of Basic Energy Sciences, Material Sciences and Engineering Division under DE-SC0001762. Additional support has been provided by the ARO under W911NF-14-2-0076.

5. References

[1] Hasan M Z and Kane C L 2010 *Rev. Mod. Phys.* **82** 3045
[2] Murakami S, Naagaosa N, Zhang S C, 2004 *Phys. Rev. Lett.* **93** 156804
[3] Kane C L and Mele E J 2005 *Phys. Rev. Lett.* **95** 146802
[4] Moore J E 2010 *Nature* **464** 194–198
[5] Fu L, Kane C L and Mele E J 2007 *Phys. Rev. Lett.* **98** 106803
[6] Hsieh D, Qian D, Wray L, Xia Y,Hor Y S, Cava R J and Hasan M Z 2008 *Nature* **452** 970
[7] Xia Y, et al. 2009 *Nat. Phys.* **5** 398
[8] Fu L and Kane C L 2008 *Phys. Rev. Lett.* **100** 096407
[9] Mourik V, Zuo K, Frolov S M, Plissard S R, Bakkers E P A M and Kouwenhoven L P 2012 *Science* **336** 1003
[10] Jiang L, Kane C L and Preskill J 2011 *Phys. Rev. Lett.* **106** 130504
[11] Wang Z, Ye T and Mani R G 2015 *Appl. Phys. Lett.* **107** 172103
[12] Mani R G, Ghenim L and Theis T N 1992 *Phys. Rev. B* **45**, 12098
[13] Ghenim L and Mani R G 1992 *Appl. Phys. Lett.* **60**, 2391
[14] Mani R G and von Klitzing K 1992 *Z. Phys. B* **92** 335
[15] Mani R G and von Klitzing K 1994 *Appl. Phys. Lett.* **64** 1262

IOP Conf. Series: Journal of Physics: Conf. Series **864** (2017) 012041 doi:10.1088/1742-6596/864/1/012041

Nano-Angle Resolved Photoemission Spectroscopy on Topological insulator Sb_2Te_3 nanowires responsible of quantum transport

José Avila[1], Chaoyu Chen[1], Yulieth C. Arango[2], Liubing Huang[3], Detlev Grützmacher[2], Hans Lüth[2], J. Grace Lu[3], Thomas Schäpers[2] & Maria C. Asensio[1*]

[1]ANTARES Beamline, Synchrotron SOLEIL, Université Paris-Saclay, L'Orme des Merisiers, Saint Aubin-BP 48, 91192 Gif sur Yvette Cedex, France.
[2]Peter Grünberg Institute (PGI-9) and JARA Jülich-Aachen Research Alliance, Research Centre Jülich GmbH, 52425 Jülich, Germany
[3]Department of Physics and Astronomy and Department of Electrophysics, University of Southern California, CA 90089, Los Angeles, USA

E-mail: asensio@synchrotron-soleil.fr

Abstract. Using high-resolution Nano-Angle Resolved Photoemission Spectroscopy (Nano-ARPES), we have determined the electronic structure of the surface and bulk states of topological insulator Sb_2Te_3 nanowires, which have been also characterized by magnetoresistance measurements. The observed Aharonov-Bohm-type oscillations could be unambiguously related to the transport by topological protected surface states directly recorded by photoemission. We have measured Nano-ARPES on individual nanowires of a few nanometers wide to provide direct evidence of the existence of the nontrivial topological surface states, as well as their doping. Our findings are consistent with theoretical predictions and confirm that the surface states of intrinsically doped unidimensional topological insulator nanowires are responsible for the quantum transport.

1. Introduction

Topological insulators (TIs) are a new type of fascinating materials that recently has attracted enormous interest. This new state of matter, known as the time-reversal-invariant Z2 TIs is created when the spin-orbit coupling (SOC) is strong enough to invert the conduction and valence bands in the material[1, 2, 3, 4]. They are characterized by an insulating bulk and a conducting surface state, which is protected by time reversal-symmetry (TRS) and it is resistant to scattering from non-magnetic impurities[5]. Their unique bandstructure with a gap less Dirac-cone (linear **E** ***vs*** **k** dispersion relation) residing in the bulk bandgap makes TIs ideal systems for a broad range of applications in condensed-matter physics, materials science, nanoelectronics, spintronics, as well as energy harvesting among others.

In the case of three dimensional (3D) TIs, such as Bi_2Se_3, Bi_2Te_3 and Sb_2Te_3 it was predicted and observed by ARPES that the surface states span the bulk energy gap and have a linear Dirac like dispersion with chiral spin-momentum locking [6, 7, 8, 9]. However, probing the surface state

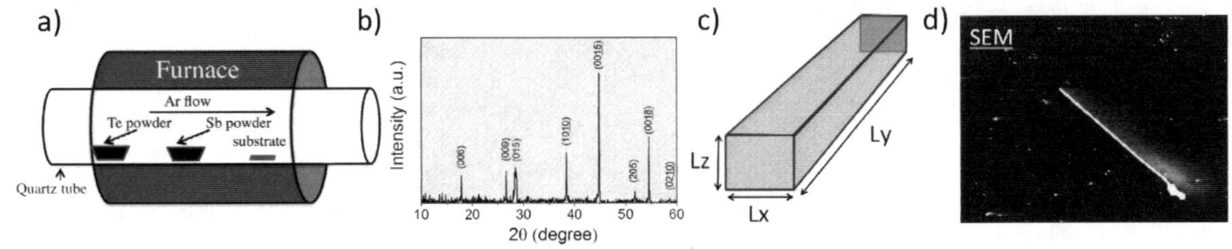

Figure 1. Synthesis and characterisation of Sb_2Te_3 nanowires.(a) Schematic of the Au-catalyzer CVD furnace used to sythetise the nanowires. **(b)** X-ray diffraction pattern of an ensemble of Sb_2Te_3 nanowires. The best fit of the peak verifies the rhombohedral structure of space group (R$\bar{3}$m) (JCPDS card 00-015-0875). **(c)**, Schematic description of the Sb_2Te_3 nanowires investigated, indicating the width, thickness and length of the nanowire as ($\mathbf{L}_x$), ($\mathbf{L}_z$), and ($\mathbf{L}_y$), respectively. **(d)** SEM image of an individual as-grown Sb_2Te_3 nanowire.

in transport experiments still poses a major challenge. In particular, the well-known Aharonov-Bohm oscillations (ABO) observed when a magnetic flux Φ_0 is penetrating the TIs nanowires, can easily be masked or perturbed by the bulk intrinsic defects or unintentional doping.

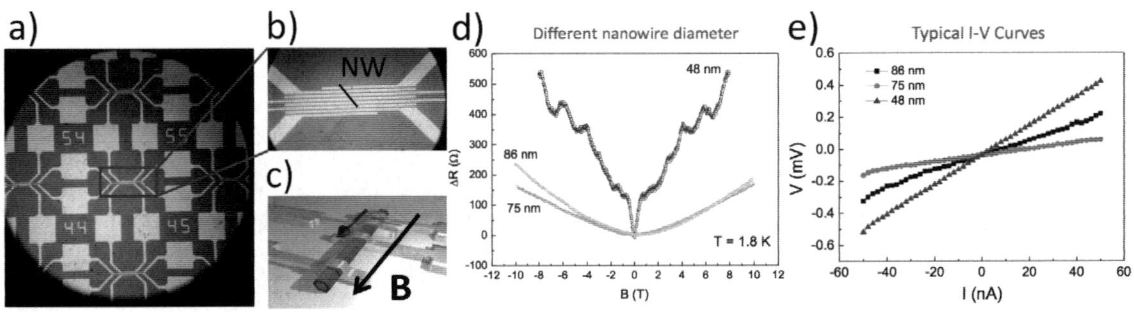

Figure 2. Aharonov-Bohm oscillations of the topological surface states. (a) Optical image of a typical circuit used to measure the nanowires (NW) transport. **(b)** Zoomed view of panel **(a)**. **(c)** Schematic diagram of a Sb_2Te_3 nanowire under a magnetic field along the wire length. **(d)** Magnetoresistance versus magnetic field for three TIs nanowires of different cross-sectional areas (width). **(e)** Typical I-V curves of the same TIs nanowires reported in panel **(d)**

As the surface contribution can be straightforward enhanced by using TIs, like nanowires, ribbons or quantum wells due to their high surface-to-volume ratio, we report here a combined study on Sb_2Te_3 nanowires of magnetotransport experiments together with Nano-ARPES, which validates that the ABO-type effect observed in TI nanowrire transport is due to the contribution of the topologically protected surface states to the quantum transport. Both the bandstructure determined by nanoARPES and the transport data confirm that the nanowires are intrinsically p-doped type.

2. Synthesis and general characterisation of the Sb_2Te_3 nanowires

Sb_2Te_3 nanowires were synthetized by Au-catalyzer chemical vapor deposition (CVD) in a horizontal tube furnace (Lindberg/BLue M) with Sb and Te powder as source materials, following the technical details from Lee et al., [10]. Nanowires were carefully characterized by scanning electron microscopy (SEM) and energy dispersive X-ray spectroscopy (EDS). Powder X-ray diffraction has been used to determiner their structural properties. Figure 1 shows the SEM image of one particular Sb_2Te_3 nanowireas well as a typical X-ray diffraction spectrum with the corresponding peak assignations, ensuring the monocrystal character of the monowires investigated. Following the nomenclature of the schematic nanowire described in Fig. 1c, we have investigated a large set of nanowires with width ($\mathbf{L}_x$), thickness ($\mathbf{L}_z$), and length ($\mathbf{L}_y$) ranging from 42 to 86 nm, from 29 to 73 nm and from 1.8 to 2.0 nm, respectively.

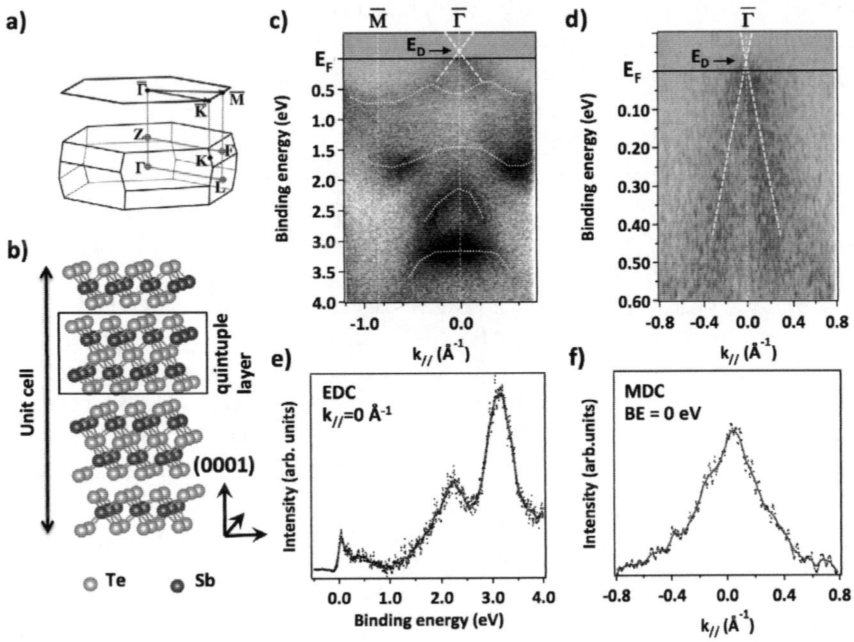

Figure 3. Electronic band structure of an individual Sb_2Te_3 nanowire. (a) Reciprocal three dimensional Brillouin zone with their respective surface projections. **(b)** Crystal lattice of the Sb_2Te_3 material composed of a typical stacks of quintuple layers of Sb and Te along the main crystal axis. **(c)** Nano-ARPES intensity map, showing the **E *vs* k** dispersion relation along the $\bar{\Gamma}\bar{M}$ symmetry direction in the Brillouin zone of a typical Sb_2Te_3 nanowire. **(d)** Zoomed view around the Fermi level of the spectrum in panel **(c)**. **(e)** An energy distribution curve taken from the nanoARPES map in panel **(c)**, at the $\bar{\Gamma}$ point. **(f)** Momentum distribution curve at the Fermi level from spectra in panel **(c)**.

3. Quantum transport measurements

Transport measurements, in principle should be able to show straightforward the low-dimensional surface topologic electronic states in TIs materials. For instance, if conduction occurs mainly through the 2D channel, the conductance would scale with the geometry of the sample surface rather than that of the bulk. Moreover, under magnetic fields, oscillations of the magnetoresistance should vary periodically with the inverse magnetic field (1/B). However, despite extensive transport experiments on bulk TIs materials, there has been no report of a conducting surface layer, and the predicted transport properties of the topological features have not been clearly identified.

Several complete transport studies have been recently reported in TIs nanowires and nanoribbons, where the ABO have been clearly observed[11, 12, 13]. However, the electronic structure of the TIs nano-objects has not been inspected, and consequently neither the surface states nor the doping of the TIs nano-objects could have been determined. Figure 1 shows the set-up of the transport measurements, as well as our main transport results. In particular Fig.2d depicts the low-temperature (T = 1.8 K) magnetoconductance versus the B field (depicted in Fig.2c as an black arrow) parallel to the nanowire axis, for three different nanowire width. Panel Fig.2e of the same figure shows the typical I-V curves of the corresponding nanowires reported in Fig.2d. In agreement with theoretical predictions, the area encircled by the TIs surface states, *(S)* (the cross-sectional area of the wire) defines the ABO period, i.e. $\Delta B = \Phi_0 / S$ [11, 12]. In our case, for the product $\Delta B x S$, values of the order of 4-5 x 10^{-15} T m^2 have been obtained, which are in excellent agreement with the theoretical expected value of the magnetic flux quantum $\Phi_0 = \mathbf{h/e}$, where **h** is Planck's constant and **e** is the electron charge.

4. Electronic bandstructure of Sb_2Te_3 nanowires

After follow a careful protocol of cleaning the nanowires under ultra high vaccum conditions, using several Ar^+ sputtering cycles, we have recorded Nano-ARPES images disclosing the electronic band structure of individual Sb_2Te_3 nanowires. They look rather similar to the bulk and TI experimental and theoretical band structured reported previously for Sb_2Te_3 bulk and films samples [8, 9, 14]. Figure 3 depicts our main results, they have been obtained using the recently developed NanoARPES set-up in the SOLEIL synchrotron[15] All the bulk bands of the Sb_2Te_3 nanowires are clearly identified and indicated by white points. The characteristic TIs surface states responsible of the quantum transport have been also observed clsoe to the Fermi level. They have been marked by dashed yellow lines in the figure. Since the Fermi level cuts the linear Dirac dispersion slightly below the Dirac point, within the gap of the bulk bands, we can confirm not only the robustness of the TI surface states in the nanowires, but also the p-doped type character of these surface states. This data clearly indicate that the nontrivial topological surface states are responsible of the quantum transport measured in the nanowires.

5. Acknowledgments

The Synchrotron SOLEIL is supported by the Centre National de la Recherche Scientifique (CNRS) and the Commissariat à l'Energie Atomique et aux Energies Alternatives (CEA), France.

6. References

[1] C. L. Kane, E. J. Mele, 2005 *Phys. Rev. Lett.* **95** 146802.
[2] B.A. Bernevig, T. L. Hughes, S. C. Zhang, 2006 *Science* **314** 1757-1761.
[3] L. Fu, C. L. Kane, E. J. Mele, 2007 *Phys. Rev. Lett. 98* 106803.
[4] H. J. Zhang, C. X. Liu, X. L. Qi, X. Dai, Z. Fang, S. C. Zhang, 2009 *Nat. Phys.* **5** 438-442.
[5] M. Z. Hasan and C. L. Kane, 2010 *Rev. Mod. Phys.* **82** 3045-3067.
[6] Y. Xia *et al.*, 2009 *Nat. Phys.* **5** 398.
[7] Y. L.Chen *et al.*, 2009 *Science* **325** 178.
[8] C. L. Kane, E. J. Mele, 2009 *Nat. Phys.* **5** 438.
[9] D. Hsieh *et al.*, 2009 *Phys. Rev. Lett.* **103** 146401.
[10] J. S. Lee *et al.*, 2008 *J. Am. Chem. Soc.* **130** 6252
[11] H. L. Peng *et al.*, 2010 *Nat. Mater.* **9** 225.
[12] F. Xiu *et al.*, 2011 *Nat. Nanotechnol.* **6** 216.
[13] L. A. Jauregui *et al.*, 2016 *Nat. Nanotechnol.* DOI: 10.1038/NNANO.2015.293.
[14] L. Plucinski *et al.*, 2013 *J. Appl. Phys.* **113** 053706.
[15] Avila J and Asensio M C. 2014 *Synchrotron Radiation News* **27** 24-30

IOP Conf. Series: Journal of Physics: Conf. Series **864** (2017) 012042 doi:10.1088/1742-6596/864/1/012042

Heat capacity and microhardness of the topological crystalline insulator $Pb_{1-x}Sn_xTe$ near the band inversion composition

E Rogacheva, O Nashchekina, A Nikolaenko and Yu Menshov
National Technical University "Kharkov Polytechnic Institute", Kharkov, UKRAINE
E-mail: rogachova.olena@gmail.com

Abstract. The goal of the present work is to reveal effects accompanying the band inversion in $Pb_{1-x}Sn_xTe$ solid solutions by measuring heat capacity and microhardness. The objects of the study are $Pb_{1-x}Sn_xTe$ alloys with Sn concentrations in the range of x = (0.59 – 0.68), near the composition corresponding to the transition to a gapless state close to room temperature. It was established that in the $Pb_{1-x}Sn_xTe$ solid solutions, the transition to the bulk gapless state with the band inversion is manifested through the appearance of peaks in the dependences of specific heat and microhardness on composition at a fixed temperature.

1. Introduction

Narrow band gap semiconductor $Pb_{1-x}Sn_xTe$ solid solutions are materials widely used in IR - optoelectronics, thermoelectricity and other fields of science and technology [1-3]. The valence band in both SnTe and PbTe consists of two subbands: a non-parabolic light-hole band and a parabolic heavy-hole band. A peculiar feature of the energy band structure of $Pb_{1-x}Sn_xTe$ alloys is the inversion of the conduction and valence bands which occurs at the certain concentration x and the realization of so-called gapless state (GLS) with zero band gap [3-7]. With increasing temperature, the band gap of $Pb_{1-x}Sn_xTe$ increases, the distance between light-hole and heavy-hole subbands in the valence band decreases and the position of the GLS point is shifted to higher concentrations of Sn. If the Sn concentration increases at a fixed temperature, the band gap decreases, then at a certain Sn concentration the GLS occurs, and the distance between the light-hole and heavy-hole subbands starts to grow.

Theoretical and practical interest in studying the $Pb_{1-x}Sn_xTe$ solid solutions has grown sharply in recent years after the prediction and subsequent experimental confirmation of the fact that those solid solutions exhibit properties of topological crystalline insulators (TCIs) in which metallic surface states are protected by the mirror symmetry of the crystal structure [8-10]. Similar to 3D-topological insulators, in the TCIs, the Dirac-cone surface states are produced by bulk band inversion caused by a strong spin-orbit coupling. A topological quantum phase transition between the TCI and trivial phases detected in the $Pb_{1-x}Sn_xTe$ solid solutions using angle-resolved photoemission spectroscopy was close to the band inversion point in the $Pb_{1-x}Sn_xTe$ energy band diagram [8-10]. This fact stimulates interest in studying the properties of the $Pb_{1-x}Sn_xTe$ solid solutions near the composition corresponding to the bulk band inversion in detail.

It can be expected that in the vicinity of the GLS, anomalies in the concentration dependences of both electronic and lattice properties will be observed. By now very few results of detailed experimental studies of electronic properties near the GLS have been reported, and, to the best of our knowledge, there are no data on thermal conductivity, heat capacity, and microhardness in the vicinity of the GLS. Some authors [11-14] tried to detect changes in the local symmetry of atoms and the electron structure in the concentration interval corresponding to the GLS. In [11,12], $Pb_{1-x}Sn_xTe$ solid solutions ($0<x<1$) were studied by Mössbauer's spectroscopy using ^{119}Sn isotope with a view to observing changes in the phonon spectrum accompanying a change in the electron spectrum resulting from a strong electron-phonon interaction near the GLS. The authors detected minima in the dependences of quadrupole splitting, isomer shift and Mössbauer's coefficient on the $Pb_{1-x}Sn_xTe$ solid solution composition in the vicinity of the GLS. They attributed the presence of the concentration-dependent anomalies [11,12] to a decrease in the frequency of optical phonons, or, in other words, the $Pb_{1-x}Sn_xTe$ phonon spectrum softening, and, as a result, a reduction in the symmetry of the Sn atom environment and strengthening of the electron-phonon interaction in the GLS.

However, the authors of [13] pointed out some experimental imperfections of the studies carried out in [11,12] and conducted a similar study of $Pb_{1-x}Sn_xTe$ solid solutions ($0<x<1$) by absorptive Mössbauer's spectroscopy using ^{119}Sn and $^{129m}Te(^{129}I)$ isotopes in order to detect possible changes in the local symmetry and electron structure of cation and anion nodes, as well as the intensity of the electron-phonon interaction near the GLS. According to the results reported in [13], in the vicinity of the GLS in $Pb_{1-x}Sn_xTe$ solid solutions, no anomalous changes in the Mössbauer's spectra width both at 80K and at 295K and no anomalous changes in the electron structure of atoms in the cation and anion sublattices are observed. That is why it was concluded in [13] that Mössbauer's spectroscopy with ^{119}Sn and $^{129m}Te(^{129}I)$ isotopes does not reveal any changes either in the local symmetry of nodes or in the electron structure of atoms or in the intensity of the electron-phonon interaction in the $Pb_{1-x}Sn_xTe$ solid solutions near the GLS. The same results were obtained later in [14] for ^{67}Cu и ^{67}Ga isotopes at 80K.

The goal of the present work was to investigate the behavior of the concentration dependences of specific heat C_p and microhardness H in the vicinity of the band inversion point in the $Pb_{1-x}Sn_xTe$ solid solutions. The objects of the study were $Pb_{1-x}Sn_xTe$ alloys with x in the range of $x = (0.59 - 0.68)$, near the composition corresponding to the transition to the GLS close to room temperature.

2. Experimental

Polycrystalline samples of $Pb_{1-x}Sn_xTe$ solid solutions were synthesized from high-purity elements in evacuated quartz ampoules and annealed for 300 h at 820 K. All samples exhibited *p*-type conductivity. The measurements of $C_p(T)$ were performed on the samples prepared by hot pressing from the polycrystalline bulks and subsequent annealing at 820 K during 200 h. The specimens for measuring C_p were prepared in the form of cylinders 15 mm in diameter and 10 mm long. For all samples, the heat capacity was measured in the range of 170 – 570 K using an adiabatic calorimeter apparatus IT-C-400 in the regime of dynamic heating. For each sample, at least four C_p measurements were taken and their results were averaged. The variation in the thus-obtained C_p values for each sample did not exceed 2 %. On the basis of the measured $C_p(T)$ dependences, the $C_p(T)$ isotherms were plotted.

The microhardness H was measured using a PMT-3 apparatus with a pyramidal diamond indentor at a load of 0.49 N. The H values were obtained as the mean value of the measurement of 30 indentations, the relative standard error being 0.5 - 1%.

3. Results and Discussion

In figure 1a, the temperature dependences of C_p for several $Pb_{1-x}Sn_xTe$ samples are presented. For the rest of the prepared samples, the character of the $C_p(T)$ dependences was similar. It is seen that all the $C_p(T)$ dependences have a minimum near 400 K. The presence of the minimum distinguishes the $C_p(T)$ dependences for the $Pb_{1-x}Sn_xTe$ solid solutions in the vicinity of the GLS from those for the constituents PbTe and SnTe, for which the lattice specific heat remains practically constant above the Debye temperature. To all appearances, such behavior of the $C_p(T)$ dependences is connected with the specificity of the GLS and requires more detailed studies for the correct interpretation of the experimental results.

On the basis of the temperature dependence of C_p we plotted specific heat isotherms, which are presented in figure 1b. As is seen, there is the pronounced peak in the C_p-isotherms near the composition $x = 0.62$, corresponding to the transition into the GLS near room temperature [1]. That is why the observed anomaly in the C_p concentration dependences can be viewed as a critical phenomenon accompanying a second-order phase transition connected with the transition into the GLS with band inversion.

The minimum in microhardness (figure 1c), which is also observed at the critical composition, indicates a reduction in the level of strain in the crystal lattice and leads to a suggestion about self-organization processes occurring in the system when it approaches the critical composition.

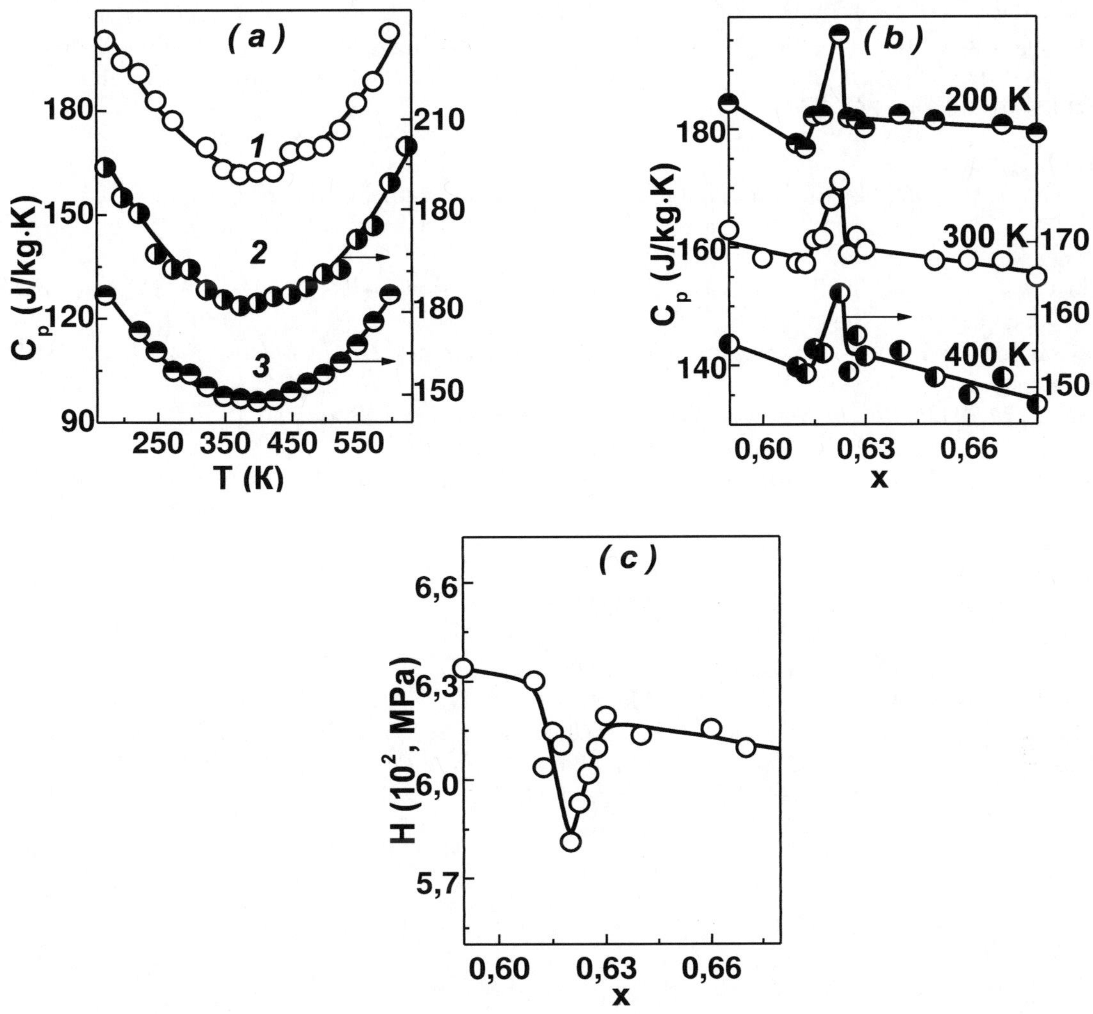

Figure 1. The temperature dependences of the specific heat C_p (a), the isotherms of C_p (b), and room-temperature dependence of the microhardness *H* (c) on the $Pb_{1-x}Sn_xTe$ solid solution composition *x*. a: 1 – x = 0.6225; 2 - x = 0.6; 3 - x = 0.68

The data obtained in this work show that, under the transition into the GLS, a change in the local symmetry of atoms and electron structure in the concentration interval corresponding to the GLS occurs with a high probability which is consistent with the conclusions of the authors of [11,12] about the $Pb_{1-x}Sn_xTe$ phonon spectrum softening leading to the reduction in the symmetry of the Sn atom environment and intensified electron-phonon interaction in the GLS.

4. Conclusions

It was established that in the temperature interval 170-570 K, the temperature dependences of the specific heat C_p of the $Pb_{1-x}Sn_xTe$ solid solutions with x close to the concentration corresponding to the transition into the GLS, represent curves with a minimum.

It was shown that in the $Pb_{1-x}Sn_xTe$ solid solutions, the transition to the bulk gapless state with the band inversion which is accompanied by the topological quantum phase transition from a TCI phase to a trivial phase [8-10] is manifested through the appearance of peaks in the dependences of specific heat and microhardness on composition at a fixed temperature

It is suggested that the transition into the GLS is a second-order phase transition, which is accompanied by changes not only in the electron but also in the phonon spectrum and, with a high probability, by ordering processes.

Acknowledgement
This work was supported by the Ukrainian Ministry of Education and Science (Project # M3925).

References

[1] Abrikosov N H and Shelimova L E 1975 *Semiconducting materials based on A^4B^6 compounds* (Moscow, Nauka) (in Russian)
[2] Ravich Yu I, Efimova B A, Smirnov I A 1970 *Semiconducting Lead Chalcogenides* (New York: Plenum Press)
[3] Rowe D M 1995 *CRC Handbook of/ Thermoelectrics* (CRC Press, Boca Raton)
[4] Zogg H, Fach A, Masek J & Blunier S 1994 *Opt. Eng.* **33** 1440
[5] Tsidil'kovskii I M 1986 *Gapless semiconductors - a new class of substances* (Moscow, Nauka) (in Russian)
[6] Dixon J R and Bis R F 1968 *J. Phys. Rev.* **176** 942
[7] Dimmock J O, Melngailis I and Strauss A J 1966 *Phys. Rev. Lett.* **16** 1193
[8] Gao X and Daw M S 2008 *Phys. Rev. B* **77** 033103
[9] Su-Yang Xu *et al* 2012 *Nature Communications* **3** 1192
[10] Chenhui Yan *et al* 2014 *Phys. Rev. Lett.* **112** 186801
[11] Nikolaev I N, Shotov A P, Volkov A F, Mar'in V P 1975 *JETP Letters* **21** 65
[12] Nikolaev N N, Potapov V P, Shotov A P, Yurchakevich E E 1977 *JETP Letters* **25** 168
[13] Masterov V F, Nasredinov F S, Nemov S A 1996 *Solid State Physics* **38** 2973 (in Russian)
[14] Nemov S A, Seregin N P 2002 *Semiconductors* **36** 852

IOP Conf. Series: Journal of Physics: Conf. Series **864** (2017) 012043 doi:10.1088/1742-6596/864/1/012043

Electromechanical coupling in suspended nanomechanical resonators with a two-dimensional electron gas

A A Shevyrin[1,2,*], A G Pogosov[1,2], A K Bakarov[1,2] and A A Shklyaev[1,2]

[1] Rzhanov Institute of Semiconductor Physics SB RAS, Lavrentyeva avenue, 13, Novosibirsk, 630090, Russia

[2] Novosibirsk State University, Pirogova street, 2, Novosibirsk, 630090, Russia

E-mail: *shevandrey@isp.nsc.ru

Abstract. A physical model describing the piezoelectric-effect-mediated influence of bending of a thin suspended cantilever with a two-dimensional electron gas on the conductivity is proposed. The model shows that the conductivity change is almost entirely caused by the rapid change in mechanical stress near the boundary of suspended and non-suspended areas, rather than by the stress itself. An experiment confirming that the electromechanical coupling is associated with the piezoelectric effect is performed. The experimentally measured conductance sensitivity to the cantilever's vibrations agree with the developed physical model.

1. Introduction

The typically studied low-dimensional electron systems are created from a two-dimensional electron gas (2DEG) contained in a semiconductor bulk (such as a GaAs/AlGaAs heterostructure). However, a heterostructure with a sacrificial layer can be used as a base for creation of thin suspended membranes with a 2DEG by means of selective etching [1-3]. The nanostructures fabricated from these membranes are mechanically moveable and can be used as nanomechanical resonators. Experience shows that vibration of the nanostructures with electron gas affects electron transport and locally changes 2DEG conductivity. These suspended nanostructures can be considered as nanoelectromechanical systems, whose mechanical motion can be detected by means of electrical measurements. A series of papers describe non-trivial hybrid electromechanical devices combining a nanoscale mechanical resonator and such mesoscopic objects as single-electron transistors [1] and quantum point contacts [2]. However, the physical mechanism of electromechanical coupling in these systems remains essentially unclear, although it is often mentioned that the coupling is possibly associated with the piezoelectric effect. Moreover, there is no common physical model predicting the magnitude of the effect. In the present paper we propose a model describing the coupling and show that its predictions agree with the experimental data.

2. Physical model

Consider a nanomechanical resonator having the form of a thin cantilever with dimensions $L \times W \times t$ performing small flexural vibrations at the fundamental mode frequency $\Omega_0/2\pi$. Introduce a coordinate system with x-axis directed along the cantilever and z-axis coinciding with the vibrations direction. Let the cantilever contain a 2DEG at the neutral surface (in the middle of the cantilever). The cantilever's shape $U(x,t)$ is known from the Euler-Bernoulli theory [4]. The mechanical stress tensor can be written using the Voigt notations [5] as

$$\sigma_\alpha = \left(-Ez \times \partial^2 U/\partial x^2 \quad 0 \quad 0 \quad 0 \quad 0 \quad 0\right)^T, \tag{1}$$

where E is the Young's modulus corresponding to the direction along the cantilever, and z is the distance from the neutral surface. Due to the piezoelectric effect, the stress leads to electrical polarization $P_\alpha = d_{\alpha\beta}\sigma_\beta$, where $d_{\alpha\beta}$ is the piezoelectric matrix. Consider the case of z -axis coinciding with [001]-direction in an $Al_{0.33}Ga_{0.67}As$ crystal and denote the angle of rotation from [100] direction to x-axis as θ. Then the piezoelectric matrix can be written in the following form:

$$d_{\alpha\beta} = \frac{d_{14}}{2}\begin{pmatrix} 0 & 0 & 0 & 2\cos 2\theta & 2\sin 2\theta & 0 \\ 0 & 0 & 0 & -2\sin 2\theta & 2\cos 2\theta & 0 \\ \sin 2\theta & -\sin 2\theta & 0 & 0 & 0 & 2\cos 2\theta \end{pmatrix}, \tag{2}$$

where d_{14} is the piezoelectric constant equal to -3.04 pm/V for the considered crystal. Electrical polarization P_α for the given stress tensor has only one non-zero component $P_z = -d_{14}/2 \times Ez\sin 2\theta \times \partial^2 U/\partial x^2$. This non-uniform polarization can be equivalently represented by the volume bound charge $\rho_\mathrm{b} = -\mathrm{div}\mathbf{P} = -\mathrm{d}P/\mathrm{d}z$ which is compensated by the surface bound charges $\sigma_\mathrm{b} = -\rho_\mathrm{b}t/2$ on the top and the bottom surfaces of the cantilever.

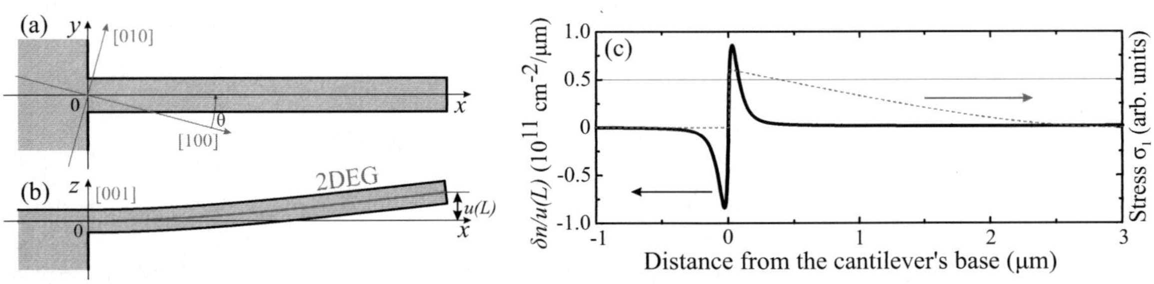

Figure 1. (a), (b) The bent cantilever and its orientation with respect to the crystallographic axes. (c) The calculated electron density change δn per unit displacement of the cantilever' free end $u(L)$. The results are shown for the cantilever with length $L = 3$ μm and width $t = 166$ nm.

The 2DEG reacts on appearance of these bound charges by changing the electron density distribution to maintain constant electrochemical potential:

$$-e(\delta\phi_\mathrm{b} + \delta\phi_\mathrm{2DEG}) + \delta\mu_\mathrm{ch} = 0. \tag{3}$$

Here e is the elementary charge, $\delta\phi_\mathrm{b}$ and $\delta\phi_\mathrm{2DEG}$ are the electrical potentials created by the piezoelectric bound charges and the changed 2DEG density δn, respectively and $\delta\mu_\mathrm{ch} = \delta n \times \pi^{-}\hbar/m^*$ is the chemical potential change (m^* is the effective electron mass). To estimate the electron density change, we neglect the term $\delta\mu_\mathrm{ch}$ and use the model of pure electrostatic screening [6]. Considering the influence of a point charge located at the distance r from the 2DEG plane, it can be shown that such approach is reasonable when $r \gg a_B$, where $a_B \approx 13$ nm is the Bohr radius in GaAs. Since $t \gg a_\mathrm{B}$ in our case, the model of pure electrostatic screening takes into account the most of the induced bound charges. The simplified equation $\delta\phi_b + \delta\phi_{2DEG} = 0$ can be solved using the method of images. To simplify the estimates we neglect the small bending of the cantilever and consider the 2DEG as an infinite zero-potential plane sandwiched between two $t/2$-thick insulating layers with dielectric constant $\varepsilon = 13$. However, we save the calculated bound charges ρ_b and σ_b at $0 < x < L$. The calculated electron density change δn per unit displacement of the cantilever free end $u(L)$ is shown in Fig. 1 (c). It can be seen that $|\delta n|$ decreases with the distance from the clamped end much more rapidly than the mechanical stress. Moreover, almost the same density distribution can be obtained for the mechanical stress in the form of the Heaviside step function $\sigma_1(x) = \sigma_1(0)\Theta(x)$. This fact shows that the change in the electron density is determined mainly not by the stress itself, but rather by the rapid change in mechanical stress occuring near the clamping point. It is clear that actual characteristic distance at which the stress becomes negligible at $x < 0$ is the cantilever thickness t.

3. Experiment

The experimental samples are created from the [001]-oriented GaAs/AlGaAs heterostructure with a 2DEG described in detail elswhere [3]. The 2DEG density is $n \approx 6.7 \times 10^{11}$ cm^{-2} and the mobility is $\mu \approx 1.2 \times 10^6$ cm^{2}V^{-1}s^{-1}. Lateral geometry of the samples is defined by means of electron beam lithography followed by reactive ion etching. The sacrificial layer is selectively removed from-under the nanostructures by means of wet etching in HF water solution. Each experimental sample represents a suspended conductive cantilever having a length of $L = 3$ μm, a width of $W = 2$ μm and a thickness of $t = 166$ nm (see Fig. 2 (a)). The cantilever is perforated by a series of through holes arrayed in a line. The regions between the holes are non-conductive due to the edge depletion, except for a single constriction near the cantilever's base where the inter-hole distance is enlarged up to 600 nm. This is the only conductive channel connecting the source and the drain areas. The measured constriction's resistance is $1/G_0 \approx 1$ kΩ (after substraction of the resistance of the contacts and the mesa). Each experimental chip contains two distanced identical cantilevers oriented in the perpendicular directions [110] and [$\bar{1}$10].

The cantilevers' flexural vibrations are driven in the direction perpendicular to the surface via

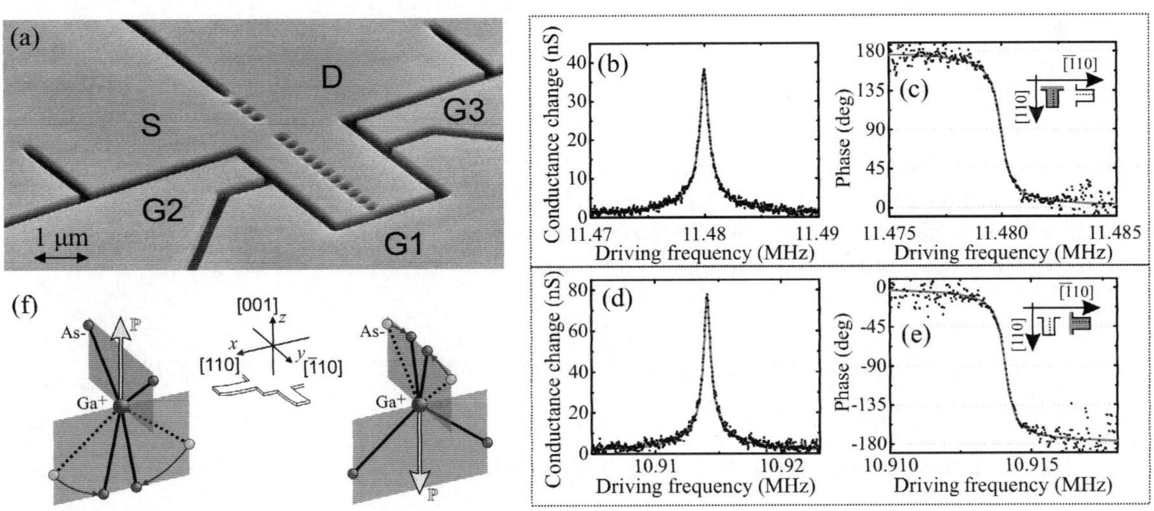

Figure 2. (a) A scanning electron microscope image of an experimental sample. Notations: S - source, D - drain, G1, G2, G3 - gates. Amplitudes (b,e) and phases (c,d) of the conductance change as a function of the driving frequency. (f) Identical stresses applied in the perpendicular directions lead to opposite electrical polarizations.

electrostatic capacitive actuation [3] by means of applying a voltage $V_{dc} + V_{ac}\cos\Omega t$ to the gate G1 surrounding the cantilever's free end. The force in the considered direction arises because the substrate underlying the cantilever has a high dielectric constant and acts as a bottom gate [3]. Amplitude δG_0 and phase φ of the change in the constriction conductance are obtained using the heterodyne down-mixing method [3]. During the experiment, the samples are placed in a vacuum chamber and cooled down to the liquid helium temperature 4.2 K.

Fig. 2 (b-e) shows δG_0 and φ as a function of the driving frequency for both cantilevers. The measurements are performed at $V_{dc} = 2$ V and $V_{ac} = 50$ mV. These curves (obtained by means of the electrical measurements) have the form characteristic for amplitude-frequency and phase-frequency dependences, which can be obtained for a classical driven linear oscillator. This fact demonstrates linearity of the electromechanical coupling. The measured resonant frequencies $\Omega_0/2\pi$ are 11.455 MHz and 10.889 MHz, while quality factors Q are 18 000 and 23 400.

The measured phase-frequency curves show that identical vibrations of the orthogonal cantilevers lead to the conductance responses differing by 180°. In other words, identical bending of the cantilevers leads to opposite changes in the conductance. This anisotropy is a qualitative evidence of the fact that electromechanical coupling has piezoelectric nature. Indeed, for the piezoelectric matrix determined by equation (2), we can expect opposite electrical responses for the perpendicularly oriented cantilevers (see Fig. 2 (f)).

The vibrations' amplitude, as shown in [3], can be estimated as $u(L) = C' V_{dc} V_{ac} Q/(2m\Omega_0^2)$, where $m = 0.24\rho t W L$ is the cantilever's effective mass, ρ is the mass density, $C' = -0.39\varepsilon_0 W L/d_0^2$ and $d_0 = 400$ nm is the distance between the cantilever and the substrate equal to the sacrificial layer's thickness. These estimates give the amplitudes of 20 and 29 nm and the relative conductance sensitivities to the vibrations $(1/u(L))\delta G_0/G_0$ of 0.56×10^{-6} and -0.86×10^{-6} nm^{-1} for the [110]- and [$\bar{1}$10]-oriented cantilevers, respectively. These values can be compared to the relative change in electron density $(1/u(L))\delta n/n$ predicted by the developed physical model. The value of δn should be taken at the distance $x = 1.3$ μm between the constriction's center and the boundary between suspended and non-suspended areas whose position can be easily determined with a scanning electron microscope. The determined value $(1/u(L))\delta n/n = 0.55 \times 10^{-6}$ nm^{-1} agrees with the experimental data and confirms the fact that the developed physical model adequately describes the experiment.

4. Conclusion

A physical model is developed, which predicts the change in the conductivity of a 2DEG contained in a vibrating nanomechanical resonator. The model shows that the conductivity change is determined mainly by rapid change in mechanical stress near the boundary of suspended and non-suspended areas, rather than by the stress itself. An experiment is described, that shows piezoelectric nature of the electromechanical coupling in nanoelectromechanical systems with a 2DEG. The experimental result agree with the model predictions.

Acknowledgments

The work is supported by RFBR Grants 16-32-60130, 15-02-05774, and 16-02-00579.

References

[1] Okazaki Y, Mahboob I, Onomitsu K, Sasaki S and Yamaguchi H 2016 *Nat. Commun.* **7** 11132
[2] Cleland A N, Aldridge J C, Driscoll D C, and Gossard A C 2002 *Appl. Phys. Lett.* **81** 1699
[3] Shevyrin A A, Pogosov A G, Budantsev M V, Bakarov A K, Toropov A I, Rodyakina E E and Shklyaev A A 2016 *Appl. Phys. Lett.* **106** 183110
[4] Landau L D and Lifshitz E M 1970 *Theory of Elasticity, A Course of Theoretical Physics Vol 7* (New York: Pergamon Press)
[5] Nye J F 1985 *Physical Properties of Crystalls* (London: Oxford University Press)
[6] Davies H and Larkin I A 1994 *Phys. Rev. B* **49** 4800

IOP Conf. Series: Journal of Physics: Conf. Series **864** (2017) 012044 doi:10.1088/1742-6596/864/1/012044

Impact of a polystyrene-based passivating interlayer on hybrid polymer/ZnO nanowire heterojunctions

S Bley[+], A Castro-Carranza, E Tansey, W Seemann and J Gutowski

Institute of Solid State Physics, Semiconductor Optics Group, University of Bremen, Bremen 28359, Germany

[+]E-Mail: bley@ifp.uni-bremen.de

Abstract. To improve the quality of hybrid p-polymer/n-zinc oxide (ZnO) nanowire heterojunctions a thin polystyrene (PS)-based passivating interlayer is deposited on plain aqueous-chemically grown ZnO nanowires by spin coating. The structural investigation of the PS-coated nanowires via scanning electron microscopy shows that the concentration of PS in the solution affects the thickness of the PS layer. For vertically aligned nanowires a concentration of 0.125 g in 10 ml toluene is needed to get thin homogeneous passivating layers. Photoluminescence measurements of the PS-coated nanowires show a decrease of the deep-level emission depending on the PS concentration which indicates a passivation on the nanowire surface without affecting the general optical properties of the ZnO. Current-voltage characteristics of these samples show a decreased hysteresis and an increased current with decreasing PS concentration/PS layer thickness. However, a reasonably thin layer reduces the tunnelling through trap states of the ZnO nanowires.

1. Introduction

Developing low-cost and efficient optoelectronic devices for applications as light emitting diodes or energy harvesting devices like solar cells is presently of huge interest.

Hybrid inorganic-organic core-shell structures, especially hybrid inorganic ZnO/organic polymer core-shell nanowires, have recently attracted considerable attention since they provide new functionalities that cannot be achieved with either the organic or inorganic part alone. ZnO is a low-cost II-VI material with a direct band gap of 3.37 eV (room temperature), high electron mobility and high absorbance in the UV spectral range [1]. In particular, ZnO nanowires show that surface-dependent optical and electrical properties are dominant due to the large surface-to-volume ratio which can be systematically tailored by surface-functionalization.

For p-doping of ZnO no reliable and reproducible procedure exists. However, the coating of n-doped ZnO nanowires with p-conductive polymers can result in p-n heterojunctions for diodes and solar cells. Polymers like poly(3,4-ethylenedioxythiophene-polystyrene sulfonate) (PEDOT:PSS) and poly({4,8-bis[(2-ethylhexyl)oxy]benzo[1,2-b:4,5-b′]dithiophene-2,6-diyl}{3-fluoro-2-[(2-ethylhexyl)carbonyl]thieno[3,4-b]thiophenediyl}) (PTB7) are well suited for such structures. Both p-type polymers are well-known for their flexibility, high conductivity, stability and highly absorbent properties, respectively [2]. A controlled deposition of the polymer shell with a thickness control in the nm range is required to form an efficient p-n heterojunction. The polymer deposition via spin coating is a suitable method to provide conformal coating of the nanowires in the nanometer range [3].

However, wet-chemical methods lead to defects at the interface. The quality of the interface has a huge impact on the recombination processes occurring in the heterojunctions. Defects at the interface can act as trap states leading to recombination of charge carriers that lowers the quality of a diode [4]. To decrease the number of defects at the interface, a thin insulating passivation layer with high transparency is necessary. Polystyrene (PS) is a promising candidate which shows a high transparency in the visible spectral range [5].

In this work aqueous-chemically grown (ACG), vertically aligned ZnO nanowires are coated with a PS layer by spin coating to study the impact of this passivating interlayer on the hybrid p-n heterojunctions. The concentration of the PS solution is systematically modified to study its impact on the layer thickness and accordingly on the defect state passivation of the heterojunction interface. Scanning electron microscopy (SEM) images serve to investigate the structural PS layer properties on the nanowire surface. The optical properties (especially the deep level emission) of the PS-coated ZnO nanowires are studied by photoluminescence spectroscopy. Electrical measurements show the I-V-characteristics of the passivated heterojunctions.

2. Experimental section

2.1. ZnO NW

The vertically aligned nanowires (diameter range 50 - 300 nm, length ~ 2 μm) were grown by a hydrothermal method [6] at 80 °C for 3 h on a ZnO seed layer on top of a conductive ITO (indium doped tin oxide) glass as substrate.

2.2. PS deposition via spin coating

For the polystyrene solution two different PS concentrations were used. 0.5 g and 0.125 g of polystyrene were dissolved in 10 ml toluene at 80 °C for 2 h.

The spin coater model WS-650-MZ-23NPP from Laurell was used with an acceleration rate of 1200 rpm/s, a rotation speed of 5000 rpm and a deposition time of 40 sec. Afterwards, the samples were annealed for ~ 5 min to evaporate toluene residuals.

2.3. Structural, optical and electrical characterization

Structural characterization was performed by the FIB/SEM dual system Nova 200 (FEI, acceleration voltage 5 kV, FIB – focused ion beam).

Optical studies of the core-shell nanowires were performed using a HeCd cw-laser IK3552R-G (Kimmon, output power 60 mW, excitation wavelength 325 nm) at room temperature. The overview spectra were recorded using a fiber-coupled spectrometer (AvaSpec-ULS3648TEC, Avantes, spatial resolution ~ 2 nm).

The electrical characterization of the structures was done using a potentiostat-galvanostat Autolab PGSTAT302N and obtained by contacting the coated ZnO NWs and their respective ITO substrates with gold plated round probe tips. A bias voltage of +/- 5 V was used.

3. Results and Discussion

3.1. Structural analysis

Figure 1 shows SEM images of uncoated ZnO NWs (figure 1(a), reference sample, PS-coated ZnO NWs obtained from a solution of 0.5 g polystyrene dissolved in 10 ml toluene (figure 1(b)) and PS-coated ZnO NWs from a solution of 0.125 g polystyrene dissolved in 10 ml toluene (figure 1(c)).

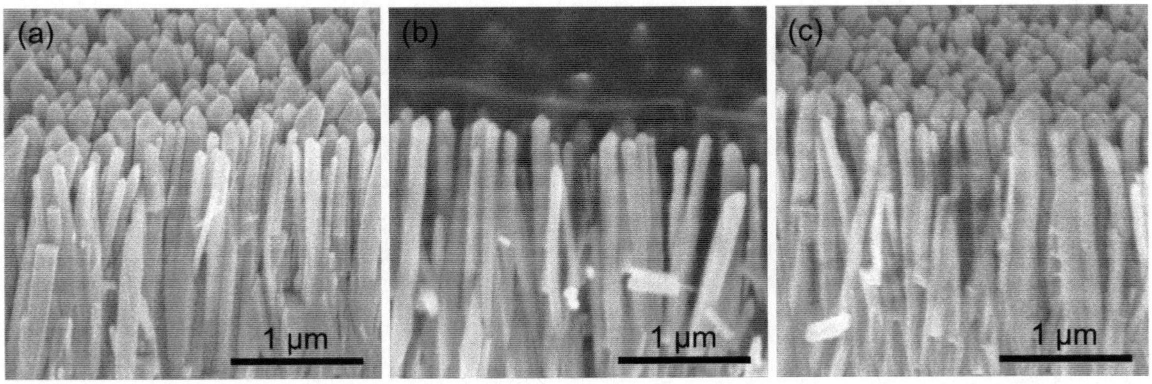

Figure 1: SEM images of (a) uncoated ZnO NWs, (b) PS-coated ZnO NWs (0.5 g PS in solution, see text) and (c) PS-coated ZnO NWs (0.125 g PS).

The PS layer obtained by using the highest PS amount in solution is as a result thick enough to act like an insulating capping layer on the NW tips. It seems that it is unable to cover the nanowires entirely. This may hinder the charge carrier transport in the p-n heterojunction.

However, the coated ZnO NWs obtained by using the lowest PS amount in solution show a very thin and homogenous coating in the small nm range, covering the nanowires completely as observed in figure 1(c).

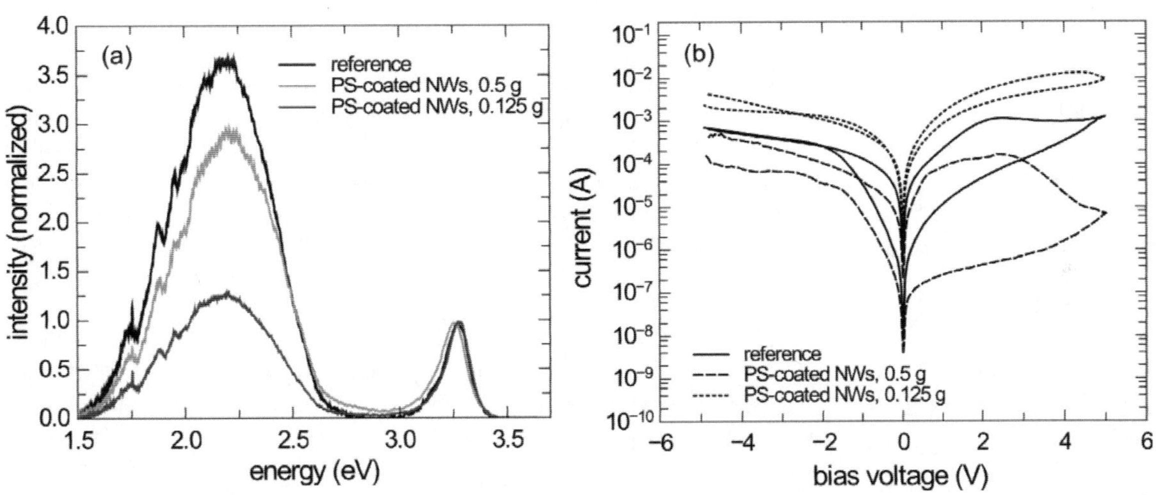

Figure 2: Overview PL spectra (a) and I-V curves (b) of uncoated ZnO NWs, PS-coated ZnO NWs (0.5 g PS amount) and PS-coated ZnO NWs (0.125 g PS amount).

3.2. Optical analysis

In figure 2(a) the PL overview spectra of the uncoated and coated ZnO NWs are depicted. All spectra are normalized with regard to the band of the near-band-edge emission (NBE) at about 3.3 eV to allow for a direct comparison of the relative intensities of the deep-level emission (DLE) which corresponds to the defect luminescence of the ZnO NWs.

In general, the overall optical properties of the nanowires are not significantly affected by the different PS coatings. However, the relative intensities of the observed bands strongly differ for various coating approaches. The black line shows the PL intensity of the uncoated nanowires. Compared to that of the NBE band, the DLE intensity is much higher which is attributed to a large number of defects. The gray curve belongs to the PS-coated NWs with an amount of 0.5 g PS in solution. The DLE band intensity is only slightly reduced. The dark gray line corresponds to the spectrum of the PS-coated NWs with an amount of 0.125 g PS in solution. Compared to the reference the DLE band intensity is strongly reduced by a factor of about 3.

This strong reduction of the DLE band is related to the passivation of the nanowire surface being achieved by an appropriate PS layer. Because of its high transparency, the absorption properties of the polystyrene should not seriously affect the luminescence in this spectral range. For thick PS layers no significant decrease of the DLE band is visible. The thick layer in combination with the wet chemical deposition method may even induce new surface defects on the nanowires.

3.3. Electrical analysis

The I-V curves of the uncoated and coated ZnO NWs are displayed in figure 2(b). The samples were scanned from negative to positive voltage and back to observe the hysteresis effects.

The black curve shows the I-V characteristics of the uncoated reference sample. The relatively low current and the hysteresis indicate that the carrier transport is hindered by surface defects in the ZnO NWs. The black dashed I-V curve belongs to the PS-coated NWs obtained for 0.5 g polystyrene. In comparison to the reference, the current is decreased and the hysteresis increased. The thick polystyrene layer is obviously forming an additional dielectric layer that hinders the carrier transport. The I-V characteristic of the PS-coated NWs obtained for 0.125 g polystyrene is shown as a dotted line. The current is increased and the hysteresis is strongly reduced.

In accordance with the interpretation of the optical measurements, the thin polystyrene layer is passivating the surface which leads to a reduction of surface defects that act as trap states for electrons and therefore, the carrier transport is enhanced.

4. Conclusions

In conclusion, the structural, optical, and electrical properties of PS-coated ZnO nanowire structures have been investigated by SEM, PL and electrical measurements. The achieved results confirm a successful passivation of the interface of ZnO/polymer core-shell heterostructures by deposition of an ultra-thin PS interlayer. The amount of the polystyrene in the solution strongly influences the thickness of the PS layer. The use of small amounts of polystyrene leads to smooth interfaces and an improved carrier transport in the hybrid devices.

Acknowledgements

The authors acknowledge funding by the DFG research group FOR 1616 “Dynamics and Interactions of Semiconductor Nanowires for Optoelectronics” and by the combined University of Bremen/FP7-PEOPLE-2012 COFUND Marie-Curie Fellowship "Bremen Trac", project no. 600411.

References

[1] Bunn C W 1935 *Proc. Phys. Soc.* **47(5)** 835
Mead C A 1955 *Phys. Lett.* **18(3)** 218
Janotti A and Van de Walle C G 2009 *Rep. Prog. Phys.* **72(12)** 126501
[2] Lövenich W 2014 *Polym. Sci. Ser. C* **56** 135
Lu L, and Yu L. 2014 *Advanced Materials* **26(26)** 4413-4430
[3] Hall D B, Underhill P and Torkelson J M 1998 *Polymer Engineering & Science* **38(12)** 2039-2045
[4] Milnes A G and Feucht D L 1972 *Heterojunctions and metal-semiconductor junctions* (Academic Press, New York)
[5] Vala Jr. M T, Haebig J and Rice S A 1965 *The Journal of Chemical Physics* **43** 886-897.
Ishii T, Handa T and Matsunaga S 1977 *Makromol. Chem.* **178(8)** 2351-2366
[6] Ma T, Guo M, Zhang M, Zhang Y and Wang X 2007 *Nanotechnology*, **18(3)**, 035605

Graphical Representation and Origin of Piezoresistance Effect in Germanium

K Matsuda[1], S Nagaoka[2] and Y Kanda[3]

[1] Department of Nano and Bio Material Engineering, Tokushima Bunri University, Sanuki, Kagawa 769-2101, Japan
[2] Department of Electric Systems Engineering, National Institute of Technology Kagawa College, Mitoyo, Kagawa 769-1101, Japan
[3] Research Institute of Industrial Technology, Toyo University, Kawagoe, Saitama 350-8585, Japan

E-mail: kmatsuda@fst.bunri-u.ac.jp

Abstract. The longitudinal and transverse piezoresistance coefficients of Ge at room temperature are represented graphically as a function of the crystal directions for orientation (001), (110) and (211) planes. Many valley model of conduction band and stress decoupling decoupling of the degenerate valence band into two bands of prolate and oblate ellipsoidal energy surface are shown to explain origin of the piezoresistance. One this basis, comparison between piezoresistance coefficient and theoretical model is discussed.

1. Introduction

Sixty years has passed since Smith [1] reported the piezoresistance (PR) effect, and Dresselhaus, Kip and Kittel (D-K-K) investigated the cyclotron resonance of electrons and holes, in Ge and Si crystals [2]. Origin of the PR effect in n-type Si was explained by carrier redistribution between the many valleys and the deformation of the valleys [3]-[5]. The PR of p-type Si has been also reported by several researchers [6]-[10].

Recently, Ge has been used in the framework of heterostructure microelectronics and high speed RF devices. However, in spite of the growing utility of Ge in these devices, and in spite of many studies for strained Si, origin of PR property of Ge as the counterpart of Si has not been fully given since the pioneering work of Smith and D-K-K.

In the present paper, typical PR coefficients of Ge are represented graphically as a functions of crystal direction for common orientations. By analogy to the theory for Si, origin of PR in Ge is explored theoretically. Correlations between the hole effective masses and the longitudinal and transverse PR coefficients of p-type Ge are discussed. The presented results should be of value for researchers to design heterostructure devices.

2. Dependence on Crystallographic Axis

According to the phenomenological description, the fractional change in resistivity with small stress is expressed with the PR coefficients;

$$\frac{\Delta\rho_i}{\rho} = -\frac{\Delta\sigma_i}{\sigma} = \sum_{J=1}^{6} \pi_{ij} X_j \tag{1}$$

where X_j is the component of the stress tensor in six-component vector notation and π_{ij} is the component of the PR tensor. In crystals with cubic symmetry such as silicon and germanium, independent components of the PR tensor are π_{11} , π_{12} and π_{44}.

The transformation is given by direction cosines. Here, three typical PR coefficients will be considered. The first one is a longitudinal PR coefficient when current and field are in direction of the uniaxial stress noted by π_l . The second one is the transverse PR coefficient when the current and field are perpendicular to the stress noted by π_l . And the third one is the shear PR coefficient π_s .

$$
\begin{aligned}
\pi'_{11} &= \pi_l = \pi_{11} - 2(\pi_{11} - \pi_{12} - \pi_{44})(l_1^2 m_1^2 + m_1^2 n_1^2 + n_1^2 l_1^2) \\
\pi'_{12} &= \pi_t = \pi_{12} + 2(\pi_{11} - \pi_{12} - \pi_{44})(l_1^2 l_2^2 + m_1^2 m_2^2 + n_1^2 n_2^2) \\
\pi'_{66} &= \pi_s = \pi_{44} + 2(\pi_{11} - \pi_{12} - \pi_{44})\left[(l_1 l_2)^2 + (m_1 m_2)^2 + (n_1 n_2)^2\right]
\end{aligned} \tag{2}
$$

In an (lmn) plane, the graph can be obtained by making the axis 3' normal to the plane, that is, $[l_3 m_3 n_3] = (l+m+n)^{-1/2}[lmn]$

The graphs of room temperature as a function of crystal direction for orientations in the (001), (011), and (211) planes are shown in Fig.1and 2. The graphs are shown based on the data of Smith. The upper halves of the graphs show positive values of the PR coefficients, that is the resistance increases with the tensile stresses. The lower halves in the figures show negative values of the PR coefficients, that is, the resistive decreases with tensile stresses.

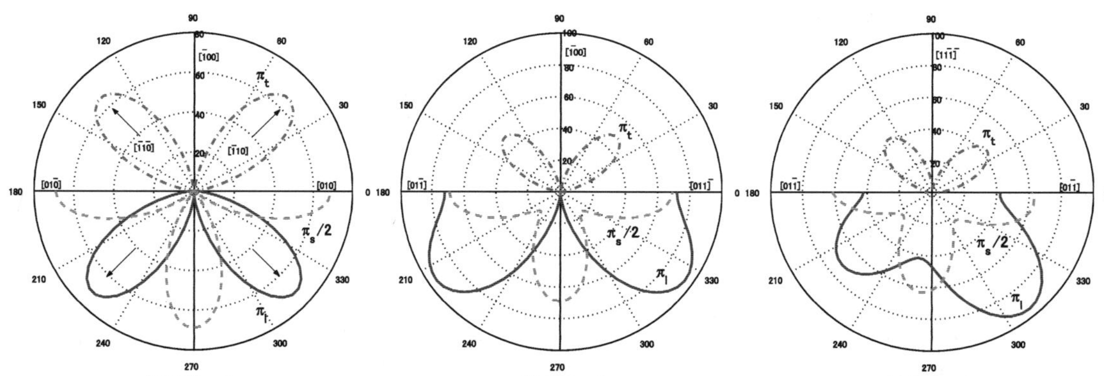

(001) plane (011) plane (211) plane

Fig1. Room temperature PR coefficients of n-Ge $(10^{-12}\ \mathrm{cm^2/dyne})$.

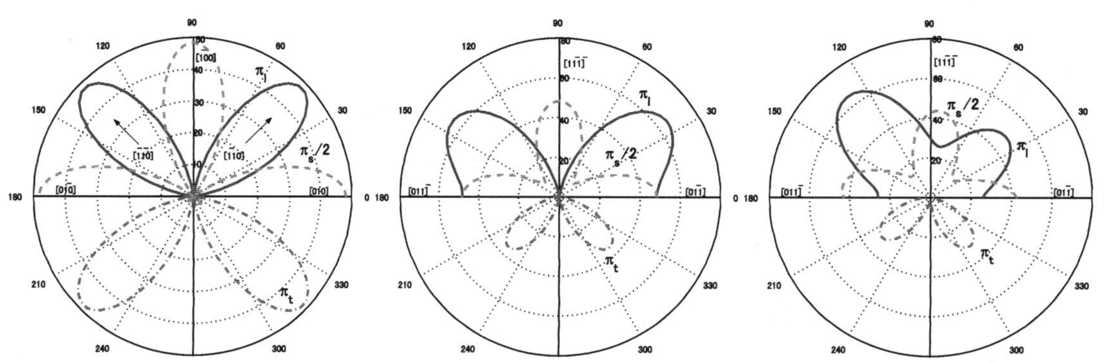

(001) Plane (011) plane (211) plane

Fig2. Room temperature PR coefficients of p-Ge $(10^{-12}\ \mathrm{cm^2/dyne})$.

3. Origin of Piezoresistance Effect

3.1. Many-Valley Model

According to Herring and Vogt [4], the energy surfaces at the band edges are ellipsoid and the carriers transfer between band edges which are shifted individually by stress. Assuming that mobilities ($\mu^{(i)}$) are independent of stress, the change in conductivity under stress is given by

$$\Delta\sigma = \mathrm{ne}\left(\frac{-1}{kT\nu}\right)\sum_{i=1}^{\nu}\mu^{(i)}\left(\Delta E^{(i)} - \frac{1}{\nu}\sum_{j=1}^{\nu}\Delta E^{(j)}\right) \tag{3}$$

where ν is the total number of valleys and $\Delta E^{(i)}$ is the energy shift at the *i*-th ellipsoid band-edge [3].

3.2. Conduction Band

The band-edge shift of each valley in C.B. are given by the deformation potential analysis (Table 1) [3]. Since all band edges in C.B. of Ge shift in same way under $\langle 100\rangle$ stress, conductivity does not change, that is π_{11} and π_{12} vanish. Then we consider the change in conductivity under $\langle 111\rangle$ stress. Conversion of the mobility tensor of each valley gives its components for $\langle 111\rangle$, $\langle 1\bar{1}0\rangle$ and $\langle 11\bar{2}\rangle$ directions (Table 1).

Table 1 Components of mobility and band edge energy shift for valleys, where $\Delta = s_{11} + 2s_{12}$, $\Xi_d + \Xi_u/3$ and Ξ_u are the deformation potentials for dilation and pure shear, respectively.

Mobility / Valley	$\mu^{(i)}_{\langle 111\rangle}$	$\mu^{(i)}_{\langle 1\bar{1}0\rangle}$	$\mu^{(i)}_{\langle 11\bar{2}\rangle}$	Band edge shift $\Delta E^{(i)}$
$L_1(\langle 111\rangle)$	$\mu_{//}$	$\mu_{\perp}$	$\mu_{\perp}$	$(\Xi_d+\Xi_u/3)\Delta\cdot X+\Xi_u s_{44}X/3$
$L_1(\langle 1\bar{1}\bar{1}\rangle)$	$(\mu_{//}+8\mu_{\perp})/9$	$(2\mu_{//}+\mu_{\perp})/3$	$(2\mu_{//}+7\mu_{\perp})/9$	$(\Xi_d+\Xi_u/3)\Delta\cdot X-\Xi_u s_{44}X/9$
$L_1(\langle \bar{1}\bar{1}1\rangle)$	$(\mu_{//}+8\mu_{\perp})/9$	$(2\mu_{//}+\mu_{\perp})/3$	$(2\mu_{//}+7\mu_{\perp})/9$	$(\Xi_d+\Xi_u/3)\Delta\cdot X-\Xi_u s_{44}X/9$
$L_1(\langle 11\bar{1}\rangle)$	$(\mu_{//}+8\mu_{\perp})/9$	$\mu_{\perp}$	$(8\mu_{//}+\mu_{\perp})/9$	$(\Xi_d+\Xi_u/3)\Delta\cdot X-\Xi_u s_{44}X/9$

Substituting Eq.(3) for the Eqns. in Table 1 and using Eqns. (1) and (2), the expression of π_{44} of n-Ge can be obtained as,

$$\pi_{11} = 0, \qquad \pi_{12} = 0, \qquad \pi_{44} = -\frac{K-1}{2K+1}\cdot\frac{\Xi_u s_{44}}{3kT} \tag{4}$$

where $K = \mu_{\perp}/\mu_{//} \cong m_{//}/m_{\perp}$ is the mobility anisotropy factor of conduction band edge..

3.3. Valence Band

By an analogy to many-valley theory, we can consider the hole transfer between the two bands (i.e. the heavy hole band and the light hole band) which are decoupled by the stress application. The energy surfaces of the two bands under a uniaxial stress are prolate and oblate ellipsoidal (Table 2) [8][9].

Since the density-of-state effective masses for HH and LH are compatible under stress, the change of conductivity of Eq.(3) can be written as,,

$$\Delta\sigma \cong \mathrm{ne}\left(\mu^{(HH)} - \mu^{(LH)}\right)\cdot\frac{\Delta \mathrm{E}}{2kT} \tag{5}$$

where $\mu = e^2\tau/m$. It is noteworthy that the change of conductivity is proportional to the difference between HH and LH mass values [10][11]. Substituting Eq.(1) for above equation and Eqns. in Table 2, the PR coefficients of p-Ge can be expressed as,

$$\pi_{11} = \frac{B}{A} \cdot \frac{b(s_{11} - s_{12})}{kT}, \quad \pi_{12} = -\frac{B}{2A} \cdot \frac{b(s_{11} - s_{12})}{kT}, \quad \pi_{44} = \frac{N}{3A} \cdot \frac{ds_{44}}{2\sqrt{3}kT} \tag{6}$$

where $N/3 = \sqrt{B^2 + C^2/3}$.

Table 2. Longitudinal and transverse inverse masses of two bands under $\langle 100 \rangle$ and $\langle 111 \rangle$ stress, and band-edge shift of V.B., where *A*, *B* and *N* are inverse mass parameters, *b* and *d* are the deformation potentials, and upper and lower signs denote the light and the heavy hole bands, respectively.

Inverse mass of ellipsoidal energy surface				Band edges shift ΔE	
$\langle 100 \rangle$ stress		$\langle 111 \rangle$ stress		$\langle 100 \rangle$ stress	$\langle 111 \rangle$ stress
Long.	*Trans.*	*Long.*	*Trans.*		
$A \pm B$	$A \mp B/2$	$A \pm N/3$	$A \mp N/6$	$b(s_{11} - s_{12})X$	$\frac{1}{2\sqrt{3}} ds_{44} X$

4. Discussions

In a quantitative argument, the following data are used: $A = -13.1$, $B = 8.3$, $C = 12.5$; $\Xi_u = 18.7$, $b = 2.1$ eV, $K = 19.3$ and $s_{11} = 0.964$, $s_{12} = -0.260$, $s_{44} = 1.49 \times 10^{-12}$ cm²/dyne [7]. The PR coefficients calculated by the many valley model are summarized in Table 3 as a comparison with the experimental data [1].

Table 3. Theoretical and experimental PR coefficients of Ge at 300 K ($\times 10^{-12}$cm^{-2}/dyne). The figures in parentheses indicate the values estimated by high temperature approximation.

	n-Ge			p-Ge		
	π_{11}	π_{12}	π_{44}	π_{11}	π_{12}	π_{44}
Theory	0	0	- 165	62.7(~0)	- 31.3(~0)	148(~97)
Experiment	- 2.3	- 3.2	- 138	- 3.7	3.2	96.7

Agreement between the theoretical and the experimental values is good but for the unusually large π_{11} and π_{12} of p-Ge. One of the reasons for the discrepancies is due to the low temperature approximation ($\Delta E > 2kT$). Since the magnitude relation of the transverse mass values of two bands is reversed at low temperature [11], the difference between them should be smaller at a high temperature. If the assumption on the DOS effective masses is incorrect in high temperature region, the magnitude of band parameters "*B*" and "*N*" in Eq. (6) will decrease, and π_{11}, π_{12} and π_{44} will become close to the experimental values. The discrepancy of π_{11} and π_{12} in n-Ge should be attributed to the destruction of the rotational symmetry of the ellipsoidal valleys by stress [12].

References

[1] Smith C S 1954 *Phys. Rev.* **94** 42
[2] Dresselhous G, Kip F and Kittel C 1955 *Phys. Rev.* **98** 368
[3] Kanda Y 1982 *IEEE Transaction on Electron Devices* **ED-29**(1) 64
[4] Herring C and Vogt E 1956 *Phys. Rev.* **101** 944
[5] Matsuda K, Suzuki K, Yamamura K and Kanda Y 1993 *J. Appl. Phys.* **73**(4) 1838
[6] Pikus G E and Bir G L 1960 *Soviet Physics –Solid State* **1** 1502
[7] Kanda Y 1987 *Jpn. J. Appl. Phys.* **26**(7) 1031
[8] Hensel J C and Feher G 1963 *Phys. Rev.* **129**(3) 1041
[9] Suzuki K, Hasegawa H and Kanda Y 1984 *Jpn. J. Appl. Phys.* **23**(11) L871
[10] Kanda Y and Matsuda K 2004 *Proceedings ot the 27th ICPS* (Arizona USA) 79
[11] Matsuda K 2004 *Journal of Computational Electronics* **3** 273
[12] Kanda Y and Suzuki K 1991 *Phys. Rev.* **B** 43(8) 6754

IOP Conf. Series: Journal of Physics: Conf. Series **864** (2017) 012046 doi:10.1088/1742-6596/864/1/012046

Insight into phonon scattering in Si nanowires through high-field hole transport: Impacts of boundary condition and comparison with bulk phonon approximation

H Tanaka, J Suda and T Kimoto

Department of Electronic Science and Engineering, Kyoto University, Kyoto, Japan

E-mail: `tanaka@semicon.kuee.kyoto-u.ac.jp`

Abstract. The impact of how to model phonon scattering on hole transport in Si nanowires was studied based on Boltzmann's transport equation. Boundary conditions for atomistic description of phonons in nanowires and approximation by bulk acoustic and optical phonons were analyzed in terms of their impacts on high-field hole transport. The boundary conditions for phonons influence the drift velocity and momentum relaxation time, especially at low electric field, but the energy relaxation time hardly depends on the boundary conditions. The impacts by the change of boundary conditions can be approximated by the change of the strength of acoustic phonon scattering in bulk phonon picture, though the behavior of energy relaxation and distribution function of holes can not be reproduced by bulk phonon approximation.

1. Introduction

Silicon (Si) nanowires (NWs) are a promising candidate for the channel material of future CMOS devices owing to their high electrostatic controllability [1]. To estimate their carrier transport capability, how to model phonon scattering in NWs is important. Compared with the quantum-confinement effects on carriers in NWs, which have been discussed by various authors [2, 3], the impacts of phonon confinement [4, 5] on carrier transport have not been sufficiently clarified.

In this paper, we calculated the hole transport in Si NWs by solving Boltzmann's transport equation in both low and high electric field regions, taking account of various models and boundary conditions for phonon scattering, and their impacts on hole transport were analyzed.

2. Calculation method

2.1. Model of phonon scattering

The phonon dispersion and phonon modes in NWs were computed by a valence force field (VFF) model [6], and the electron-phonon interaction was treated atomistically [7]. Here, the boundary condition (BC) for VFF phonons was varied from free to "free+4 nm" (a NW surrounded by Si with about 4-nm thickness whose outer surface has free BC), and from fixed to "fixed+4 nm". The schematic images of these BCs are presented in Fig. 1.

Approximation by bulk phonons as elastic acoustic phonon scattering and inelastic optical phonon ($\hbar\omega_{\mathrm{op}} = 63$ meV) scattering was also used for comparison. The expression used for bulk phonon scattering in NWs is similar to that in [8], except the substitution of atom-averaged tight-binding probability density ((probability at an atom in a unit cell)/$N_{\rm cell} \times 8/a_0^3$) for squared envelope function ($|\psi_n(y,z)|^2/L$), where L, $N_{\rm cell}$, and a_0 are the total system length, the number of unit cells (as a NW) in L, and the lattice constant of bulk, respectively.

2.2. Model of hole transport
The valence band structure and wave functions were calculated by an $sp^3d^5s^*$ tight-binding approximation [9]. Considering the scattering rate of phonon scattering by Fermi's golden rule, the hole distribution functions under various electric fields were computed by solving the Boltzmann's transport equation. At high field, a nonlinear Boltzmann's transport equation was solved iteratively [10, 11]. The transport orientation and sidewalls of the targeted NW were $[110]/(1\bar{1}0)/(001)$, and the cross-section was a square of 1.9 nm × 1.9 nm. Hole density was fixed at 10^{17} cm^{-3}, and the temperature was assumed to be 300 K.

3. Results and discussion
3.1. Impacts of boundary conditions for phonons
The low-field phonon-limited mobility calculated with various BCs for VFF phonons is presented in Fig. 2(a). Free BC gives the lowest mobility, and fixed BC the highest, since the fixed BC pushes up the low-energy phonons to higher energy as shown in Fig. 2(b). For both BCs at the surface, surrounding Si thickness of 4 nm gives similar mobility, which means that surrounding Si thickness of 4 nm is sufficient to realize a "quasi-bulk" BC.

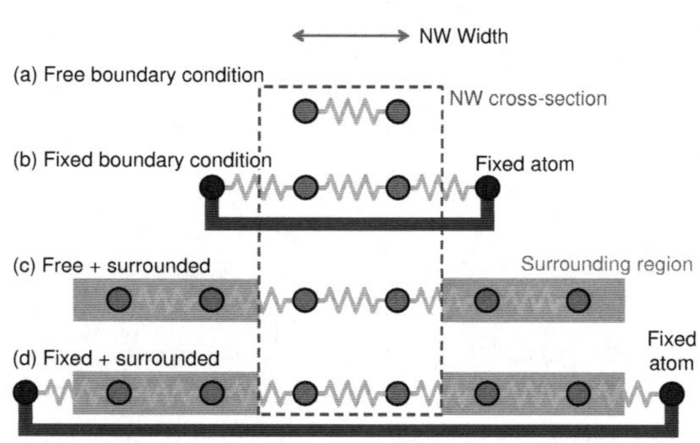

Figure 1. Schematic image of BC models for calculation of phonons in NWs by VFF.

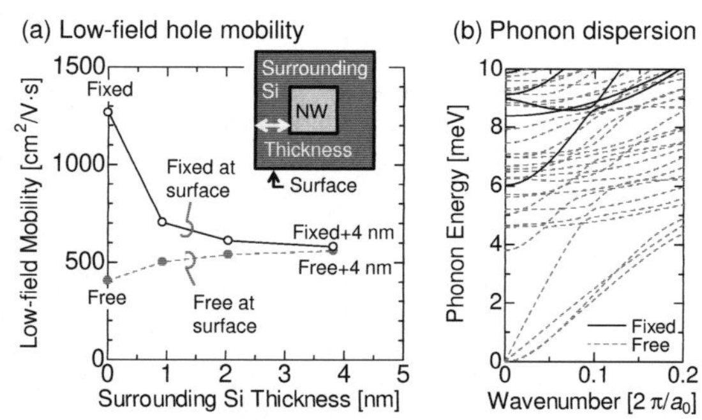

Figure 2. (a) Low-field phonon-limited hole mobility in the Si NW with various BCs. (b) Phonon dispersions in the low-energy region by VFF for the Si NW with free and fixed BCs.

Fig. 3(a) shows the electric field F dependence of average hole drift velocity $\langle v_{\rm d}\rangle$. Compared to the large difference at low field, the BC dependence of $\langle v_{\rm d}\rangle$ is weaker at high field. An averaged effective mass by non-equilibrium hole distribution $m_{\rm avg}$ and average momentum relaxation time $\langle \tau_{\rm m}\rangle$ defined as $\langle v_{\rm d}\rangle = F \times e\langle \tau_{\rm m}\rangle/m_{\rm avg}$ are shown in Figs. 3(b) and (c), respectively. $m_{\rm avg}$ increases with F due to nonparabolicity of the valence band structure [11], and depends on BCs only weakly. The increase of $m_{\rm avg}$ and the decrease of $\langle \tau_{\rm m}\rangle$ cause the negative differential resistance behavior of $\langle v_{\rm d}\rangle$[10, 11]. BC dependence of $\langle \tau_{\rm m}\rangle$ steeply decreases at high field, as optical phonon scattering dominates, which is less affected by BCs.

To analyze the energy relaxation process, the relationship between average hole energy increase ΔE (from equilibrium) and energy relaxation time $\langle \tau_E\rangle$ defined as $1/\langle \tau_E\rangle = e\langle v_{\rm d}\rangle F/\Delta E$ ($e\langle v_{\rm d}\rangle F$ is the energy a hole obtains per unit time from the electric field) is depicted in Fig. 4. Fig. 4 indicates that $\langle \tau_E\rangle$ is almost independent on the BCs. This result originates from the small

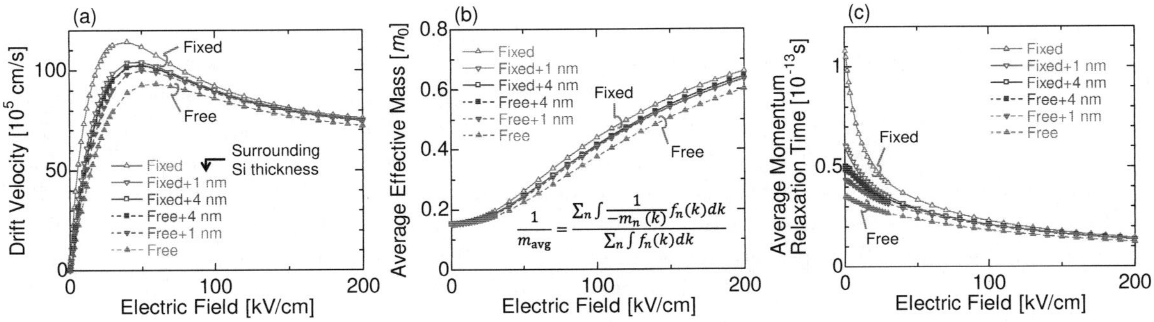

Figure 3. Field dependence of (a) average hole drift velocity $\langle v_{\rm d}\rangle$, (b) average effective mass $m_{\rm avg}$, and (c) average momentum relaxation time $\langle \tau_{\rm m}\rangle$ in the Si NW with various BCs for VFF phonons.

impact of BCs on the phonons with large energy, which dominate energy relaxation. Thus, the smaller $m_{\rm avg}$ in free BC (Fig. 3(b)), which means weaker "heating" of holes, can be attributed to the smaller energy increase originating from smaller $\langle v_{\rm d}\rangle$ ($\propto$ average energy gain per unit time), not from faster energy relaxation.

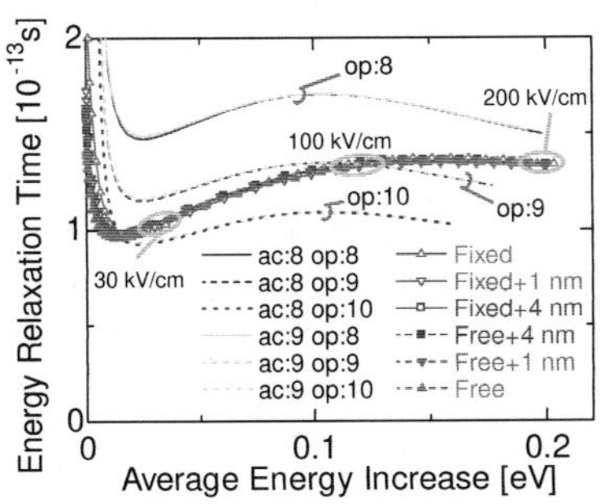

Figure 4. Energy relaxation time $\langle \tau_E\rangle$ as functions of average hole energy increase ΔE (including results by bulk phonon approximation using various $D_{\rm ac}$ [eV] and $D_{\rm op}$ [10^{10} eV/m]).

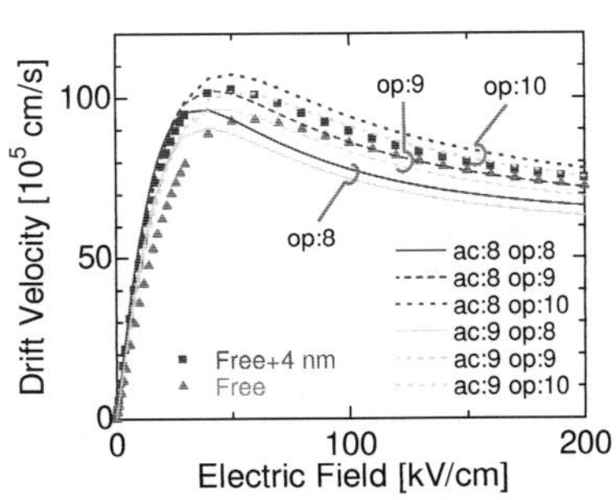

Figure 5. Field dependence of average hole drift velocity $\langle v_{\rm d}\rangle$ calculated using various $D_{\rm ac}$ [eV] and $D_{\rm op}$ [10^{10} eV/m].

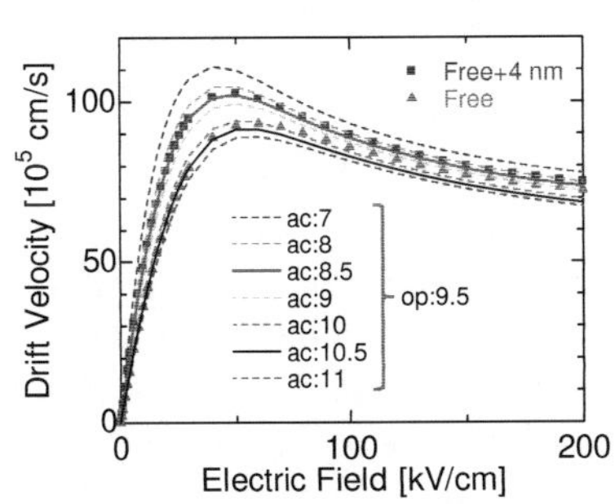

Figure 6. Field dependence of average hole drift velocity $\langle v_{\rm d}\rangle$ calculated using various $D_{\rm ac}$ [eV] and constant $D_{\rm op}$ of 9.5×10^{10} eV/m.

3.2. Comparison with bulk phonon approximation

In Fig. 4, the results by bulk phonon approximation are shown, which were obtained using an acoustic phonon deformation potential ($D_{\rm ac}$) of 8 or 9 eV and an optical phonon deformation potential ($D_{\rm op}$) of 8, 9 or 10 $\times$ 10^{10} eV/m. The $\langle \tau_E\rangle$ is almost independent of $D_{\rm ac}$, and this suggests that $D_{\rm op}$ can be fitted to $\langle \tau_E\rangle$. However, ΔE-$\langle \tau_E\rangle$ curves cannot be reproduced by any $D_{\rm op}$. The trend of the larger $D_{\rm op}$ required to reproduce $\langle \tau_E\rangle$ by VFF phonons at smaller ΔE is explained by the contribution of acoustic phonons to energy relaxation, which is influential at small ΔE and is ignored in the bulk phonon approximation. The field dependence of $\langle v_{\rm d}\rangle$ for these deformation potentials is plotted in Fig. 5 with VFF results for free and quasi-bulk (free+4 nm) BCs. Note that $\langle v_{\rm d}\rangle$ at high field is larger for larger $D_{\rm op}$, owing to faster energy relaxation and resulting smaller $m_{\rm avg}$.

From Figs. 4 and 5, we tried to determine D_{ac} and D_{op} values that can approximate the high-field hole transport in the Si NW. Here, we assumed $D_{op} = 9.5 \times 10^{10}$ eV/m, which reproduces $\langle \tau_E \rangle$ in the 30–100 kV/cm range, and D_{ac} was fitted to $\langle v_d \rangle$. In Fig. 6, drift velocities calculated using various D_{ac} and constant D_{op} of 9.5×10^{10} eV/m are shown. D_{ac} values of 10.5 and 8.5 eV almost reproduced the F-$\langle v_d \rangle$ curves for free and quasi-bulk BCs, respectively. The factor of (10.5/8.5) means that the free BC of phonons can have an equivalent impact to about 25% increase of D_{ac} compared to quasi-bulk. This increase ratio (10.5/8.5) is close to (but smaller than) the experimentally reported increase of D_{ac} (12/9) for electrons in Si MOS interface [12] compared to bulk Si, which may validate the use of free BC to imitate the Si/oxide interface. Even though $\langle v_d \rangle$ was well reproduced around 30 kV/cm, the distribution functions are different between VFF and bulk phonons as shown in Fig. 7, resulting in different ΔE and m_{avg}.

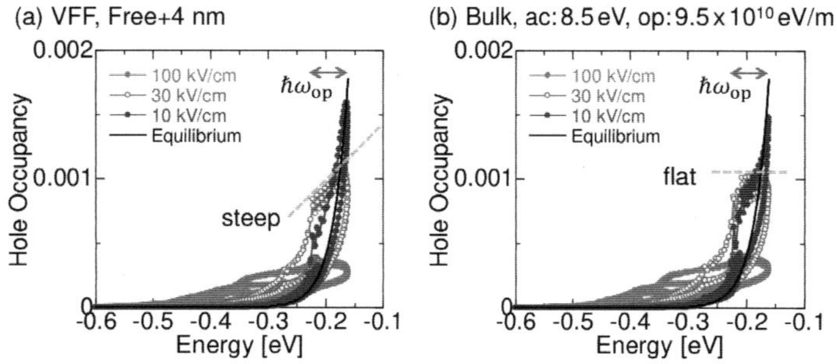

Figure 7. Calculated hole distribution functions of the Si NW under various electric fields. The case with VFF phonons with quasi-bulk BC (a) shows larger occupation near the band edge, while the results for bulk phonon approximation (b) show flatter occupation below $\hbar\omega_{op}$ from the band edge.

4. Conclusion

The boundary condition for phonons in Si NWs has strong impact on low-field hole mobility, but less at high field. The energy relaxation time is almost independent of boundary conditions. The free boundary condition in the Si NW may be described by an acoustic phonon deformation potential of about 25% larger than the quasi-bulk boundary condition. The approximation assuming elastic acoustic phonon and inelastic optical phonon scattering cannot reproduce the behavior of energy relaxation. Our results suggest that taking into account of energy relaxation by acoustic phonons in Si NWs is important to describe energy relaxation process, which is crucial in scaled MOSFETs under quasi-ballistic regime [13], and that the enhanced phonon scattering near the MOS interface may be described by a free boundary condition for phonons.

Acknowledgments

This work was supported by Japan Society for the Promotion of Science (JSPS) KAKENHI Grant Number 15J03785.

References

[1] Park J T and Colinge J 2002 *IEEE Trans. Electron Devices* **49** 2222–9
[2] Neophytou N, Paul A, Lundstrom M and Klimeck G 2008 *IEEE Trans. Electron Devices* **55** 1286–97
[3] Morioka N, Yoshioka H, Suda J and Kimoto T 2011 *J. Appl. Phys.* **109** 064318
[4] Hattori J, Uno S, Mori N and Nakazato K 2010 *Jpn. J. Appl. Phys.* **49** 04DN09
[5] Paul A, Luisier M and Klimeck G 2011 *J. Appl. Phys.* **110** 094308
[6] Sui Z and Herman I P 1993 *Phys. Rev. B* **48** 17938–53
[7] Zhang W, Delerue C, Niquet Y M, Allan G and Wang E 2010 *Phys. Rev. B* **82** 115319
[8] Ramayya E B, Vasileska D, Goodnick S M and Knezevic I 2008 *J. Appl. Phys.* **104** 063711
[9] Niquet Y M, Rideau D, Tavernier C, Jaouen H and Blase X 2009 *Phys. Rev. B* **79** 245201
[10] Li J, Mugny G, Niquet Y M and Delerue C 2015 *Appl. Phys. Lett.* **107** 063103
[11] Tanaka H, Suda J and Kimoto T 2016 *Silicon Nanoelectronics Workshop (SNW2016)* pp 192–3
[12] Takagi S, Hoyt J L, Welser J J and Gibbons J F 1996 *J. Appl. Phys.* **80** 1567–77
[13] Datta S 2005 *Quantum Transport: Atom to Transistor* (New York: Cambridge University Press)

IOP Conf. Series: Journal of Physics: Conf. Series **864** (2017) 012047 doi:10.1088/1742-6596/864/1/012047

Magneto-transport characteristics of a 2D electron system driven to negative magneto-conductivity by microwave photoexcitation

R. G. Mani, A. Kriisa

Department of Physics and Astronomy, Georgia State University, Atlanta, Georgia 30303, USA

E-mail: `rmani@gsu.edu`

Abstract. The magneto-transport characteristics of the negative conductivity/resistivity state in the microwave photo-excited two-dimensional electron system (2DES) is examined through a numerical solution of the associated boundary value problem. The results suggest, surprisingly, that a bare negative diagonal conductivity/resistivity state in the 2DES under photo-excitation should yield a positive diagonal resistance along with a sign reversal in the Hall voltage.

1. Introduction

Microwave-induced zero-resistance states arise from "1/4-cycle-shifted" microwave radiation-induced magnetoresistance oscillations in the high mobility GaAs/AlGaAs system as a reduction in T leads to the saturation of the resistance minima at zero resistance.[1, 2, 3, 4, 5, 6, 7, 8, 9, 10, 11, 12, 13, 14, 15, 16, 17, 18, 19, 20, 21]. These zero-resistance states exhibit activated transport similar to the quantum Hall situation although the Hall resistance, R_{xy}, does not exhibit plateaus or quantization in this instance.[1] The observations have led to some theoretical interest in understanding associated phenomena.[22, 23, 25, 26, 27, 28, 29, 30, 31, 32, 33, 34, 35, 36, 37, 38]

Some theories have utilized a two step approach to model the observations. (There do exist alternate approaches which directly realize zero-resistance states, see for example [27, 33, 36]) In the first step, theory identifies a mechanism that helps to realize large oscillations in the diagonal magneto-photo-conductivity/resistivity, where the minima of the oscillatory diagonal conductivity/resistivity can even take on negative values.[22, 23, 25, 29] The next step invokes an instability in the zero-current-state at negative resistivity (and conductivity),[24] which favors the appearance of current domains with a non-vanishing current density,[24] and zero-resistance.

Naively, one believes that negative magneto-resistivity/conductivity should lead to observable negative magneto-resistance/conductance, based on expectations for the zero-magnetic-field situation. At the same time, the existence of the magnetic field is no doubt an important additional feature because $\rho_{xy} >> \rho_{xx}$ for $\omega_c \tau >> 1$, the typical situation in these experiments. Here, ρ_{xy} and ρ_{xx} are the Hall off-diagonal and diagonal resistivities, respectively, ω_c is the cyclotron frequency and τ is the relaxation time. Thus, one wonders if the magnetic field, and the extremely strong Hall effect, are sufficiently important to overcome nominal expectations, based on the zero-magnetic-field analogy, for an instability in a negative magneto-

IOP Conf. Series: Journal of Physics: Conf. Series **864** (2017) 012047 doi:10.1088/1742-6596/864/1/012047

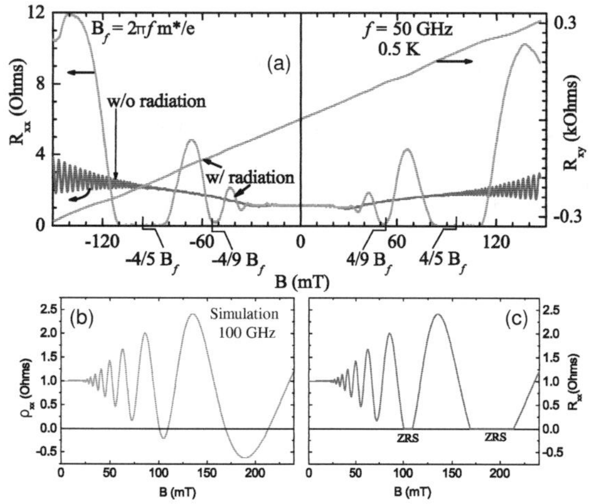

Figure 1. (a) The dark- and photo-excited-diagonal- (R_{xx}) and the photo-excited-Hall-resistance (R_{xy}) vs. the magnetic field B for a GaAs/AlGaAs device. R_{xx} exhibits a non-vanishing resistance in the dark. Under photo-excitation (red traces), R_{xx} exhibits large oscillations with vanishing resistance in the vicinity of $\pm(4/5)B_f$, where $B_f = 2\pi f m^*/e$, without concurrent Hall quantization. (b) Theory predicts negative diagonal resistivity, i.e., $\rho_{xx} < 0$, at the oscillatory minima, observable here in the vicinity of $B \approx 0.19$ Tesla and $B \approx 0.105$ Tesla. (c) Theory asserts that current domain formation leads to zero-resistance states (ZRS).

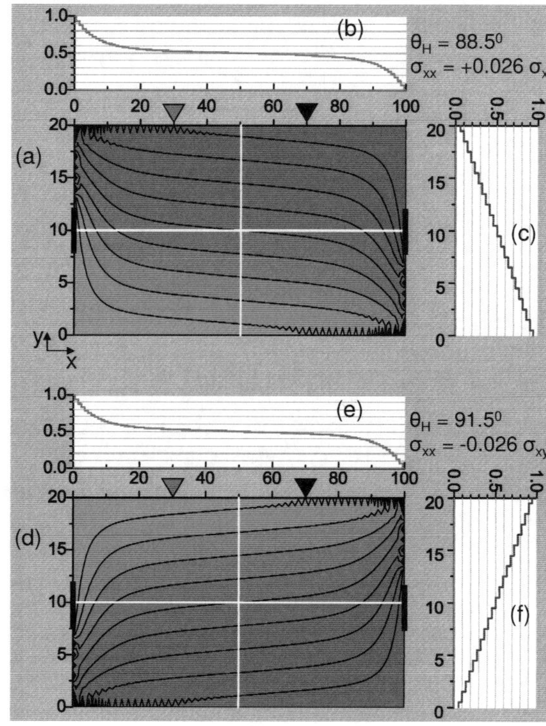

Figure 2. Numerical simulation of the potential within a Hall bar device. Panel (a) shows the potential profile at $\sigma_{xx} = +0.026\sigma_{xy}$. Panel(b) shows the potential variation from the left to the right end at $y = 10$. Panel(c) suggests a large Hall voltage between the bottom and top edges. Panel (d) shows the potential profile at $\sigma_{xx} = -0.026\sigma_{xy}$. Here, the potential profile is reflected with respect panel (a) about the line at $y = 10$ when the σ_{xx} shifts from a positive to a negative value. Panel (e) shows that the potential still decreases from left to right, implying a positive R_{xx}. Panel (f) shows a sign reversal in the Hall voltage in going from positive conductivity to negative conductivity

conductivity/resistivity state. Indeed, a related question of interest is: what are the magneto-transport characteristics of a bare negative conductivity/resistivity state?

To address this last question, we examine here the transport characteristics of the photo-excited 2DES at negative diagonal conductivity/resistivity through a numerical solution of the associated boundary value problem. The results suggest, rather surprisingly, that negative conductivity/resistivity in the 2DES under photo-excitation should generally yield a positive diagonal resistance, i.e., $R_{xx} > 0$. The simulations also identify an associated, unexpected sign reversal in the Hall voltage under these conditions.

2. Experiments and results

Figure 1(a) exhibits measurements of R_{xx} and R_{xy} at $T = 0.5K$ with and without microwave photoexcitation. The green curve represents the R_{xx} in the absence of photo-excitation (w/o radiation). Microwave photo-excitation of this specimen at 50GHz, see orange traces in Fig. 1, produces radiation-induced magnetoresistance oscillations in R_{xx}, and these oscillations grow in amplitude with increasing $|B|$. At the deepest minimum, near $|B| = (4/5)B_f$, where $B_f = 2\pi f m^*/e$, the R_{xx} saturates at zero-resistance.[1]

Both the displacement theory,[22] and the inelastic model,[29] for the radiation-induced magnetoresistivity oscillations suggest that the magnetoresistivity can take on negative values over the B-spans where experiment indicates zero-resistance states. For illustrative purposes, such theoretical expectations are sketched in Fig. 1(b), which presents the simulated ρ_{xx} at $f = 100$ GHz. This figure shows that the deepest ρ_{xx} minima at $B \approx 0.19$ Tesla and $B \approx 0.105$ Tesla exhibit negative resistivity, similar to theoretical predictions.[22, 23, 25, 29] Theory states,[24] that the only time-independent state of a negative resistivity/conductivity includes a current which almost everywhere has a magnitude j_0 fixed by the condition that nonlinear dissipative resistivity equals zero. As a consequence, the ρ_{xx} curve of Fig. 1(b) turns into the magnetoresistance, R_{xx}, trace shown in Fig. 1(c), with zero-resistance over the B-domains that exhibited negative resistivity in Fig. 1(b).

3. Numerical simulation

Hall effect devices can be numerically simulated on a grid/mesh,[39, 40, 41, 42], by enforcing $\nabla.\overrightarrow{j} = 0$, where $\overrightarrow{j}$ is the 2D current density with components j_x and j_y, $\overrightarrow{j} = \sigma\overrightarrow{E}$, and σ is the conductivity tensor.[39] Enforcing $\nabla.\overrightarrow{j} = 0$ is equivalent to solving the Laplace equation $\nabla^2 V = 0$, which may be carried out in finite difference form using a relaxation method, subject to the boundary conditions that current injected via current contacts is confined to flow within the conductor. Simulations were carried out using a 101×21 point grid with 6 points wide current contacts at the ends. For the sake of simplicity, the negative current contact is set to ground potential, i.e., $V = 0$, while the positive current contact is set to $V = 1$.

Fig. 2(a) shows the potential profile at positive conductivity $\sigma_{xx} = +0.026\sigma_{xy}$. The salient feature here is that the equipotential contours are nearly parallel to the long axis of the Hall bar, see Fig. 2(b). Concurrently, Fig. 2(c) suggests a large Hall voltage between the bottom and top edges. Here the Hall voltage decreases from the bottom- to the top- edge.

Fig. 2(d) shows the potential profile at $\sigma_{xx} = -0.026\sigma_{xy}$, i.e., the negative conductivity case. The important feature here is the reflection of the potential profile with respect Fig. 2(a) about the line at $y = 10$ when the σ_{xx} shifts from a positive ($\sigma_{xx} = +0.026\sigma_{xy}$) to a negative ($\sigma_{xx} = -0.026\sigma_{xy}$) value. Fig. 2(e) shows, remarkably, that in the negative σ_{xx} condition, the potential still decreases from left to right, implying $V_{xx} > 0$ and $R_{xx} > 0$ even in this $\sigma_{xx} \leq 0$ condition. Fig. 2(f) shows that for $\sigma_{xx} = -0.026\sigma_{xy}$, the potential *increases* from the bottom edge to the top edge, in sharp contrast to Fig. 2(c). Thus, these simulations show that the Hall voltage undergoes sign reversal when $\sigma_{xx} \leq 0$, although the diagonal voltage (and resistance) exhibits positive values. Thus, a negative diagonal conductivity state is expected to exhibit a positive diagonal resistance along with a sign reversal in the Hall effect.

4. Acknowledgments

Magnetotransport measurements at Georgia State University are supported by the U.S. Department of Energy, Office of Basic Energy Sciences, Material Sciences and Engineering Division under DE-SC0001762. Additional support is provided by the ARO under W911NF-07-01-015.

References

[1] Mani R G, Smet J H, von Klitzing K, Narayanamurti V, Johnson W B and Umansky V 2002 *Nature (London)* **420** 646
[2] Zudov M A, Du R R, Pfeiffer L N and West K W 2003 *Phys. Rev. Lett.* **90** 046807
[3] Mani R G, Narayanamurti V, von Klitzing K, Smet J H, Johnson W B and Umansky V 2004 *Phys. Rev. B* **69** 161306(R); 2004 *Phys. Rev. B* **70** 155310
[4] Mani R G, Smet J H, von Klitzing K, Narayanamurti V, Johnson W B and Umansky V 2004 *Phys. Rev. Lett.* **92** 146801; 2004 *Phys. Rev. B* **69** 193304
[5] Mani R G 2004 *Physica E (Amsterdam)* **22** 1-6; 2004 *Physica E Amsterdam* **25** 189-197; 2004 *Int. J. Mod. Phys. B* **18** 3473-3480; 2004 *Appl. Phys. Lett.* **85** 4962-4964; 2004 *Phys. Rev. B* **72** 075327; 2007 *Appl. Phys. Lett.* **91** 132103; 2007 *Solid State Comm.* **144** 409-412; 2008 *Physica E (Amsterdam)* **40** 1178-1181; 2008 *Appl. Phys. Lett.* **92**, 102107
[6] Smet J H, Gorshunov B, Jiang C, Pfeiffer L, West K, Umansky V, Dressel M, Meisels R, Kuchar F and von Klitzing K 2005 *Phys. Rev. Lett.* **95** 116804
[7] Studenikin S A, Sachrajda A S, Gupta J A, Wasilewski Z R, Fedorych O M, Byszewski M, Maude D K, Potemski M, Hilke M, West K W and Pfeiffer L N 2007 *Phys. Rev. B* **76** 165321
[8] Raichev O E 2008 *Phys. Rev. B* **78** 125304
[9] Mani R G, Johnson W B, Umansky V, Narayanamurti V and Ploog K 2009 *Phys. Rev. B* **79** 205320
[10] Fedorych O M, Potemski M, Studenikin S A, Gupta J A, Wasilewski Z R and Dmitriev I A 2010 *Phys. Rev. B* **81** 201302(R)
[11] Mani R G, Gerl C, Schmult S, Wegscheider W and Umansky V 2010 *Phys. Rev. B* **81**, 125320
[12] Wiedmann S, Gusev G M, Raichev O E, Bakarov A K and Portal J C 2010 *Phys. Rev. Lett.* **105** 026804
[13] Mani R G, Ramanayaka A N and Wegscheider W 2011 *Phys. Rev. B* **84** 085308
[14] Bogan A, Hatke A T, Studenikin S A, Sachrajda A, Zudov M A, Pfeiffer L N and West K W 2012 *Phys. Rev. B* **86** 235305
[15] Ramanayaka A N, Mani R G, Iñarrea J and Wegscheider W 2012 *Phys. Rev. B* **85** 205315
[16] Mani R G, Ramanayaka A N, Ye T, Heimbeck M S, Everitt H O and Wegscheider W 2013 *Phys. Rev. B* **87** 245308
[17] Konstantinov D, Monarkha Y and Kono K 2013 *Phys. Rev. Lett.* **111** 266802
[18] Mani R G and Kriisa A, 2013 *Sci. Rep.* **3**, 3478
[19] Ye T, Liu H-C, Wegscheider W and Mani R G 2014 *Phys. Rev. B* **89** 155307
[20] Liu H-C, Ye T, Wegscheider W and Mani R G 2015 *J. Appl. Phys.* **117** 064306
[21] Ye T, Liu H-C, Wang Z, Wegscheider W and Mani R G 2015 *Sci. Rep.* **5** 14880
[22] Durst A C, Sachdev S, Read N and Girvin S M 2003 *Phys. Rev. Lett.* **91** 086803
[23] Ryzhii V and Suris R 2003 *J. Phys. Cond. Matt.* **15** 6855
[24] Andreev A V, Aleiner I L and Millis A J 2003 *Phys. Rev. Lett.* **91**, 056803 (2003).
[25] Lei X L and Liu S Y 2003 *Phys. Rev. Lett.* **91** 226805
[26] Koulakov A A and Raikh M E 2003 *Phys. Rev. B* **68** 115324
[27] Iñarrea J and Platero G 2005 *Phys. Rev. Lett.* **94** 016806
[28] Lei X L and Liu S Y 2005 *Phys. Rev. B* **72** 075345
[29] Dmitriev I A, Vavilov M G, Aleiner I L, Mirlin A D and Polyakov D G 2005 *Phys. Rev. B* **71** 115316
[30] Iñarrea J and Platero G 2007 *Phys. Rev. B* **76** 073311
[31] Wang S and Ng T-K 2008 *Phys. Rev. B* **77** 165324
[32] Iñarrea J, Mani R G and Wegscheider W 2010 *Phys. Rev. B* **82** 205321
[33] Mikhailov S A 2011 *Phys. Rev. B* **83** 155303
[34] Chepelianskii A D, Laidet J, Farrer I, Beere H E, Ritchie D A and Bouchiat H 2012 *Phys. Rev. B* **86** 205108
[35] Lei X L and Liu S Y 2012 *Phys. Rev. B* **86** 205303
[36] Zhirov O V, Chepelianskii A D and Shepelyansky D L 2013 *Phys. Rev. B* **88** 035410
[37] Yar A and Sabeeh K 2015 *J. Phys.:Cond. Matt.* **27** 435007
[38] Ibarra-Sierra V G, Sandoval-Santana J C, Cardoso J L and Kunold A 2015 *Annal. Phys.* **362** 83
[39] Mani R G 1996 *J. Phys. Soc. Jpn.* **65**, 1751-1759
[40] Mani R G and von Klitzing K 1994 *Appl. Phys. Lett.* **64** 1262-1264
[41] Mani R G and von Klitzing K 1993 *Z. Phys. B* **92** 335-339
[42] Oswald M, Oswald J and Mani R G 2005 *Phys. Rev. B* **72**, 035334

IOP Conf. Series: Journal of Physics: Conf. Series **864** (2017) 012048 doi:10.1088/1742-6596/864/1/012048

Comparative study of microwave radiation-induced magneto-resistance oscillations induced by circularly- and linearly- polarized microwaves

Tianyu Ye, Han-Chun Liu, Ramesh G. Mani
Department of Physics and Astronomy, Georgia State University, Atlanta, Georgia 30303, USA

W. Wegscheider
Laboratorium für Festkörperphysik, ETH Zürich, 8093 Zürich, Switzerland

E-mail: tye2@student.gsu.edu

Abstract. A systematic comparative study of radiation-induced magneto-resistance oscillations using circularly polarized- and linearly polarized microwaves was carried out on the high mobility GaAs/AlGaAs heterostructure two dimensional electron system (2DES). The results showed that, the sinusoidal sensitivity in the amplitude of the radiation-induced magneto-resistance oscillations observed under launcher rotation for linearly polarized microwaves, is remarkably absent in the similar experiment carried out with circularly polarized microwaves.

1. Introduction

Microwave radiation induced zero resistance states and associated microwave radiation-induced magnetoresistance oscillations in the 2D electron system [1, 2, 3, 4, 5, 6, 7, 8, 9, 10, 11] have been a topic of interest over the past decade. The role of the microwave polarization in such experiments is thought to be a discriminating feature among existing theories, see discussion in [12], and therefore, the effect of rotation of the polarization of linearly polarized microwaves has been heavily investigated.[12, 13, 14, 15, 16, 17, 18] Associated experiments have shown, remarkably, that the amplitude of the radiation-induced magnetoresistance oscillations varies sinusoidally with the linear polarization angle, following a cosine-square function.[13] So far as circularly polarized microwave photo-excitation is concerned, only one experimental study[19] examined the magnetotransport response for such polarization and it reported immunity of the radiation-induced magnetoresistance oscillations to the polarization orientation for both circularly polarized and linearly polarized radiation. On the theoretical side, Lei and Liu examined radiation-induced magnetoresistance oscillations under a variety of polarization conditions, and found that the amplitude of the magnetoresistance oscillations differs with the type of polarization of the radiation. [20, 21]

Here, we summarize a systematic comparative study of radiation-induced magnetoresistance oscillations using circularly polarized- and linearly polarized- microwaves, measured in the same sample, in a single cooldown, under nearly the same experimental conditions. The results show a striking sensitivity in the amplitude of the radiation-induced magnetoresistance

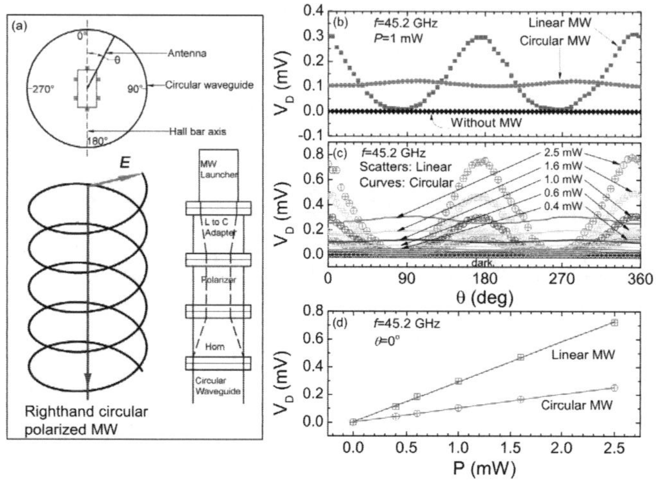

Figure 1. (a) Schematic diagram of the polarization-dependence measurement-geometry. Top-left: Hall bar sample is located at the waveguide center. Bottom-left: Right-handed circularly polarized microwaves. Bottom-right: The linear-to-circularly-polarized microwave converter. (b) Received microwave power at the bottom of the sample holder as a function of launcher antenna angle for linearly and circularly polarized microwaves. (c) Same as panel (b) but at different microwave powers. (d) Microwave power detector response as a function of source microwave power for linearly and circularly polarized microwaves at 0° polarization angle.

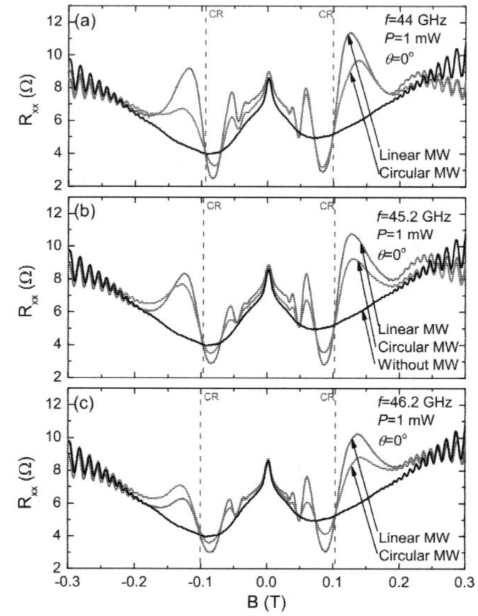

Figure 2. (a) Magnetoresistance measurements of a GaAs/AlGaAs 2DES sample with microwave excitation at (a) 44 GHz, (b) 45.2 GHz, and (c) 46.2 GHz. Dashed lines mark the nominal magnetic field for cyclotron resonance (CR).

oscillations under launcher rotation for linearly polarized microwaves, which is absent in the similar experiment carried out with circularly polarized microwaves.

2. Experiments and results

Experiments were carried out on Hall bars fabricated from high mobility GaAs/AlGaAs heterojunctions. A long cylindrical waveguide sample holder with the sample mounted at the end was inserted into a variable temperature insert (VTI), inside the bore of a superconducting solenoid. A temperature of 1.5 K was realized by pumping on liquid helium. The specimens reached the high mobility condition after brief illumination with a red light-emitting-diode. A commercially available microwave synthesizer was used to provide microwave excitation via a launcher at the top of the sample holder. For the linearly polarized microwaves, the microwave launcher was connected with the circular waveguide through a rectangular to circular adaptor. For the circularly polarized microwaves, a commercially available circular polarizer and horn assembly was inserted between the adapter and the circular waveguide, see Fig. 1 (a). This assembly converts the linearly polarized microwaves to right-hand circularly polarized microwaves within the frequency band $43 \leq f \leq 50$ GHz.

In order to determine and test the microwave polarization, the launcher was rotated as a power sensor was located at the bottom of the sample holder (Fig. 1b). Here, the polarization angle is defined as the angle between the antenna and a reference mark on the circular waveguide. (During magnetotransport measurements, the Hall bar sample is located at the center of the

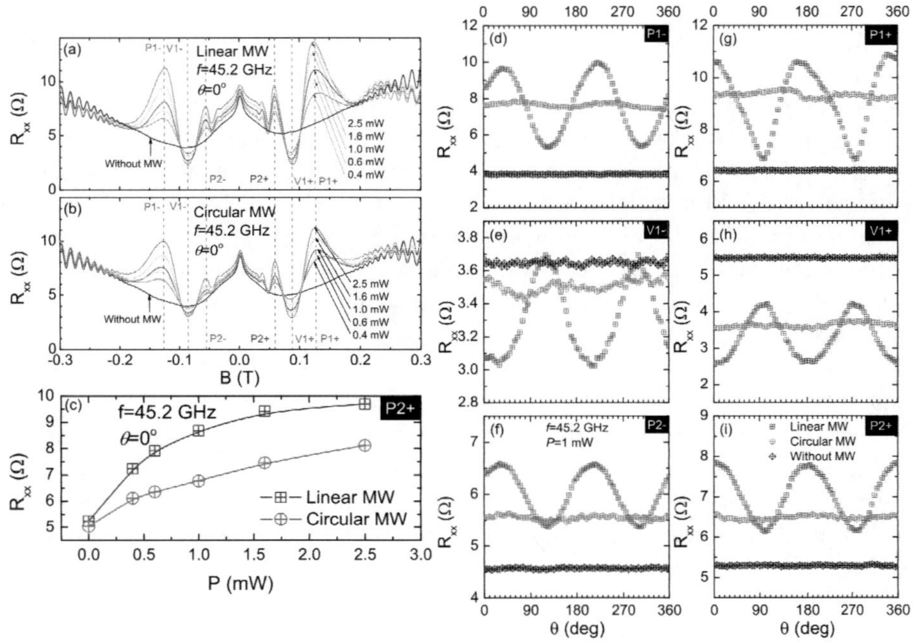

Figure 3. (Color online) Microwave power dependence of R_{xx} for (a) linearly polarized and (b) circularly polarized microwave radiation. (c) Diagonal resistance as a function of microwave power for linearly and circularly polarized microwaves at the peak labelled P2+. Diagonal resistance as a function of polarization angle for linearly polarized (blue symbols) microwaves, circularly polarized (red symbols) microwaves, and without microwave (black symbols) at the magnetic field corresponding to (d)P1-, (e)V1-, (f)P2-, (g)P1+, (h)V1+ and (i)P2+.

circular waveguide with its long axis aligned with the reference mark on the circular waveguide). Fig. 1 (b) exhibits the measurement for both linearly and circularly polarized microwaves with $f = 45.2$ GHz with a microwave diode detector attached at the bottom end of waveguide sample holder. Black line shows the microwave detector response in the absence of microwaves. The green curve shows the sinusoidal change in the microwave intensity with linearly polarized microwaves as the highest intensity appears at around zero degree, which means that the antenna in the launcher is parallel to the antenna in the detector. The red curve exhibits the microwave intensity with circularly polarized microwaves. Here, the power dependence was also checked by measuring microwave intensity versus polarization angle at microwave powers ranging from 0 to 2.5 mW, see Fig. 1 (c). It is clear from Fig. 1 (d), that at the bottom end of circular waveguide, at fixed polarization angle, the microwave power changes linearly with the source power, for both linearly polarized or circularly polarized microwaves.

Figure 2 exhibits microwave radiation-induced magnetoresistance oscillations for both linearly and circularly polarized microwaves at several frequencies. For both types of polarization, the magnetoresistance oscillations are periodic in $1/B$ and exhibit the 1/4-cycle phase shift. However, the oscillatory amplitude differs for the two types of polarizations. At every frequency, the amplitude of oscillations for circularly polarized microwaves is always smaller than the amplitude for linearly polarized microwaves, except for the first peak on the left in the 46.2 GHz trace. Here, although the height of oscillatory peak for circularly polarized microwaves is higher than the oscillatory peak for linearly polarized microwaves, the overall amplitude (from top of the peak to bottom of the valley) is the same for both polarizations.

Figure 3 (a) and (b) exhibit the microwave power dependence of the diagonal resistance R_{xx} both linearly and circularly polarized microwaves. As the microwave power increases, the amplitude of the microwave induced oscillations increases in both cases. However, the increase in the oscillatory response at the peaks is nonlinear with the power, see Fig. 3 (c). Here, it is worth noting that the two R_{xx} vs P curves start to split more as the microwave power increases. Fig. 3 (d) to (i) shows the launcher angle dependence of R_{xx} for both polarizations at the magnetic fields corresponding to P1-, V1-, P2- and P1+, V1+, P2+ (marked by dashed lines in Fig. 3 (a) and (b)). The angular dependence of R_{xx} for the linearly polarized microwaves (blue curves) is sinusoidal at all the peaks and valleys. In contrast, however, the R_{xx} traces for the circularly polarized radiation always exhibit launcher angle insensitivity while being shifted away from the dark curve.

3. Summary

We carried out a comparative study of microwave radiation-induced magnetoresistance oscillations induced by linearly- and circularly polarized radiation. For the linearly polarized radiation, the magneto-resistive response is a strong sinusoidal function of the linear polarization angle, θ. For circularly polarized radiation, the oscillatory magneto-resistive response is hardly sensitive to θ. Finally, for circular polarized radiation, the magneto-resistive response for the cyclotron resonance active and inactive conditions is approximately the same over the entire field range.

4. Acknowledgments

Magnetotransport measurements at Georgia State University are supported by the U.S. Department of Energy, Office of Basic Energy Sciences, Material Sciences and Engineering Division under DE-SC0001762. Additional support is provided by the ARO under W911NF-07-01-015.

References

[1] Mani R G, Smet J, von Klitzing K, Narayanamurti V, Johnson W and Umansky V 2002 *Nature* **420**, 646-650
[2] Zudov M A, Du R R, Pfeiffer L N and West K W 2003 *Phys. Rev. Lett.* **90**, 046807
[3] Mani R G, Narayanamurti V, Klitzing K, Smet J, Johnson W and Umansky V 2004 *Phys. Rev. B* **69** 161306
[4] Mani R G, Smet J H, von Klitzing K, Narayanamurti V, Johnson W B and Umansky V 2004 *Phys. Rev. Lett.* **92**, 146801
[5] Mani R G 2004 *Physica E.* **25**, 189
[6] Mani R G 2008 *Appl. Phys. Lett.* **92**, 102107
[7] Mani R G, Gerl C, Schmult S, Wegscheider W and Umansky V 2010 *Phys. Rev. B* **81**, 125320
[8] Mani R G, Ramanayaka A N, Ye T, Heimbeck M S, Everitt H O, and Wegscheider W 2013 *Phys. Rev. B* **87**, 245308
[9] Mani R G and Kriisa A, 2013 *Sci. Rep.* **3**, 3478
[10] Ye T, Mani R G and Wegscheider W 2013 *Appl. Phys. Lett.* **102**, 242113
[11] Ye T, Mani R G and Wegscheider W 2013 *Appl. Phys. Lett.* **103**, 192106
[12] Mani R G, Ramanayaka A N and Wegscheider W 2011 *Phys. Rev. B* **84**, 085308
[13] Ramanayaka A N, Mani R G , Iñarrea J and Wegscheider W, 2012 *Phys. Rev. B* **85**, 205315
[14] Lei X L and Liu S Y, 2012 *Phys. Rev. B* **86**, 205303
[15] Iñarrea J 2013 *J. Appl. Phys.* **113**, 183717
[16] Ye, T, Liu H C, Wegscheider W and Mani R G 2014 *Phys. Rev. B* **89**, 155307
[17] Ye T, Wegscheider W and Mani R G 2014 *Appl. Phys. Lett.* **105**, 191609
[18] Ye T, Liu H C, Wang Z, Wegscheider W and Mani R G 2015 *Sci. Rep.* **5**, 14880
[19] Smet J H et al 2005 *Phys. Rev. Lett.* **95**, 116804
[20] Lei X L and Liu S Y 2005 *Phys. Rev. B* **72**, 075345
[21] Lei X L and Liu S Y,Infrared, Millimeter, and Terahertz Waves (IRMMW-THz), 2013 38th Intl. Conf. pp. 1-6 Sept. 2013.

IOP Conf. Series: Journal of Physics: Conf. Series **864** (2017) 012049 doi:10.1088/1742-6596/864/1/012049

The electrical and thermal transport properties of hybrid zigzag graphene-BN nanoribbons

Song Gao[1], Wei Lu[1], Guo-Hui Zheng[1], Yalei Jia[1], and San-Huang Ke[1,2,*]

[1]MOE Key Labortoray of Microstructured Materials, School of Physics Science and Engineering, Tonji University, 1239 Siping Road, Shanghai 200092, P.R. China

[2]Beijing Computational Science Research Center, 10 Dongbeiwang West Road, Beijing 100094, P.R. China

E-mail: *shke@tongji.edu.cn

Abstract. The electron and phonon transport in hybrid graphene-BN zigzag nanoribbons are investigated by the nonequilibrium Green's function method combined with density functional theory calculations. A 100% spin-polarized electron transport in a large energy window around the Fermi level is found and this behavior is independent of the ribbon width as long as there contain 3 zigzag carbon chains. The phonon transport calculations show that the ratio of C-chain number to BN-chain number will modify the thermal conductance of the hybrid nanoribbon in a complicated manner.

1. Introduction

Graphene and monolayer hexagonal boron nitride (h-BN) have been attracted much attention for their potential applications in next-generation electronics[1, 2, 3, 4, 5] after they were successfully synthesized in experiments by different methods[6, 7, 8, 9]. On the other hand, the experimental realization of hybrid graphene and h-BN provides a new way to modify the electronic and transport properties of graphene-based materials. In the past few years the electronic and transport properties of graphene nanoribbons (GNRs) modified with BN domains or BN nanoribbons (BNNRs) were investigated extensively and several approaches have been proposed to control the spin transport in GNR/BNNR hybrid systems. For instance, Dong *et al.* found that a BNNR seamlessly connected between two pieces of GNRs can result in highly spin-polarized transport[10]. He *et al.* found that the currents with different spin configurations display different behavior in hybridized zigzag $BNNR_m$-GNR_n systems with certain n [11], though they did not find a way to achieve 100% spin polarization of current. While the electronic properties of BNNRs and GNRs have been extensively explored and the thermal transports in BNNRs and GNRs have also got much attention[12, 13, 14], the investigation of thermal transport properties of hybrid GNR/BNNR systems is still limited to our best knowledge. In this work, we investigate the electron and phonon transport properties of parallel hybrid zigzag-edged GNR/BNNR (zGNR/zBNNR) systems with different widths by performing first-principles calculations based on the nonequilibrium Green's function (NEGF) method combined with density functional theory (DFT).

2. Models and Computation

We first consider a zGNR consisting of 8 zigzag carbon chains (8-zGNR). Then we gradually substitute these carbon chains with BN chains to construct parallel zGNR/zBNNR hybrid nanoribbon systems. The connection is made via the B atoms (system A) or N atoms (system B). The two systems are denoted by

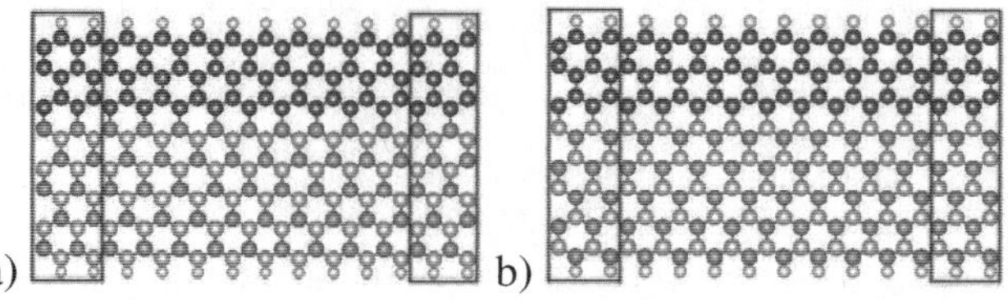

Figure 1. (Color online) Structures of 3-zGNR/5-zBNNR hybrid systems with the connection via (a) B atoms and (b) N atoms. The brown, pink, gray, and green balls denote C, H, N, and B atoms, respectively.

mBn and mNn, respectively, with m and n being the numbers of C chains and BN chains, respectively. The two edges of the hybrid nanoribbons are saturated by H atoms. As an example, the structures of the 3B5 and 3N5 nanoribbons are shown in Fig.1. The majority spin of the spin-polarized edge state localized around the zGNR edge is denoted by α-spin.

To investigate the electron and phonon transport through the nanoribbon systems, we adopt the NEGF method combined with first-principles DFT calculations[15]. The periodic nanoribbon systems are divided artificially into three subsystems: left lead, scattering region, and right lead (see Fig.1). For the electron transport, the self-consistent Kohn-Sham Hamiltonian of the scattering region and the self-energies of two semi-infinite leads are used to construct a single-particle Green's function from which the transmission coefficient at any energy is calculated. The computational techniques have been described in details previously[15]. For the phonon transport calculations, the similar NEGF framework is adopted, where the phonon Green's function of the scattering region can be expressed in terms of the force constant matrix $K_{ij} = \partial^2 E_{total}/\partial u_i \partial u_j$ and the phonon self-energy Σ_L^p (Σ_R^p) of the left (right) thermal lead:

$$G^p = [(\omega + i\eta)^2 I - K - \Sigma_L^p - \Sigma_R^p]^{-1}. \quad (1)$$

Then the phonon transmission function can be determined by the Caroli equation: $T(\omega) = Tr[\Gamma_L^p(\omega)G^p(\omega)\Gamma_R^p(\omega)G^{p\dagger}(\omega)]$, and the lattice thermal conductance can be calculated by

$$\kappa_p = \frac{1}{2\pi k_B T^2}\int_0^\infty (\hbar\omega)^2 \frac{e^{\hbar\omega/k_B T}}{(1-e^{\hbar\omega/k_B T})^2} T(\omega)d\omega. \quad (2)$$

For the DFT electronic structure calculations, we employ a double-ζ plus polarization (DZP) numerical basis set for all atomic species[16, 17]. The optimized Troullier-Martins pseudopotentials[18] are adopted to describe the atomic cores and the generalized gradient approximation (GGA)[19] is used for the electron exchange and correlation. The atomic structures of the systems studied are fully optimized by minimizing the forces on each single atom to be smaller than 0.005 eV/Å.

3. Results and Analyses

We first present in Figs.2 (a) and (b) the transmission functions of pristine 8-zGNR with ferromagnetic (FM) and antiferromagnetic (AFM) spin configurations, respectively. These results agree well with a previous work about the transport in pristine 8-zGNRs[20]. Now, let us look at the transmission function of 3B5, as shown in Fig.2(c). One can see that the α-spin shows a similar behavior to the FM 8-zGNR while the β-spin shows a similar behavior to the AFM 8-zGNR. To understand better this result, we calculate the band structure of 3B5 (Fig.2(g)) and compare it to that of FM and AFM 8-zGNR (Figs.2(e) and(f)). It can be seen that the α-spin bands and β-spin bands around the Fermi level show indeed similar behavior to those of the FM and AFM 8-zGNR, respectively. As a result, the 3B5 system possesses a complete (100%) spin polarization in a large energy range around the Fermi level. Next, let us look at the transmission function of 3N5, as shown in Fig.2(d). It is quite different from that of 3B5: Now the α-spin shows a similar behavior to the AFM 8-zGNR while the β-spin shows a similar behavior to

IOP Conf. Series: Journal of Physics: Conf. Series **864** (2017) 012049 doi:10.1088/1742-6596/864/1/012049

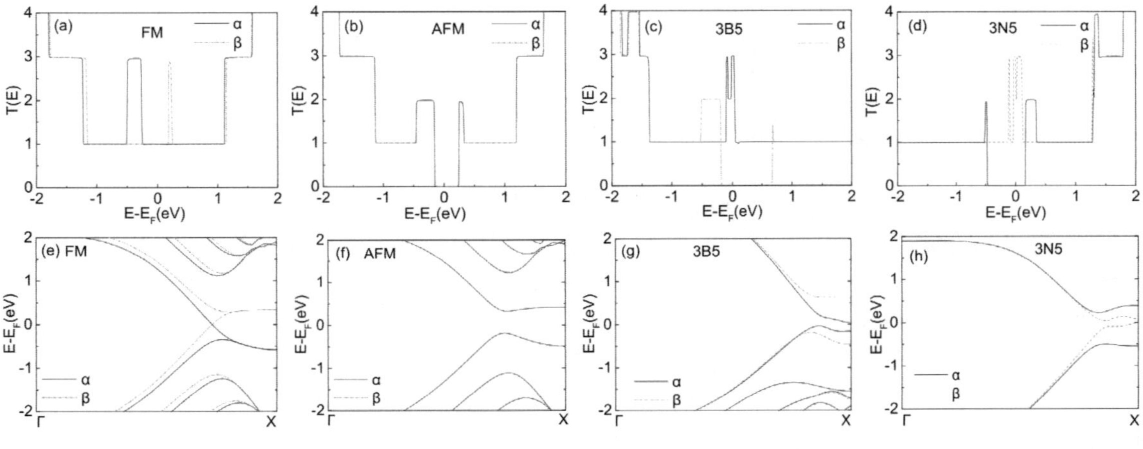

Figure 2. (Color online) Transmission functions of (a) pristine 8-zGNR with FM spin configuration, (b) pristine 8-zGNR with AFM spin configuration, (c) 3B5, and (d) 3N5. (e)-(f) Band structures of the FM 8-zGNR, AFM 8-zGNR, 3B5, and 3N5, respectively.

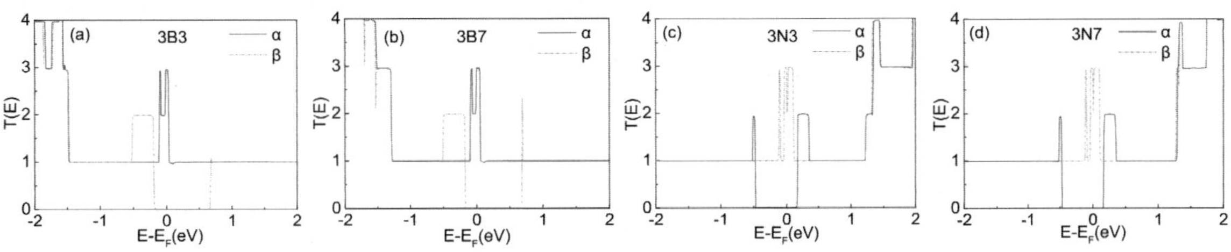

Figure 3. (Color online) Transmission functions of (a) 3B3, (b) 3B7, and (c) 3N3, (d) 3N7.

FM 8-zGNR, just being opposite to 3B5. This can also be seen in the band structure of 3N5 shown in Fig.2(h). The result is also a 100% spin polarization around the Fermi level.

Our results show that both the 3B5 and 3N5 systems can provide a large energy window around the Fermi level where the electron transport is 100% spin polarized. To check whether or not this behavior is general for hybrid nanoribbons consisting of 3 zigzag carbon chains, we further calculate other systems with different ribbon widths: 3B3, 3B7, and 3N3, 3N7. The transmission functions are given in Fig.3. In all the cases there exists an energy window of 100% spin polarization around the Fermi level, indicating that this behavior is general as long as the system contains 3 zigzag carbon chains. Our further calculations show that this behavior is weak or absent for other hybrid nanoribbons whose number of carbon chain is not three.

Finally, let us look at the phonon transport properties of the hybrid nanoribbons. Fig.4(a) shows the calculated phonon transmission functions (T(E)) of the 8-zGNR (8CC), 8-zBNNR (8BN), 3B5, and 3N5. We note that T(E) of 8CC is overall larger than that of 8BN and those of 3B5 and 3N5 are basically in between. To have a systematic comparison, we give the thermal conductance as a function of the number of carbon chain for system A (Fig.4(b)) and system B (Fig.4(c)) under a series of temperatures. At low temperatures (¡ 200K) the variation of thermal conductance is very flat for both system A and B. When the temperature is higher than 300K, the κ_p of system A first increases and then decreases for 8BN to 3B5, and then increases linearly for 3B5 to 8CC. As a result, the 3B5 has the minimum thermal conductance. System B has the similar behavior from 8BN to 4N4, but the increase after 4N4 is not linear, being different from system A. The ratio of the C-chain number to the BN-chain number provide a way to modifying the thermal transport properties of the hybrid nanoribbons.

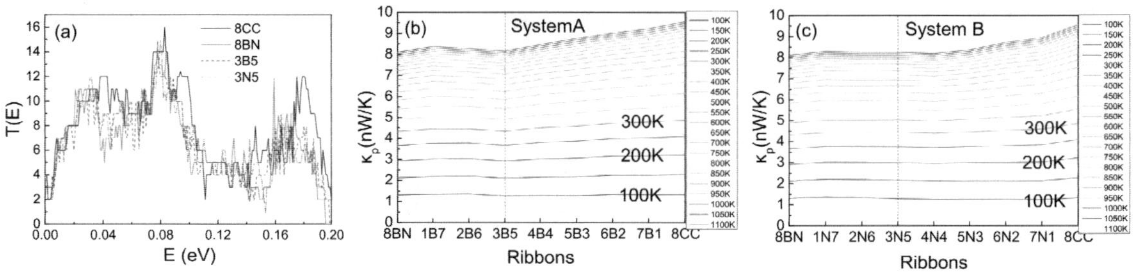

Figure 4. (Color online) (a) Phonon transmission functions of 8CC, 8BN, 3B5, and 3N5. (b) and (c) Thermal conductance of system A and B, respectively.

4. Conclusions

By performing first-principles calculations we have studied the electron and phonon transport properties of hybrid zGNR/zBNNR nanoribbons with different connection atoms and ribbon widths, which can be now fabricated in experiments. It is found that nanoribbons consisting of 3 zigzag carbon chains will provide a large energy window around the Fermi level where the electron transport is 100% spin polarized, regardless of the ribbon width and connection atom. The phonon transport calculations show that the ratio of C-chain number to BN-chain number will modify the thermal conductance of the hybrid ribbon under different temperatures in a complicated manner.

5. Acknowledgments

This work was supported by the National Natural Science Foundation of China under Grants No. 11174220 and 11374226.

References

[1] A. K. Geim, 1992 Science **324**, 1530.
[2] A. H. Castro Neto, F. Guinea, K. S. Novoselov and A. K. Geim, 2009 Rev. Mod. Phys **81**, 183.
[3] A. Nad, K. Raidongia, K. P. S. S. Hembram, R. Datta, U. V. Waghmare, and C. N. R. Rao, 2010 ACS Nano **4**, 1539.
[4] D. Golberg, Y. Bando, Y. Huang, T. Terao, C. Tang, and C. Zhi, 2010 ACS Nano **4**, 2979.
[5] M. L. Hu, Z. Z. Yu, K. W. Zhang, and L. Z. S. J. X. Zhong, 2011 J. Phys. Chem. C **115**, 8260.
[6] K. S. Novoselov, A. K. Geim, S. V. Morozov, D. Jiang, Y. Zhang, S. V. Dubonos, I. V. Grigorieva, and A. A. Firsov, 2004 Science **306**, 666.
[7] W. Q. Han, L. Wu, Y. Zhu, K. Watanabe, and T. Taniguchi, 2008 Appl. Phys. Lett. **93**, 223103.
[8] C. Jin, F. Lin, K. Suenaga, and S. Iijima, 2009 Phys. Rev. Lett. **102**, 195505.
[9] N. Alem, R. Erni, C.kisielowski, M. D. Rossell, W. Gannett, and A. Zettl, 2009 Phys. Rev. B **80**,155425.
[10] J. C. Dong and H. Li, 2012 J. Phys. Chem. C **116**, 17259.
[11] J. He, K. Q. Chen, Z. Q. Fan, L. M. Tang, and W. P. Hu, 2010 Appl. Phys. Lett. **97**, 193305.
[12] Y.-C. Chen, S.-C. Lee, T.-H. Liu, C.-C. Chang, I.J. Therm. 2015 Science **94**, 72e78.
[13] L. Duclaux, B. Nysten, J.P. Issi, A.W. Moore, 1992 Phys. Rev. B **46**, 3362.
[14] D.L. Nika, E.P. Pokatilov, A.S. Askerov, A.A. Balandin, 2009 Phys. Rev. B **79**, 155413.
[15] S.-H. Ke, H. Baranger, and W. T. Yang, 2004 Phys. Rev. B **70**, 085401.
[16] D. Sánchez-Portal, P. Ordejon, E. Artacho, and J. M. Soler, Int. 1997 J. Quantum Chem.**65**, 453.
[17] J. M. Soler, E. Artacho, J. D. Gale, A. Garcia, J. Junquera, P. Ordejón, and D. Sónchez-Portal, 2002 J. Phys.:Condens. Matter **14**, 2745.
[18] N. Troullier and J. L. Martins, 1991 Phys. Rev. B **43**, 1993.
[19] J. P. Perdew, K. Burke, and M. Ernzerhof, 1996 Phys. Rev. Lett. **77**, 3865.
[20] S. M.-M. Dubois, Z. Zanolli, X. Declerck, and J.-C. Charlier, 2009 Eur. Phys. J. B **72**, 1.

IOP Conf. Series: Journal of Physics: Conf. Series **864** (2017) 012050 doi:10.1088/1742-6596/864/1/012050

Extraction of overlapping radiation-induced magnetoresistance oscillations and bell-shaped giant magnetoresistance in the GaAs/AlGaAs 2DES using a multiconduction model

R L Samaraweera[1], H C Liu[1], Z Wang[1], W Wegscheider[2] and R G Mani[1]

[1] Department of Physics and Astronomy, Georgia State University, Atlanta, GA 30303
[2] Laboratorium für Festkörperphysik, ETH-Zürich, 8093 Zürich, Switzerland

mani.rg.gsu@gmail.com

Abstract. We present an experimental study aimed at extracting the microwave radiation-induced magnetoresistance oscillations from the bell-shape giant magnetoresistance in high mobility GaAs/AlGaAs devices using a multi-conduction model. The results show that the multi-conduction model describes the observed giant magnetoresistance effect and the model helps to extract radiation-induced magnetoresistance oscillations, over a wider parameter space.

1. Introduction

Interest in the GaAs/AlGaAs heterostructure two-dimensional electron systems (2DES) has been motivated by the enhanced physical and electrical properties of this system, including 2D electron mobilities well above 10^7 cm^2 /Vs. This 2DES shows interesting physical phenomena such as novel microwave induced zero-resistance states and associated radiation- induced magnetoresistance oscillations, [1-14] and Giant Magneto-Resistance (GMR) [14-18]. Photo-excited magnetotransport has been studied over the past decade, while negative GMR, which is proportional to the square of magnetic field, B^2, has been known since the initial report in 1983 [15]. Studies of negative GMR have revealed the influence of various factors [15-18]. Typically, the bell-shape negative GMR effect is fit with a parabolic magnetic –field-dependence (B^2- fit) model [17,18]. However, from the experimental perspective, there is more to the bell-shape negative GMR than the initial parabolic term.

In this study, we examine the overlap of microwave induced magnetoresistance oscillations with the bell-shape positive and negative-GMR in GaAs/AlGaAs 2DES. The aim of this study is to develop a technique to separate coexisting GMR and microwave induced magnetoresistance oscillations's and extract associated features. We utilized the multi-conduction model to fit the GMR effect [16]. Subtracting the multi-conduction fit results from the experimental data helps to isolate the microwave induced magnetoresistance oscillations.

2. Experimental

Electrical measurements were carried out at $T \leq 1.7$ K using low frequency lock-in-based techniques on a 2DES with mobility and electron density of 10^7 cm^2/V s and 2.4 x 10^{11} cm^{-2} respectively. Both a constant I_{ac} = 2 μA and a variable I_{dc} were applied along the Hall bar direction. The magnetoresistance $R_{xx}= V_{xx}/I_{ac}$. Microwaves at frequency f=70.1GHz were conveyed to the sample via a waveguide.

3. Results and Data fit

The solid lines of Fig. 1(a) exhibit the R_{xx} vs. B at 1.7 K under microwave excitation at f=70.1 GHz. Microwave induced magnetoresistance oscillations are evident in the range $-0.2 < B < 0.2$ Tesla. In addition, a close study of the data reveals sharp negative-magnetoresistance in the range of $-0.01 \leq B \leq 0.01$ Tesla, with a peak at B=0 Tesla, which looks like the weak-Localization (WL) effect [19,20]. Above 0.05 T, Fig. 1(a) shows positive magnetoresistance. In Fig. 1 (a), the amplitude of the microwave induced magnetoresistance oscillations decreases with decreasing MW power. Figure 1(b) illustrates that the amplitude of microwave induced magnetoresistance oscillations decreases with increasing the I_{dc}. Also, a large bell-shaped negative GMR develops with increasing I_{dc}. The WL-peak is unaffected by either P or I_{dc}.

In order to extract microwave induced magnetoresistance oscillations from the overlapping magnetoresistive effects, we introduced a fitting model that addresses the WL-effect and the bell-shaped GMR effect. Weak localization effect is modelled using the 2D WL theory by neglecting spin-

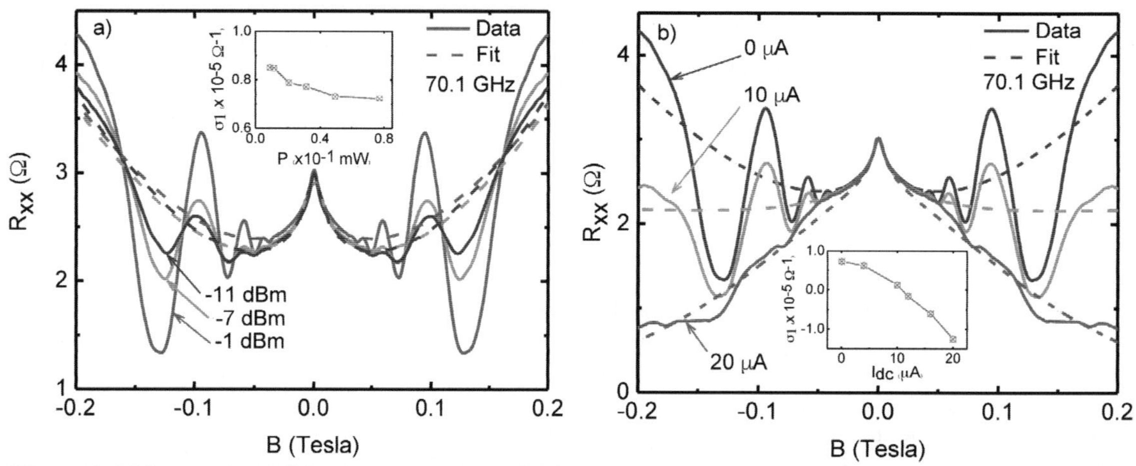

Figure 1. This panel exhibits the R_{xx} vs. B (solid lines) under photo-excitation at f=70.1 GHz, at T = 1.7 K, also shown are the fits (dashed-lines) to the data using multi-conduction model, see text. a) Exhibits the R_{xx} at various MW-powers, P ranging from 0.1 mW to 0.8 mW, at I_{dc}= 0 μA. Insert shows the fit extracted σ_1 as a function of P. b) Exhibits the R_{xx} at different dc-bias, I_{dc} from 0 μA to 20 μA, the insert illustrates the fit extracted σ_1 as a function of I_{dc}

orbit and spin scattering [19-20]. The bell-shape negative GMR effect is addressed with the multi-conduction model [16]. To account for the experimentally observed bell-shape GMR, an additional conductivity term described by conductivity, σ_1 and mobility, μ_1 is introduced to the multi-conduction model, eqns. 3 and 4 in Ref 16. Thus, the total diagonal conductivity, σ_{xx} and off-diagonal conductivity, σ_{xy} are given as follows,

$$\sigma_{xx} = \left\{ \sigma_0 / [1 + (\mu_0 B)^2] \right\} + \left\{ \sigma_1 / [1 + (\mu_1 B)^2] \right\} \tag{1}$$

$$\sigma_{xy} = \left\{ \sigma_0 \mu_0 B / [1 + (\mu_0 B)^2] \right\} + \left\{ \sigma_1 \mu_1 B / [1 + (\mu_1 B)^2] \right\} \tag{2}$$

The zeroth conductivity terms, in eqns. (1) and (2) represents the high mobility electrons in 2D-electron system, thus $\sigma^0_{xx}=n_0 e\mu_0$ where n_0, electron density and μ_0, electron mobility of the 2DES. The number of free parameters in the fit model has been reduced to two, by holding constant the values of n_0 and μ_0, respectively.

Dashed lines in Fig. 1 exhibit the fitting results to the experimental data. Also, fit extracted μ_l= 1.9 x 10^4 cm^2/V s is unaffected by P in Fig. 1(a). However, the fit result shows I_{dc} dependency of μ_l: as I_{dc} changes from 0 μA to 20 μA, μ_l increases from 1.9 to 6.2 x 10^4 cm^2/V s.

Insets in Fig. 1 exhibit the fit extracted additional conductivity term σ_l, which captures the bell-shape GMR effect observed in the experimental data. As shown in Fig. 1(a) (inset), the σ_l is not significantly affected by the applied P, however, it slightly decreases at higher power. In contrast, in Fig. 1(b) (inset), the σ_l shows significant change as a function of I_{dc}. One can observe a clear transition of σ_l from positive to a negative value with increasing dc-bias from 0 μA to 20 μA. The transition occurs in the range of 10 μA < I_{dc}< 12 μA.

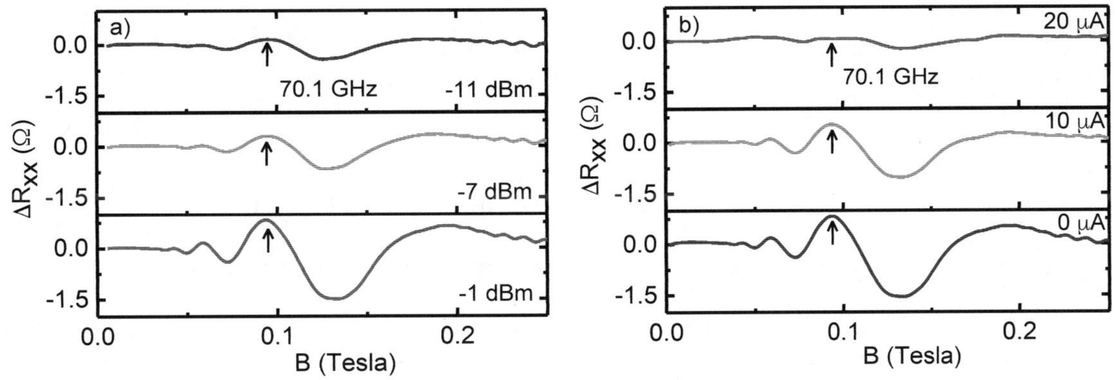

Figure 2. This figure shows the extracted microwave induced magnetoresistance oscillations, ΔR_{xx}, after subtracting the fit results from the experimental data, in the positive magnetic field. a) Exhibits the influence of power, P on the microwave induced magnetoresistance oscillations, b) Exhibits the effect of I_{dc} on the microwave induced magnetoresistance oscillations, at given conditions as indicated.

In Figure 2, ΔR_{xx} is obtained by subtracting the GMR fit results from the experimental data. One can observe clear microwave induced magnetoresistance oscillations without indication of other magnetoresistance effects. In Fig. 2(a), the height of the microwave induced magnetoresistance oscillations peak indicated by arrow (↑) has decreased by about 80% upon reducing the MW-power by a factor of eight. On the other hand, as shown in Fig. 2(b), the microwave induced magnetoresistance oscillations peaks disappear at I_{dc}=20 μA. Also, the peaks do not show any phase shift as a function of either MW-power or dc-bias.

4. Discussion

Magnetoresistance in the 2DES, under the photo-excitation and applied dc-bias, over the examined magnetic field range can arise from a number of mechanisms such as the weak localization effect, microwave-induced magnetoresistance oscillations, and bell-shape GMR. One might also examine the interplay of these effects with an applied I_{dc}. The weak localization effect observed at low magnetic fields is mainly a low-temperature disorder induced effect [16,19,20]. On the other hand, microwave radiation-induced magnetoresistance oscillations have been studied as a function of radiation frequency, power, polarization angle, etc. [1-14]. Studies report a nonlinear growth of microwave induced magnetoresistance oscillations with the radiation power P [11]. In this study, we found an increase in microwave induced magnetoresistance oscillations amplitude as a function of the microwave power by examining the ΔR_{xx} obtained by subtracting the fit result from the R_{xx}. Experimental studies have reported the effect of an applied I_{dc}, and the interplay of dc-bias and microwaves in the magneto transport of GaAs/AlGaAs 2DES [22-24]. In this study, quenching of microwave induced magnetoresistance oscillations was observed at about I_{dc} = 20 μA. Our results indicate a non-linear behaviour in the fit extracted σ_l. Also, bell-shape GMR effect, parameterized by σ_l, varies from positive to negative value as a function of dc-bias.

5. Conclusion

In summary, this study has shown that the multiconduction model along with weak-localization term provides a good fit to the non-oscillatory experimental data, as it also captures the essential features in bell-shape GMR effect, over various experimental parameters. In addition, the model serves to extract coexisting effects such as, radiation induced magnetoresistance oscillations, from the bell-shape GMR.

Acknowledgement
This work was supported by the U.S. Department of Energy, Office of Basic Energy Sciences, Material Science and Engineering Division under DE-SC0001762, and by the Army Research Office under W911NF-14-2-0076 and W911NF-15-1-0433.

References

[1] Mani R G, Smet J H, von Klitzing K, Narayanamurti V, William B, Johnson W B and Umansky V 2002 *Nature* **420** 646-650
[2] Zudov M A, Du R R, Pfeiffer L N and West K W 2003 *Phys. Rev. Lett.* **90** 046807
[3] Mani R G 2008 *Appl. Phys. Lett.* **92** 102107
[4] Mani R G, Narayanamurti V, von Klitzing K, Smet J H, Johnson W B and Umansky V 2004 *Phys. Rev. B* **69** 161306
[5] Mani R G, Smet J H, von Klitzing K, Narayanamurti V, Johnson W B and Umansky V 2004 *Phys. Rev. Lett.* **92** 146801
[6] Ramanayaka A N, Mani R G and Wegscheider W 2011 *Phys. Rev. B* **83** 165303
[7] Liu H C, Tianyu Ye,Wegscheider W and Mani R G 2015 *J. of App. Phys.***117** 064306
[8] Mani R G, Ramanayaka A N and Wegscheider W 2011 *Phys. Rev. B* **84** 085308
[9] Ye T, Liu H-C, Wang Z, Wegscheider W, Mani R G, 2015 *Sci. Rep.* 5, 14880
[10] Mani R G, Gerl C, Schmult S, Wegscheider W and Umansky V 2010 *Phys. Rev. B* **81** 125320
[11] Inarrea J, Mani R G and Wegscheider W 2010 *Phys. Rev. B* **82** 205321
[12] Ramanayaka A N, Mani R G, Inarrea J and Wegscheider W, 2012 *Phys. Rev. B* **85**, 205315
[13] Mani R G and Kriisa A, 2013 *Sci Rep.* **3**, 3478
[14] Mani R G, Ramanayaka A N, Ye T, Heimbeck M S, Everitt H O and Wegscheider W, 2013 *Phys. Rev. B* **87**, 245308
[15] Paalanen M A, Tsui D C and Hwang J C M 1983 *Phys. Rev. Lett.* **51** 2226–2229
[16] Mani R G, Kriisa A and Wegscheider W 2013 *Sci. Rep.* **3** 2747
[17] Bockhorn L, Barthold P, Schuh D, Wegscheider W, and Haug R J 2011 *Phys. Rev. B* **83** 113301-1–4
[18] Bockhorn L, Gornyi I V, Schuh D,Reichi C, Wegscheider W and Haug R J 2014 *Phys. Rev. B* **90** 165434
[19] Hikami S, Larkin A I and Nagoka Y 1980 *Prog. Theor. Phys.* **63** 707-710
[20] Bergmann G, 1984 *Phys. Repts.* **107** 1–58
[21] Bayot V, Piraux L, Michenaud J P, Issi J P, Lelaurain M and Moore A 1990 *Phys. Rev. B* **41** 17 11770-11779
[22] Zhang W, Zudov M A, Pfeiffer L N and West K W 2007 *Phys. Rev. Lett.* **98** 106804 1-4
[23] Bykov A A, Zhang J, Vitkalov S, Kalagin A K and Bakarov A K 2005 *Phys. Rev. B* **72** 245307 1-5
[24] Jesús Iñarrea 2009 *Phys. Rev. B* **80** 193302 1-4

IOP Conf. Series: Journal of Physics: Conf. Series **864** (2017) 012051 doi:10.1088/1742-6596/864/1/012051

The 2DEG mobility enhancement for low- and high-electric fields in a new type of AlGaAs/InGaAs heterostructures with donor-acceptor doping

D.Yu. Protasov‡§, D.V. Gulyaev‡, A.K. Bakarov‡, A.I. Toropov‡ and K.S. Zhuravlev‡†

‡ Rzhanov Institute of Semiconductor Physics, Siberian Branch of Russian Academy of Sciences, 13, Lavrentiev avenue, Novosibirsk, 630090, Russia

§ Novosibirsk State Technical University, 20, K.Marx Avenue, Novosibirsk, 630073, Russia

† Novosibirsk State University, 2, Pirogov Street, Novosibirsk, 630090, Russia

e-mail: protasov@isp.nsc.ru

Abstract. The splitting of the delta-layers in the DA-pHEMT heterostructures has resulted in the increase of the spacer's effective thickness and growth of the low-field 2DEG mobility from 4000÷5000 cm^2 V^{-1} s^{-1} up to 6500 cm^2 V^{-1} s^{-1} at the temperature of 300 K and 2DEG density of 4.0×10^{12} cm^{-2}. The 2DEG mobility in the δ-splitted DA-pHEMT heterostructures almost coincides with the mobility in standard pHEMT heterostructures, but the 2DEG density in the DA-pHEMT heterostructures is approximately twice higher. The additional potential barrier in the DA-pHEMT heterostructures formed by the acceptors causes the reduction of the real-space transfer effect. Therefore, the drift saturation velocity in these heterostructures is higher than the drift saturation velocity in standard pHEMT heterostructures.

1. Introduction

Today the fastest semiconductor transistors are those based on quantum wells (QWs). It has been theoretically predicted that electron and phonon quantization leads to a decrease in the energy relaxation rate of hot electrons and, therefore, allows enhancing ultimate parameters of high-electron mobility transistors (HEMT) [1]. In rather shallow QWs these effects are masked by the real space transfer effect (RST) [2] i.e. the hot electrons escape from a QW into the barriers when the mean electron energy accelerated by a high electric field becomes comparable with the band-gap discontinuity.

Recently we have proposed to increase the effective depth of a QW by modulation doping of QW barriers by donors and acceptors. In this case the height of the barriers is approximately comparable with the band-gap value in the AlGaAs of about 1.5 eV, whereas the height of the barriers without such a doping is limited by the band-gap off-set at the AlGaAs/InGaAs heterointerface of 0.3 eV. A new type of AlGaAs/InGaAs heterostructures with donor-acceptor doped barriers (DA-pHEMT) allows us to suppress the parasitic parallel conductivity via the δ-n-layers placed in the barriers and enhance the output power density of the DA-pHEMT transistor [3]. However, the electron mobility in

IOP Conf. Series: Journal of Physics: Conf. Series **864** (2017) 012051 doi:10.1088/1742-6596/864/1/012051

DA-pHEMT heterostructures is lowered by scattering on ionized donors from the broadened δ-sublayers located in the barriers [4].

In the present paper we have investigated the influence of the δ-layers splitting on the mobility of two-dimensional electron gas (2DEG) in a low-electric field. The enhancement of the 2DEG mobility in a high electric field due to the suppression of the RST effect by additional potential barriers in DA-pHEMTs has been also studied.

2. Details of the experiments and low-field mobility calculations

The DA-pHEMT heterostructures under study were grown on the (001) GaAs substrate by the molecular-beam epitaxy using the Compact 21T machine. The heterostructure buffer consists of a 0.4 micrometer-thick GaAs layer and twelve-period GaAs/AlAs superlattice. The growth rates were equal to 0.28 nm s^{-1} at the temperature of 620 °C and 0.24 nm s^{-1} at the temperature of 500÷520 °C for the (Al)GaAs layers and InGaAs layers, respectively. The growth of the δ-layers was carried out at 530 °C. The layer consequence for aDA-pHEMT heterostructure is shown in Figure 1 (without a top n^+-GaAs contact layer).

The values of the 2DEG density and mobility were determined from the variable-field Hall effect measurement by the Van der Pauw method at 77 K and 300 K temperatures in the magnetic field range of 0÷2 T by using the mobility spectrum analysis with multi-carrier fitting [4]. Ohmic contacts to the samples were fabricated by the high-voltage discharge through the indium foil imposed on the sample.

The rectangular mesa-samples were made by photolithography and wet etching to measure drift velocity and electroluminescence (EL). The Ge/Au/Ni/Au ohmic contacts were formed by the electron-beam evaporation and 5-min annealing in a hydrogen atmosphere at 420 °C. The distance between the contacts ranged from 6 to 500 mkm for different samples. The velocity-field characteristics were determined from the pulsed current-voltage measurements with a pulse duration of 800 ns. We assume that the 2DEG density does not have any dependence upon an electric field intensity. The EL radiation emitted from the samples under the voltage pulse was measured by the spectrometer on the basis of the single diffraction monochromator equipped with a Si CCD matrix at 300 K.

AlGaAs	6 nm
AlGaAs:Be	8 nm, 5×10^{18} cm^{-3}
AlGaAs	7 nm
2ML GaAs	
δ_{Si}	6×10^{12} cm^{-2}
2ML GaAs	
AlGaAs spacer	3 nm
GaAs smooser	1.5 nm
InGaAs	14 nm, x=0.165
GaAs smooser	3.0 nm
AlGaAs spacer	3 nm
2ML GaAs	
δ_{Si}	6.5×10^{12} cm^{-2}
2ML GaAs	
AlGaAs	5 nm
AlGaAs:Be	11 nm, 4×10^{18} cm^{-3}
Buffer layers/GaAs substrate	

Fig. 1. The layer sequence and doping levels for the DA-pHEMT heterostructure

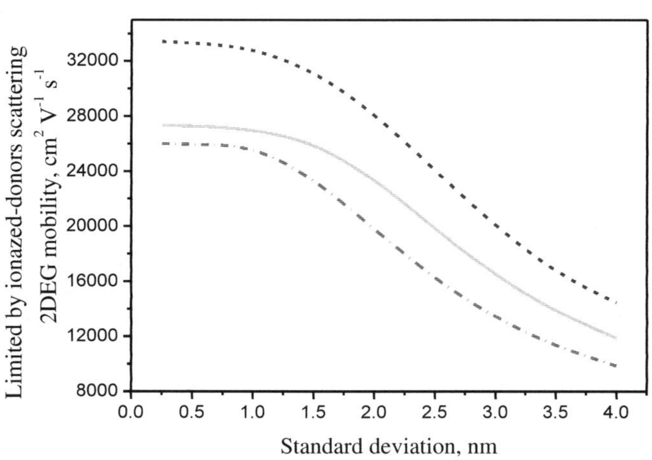

Fig. 2. The dependence of the low-temperature 2DEG mobility limited by ionized-donors scattering upon the broadening of δ-layers splitting: dash-dot red line – unsplitted δ-layers, solid green line – 1 nm, dash blue line – 2 nm distance between the δ-sublayers.

IOP Conf. Series: Journal of Physics: Conf. Series **864** (2017) 012051 doi:10.1088/1742-6596/864/1/012051

The calculations of the 2DEG mobility limited by ionized-donors scattering were made at low temperature (77 K) according to the theoretical model described in our previous paper [4]. The broadening of the δ-layers was taken into account. We assume that the Gaussian distribution with a standard deviation (σ) describes the δ-layer profile. The heterostructure band diagrams and subband wave functions $\varphi(z)$ required for 2DEG mobility calculations were determined using the self-consistent Schrodinger-Poisson solver NextNano [5].

3. Results and discussion

Splitting of the δ-layers has been proposed to reduce the Coulomb scattering on the ionized donors that dominates at low temperatures in DA-pHEMT heterostructures [4]. So far such splitting has been applied only for suppression of parasitic parallel conductivity [6, 7]. The δ-layers from each side of a QW (see Fig. 1) can be split into two δ-sublayers with the total donor density in the nearest to the QW δ-sublayers equal to the 2DEG density. The donor concentration in the δ-sublayers remote from the QW should be close to the net acceptor concentration for the best compensation. The calculations have shown that δ-layer splitting does not result in a noticeable change in the potential diagram or subband wave functions in the QW region. Nevertheless, the 2DEG mobility limited by scattering on the ionized donors has improved significantly due to the increase in the effective thickness of the spacer. The calculated 2DEG mobility at 77 K for different δ-layers splitting is presented in the Fig.2. As one can see from Fig. 2, the mobility improves from 11230 cm^2 V^{-1} s^{-1} for the zero splitting to 16600 cm^2 V^{-1} s^{-1} for the 2-nm-splitting at the 3.4-nm standard deviation of the δ-layers.

The obtained values of the 2DEG density and mobility are listed in Table 1 for standard pHEMT heterostructures, DA-pHEMT heterostructures with unsplit δ-layers and DA-pHEMT heterostructures with split δ-layers. The parasitic parallel conductivity has been observed for standard pHEMT heterostructures with the density of slightly less than 2×10^{12} cm^{-2} and mobility of about 1700 cm^2 V^{-1} s^{-1} at the room temperature. As one can see from Table 1, the 2DEG mobility for the DA-pHEMT heterostructures approximately is halved in comparison with the standard pHEMT heterostructures. This results from a higher density of the ionized donors in both δ-layers which leads to an increase in scattering on the ionized donors [4]. But the 2DEG mobility in DA-pHEMT heterostructures after splitting of the δ-layers becomes comparable with mobility in standard pHEMT heterostrutures. Such a mobility enhancement testifies the theoretical calculations shown in Fig. 2.

Table 1. The 2DEG density and mobility for the heterostructures under study at 300 K and 77 K.

	300 K		77 K	
	n, $\times10^{12}$ cm^{-2}	μ, $cm^2V^{-1}s^{-1}$	n, $\times10^{12}$ cm^{-2}	μ, $cm^2V^{-1}s^{-1}$
pHEMT	2.5	7070	2.7	19500
pHEMT	2.6	6900	-	-
DA-pHEMT	3.7	4150	-	-
DA-pHEMT	3.4	4770	3.7	11000
splitted DA-pHEMT	4.0	6600	4.0	21200
splitted DA-pHEMT	3.9	6590	4.0	19600

The electric field dependencies of drift velocity are shown in Fig. 3 for the standard pHEMT heterostructures, DA-pHEMT heterostructures with unsplit δ-layers and DA-pHEMT heterostructures with split δ-layers. As one can see from this figure, the area of negative differential resistance is observed for the standard pHEMT heterostructure, whereas this feature is absent for both DA-pHEMT heterostructures. Moreover, the drift velocity in the DA-pHEMT heterostructures is higher than the drift velocity in the standard pHEMT heterostructure. The splitting of the δ-layers in the DA-pHEMT heterostructure leads to the increase in drift saturation velocity. These peculiarities can be explained by the suppression of the RST effect in DA-pHEMT heterostructures with an additional potential barrier formed by acceptors. Therefore, hot carriers don't escape from a QW into wide gap barrier layers and possesses a higher drift velocity.

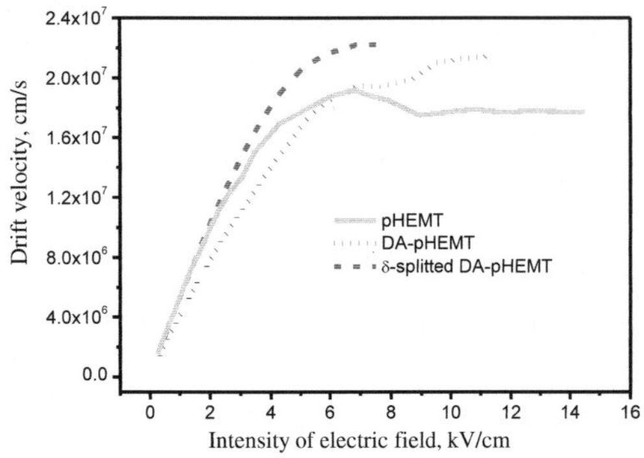

Fig. 3. The filed-dependencies of drift velocity at 300 K

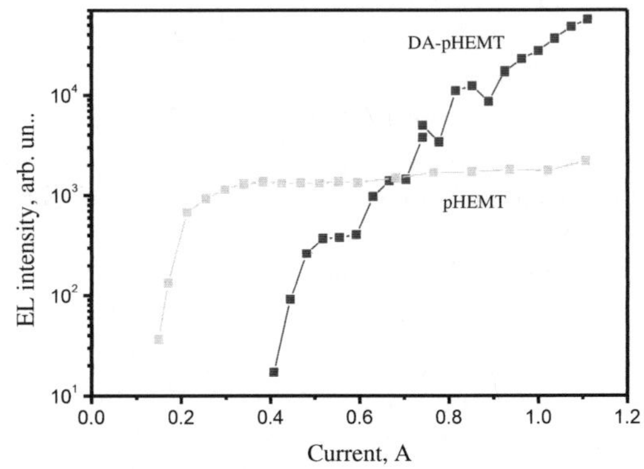

Fig. 4. The dependencies of EL intensity on current, passed through the sample, at 300 K.

The suppression of theRST effect is also confirmed by the EL measurements. The EL spectrum of a standard pHEMT heterostructure consists of many bands due to the optical transition in various heterostructure layers. The bands related to the optical transitions between the subbands in the InGaAs QW dominate in the spectrum of the DA-pHEMT heterostructures. Therefore, the electrons do not escape from the QW in a high electric field (up to few kV/cm) in the DA-pHEMT heterostructures. The EL intensity becomes saturated at the current of 0.3 A in standard pHEMT heterostructures (see Fig. 4). In case of a DA-pHEMT heterostructure the EL intensity grows up to the maximum applied current of 1.2 A. Such a difference confirms that the high potential barriers in the DA-pHEMT heterostructures suppress the RST effect and the electrons remain confined in high electric fields.

4. Conclusions

It has been shown both theoretically and experimentally that the splitting of the δ-layers leads to a significant increase in a low-field 2DEG mobility in DA-pHEMT heterostructures due to an effective spacer thickness growth and, consequently, reduction of the ionized donor scattering. There is no negative resistance area on field dependencies of drift velocity in DA-pHEMT heterostructures. Besides, the drift saturation velocity in DA-pHEMT heterostructures is higher than the drift saturation velocity in standard pHEMT heterostructures. The bands related to the optical transitions between the subbands in a InGaAs QW dominate in the EL spectra of DA-pHEMT heterostructures. Moreover, the EL intensity doesn't saturate at an electric field intensity growth in DA-pHEMT heterostructures. These features testify that the RST effect is suppressed in DA-pHEMT heterostructures by an additional potential barrier formed by acceptors.

References

[1] B.K. Ridley, Rep. Prog. Phys., **54**, (1991) 169
[2] Z.S. Gribnikov et al, Appl. Phys. Rev., **77**(4), (1995) 1337
[3] V. M. Lukashin et al, Semiconductors, **48**(5), (2014) 666
[4] D. V. Gulyaev et al, J. Phys. D: Appl. Phys., **49**, (2016) 095108
[5] www.nextnano.de
[6] K. Kalna and A. Asenov, Solid-State Electron. **48**, (2004) 1223
[7] R, Gupta et al, Microelectronics J., **37** (2006) 919

Resonant tunneling of charge carriers in InGaN/GaN superlattice

V V Kopyev, I A Prudaev, V L Oleynik

Functional Electronics Laboratory, Tomsk State University, 36 Lenin Avenue, Tomsk, 634050, Tomsk, Russia

E-mail: viktor.kopev@gmail.com

Abstract. The result of studies of resonant tunnelling of charge carriers in InGaN/GaN unipolar structure is presented. Authors show that at temperatures below 150 K the multiple negative-differential resistance regions are observed on a reverse current-voltage characteristic which is typical for inhomogeneous distribution field for sample with 6 nm barrier thickness. At GaN barrier thickness 3 and 12 nm resonant tunneling were not observed.

1. Introduction

The InGaN/GaN superlattices (SLs) and multiple quantum wells (MQWs) are widely used as active regions or buffer layers in laser diodes, ligth-emitting diodes, receivers of visible and ultraviolet ranges, etc. As a rule, the electrical characteristics of these devices are investigated in details. However, the characteristics of InGaN/GaN SLs and MQWs are not studied intensively. The author's different works showed that low-temperature transport of electrons cannot be explained by conventional quasi-equilibrium drift-diffusion model [1].

Studies of current-voltage (I-V) characteristics of structures with InGaN/GaN multiple quantum wells used for analysis of transport mechanism of charge carriers. It is well known that resonant tunneling of charge carriers in superlattice can be detected by using I-V measurements. In this case the regions of negative-differential resistance (NDR) at I-V curves are observed in the experiment [2]. Previously, such NDR regions were measured for different material systems (AlGaAs/GaAs [3], AlGaN/GaN [4, 5, 6] etc.) This paper presents the results of experimental study of resonant tunneling in InGaN/GaN superlattice.

2. Experimental details

Experiments were carried out for unipolar structures grown by a MOCVD method. on the [0001] patterned sapphire substrate. As precursors of III-group elements the following metal organic compounds were used: trimethylindium $In(CH_3)_3$, trimethylgallium $Ga(CH_3)_3$, triethylgallium $Ga(C_2H_5)_3$. As nitrogen precursor an ammonia (NH_3) was used. For donor dopant (silicon) silicomethane was used. The structure consisted of a 4-µm-thick GaN nucleation layer, a 2-µm-thick Si-doped *n*-type GaN, doped *n*-type layer with $In_xGa_{1-x}N$/GaN MQWs, and 17-nm-thick Si-doped *n*-type GaN. The MQW layers comprised ten periods of an $In_{0.15}Ga_{0.85}N$ well (2 nm) and a GaN barrier (6 nm). Additionally we investigated structures with GaN barrier thickness 3 and 12 nm. As a top contact we used Ni Schottky barrier. As an ohmic contact for buffer *n*-GaN layer we used fusing In contact of large area. Schematically investigated structure is shown in figure 1.

In the experiment we measured the current-voltage characteristics in the temperature range $T = 10 - 300$ K using a Janis cryostat and Keithley 2636A Source Meter. The measurements were carried out in a static and pulse mode to avoid overheating structures.

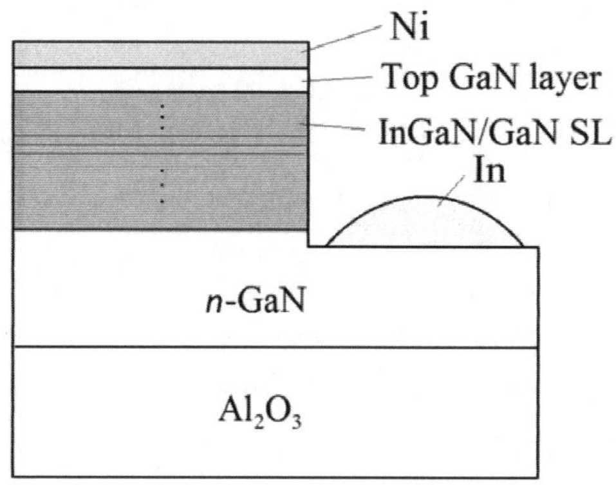

Figure 1. Schematic view of experimental samples for investigated InGaN/GaN unipolar structures.

3. Experimental results

Initially we studied dependence of barrier height on temperature using *n*-GaN/Ni structure:

$$\Phi_k = kT \cdot \ln\left(A^* T^2 / j_s \right), \tag{1}$$

where k – Boltzmann constant, T – temperature, A^* – Richardson constant, j_s – saturation current. In figure 2 the dependence of Schottky barrier height on temperature is shown. Drop height of the barrier is observed at temperatures below 200 K.

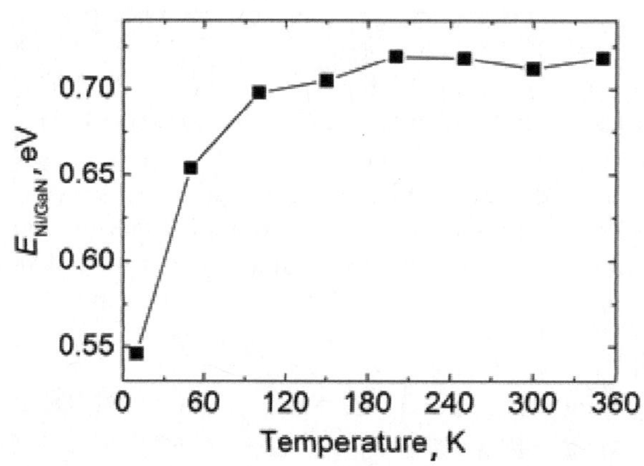

Figure 2. Dependence of Schottky barrier height on different temperature.

Then we measured I-V characteristics of unipolar samples (GaN barrier thickness at 3, 6 12 nm). And we found that at the different thickness barriers GaN experimental results for I-V curves are very different. The measurements of I-V curves showed an increasing of current limitation with barrier thickness increasing (figures 3 - 6). This limitation trend is more expressed at reverse bias for Ni/GaN Schottky barrier.

In the figures 3 and 4 it is show that structures with thickness 3 and 12 nm don't have NDR regions. This can be explained by the fact that the tunneling probability of the charge carriers is decreased with increasing GaN barrier thickness up to 12 nm. A split of discrete levels in InGaN/GaN QW in the miniband occurs with reducing GaN barrier thickness up to 3 nm where carriers are transported without tunneling.

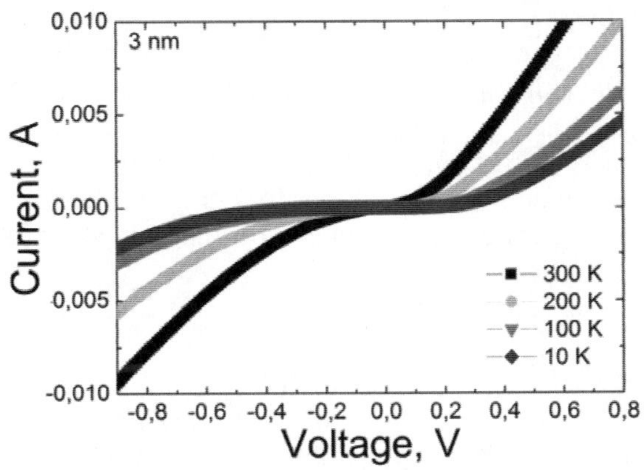

Figure 3. I–V curves of InGaN/GaN with 3 nm barrier thickness at different temperatures.

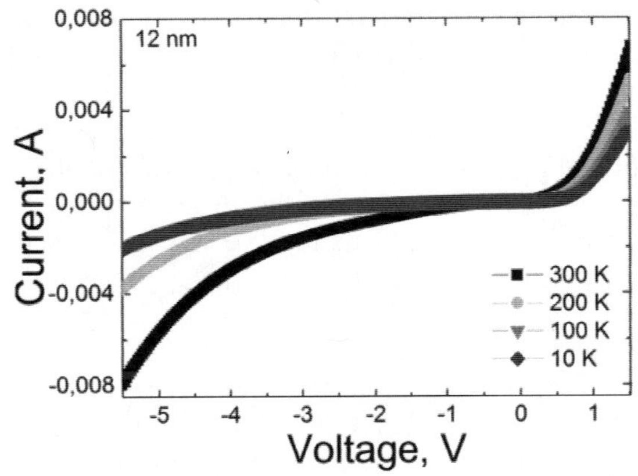

Figure 4. I–V curves of InGaN/GaN with 12 nm barrier thickness at different temperatures.

In the figures 5 and 6 experimental I-V characteristics for sample with 6 nm are shown. In the range from 0.5 to 0.7 V one or two regions of the NDR are observed on forward I-V characteristics which disappear when the temperature increases (Figure 1). Such behavior of I-V characteristic is typical for the case of homogeneous electric field distribution in superlattice [2]. At temperatures below 150 K at above 0.7 V the multiple NDR regions are observed on a reverse I-V characteristic which is typical for inhomogeneous distribution field (Figure 2).

In both figures 5 and 6 it is shown that the numbers of NDR regions are decreased by growing up the temperature. However, at room temperature (T = 300 K) those NDR regions are not observed at all.

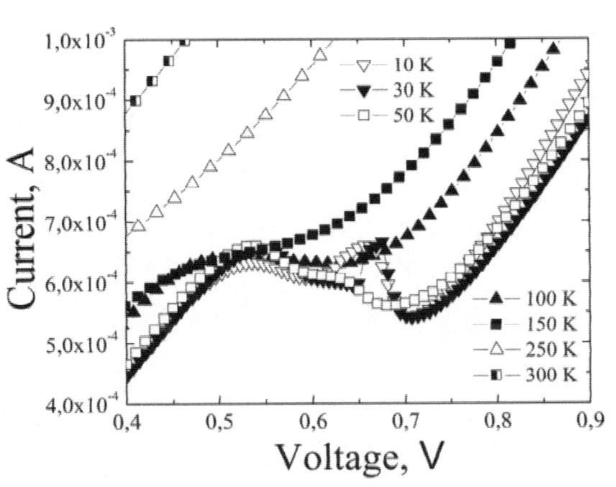

Figure 5. Forward I–V curves of InGaN/GaN with 6 nm barrier thickness at different temperatures.

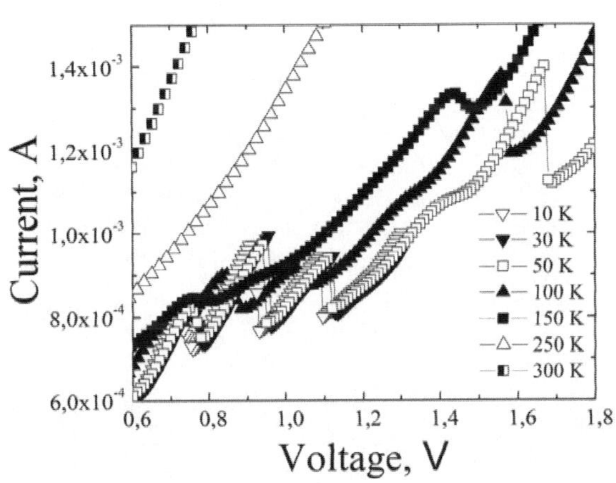

Figure 6. Reverse I–V curves of InGaN/GaN with 6 nm barrier thickness at different temperatures.

IOP Conf. Series: Journal of Physics: Conf. Series **864** (2017) 012052 doi:10.1088/1742-6596/864/1/012052

4. Conclusion

Thus, according to the results of this study, NDR regions were observed in the forward and reverse I-V curves for the InGaN/GaN SL with GaN barriers thickness of 3 nm. Found areas are observed at temperatures below 150-200 K. And at the same temperature reduction in the potential *n*-GaN/Ni barrier height occur.

Therefore it is explained the fact that NDR regions occur due to the resonant tunneling of charge carriers through the GaN barriers. By increasing the thickness of GaN barrier NDR regions do not occur. It can be associated with decreasing of tunneling probability of the charge carriers. By reducing the thickness of barriers a split of discrete levels in the miniband occurs. The detailed analysis of experimental results considers miniband vertical transport and resonant tunneling under condition of homogeneous and inhomogeneous electric field distribution in SLs [2]

In this work we have shown the result which demands additional studies. Hopefully, in the next few years, we are going to have new data that will provide answers and guidance for our next steps in the investigation of transport change carries of InGaN/GaN SL.

References

[1] Browne D A, Mazumber B, Wu Y, Speck J S 2015 J. Appl. Phys. **117** 185703
[2] Waker A 2002 Physics Reports **357** 1
[3] Andronov A A, Dodin E P, Zinchenko D I, Nozdrin Yu N 2009 Semiconductors **43** 240-247
[4] Haoran Chen, Lin'an Yang, Yue Hao 2014 Journal Of Applied Physics **116** 074510
[5] Warde E, Sakr S, Tchernycheva M Julien H 2012 Journal of Electronic Materials **41** 965-970
[6] Noriyuki Watanabe, Haruki Yokoyama, Naoteru Shigekawa 2013 Jpn. J. Appl. Phys. **52** 08JN03

Transport in vertically stacked hetero-structures from 2D materials

Fan Chen, Hesameddin Ilatikhameneh, Yaohua Tan, Daniel Valencia, Gerhard Klimeck and Rajib Rahman

Network for Computational Nanotechnology (NCN), Purdue University, West Lafayette, IN 47906, USA

E-mail: fanchen@purdue.edu

Abstract. In this work, the transport of tunnel field-effect transistor (TFET) based on vertically stacked hereto-structures from 2D transition metal dichalcogenide (TMD) materials is investigated by atomistic quantum transport simulations. WTe2-MoS2 combination was chosen due to the formation of a broken gap hetero-junction which is desirable for TFETs. There are two assumptions behind the MoS2-WTe2 hetero-junction tight binding (TB) model: 1) lattice registry. 2) The $S-Te$ parameters being the average of the $S-S$ and $Te-Te$ parameters of bilayer MoS2 and WTe2. The computed TB bandstructure of the hetero-junction agrees well with the bandstructure obtained from density functional theory (DFT) in the energy range of interest for transport. NEGF (Non-Equilibrium Green′s Function) equations within the tight binding description is then utilized for device transfer characteristic calculation. Results show 1) energy filtering is the switching mechanism; 2) the length of the extension region is critical for device to turn off; 3) MoS2-WTe2 interlayer TFET can achieve a large on-current of $1000\mu A/\mu m$ with $V_{DD}=0.3V$, which suggests interlayer TFET can solve the low ON current problem of TFETs and can be a promising candidate for low power applications.

1. Introduction

The fast growth of today′s technology has been sustained by continuous scaling of silicon-based MOSFETs. The scaling of transistors face two major challenges nowadays: the degradation of gate gate control and the fundamental thermionic limitation of the steepness of sub-threshold swing (SS). 2D materials have emerged as promising channel materials for transistors, as they can maintain excellent device electrostatics at much reduced channel length and thickness [1]. Interlayer tunnel field effect transistors (TFETs) based on vertically stacking 2D materials have also been shown to break the thermionic limitation of sub-threshold swing (SS)[2, 3]. However, experimentally demonstrated vertical TFET with small SS still suffer from low current levels[4]. TFET designs exploiting different 2D material properties[5, 6, 7, 8, 9, 10, 11] have been proposed to increase the tunneling current. Theoretical works also demonstrated the possibility of a high ON current in interlayer TFET′s [12, 13]. Yet, an understanding of device physics of interlayer TFETs such as the effect from device geometry, lattice mismatch, twisted angles[14] and Van der waals bonding in order to provide a guideline for interlayer TFETs to achieve high ON currents experimentally. In this work, an interlayer TFET based on vertically stacked MoS2 and WTe2 as shown in Fig. 4) is studied. The model assumptions are described in details and have been carefully examined. The switching mechanism and performance of the device are then investigated and evaluated.

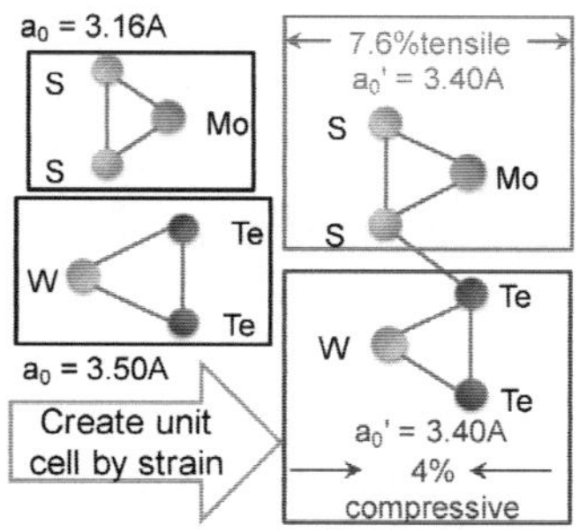

Figure 1. Assumption I : MoS2 and WTe2 layers have been strained to the same lattice constant to be registered.

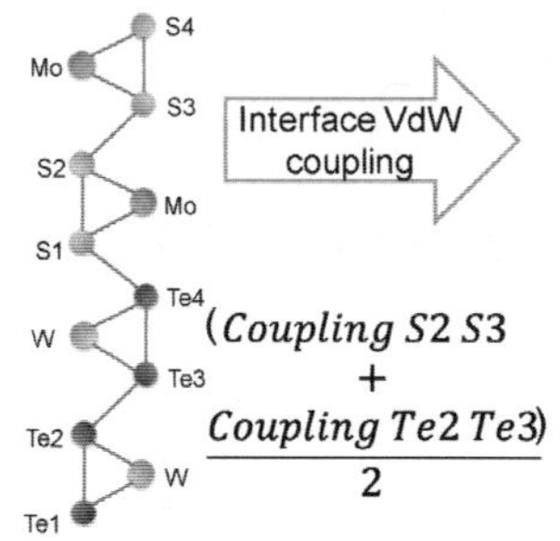

Figure 2. Assumption II : Interface VdW coupling of vertical MoS2-WTe2 hetero-junction is the average of the VdW coupling of the two materials.

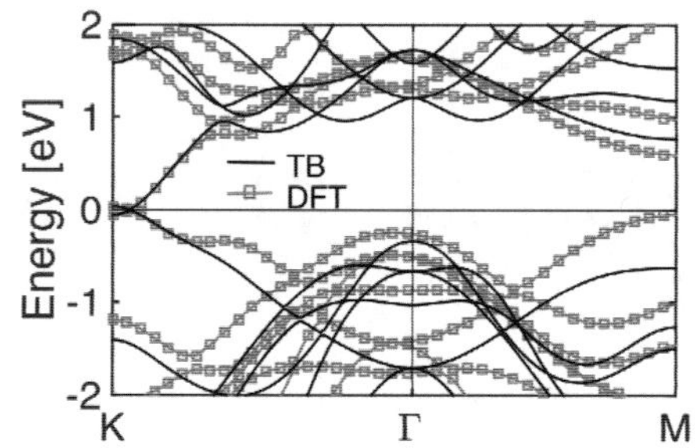

Figure 3. A comparison of band structure plots for vertical MoS2-WTe2 hetero-junction from tight binding (TB) and density function theory (DFT). They match well along the band edge at K point.

2. Method

The Transistion Metal Dichalcogenide (TMD) Hamiltonian (strained) is represented by an $sp3d5$ second nearest neighbor tight-binding (TB) model including $spin-orbit$ coupling. DFT-guided TB model is adapted to reduce computational expense and system size compared to $ab-initio$ calculations [15]. The PBE functional is used in the density functional theory (DFT) calculations since it is known to produce band gap comparable with experimental measurements for TMDs[18]. The Van der Waals correction for the interlayer coupling of multilayer MoS2 or WTe2 is included by the optB88 functional[16].The TB parameters are well calibrated to match the band structure and effective mass from DFT by a well-established mapping method [17, 15]. The TB parameters are general and capture the band structure of both the bulk and monolayer TMDs [18, 23].

There are two major assumptions to obtain the TB parameters for the hetero-junction: 1) The materials are lattice matched by applying 7.6 percent tensile strain to MoS2 and 4 percent compressive strain to WTe2 as shown in Fig. 1; 2) The interlayer coupling parameters between the Sulphur and the Telluride atoms are obtained by averaging the coupling parameters of S-S in bilayer MoS2 and and Te-Te in WTe2 as shown in Fig. 2.

A strained TB model is developed to fulfill the lattice registry. The relation of TB parameters with strain percentage is obtained by fitting to a set of strained DFT MoS2 or WTe2 monolayers. This model is then carefully examined by reproducing the DFT bandstructure under different strain. In order to obtain the DFT band structure of MoS2-WTe2 interface for benchmarking, an ionic relaxation was carried out on the supercell until all the atomic forces on each ion were less than $0.0001eV/nm$. Subsequently, the EK diagram was obtained for the relaxed configuration. This calculation was performed with a generalized gradient approximation (GGA) and PBE functional with a vdW correction as implemented in VASP (DFT-D2).

All the transport characteristics of the MoS2-WTe2 TFET have been simulated using the self-consistent Poisson-semiclassical approach to obtain the potential profile and then this potential is passed to quantum transmitting boundary method (an equivalent of non-equilibrium Green's function method for ballistic transport) in the multi-scale [19] and multi-physics [20, 21, 22] Nano-Electronic MOdeling (NEMO5) tool [23]. The default source and drain regions are doped with the doping level of $10^{20}cm^{-3}$. Equivalent oxide thickness (EOT) is set to 0.5nm.

IOP Conf. Series: Journal of Physics: Conf. Series **864** (2017) 012053 doi:10.1088/1742-6596/864/1/012053

3. Results and Discussion

The DFT band-diagram of a relaxed MoS2-WTe2 vertical hetero-junction is plotted in Fig. 3. The zero band gap in the band diagram shows MoS2 and WTe2 forms a broken bandgap junction. The conduction and valence band edges meet at K point which provides the density of states for carrier transport. Furthermore, the TB band diagram computed under the assumptions agrees well with the DFT results in the energy range that is important for transport.

The tight-binding material model developed in this manner was used to simulate quantum transport for the device shown in Fig. 4. The device geometry parameters of the MoS2-WTe2 interlayer TFET $L_{overlap}$ and L_{ext} shown in Fig. 4 are optimized to be $30nm$ and $15nm$ respectively. The extension region is critical for the device to turn off. A zero L_{ext} would cause a high leakage current such that the small SS can not be observed. The leakage current mainly comes from the edges states. Fig. 4 shows that the transfer characteristics for a supply voltage of -0.3 V. The striking feature is the large ON current of $1000\mu A/\mu m$, which is significantly higher than the homo-junction 2D TFETs simulated thus far. The SS is around $20mV/dec$ in 4 orders of magnitude of current change.

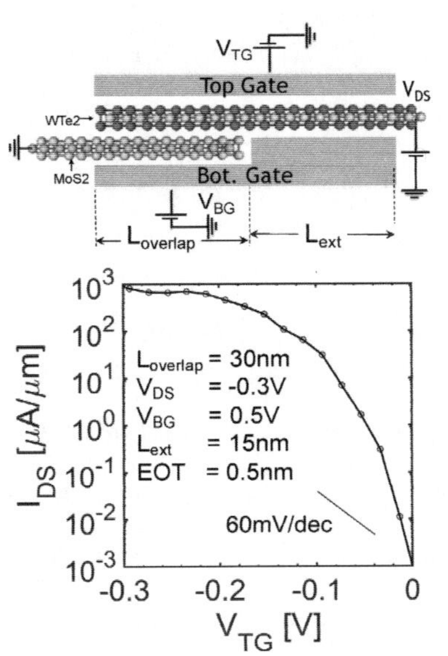

Figure 4. The device structure and the transfer characteristics $(Id - Vg)$ of a MoS2-WTe2 interlayer TFET. The device has an overlap region and an extension region denoted by $L_{overlap}$ and L_{ext}, respectively.

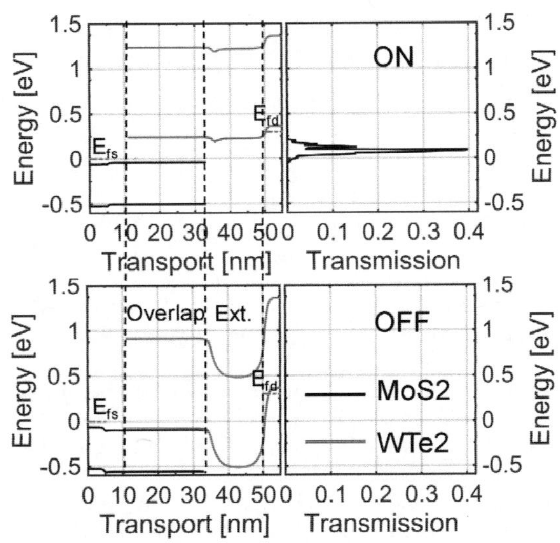

Figure 5. The band edges of a MoS2-WTe2 interlayer TFET at ON and OFF state aligned with the energy resolved transmission respectively. At ON state, WTe2 Ev is higher than MoS2 Ec. This opens a window for electrons to flow and it has a non-zero transmission. At OFF state, WTe2 Ev is lower than MoS2 Ec. No electrons can go through and it has a zero transmission.

The band edges of a MoS2-WTe2 interlayer TFET aligned with the energy resolved transmission at ON and OFF state respectively are plotted in Fig. 5. At ON state, WTe2 Ev is higher than MoS2 Ec. This opens a window for electrons to flow, and therefore achieves a high ON current. This indicates an energy filtering switching mechanism.

4. Conclusion

In conclusion, MoS2-WTe2 interlayer TFET is studied. The model assumptions have been carefully examined. The band diagram aligned with transmission shows the switching mechanism is the energy filtering. The MoS2-WTe2 interlayer TFET shows a SS smaller than $60mV/dec$. The extension region is critical for the device to turn off. An optimized device can achieve an ON current of $1000\mu A/\mu m$ with $V_{DD} = 0.3$.This suggests MoS2-WTe2 interlayer TFET as a promising candidate for low power applications.

Acknowledgments

This work was supported in part by the Center for Low Energy Systems Technology, one of six centers of STARnet, and in part by the Semiconductor Research Corporation Program through Microelectronics Advanced Research Corporation and Defense Advanced Research Projects Agency.

References

[1] Geim A and Grigorieva I 2013 *Nature* **499** 419–425 ISSN 0028-0836
[2] Szabo A, Koester S J and Luisier M *Device Research Conference (DRC), 2014 72nd Annual* (IEEE) pp 19–20 ISBN 1479954055
[3] Lam K T, Seol G and Guo J *Electron Devices Meeting (IEDM), 2014 IEEE International* (IEEE) pp 30.3. 1–30.3. 4
[4] Sarkar D, Xie X, Liu W, Cao W, Kang J, Gong Y, Kraemer S, Ajayan P M and Banerjee K 2015 *Nature* **526** 91–95
[5] Ameen T A, Ilatikhameneh H, Klimeck G and Rahman R 2015 *arXiv preprint arXiv:1512.05021*
[6] Ilatikhameneh H, Ameen T, Novakovic B, Tan Y, Klimeck G and Rahman R 2016 *arXiv preprint arXiv:1605.03979*
[7] Chen F W, Ilatikhameneh H, Ameen T A, Klimeck G and Rahman R 2016 *arXiv preprint arXiv:1607.04065*
[8] Ilatikhameneh H, Chen F W, Rahman R and Klimeck G 2015 *Computational Electronics (IWCE), 2015 International Workshop on* pp 1–3
[9] Ilatikhameneh H, Ameen T, Klimeck G, Appenzeller J and Rahman R 2015 *Electron Device Letters, IEEE* **36** 1097–1100 ISSN 0741-3106
[10] Chen F W, Ilatikhameneh H, Klimeck G, Chen Z and Rahman R 2016 *IEEE Journal of the Electron Devices Society* **4** 124–128
[11] Chen F, Ilatikhameneh H, Chu T, Rahman R, Appenzeller J, Chen Z and Klimeck G 2015 *Proc. TECHCON*
[12] Cao J, Cresti A, Esseni D and Pala M 2016 *Solid-State Electronics* **116** 1–7 ISSN 0038-1101
[13] Li M O, Esseni D, Snider G, Jena D and Xing H G 2014 *Journal of Applied Physics* **115** 074508 ISSN 0021-8979
[14] Tan Y, Chen F and Ghosh A 2016 *arXiv preprint arXiv:1606.01858*
[15] Tan Y P, Povolotskyi M, Kubis T, Boykin T B and Klimeck G 2015 *Physical Review B* **92** 085301
[16] Farmanbar M and Brocks G 2016 *Physical Review B* **93** 085304
[17] Tan Y, Povolotskyi M, Kubis T, He Y, Jiang Z, Klimeck G and Boykin T B 2013 *Journal of Computational Electronics* **12** 56–60
[18] Ilatikhameneh H, Tan Y, Novakovic B, Klimeck G, Rahman R and Appenzeller J 2015 *IEEE Exploratory Solid-State Computational Devices and Circuits* **1** 12–18 ISSN 2329-9231
[19] Chen F W, Manfra M, Klimeck G and Kubis T 2015 *International Workshop on Computational Electronics (IWCE 2015)* (http://in4.iue.tuwien.ac.at/pdfs/iwce/iwce18_2015/IWCE_2015_33-34.pdf)
[20] Miao K, Sadasivam S, Charles J, Klimeck G, Fisher T and Kubis T 2016 *Applied Physics Letters* **108** 113107
[21] Chen F, Jauregui L, Tan Y, Manfra M, Chen Y, Gerhard K and Kubis T 2015 *Applied Physics Letters* **107** 121605
[22] Chen F W, Ilatikhameneh H, Klimeck G, Rahman R, Chu T and Chen Z *2015 International Conference on Simulation of Semiconductor Processes and Devices (SISPAD),* (IEEE) pp 177–181 ISBN 1467378585
[23] Fonseca J E, Kubis T, Povolotskyi M, Novakovic B, Ajoy A, Hegde G, Ilatikhameneh H, Jiang Z, Sengupta P and Tan Y 2013 *Journal of Computational Electronics* **12** 592–600 ISSN 1569-8025

IOP Conf. Series: Journal of Physics: Conf. Series **864** (2017) 012054 doi:10.1088/1742-6596/864/1/012054

Observation of Conductance Fluctuation due to Zitterbewegung in InAs 2-dimentional Electron Gas

Y Iwasaki, Y Hashimoto, T Nakamura and S Katsumoto

Institute for Solid State Physics, University of Tokyo, 5-1-5 Kashiwanoha, Kashiwa, Chiba 277-8581, Japan

E-mail: `you.iwasaki@issp.u-tokyo.ac.jp`

Abstract. Zitterbewegung (ZB; trembling motion) is one of the peculiar phenomena predicted in the Dirac equation in that the velocity and the spatial coordinate cannot be well defined kinetic constant. While the ZB in vacuum is far out of reach, feasibility of observation in solids has been discussed though very few experiments have been reported yet. We here report the ZB in two-dimensional electron gas with strong Rashba-type spin-orbit coupling causes large fluctuation in conductance. Such an experiment was impossible until recent development of point contact injectors with spin-polarised electrons. Numerical calculations on a simple model successfully simulate the experimental observations.

1. Introduction

A characteristic oscillation in the velocity of a free electron which obeys the Dirac equation, was first found by Schrödinger [1]. The phenomenon, now is called "Zitterbewegung" (ZB; German word for trembling motion), is however far out of experimental reach in its amplitude and frequency [2]. While some successful imitations [3–5] have been reported, very few have been reported on electrons in "emulated vacuums" like graphene [6] or semiconductors [7] due to the still stringent conditions. For example in two-dimensional electron gas (2DEG) with strong Rashba-type spin-orbit coupling (SOC), the ZB should appear as a meandering of electrons with amplitudes around ten nanometers [7], which is now barely observable by nanoscale techniques.

Here we report observation of the ZB in an InAs 2DEG with strong Rashba SOC as conductance fluctuations (CFs). For the emitters and the collectors of spin-polarised electrons, we used quantum point contacts (QPCs). The amplification of the ZB here is by virtue of small scattering range of impurities. The experimental observation can be simulated by a numerical calculation based on a simple tight-binding model.

2. Experimental

Figure 1(a) is the cross section of the layered structure grown by molecular beam epitaxy. The 2DEG has the mobility $\mu = 65900$ cm^2/Vs, the sheet carrier concentration $n = 1.1 \times 10^{12}$ /cm^2 and the Rashba parameter $\alpha = 3.6 \times 10^{-11}$ eVm, which were obtained from Hall and Shubnikov-de Haas measurement. Figure 1(b) is a scanning electron micrograph image of the sample fabricated with electron-beam lithography and Ar dry-etching. Electrodes 1, 2, 4, 5, 6 are connected through QPCs while 3 and 7 are through 1.8 µm wide strips. The sample was cooled down to 70 mK in a dilution fridge with a superconducting magnet. Conventional lock-in technique was used for measuring two-wire conductances with frequencies lower than 1 kHz.

IOP Conf. Series: Journal of Physics: Conf. Series **864** (2017) 012054 doi:10.1088/1742-6596/864/1/012054

The conductances of the QPCs in the present experiment were tuned to the conductance plateau at $2e^2/h \equiv G_\mathrm{q}$, on which the emitted electrons are highly polarised as established theoretically [9] and experimentally [10]. In the two-terminal conductance measurements between the QPCs, the electrons emitted from a QPC should experience several scatterings during the traversal to the collector QPC because the mean free path 1.1 µm is a few times shorter than the length between the confronted QPCs (Fig. 1(c)). Though the opening of the QPCs is in the same order as the amplitude of the ZB, the range of scattering potential can be much shorter. Hence if a trajectory meandered by the ZB is modified by *e.g.* external magnetic field $\boldsymbol{B}$ [2], the scattering cross section should be largely modified resulting in the modification of conductance.

3. Results and Discussion

In conductance between terminals 1 and 5 (G_{1-5}), we observe reproducible CF as a function of $\boldsymbol{B}$ (Fig. 2(a)). Here "reproducible" means patterns of aperiodic variations are almost the same for independent sweeps of $\boldsymbol{B}$. Cooling the sample from 1.0 K to 70 mK enhanced the amplitude of the CF from 3 % to 15 % of the averaged conductance, suggesting that the CFs originate from some quantum effect. These characteristics are reminiscent of so called "universal conductance fluctuation" (UCF) in mesoscopic conductors though this is not the case as we see in the following.

Figure 2(a) also shows angular dependence of the CF when $\boldsymbol{B}$ was rotated in y-z plane from $\theta = 0°$ (perpendicular to the 2DEG) to 90° (paralle to the 2DEG). Though the pattern of CF changes with θ, the amplitude and the averaged frequency versus B are almost constant manifesting that the CF does not contain the portion by interference between spatial orbits in 2DEG. Particularly the CF around $\theta = 90°$ is the evidence that some mechanism through electron spin should be the origin. A dramatic change occurs when we switch to the conductance between electrodes 3 and 7 (G_{3-7}), the path of which has no QPC on it. As shown in Fig. 2(b), CF is negligible for both the parallel and the perpendicular field even if we consider the channel-averaging effect. For the perpendicular field, even a large Shubunikov-de Haas oscillation is observed just like an ordinary 2DEG. We also should add that much smaller CF is found in the transport through a single QPC in the present sample.

Because these two measurements are on the same part of the sample, we should look for the

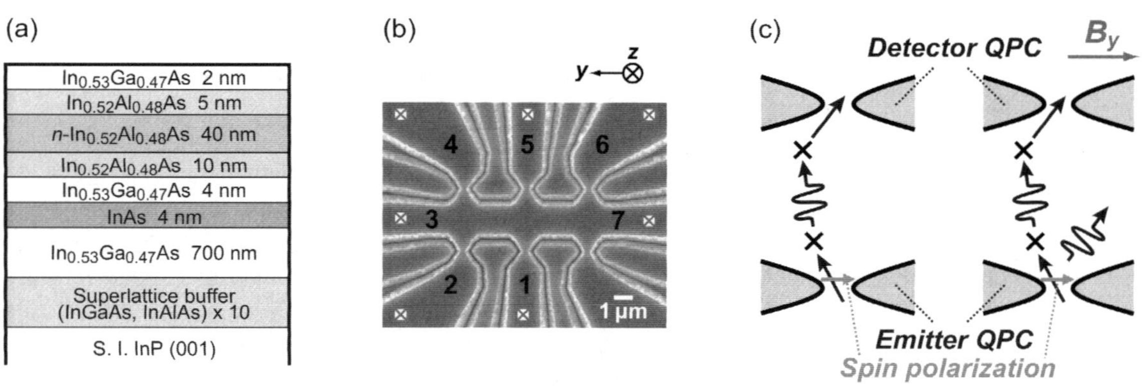

Figure 1. (a) Cross sectional view of the layered structure. (b) Scanning electron micrograph of the sample. Red shaded areas are gates of QPCs. (c) Illustrates how the ZB occurs and affects a scattering by an impurity.

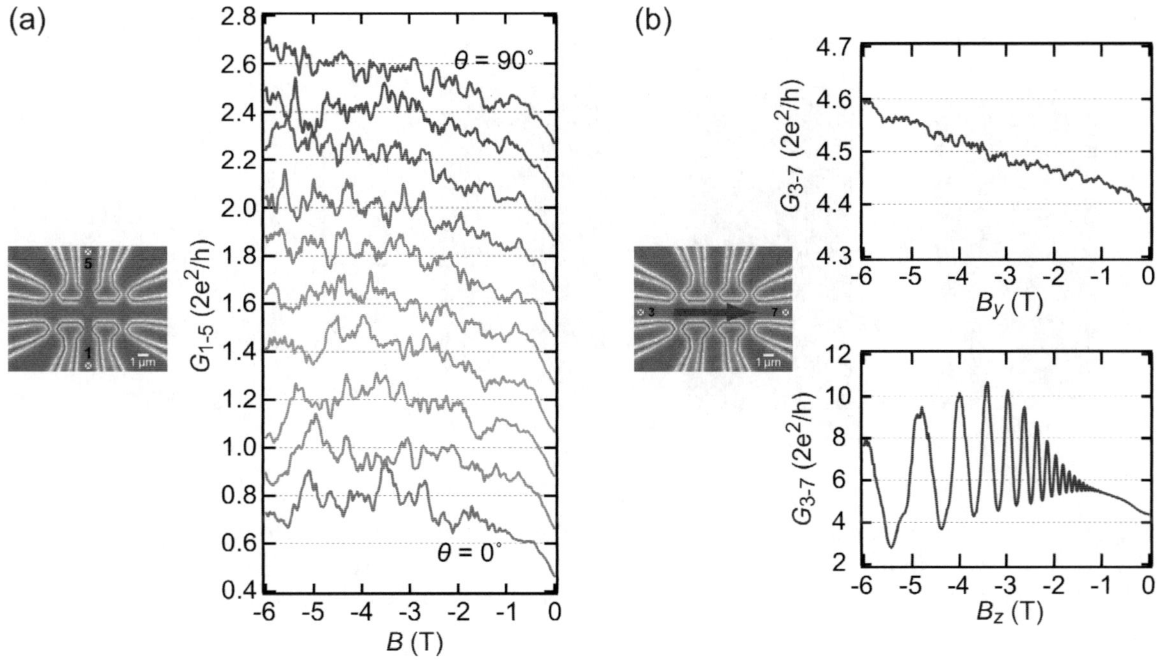

Figure 2. (a) G_{1-5} versus B with different $\boldsymbol{B}$ zenith angle θ from 0° in z-direction (bottom) to 90° in y-direction (top). The parameter step and offset in conductance are 10° and $0.2G_\mathrm{q}$/step respectively. (b) G_{3-7} versus B_y (top panel) and B_z (bottom panel). The photographs indicate the directions of electron propagations. The temperature is 70 mK.

origin of the difference in the nature of emitted electrons, which should be highly spin-polarized by the emitter QPC. Hence what we can consider as the origin of the CF is nothing but the ZB, the effect of which is amplified with the small range of the impurity scattering potentials.

In order to see the feasibility of our inference, we next conduct numerical calculations on the program package "Kwant" [11]. The parameters, the QPC width 400 nm and the Rashba strength 3.6×10^{-11} eVm are directly taken from the experiment. The effective mass $0.023m_0$ (where m_0 is the electron mass in vacuum) and g-factor -14.9 are extracted from data of bulk InAs. Fermi energy ε_F is estimated from the above parameters under the condition that the QPC conductance is $1.0G_\mathrm{q}$. The model area (3.6 µm× 3.0 µm) is divided into square unit cell (the lattice constant 25 nm). The boundary condition is absorption walls except for the QPCs. The scattering center s are introduced by adding randomly-distributed values on on-site energy with amplitudes of 0.2 eV.

First we calculate distributions of the charge density near the emitter QPC for the case of no scattering center and for the electrons spin-polarised along x-direction, the results of which are given in Figs.3(a)(b)(c). Clear meandering pattern due to the ZB is visible in Fig. 3(a). The oscillation period is modulated by applying $B_y = \pm 2.0$ T (Fig. 3(b)(c)) as predicted theoretically by Zawadzki and Rusin [2].

Next we introduce impurities and change the polarisation of emitted electrons to those along y. Figure 3(d) shows thus calculated G_{1-5}. In spite of the simpleness in the model, the calculated conductance between two QPCs really shows aperiodic fluctuation against the in-plane magnetic field, supporting the essential legitimacy of our model for the experiment.

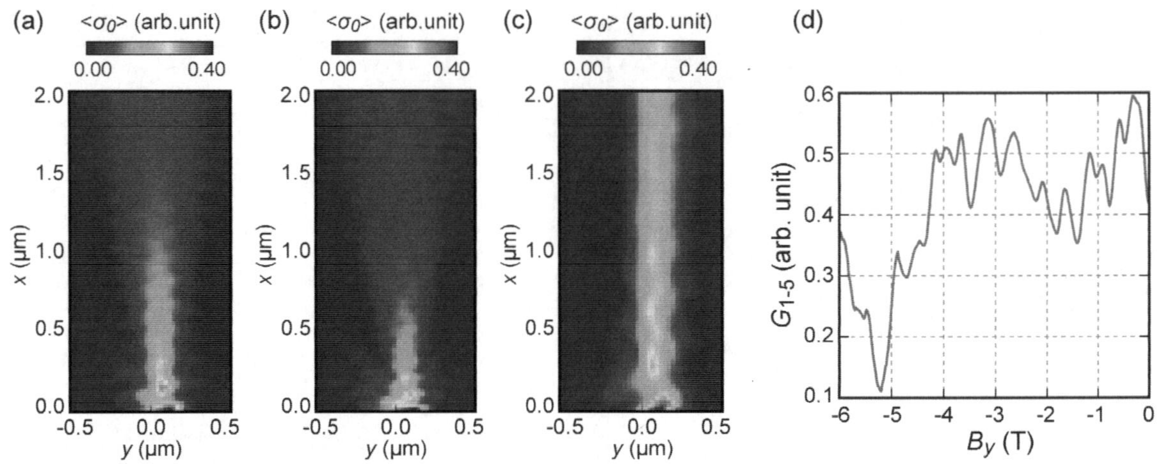

Figure 3. (a)(b)(c) Calculated space distributions of charges. From the QPC at the origin, electrons polarised along x are injected. B_y is 0.0 T, -2.0 T and $+2.0$ T for (a), (b) and (c) respectively. (d) G_{1-5} as a function of B_y. The polarisation of the electrons is along y-axis.

4. Summary

We found that the ZB appears in a 2DEG with strong Rashba SOC as a reproducible CF versus magnetic field. The small range of scattering centers amplifies the ZB through modulation of the scattering cross section. This picture is reproduced in a numerical calculation on a simple model.

5. Acknowledgment

This work was supported by Grant-in-Aid for Scientific Research on Innovative Area, "Nano Spin Conversion Science" (Grant No.26103003), also by Grant No.25247051 and by Special Coordination Funds for Promoting Science and Technology. Iwasaki was funded by Japan Society for the Promotion of Science through Program for Leading Graduate Schools (MERIT).

References

[1] Schrödinger E, 1930 *Sitz. ber. Preuss. Akad. Wiss. Phys. Math. Kl.* **24,** 418.
[2] Zawadzki W and Rusin T M 2011 *J. Phys.: Condens. Matter* **23,** 143201.
[3] Gerritsma R, Kirchmair G, Zähringer F, Solano E, Blatt R and Roos C F 2010 *Nature* **463,** 68.
[4] LeBlanc L J, Beeler M C, Jiménez-García K, Perry A R, Sugawa S, Williams R A and Spielman I B 2013 *New J. Phys.* **15,** 073011.
[5] Qu C, Hamner,C, Gong M, Zhang C and Engels P 2013 *Phys. Rev.* A **88,** 021604.
[6] Rusin T M and Zawadzki W 2009 *Phys. Rev.* B **80,** 045416.
[7] Brusheim P and Xu H Q 2007 *Phys. Rev.* B **75,** 195333.
[8] Biswas T and Ghosh T K 2012 *J. Phys.: Condens. Matter* **24,** 185304.
[9] Eto M, Hayashi T, Kurotani Y and Yokouchi H 2006 *phys. stat. sol. (c)* **3,** 4168—71.
[10] Kim S W, Hashimoto Y, Nakamura T and Katsumoto S, *Preprint* 1605.04685.
[11] Groth C W, Wimmer M, Akhmerov A R, Waintal X, 2014 *New J. Phys.* **16**, 063065.

IOP Conf. Series: Journal of Physics: Conf. Series **864** (2017) 012055 doi:10.1088/1742-6596/864/1/012055

Behaviour of Bichromatic Microwave Induced Magneto-resistance Oscillations in the High Mobility GaAs/AlGaAs 2D electron System

Binuka Gunawardana[1], Han-Chun Liu[1], Rasanga L. Samaraweera[1], C. Reichl[2], W Wegscheider[2] and R.G. Mani[1]

[1]Dept. of Physics and Astronomy, Georgia State University, Atlanta, GA 30303 USA
[2]ETH-Zurich, Zurich, Switzerland

Kgunawardana1@student.gsu.edu

Abstract: Microwave radiation-induced magneto-resistance oscillations are examined under bichromatic excitation for various frequency combinations in order to obtain a better understanding of the lineshape observed in the dual excitation experiment of the high mobility GaAs/AlGaAs 2D electron system. Here, we examine superposition- or lack thereof- in the lineshape observed in the bichromatic experiment, and report a trend observed between the monochromatic and bichromatic responses of the oscillatory diagonal resistance.

Introduction

Microwave radiation-induced magnetoresistance oscillations refer to the B^{-1}-periodic oscillatory variation in the diagonal (R_{xx}) - and off-diagonal (R_{xy}) resistance observed under microwave excitation in the 2D electron system under the influence of a transverse magnetic field, B.[1-24] Such oscillations exhibit a so-called ¼ -cycle phase shift, non-linear variation in the oscillatory amplitude with microwave power,[13,21,22] sensitivity in the oscillatory amplitude to the orientation of the linearly polarized microwaves with respect to the device,[15] etc. Many theoretical explanations have been proposed to help understand associated phenomena [25-42], including, for example, mechanisms based on (a) the scattering of electrons by phonons and impurities between Landau levels [26,30,33], (b) the periodic motion of electron orbit centers under microwave irradiation,[27,35] and (c) the re-collision of cyclotron electrons from scattering centers [42].

In order to obtain a better understanding of both the underlying physical mechanism and the experimental phenomenology, microwave radiation-induced magneto-resistance oscillations in the diagonal resistance R_{xx} are examined here under bichromatic excitation for various frequency combinations. Thus, the 2D electron system was photo-excited with microwave frequencies at both f_1 and f_2, which were chosen to realize different f_1/f_2 ratios ranging from $1.84 < f_1/f_2 < 3.4$. In addition, the microwave power was changed systematically to study the evolution of the lineshape in going from the monochromatic to the bichromatic photoexcitation situation.

1. Experiment and results

Measurements were carried out on a GaAs/AlGaAs 2D heterostructure sample with an electron density of 3.3×10^{11} cm^{-2} and mobility 14.6×10^{6} cm^2/Vs at 1.8K. Two microwave sources served to produce microwaves at two different frequencies and bichromatic photoexcitation of the specimen was realized by combining the output of these two sources with a waveguide coupler. In experiment, the sample was first illuminated with monochromatic microwaves at frequency f_1, with the other microwave source at f_2 switched off, to obtain the R_{xx} vs. B response at f_1. Then, the source at f_1 was switched off, and the source at f_2 was switched on, to obtain the R_{xx} vs. B response at f_2. Finally, the sources at both f_1 and f_2 were switched on for the bichromatic photo-excitation experiment. Traces of R_{xx} vs. B obtained for frequency combinations of f_1 = 90.6GHz and f_2 = 34.65GHz, f_1 = 90.6GHz and f_2 = 41.0GHz and f_1 = 90.6GHz and f_2 = 48.7GHz are shown in figure 1.

A study of the experimental results of Fig. 1 suggests that the bichromatic microwave response of R_{xx} seems to follow the low frequency monochromatic microwave response of R_{xx} at low magnetic fields. However, the bichromatic response tends to deviate towards the high frequency monochromatic microwave response when the magnetic field strength increases, for all the examined combinations.

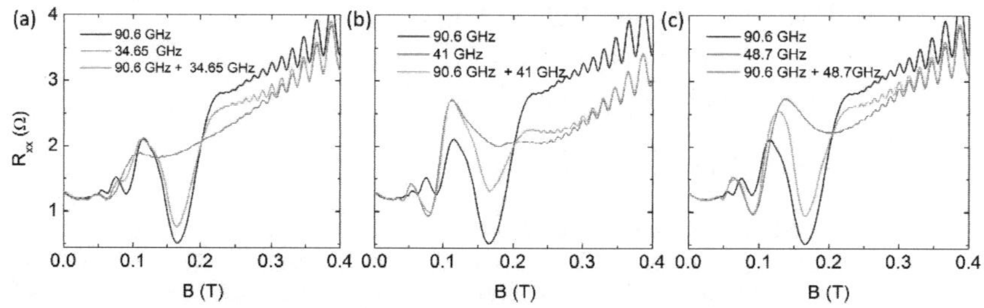

Figure 1. Monochromatic and bichromatic microwave induced magnetoresistance for high frequency $f_1 = 90.6$GHz and low frequency (a) $f_2 = 34.65$ GHz, (b) $f_2 = 41.0$ GHz and (c) $f_2 = 48.7$ GHz

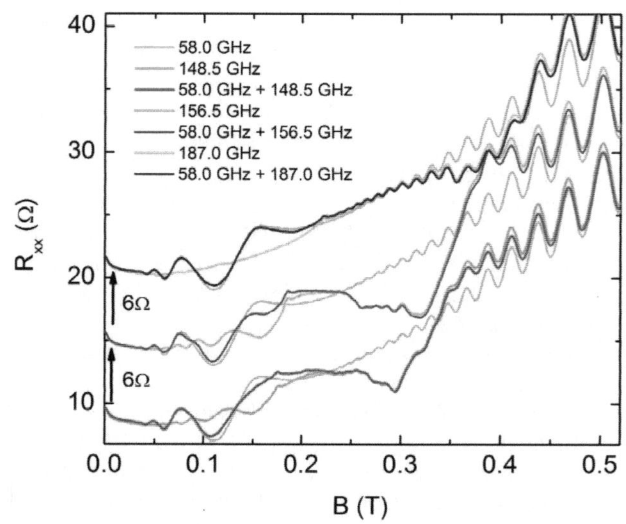

Figure 2. Monochromatic and bichromatic microwave induced oscillations in the diagonal resistance, R_{xx} for f_l = constant = 58 GHz and f_2 = 141, 161, 174, and 207 GHz. The traces are shifted by 6Ω for clarity. The yellow solid lines show the monochromatic response of R_{xx} at 58 GHz for each case.

To further examine this behavior, the sample was also illuminated with microwave frequencies which produce R_{xx} oscillations that are further apart on the magnetic field axis such as with $f_l = 58.0$GHz while f_2 = 148.5, 156.5, 187.0 GHz. The results of such measurements, shown in figure 2, indicate once again that the monochromatic low frequency R_{xx} response sets bichromatic R_{xx} response at low magnetic fields as the monochromatic high frequency R_{xx} response mostly determines the bichromatic R_{xx} response at high magnetic fields in the bichromatic experiment. .

The microwave induced magnetoresistance under bichromatic excitation with frequencies $f_l = 90.6$ GHz and $f_2 = 41.0$GHz was also examined upon changing the microwave power, P, of the low frequency excitation at f_2; the results of such experiments are shown in Fig.3. The R_{xx} at the extrema indicated by the labels 1, 2 and 3 of Fig. 3 are plotted versus the microwave power in Fig. 4, and Fig. 4 shows a nonlinear variation of the extremal resistance with P. This behavior was also observed at other frequency combinations utilized in this experiment.

2. Discussion

The bichromatic microwave response of the diagonal resistance, R_{xx} follows the low frequency monochromatic response at low magnetic fields and it deviates towards high frequency monochromatic microwave response as the magnetic field strength increases. Previous studies [24,25] had suggested that the bichromatic microwave induced magneto-resistance oscillations could be explained by averaging the two monochromatic microwave responses. This suggestion does not seem to agree with the results of this experiment since our experiment has shown that the bichromatic response depends on the magnetic field strength and the microwave frequency. Also, the bichromatic microwave R_{xx} response varies nonlinearly with the microwave power, see Fig. 4, similar to the non-linear variation observed with monochromatic microwaves.[13,21,22].

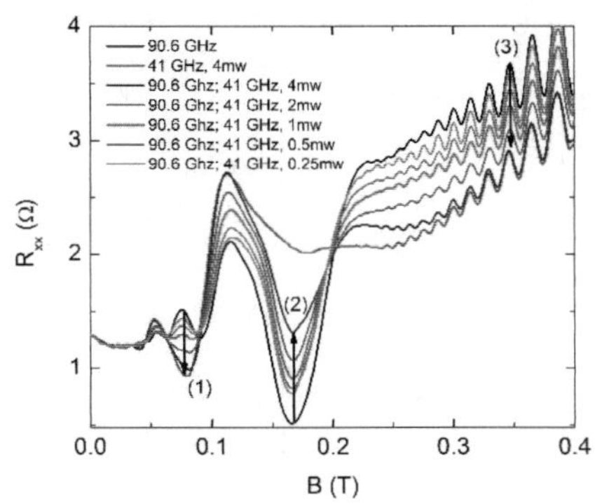

Figure 3. Monochromatic and bichromatic microwave induced magnetoresistance, R_{xx}, at frequencies 90.6 GHz and 41.0 GHz for microwave powers of 4, 2, 1, 0.5 and 0.25 microwave at the low frequency and fixed power at the high frequency. Positions (1), (2) and (3) are three extrema B-values of bichromatic R_{xx} oscillatory response.

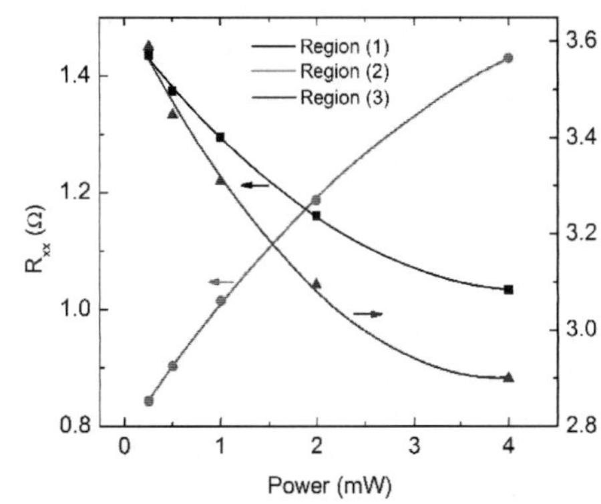

Figure 4. Microwave power variation of the diagonal resistance R_{xx} at the extremal positions 1, 2 and 3 marked in Fig. 3.

Acknowledgments
Magneto-transport studies at Georgia State University have been funded by U.S. Department of Energy, Office of Basic Energy Sciences, Material Sciences and Engineering division under the grant DESC0001762. Army Research Office under W911NF-14-2-0076 and W911NF-15-1-0433 has funded the microwave/terahertz- and 2D-material- studies.

References
[1] Mani R G, Smet J H, von Klitzing K, Narayanamurti V, Johnson W B and Umansky V 2002 *Nature (London)* **420**, 646; 2004 *Phys. Rev. Lett.* **92**, 146801; 2004 *Phys. Rev. B* **69**, 193304
[2] Mani R G, Narayanamurti V, von Klitzing K, Smet J H, Johnson W B, and Umansky V 2004 *Phys. Rev. B* **70**, 155310; 2004 *Phys. Rev. B.* **69**, 161306
[3] Mani R G, 2004 *Physica E* **22**, 1; 2004 ibid. **25**, 189
[4] Kovalev A E, Zvyagin S A ,Bowers C R, Reno J L , and Simmons J A 2004 *Solid State Commun.*

130, 379
[5] Zudov M A 2004 *Phys. Rev. B* **69**, 041304
[6] Simovic B, Ellenberger C, Ensslin K, Tranitz H P, and Wegscheider W 2005 *Phys. Rev. B* **71**, 233303
[7] Mani, R G 2005 *Phys. Rev. B* **72**, 075327; 2007 *Appl. Phys. Lett.* **91**, 132103; 2008 *Physica E* **40**, 1178
[8] Smet J.H et al. 2005 *Phys. Rev. Lett.* **95**, 116804
[9] Mani R G 2008 *Appl. Phys. Lett.* **92**, 102107
[10] Wiedmann S, Gusev G M, Raichev O E, Lamas T E, Bakarov A K and Portal J C 2008 *Phys. Rev. B.* **78**, 121301
[11] Konstantinov D and Kono K 2009 *Phys. Rev. Lett.* **103**, 266808
[12] Mani R G, Johnson W B, Umansky V, Narayanmurti V, and Ploog K 2009 *Phys. Rev. B* **79**, 205320
[13] Mani R G, Gerl C, Schmult S, Wegscheider W and Umansky V 2010 *Phys. Rev. B* **81**, 125320
[14] Ramanayaka A N, Mani R G, and Wegscheider W 2011 *Phys. Rev. B* **83**, 162303
[15] Mani R G, Ramanayaka A N, and Wegscheider W 2011 *Phys. Rev. B* **84**, 085308; Ramanayaka A N, Mani R G, Inarrea J and Wegscheider W 2012 *ibid.* **85**, 205315
[16] Mani R G, Hankinson J, Berger C and de Heer W A 2012 *Nat. Comm.* **3**, 996
[17] Mani R G, Ramanayaka A N, Ye T, Heimbeck M S, Everitt H O and Wegscheider W 2013 *Phys. Rev. B* **87**, 245308
[18] Ye T, Mani R G and Wegscheider W 2013 *Appl. Phys. Lett.* **102**, 242113; 2013 *Appl. Phys. Lett.* **103**, 192106
[19] Mani R G and Kriisa A 2013 *Sci. Rep.* **3**, 3478
[20] Mani R G, Kriisa A, and Wegscheider W 2013 *Sci. Rep.* **3**, 2747
[21] Ye T, Liu H C, Wegscheider W and Mani R G 2014 *Phys. Rev. B* **89**, 155307
[22] Ińarrea J, Mani R G and Wegscheider W 2010 *Appl. Phys. Lett.* **82**, 205321
[23] Lei X L and Liu S Y 2005 *Phys. Rev. B* **72**,075345
[24] Zudov M A, Du R R, Pfeiffer L N and West K W 2006 *Phys. Rev. Lett.* **96**, 236804
[25] Lei X L and Liu S Y 2006 *Appl. Phys. Lett.* **89**, 182117
[26] Durst A C, Sachdev S, Read N and Girvin S M 2003 *Phys. Rev. Lett.* **91**, 086803
[27] Inarrea J and Platero G 2005 *Phys. Rev. Lett.* **94**, 016806
[28] Lei X L, 2009 *Physica E* **42**, 63
[29] Andreev, A V, Aleiner I L and Millis, A J 2003 *Phys. Rev. Lett.* **91**, 056803
[30] Ryzhii V and Suris R 2003 *J. Phys. Condens. Matter* 15, 6855
[31] Koulakov A A and Raikh M E 2003 *Phys. Rev. B* **68**, 115324
[32] Ryzhii V 2003 *Phys. Rev. B* **68**, 193402
[33] Lei X L and Liu S Y 2003 *Phys. Rev. Lett.* 91, 226805
[34] Dmitriev I A, Vavilov M G, Aleiner I L, Merlin A D and Polyakov D G 2005 *Phys. Rev. B* **71**, 115316 (2005)
[35] Inarrea J and Platero G 2011 *Phys. Rev. B* **76**, 073311
[36] Chepelianskii A D, Pikovsky A S and Shepelyansky, D L 2007 *Eur. Phys. J. B* **60**, 225
[37] Finkler I G and Halperin B I 2009 *Phys. Rev. B* **79**, 085315
[38] Chepelianskii A D and Shepelyansky D L 2009 *Phys. Rev. B* **80**, 241308
[39] Mikhailov S A 2011 *Phys. Rev. B* **83**, 155303 (2011)
[40] Inarrea J 2011 *Appl. Phys. Lett.* **99**, 232115 (2011)
[41] Lei X L and Liu S L 2012 *Phys. Rev. B* **86**, 205303 (2012)
[42] Beltukov Y M and Dyakonov M I 2016 *Phys. Rev. Lett.* **116**, 176801

IOP Conf. Series: Journal of Physics: Conf. Series **864** (2017) 012056 doi:10.1088/1742-6596/864/1/012056

Evolution of the frequency-dependent polarization-angle phase-shift in the microwave radiation-induced magnetoresistance oscillations

Han-Chun Liu[1], Rasanga L. Samaraweera[1], C. Reichl[2], W. Wegscheider[2] and R. G. Mani[1]

[1]Department of Physics and Astronomy, Georgia State University, Atlanta, Georgia 30303
[2]Laboratorium für Festkörperphysik, ETH Zürich, CH-8093 Zürich, Switzerland

E-mail: hliu20@student.gsu.edu

Abstract. We report the evolution of the phase shift, θ_0, extracted from traces of the diagonal resistance, R_{xx}, vs. the linear polarization angle, θ, at oscillatory extrema of the microwave radiation induced magnetoresistance oscillations over the $36 \leq f \leq 40$ GHz band in GaAs/AlGaAs system. A reference phase shift for the linear polarization angle in the vicinity of the specimen is obtained with the help of a sensitive carbon resistor. We fit an empirical cosine square law to the sinusoidal responses of R_{xx} vs. θ to extract the phase shift θ_0. The quasi-continuous variation θ_0 vs. f trace suggests a preferable polarization orientation for the specimen, and the $f-$ and $B-$ independence of overall average of θ_0.

1. Introduction

Microwave radiation-induced zero-resistance states and associated magnetoresistance oscillations (MRIMOs)[1, 2], which appear in the 2D electron system subjected to crossed electric and magnetic fields, have revealed a "1/4-cycle phase-shift" in its periodicity in $1/B$ along with a multitude of interesting properties after nearly a decade research[3, 4, 5, 6, 7, 8, 9, 10, 11, 12, 13, 14, 15, 16, 17, 18, 19]. At the same time, a number of theoretical works have attempted to provide plausible explanations of MRIMOs[20, 21, 22, 23, 24, 25, 26, 27, 28, 29, 30, 31, 32, 33, 34, 35], based on the displacement model[20, 21, 22, 25], the nonparabolicity model[23], the inelastic model[26] and the radiation-driven electron-orbit model[24, 27].

Recently, the sensitivity of the amplitude of MRIMOs to linearly polarized radiation polarization was reported[12, 14, 18]. The results demonstrated a sinusoidal variation in the amplitude of the oscillatory magnetoresistance with microwave polarization angle at moderate radiation power. This sinusoidal variation can be described by an empirical cosine square function, $R_{xx}(\theta) = A \pm Ccos^2(\theta - \theta_0)$, where R_{xx} is diagonal magnetoresistance, θ is microwave polarization angle, θ_0 is phase shift, and the plus and minus signs correspond to the oscillatory maxima and minima, respectively. Previous works examined the dependence of θ_0 on microwave frequency, f, and B at a set of discrete frequencies. In this approach, however, the evolution of θ_0 between the measured f was unknown. In this report, therefore, the quasi-continuous evolution of the phase shift θ_0 extracted from R_{xx} vs. θ traces of the radiation-induced magnetoresistance oscillations at a set of six oscillatory extrema, and the reference phase shift reported by a

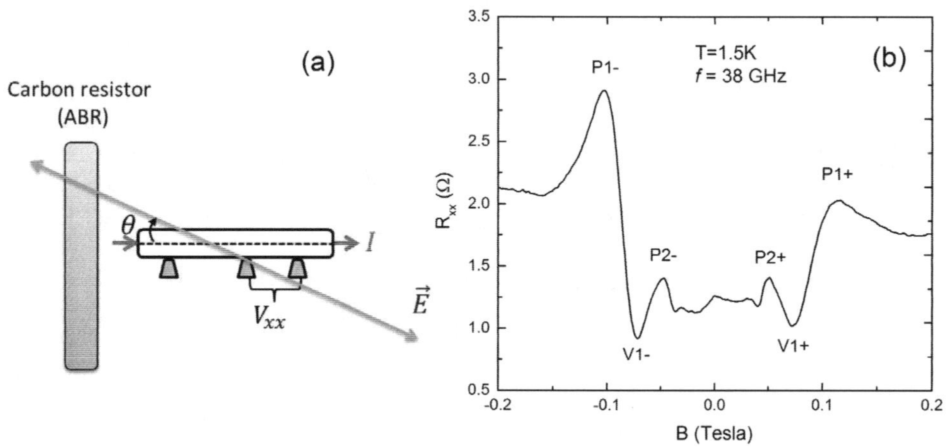

Figure 1. (Color online) (a) This schematic shows relative device arrangement of GaAs/AlGaAs heterojunction and the carbon resistor (ABR) subjected to rotatable linearly polarized radiation. The polarization angle, θ, is defined with respect to Hall bar axis clockwise. Diagonal voltage, V_{xx}, is collected via four-terminal lock-in technique. (b) The trace of R_{xx} vs. B shows strong MRIMOs at $f = 38\ GHz$. The oscillatory extrema of MRIMOs are assigned to $P1+$, $V1+$, $P2+$, $P2-$, $V1-$, and $P1-$ respectively over $36 \leq f \leq 40\ GHz$ band.

sensitive carbon resistor, are examined over the frequency band $36 \leq f \leq 40\ GHz$ for Hall-bar section. The results indicate that θ_0 results from the intrinsic response of specimen and the overall average of θ_0 is independent upon f and B.

2. Experiment and Results

Magnetotransport in 200-μm-wide Hall-bar type GaAs/AlGaAs heterojunction devices with electron density $n = 3.3 \times 10^{11} cm^{-2}$ and mobility $\mu = 14 \times 10^6 cm^2 \cdot V^{-1} \cdot s^{-1}$ at $T = 1.5K$ were measured using the four-terminal lock-in technique. The diagonal voltage, V_{xx}, was collected vs. the polarization angle, θ, which is defined as the angle between the Hall bar axis and linear-polarized microwave electric field as in Fig. 1(a). A carbon resistance sensor with a strong negative temperature coefficient, i.e., $dR_{ABR}/dT \leq 0$, at liquid helium temperatures, i.e., an Allen-Bradley resistor (ABR), was placed next to the Hall bar specimen for the sake of *in-situ* measurement of θ_0 and the independent detection of microwave polarization rotation at the sample location. Typical MRIMOs in R_{xx} vs. B at $f = 38\ GHz$ are shown in Fig. 1(b). We define the first (second) peak and the first valley on the positive side of B as $P1+$ ($P2+$) and $V1+$ respectively. Likewise, $P1-$, $V1-$ and $P2-$ are labels utilized to identify extrema on the negative side of the B axis. For the R_{xx} vs. θ measurements, we swept f from 36 to 40 GHz with the polarization angle held constant between $0^\circ \leq \theta \leq 360^\circ$ at 10° intervals.

To rule out the possibility that the θ_0 observed in the specimen is due to uncharacterized incident polarization rotation, the carbon resistor (ABR) serves as a radiation polarization sensor to determine the polarization orientation near the sample. Fig. 2(a) displays normalized R_{ABR} color plot of f vs. θ over $36 \leq f \leq 40$ GHz with $0^\circ \leq \theta \leq 360^\circ$ at $V1-$. High normalized R_{ABR} occurring near 0° matches our expectations for the incident microwave polarization. A nice sinusoidal curve of normalized R_{ABR} vs. θ at $f = 38.3\ GHz$ is fit to the empirical cosine square function to extract $\theta_{0,ABR}$; see Fig. 2(b). The extracted θ_0 over $36 \leq f \leq 40\ GHz$ are

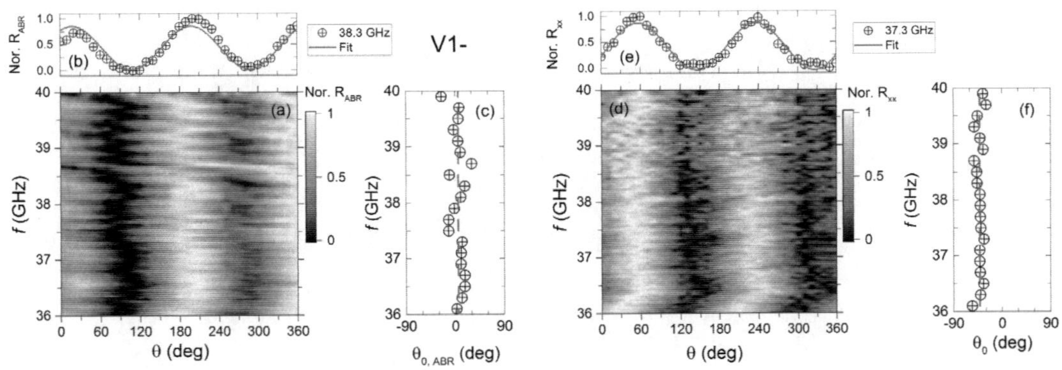

Figure 2. (Color online) This figure shows the normalized R_{ABR} (left color plot: a) and R_{xx} (right color plot: d) over $36 \leq f \leq 40$ GHz band with $0° \leq \theta \leq 360°$ at $V1-$. Panel (b) and (e) illustrate nice sinusoidal variation in normalized resistance fit to the empirical cosine square function with θ at given f. Panel (c) and (f) demonstrate the plot of f vs. θ_0 with the average θ_0, which are extracted from the fit at each f. The figure (c) and (f) imply a constant average phase shift at $V1-$.

plotted in Fig 2(c) with an average $\theta_{0,ABR}$ indication at 5.3° (red dashed line), which proves that the incident microwave polarization orientation has not changed near the specimen. Similarly, Fig. 2(d) exhibits normalized R_{xx} color plot of f vs. θ. Then, we fit cosine square function to the sinusoidal curve of normalized R_{xx} vs. θ (Fig. 2(e)) to obtain θ_0 within $36 \leq f \leq 40\ GHz$ with average $\theta_0 = -41.3°$ (Fig. 2(f)). Clearly, the average θ_0 shows a recognizable difference with respect to $\theta_{0,ABR}$.

3. Discussion

Although the microwave launcher and the Hall bar axis were pre-aligned before carrying out the magnetotransport measurements in previous microwave polarization angle dependence studies with the aid of a microwave power detector, there remained the possibility that the observed θ_0 was due to the reflection caused by metallic contacts and bonded gold wires. However, the response to incident microwaves reported by the carbon resistor collected here simultaneously with magnetotransport measurements suggests that the average $\theta_{0,ABR}$ is extremely similar to the standard error obtained by microwave power detector in the prealignment process. This feature affirms that the observed θ_0 results from intrinsic sample properties. Plus, according to the average θ_0 examined at remaining oscillatory extrema, which are all located at $\approx -42°$ (not shown here), while the extracted $\theta_{0,ABR}$ and θ_0 manifest similar standard errors, the average θ_0 is independent upon f and B. Thus, our results imply that the GaAs/AlGaAs Hall-bar device possesses an intrinsic preferred polarization angle for strongest MRIMOs response.

4. Conclusion

We suggest here that the θ_0 is an intrinsic property of the GaAs/AlGaAs 2D specimen with the help of a sensitive carbon resistor. In addition, the similar average θ_0 values examined at different oscillatory extrema suggest that the average θ_0 is independent of f and B, and the sample exhibits a preferred polarization orientation to incident microwaves.

Acknowledgments
Magnetotransport studies and H.-C. Liu are supported by the US Department of Energy, Office of Basic Energy Sciences, Material Sciences and Engineering Division under DE-SC001762. R. L. Samaraweera is supported by the ARO under W911NF-10-1-0450. High quality MBE GaAs/AlGaAs heterostructure wafers have been kindly provided by CR and WW.

References
[1] Mani R G, Smet J H, von Klitzing K, Narayanamurti V, Johnson W B and Umansky V 2002 *Nature (London)* **420** 646
[2] Zudov M A, Du R R, Pfeiffer L N and West K W 2003 *Phys. Rev. Lett.* **90** 046807
[3] Mani R G, Narayanamurti V, von Klitzing K, Smet J H, Johnson W B and Umansky V 2004 *Phys. Rev. B* **69** 161306(R); 2004 *Phys. Rev. B* **70** 155310
[4] Mani R G, Smet J H, von Klitzing K, Narayanamurti V, Johnson W B and Umansky V 2004 *Phys. Rev. Lett.* **92** 146801; 2004 *Phys. Rev. B* **69** 193304
[5] Mani R G 2004 *Physica E (Amsterdam)* **22** 1-6; 2004 *Physica E Amsterdam* **25** 189-197; 2004 *Int. J. Mod. Phys. B* **18** 3473-3480; 2004 *Appl. Phys. Lett.* **85** 4962-4964; 2004 *Phys. Rev. B* **72** 075327; 2007 *Appl. Phys. Lett.* **91** 132103; *Solid State Comm.* 2007 **144** 409-412; 2008 *Physica E (Amsterdam)* **40** 1178-1181
[6] Smet J H, Gorshunov B, Jiang C, Pfeiffer L, West K, Umansky V, Dressel M, Meisels R, Kuchar F and von Klitzing K 2005 *Phys. Rev. Lett.* **95** 116804
[7] Studenikin S A, Sachrajda A S, Gupta J A, Wasilewski Z R, Fedorych O M, Byszewski M, Maude D K, Potemski M, Hilke M, West K W and Pfeiffer L N 2007 *Phys. Rev. B* **76** 165321
[8] Raichev O E 2008 *Phys. Rev. B* **78** 125304
[9] Mani R G, Johnson W B, Umansky V, Narayanamurti V and Ploog K 2009 *Phys. Rev. B* **79** 205320
[10] Fedorych O M, Potemski M, Studenikin S A, Gupta J A, Wasilewski Z R and Dmitriev I A 2010 *Phys. Rev. B* **81** 201302(R)
[11] Wiedmann S, Gusev G M, Raichev O E, Bakarov A K and Portal J C 2010 *Phys. Rev. Lett.* **105** 026804
[12] Mani R G, Ramanayaka A N and Wegscheider W 2011 *Phys. Rev. B* **84** 085308
[13] Bogan A, Hatke A T, Studenikin S A, Sachrajda A, Zudov M A, Pfeiffer L N and West K W 2012 *Phys. Rev. B* **86** 235305
[14] Ramanayaka A N, Mani R G, Iñarrea J and Wegscheider W 2012 *Phys. Rev. B* **85** 205315
[15] Mani R G, Ramanayaka A N, Ye T, Heimbeck M S, Everitt H O and Wegscheider W 2013 *Phys. Rev. B* **87** 245308
[16] Konstantinov D, Monarkha Y and Kono K 2013 *Phys. Rev. Lett.* **111** 266802
[17] Ye T, Liu H-C, Wegscheider W and Mani R G 2014 *Phys. Rev. B* **89** 155307
[18] Liu H-C, Ye T, Wegscheider W and Mani R G 2015 *J. Appl. Phys.* **117** 064306
[19] Ye T, Liu H-C, Wang Z, Wegscheider W and Mani R G 2015 *Sci. Rep.* **5** 14880
[20] Durst A C, Sachdev S, Read N and Girvin S M 2003 *Phys. Rev. Lett.* **91** 086803
[21] Ryzhii V and Suris R 2003 *J. Phys. Cond. Matt.* **15** 6855
[22] Lei X L and Liu S Y 2003 *Phys. Rev. Lett.* **91** 226805
[23] Koulakov A A and Raikh M E 2003 *Phys. Rev. B* **68** 115324
[24] Iñarrea J and Platero G 2005 *Phys. Rev. Lett.* **94** 016806
[25] Lei X L and Liu S Y 2005 *Phys. Rev. B* **72** 075345
[26] Dmitriev I A, Vavilov M G, Aleiner I L, Mirlin A D and Polyakov D G 2005 *Phys. Rev. B* **71** 115316
[27] Iñarrea J and Platero G 2007 *Phys. Rev. B* **76** 073311
[28] Wang S and Ng T-K 2008 *Phys. Rev. B* **77** 165324
[29] Iñarrea J, Mani R G and Wegscheider W 2010 *Phys. Rev. B* **82** 205321
[30] Mikhailov S A 2011 *Phys. Rev. B* **83** 155303
[31] Chepelianskii A D, Laidet J, Farrer I, Beere H E, Ritchie D A and Bouchiat H 2012 *Phys. Rev. B* **86** 205108
[32] Lei X L and Liu S Y 2012 *Phys. Rev. B* **86** 205303
[33] Zhirov O V, Chepelianskii A D and Shepelyansky D L 2013 *Phys. Rev. B* **88** 035410
[34] Yar A and Sabeeh K 2015 *J. Phys.:Cond. Matt.* **27** 435007
[35] Ibarra-Sierra V G, Sandoval-Santana J C, Cardoso J L and Kunold A 2015 *Annal. Phys.* **362** 83

IOP Conf. Series: Journal of Physics: Conf. Series **864** (2017) 012057 doi:10.1088/1742-6596/864/1/012057

Remotely sensed in microwave irradiated GaAs/AlGaAs two-dimensional electron system

A Kriisa,[1] H-C Liu,[1] R L Samarweera,[1] M S Heimbeck,[2] H O Everitt,[2,3] W Wegscheider[4] and R G Mani[1]

[1] Dept. of Physics and Astronomy, Georgia State University, Atlanta, GA 30303
[2] Army Aviation & Missile RD&E Center, Redstone Arsenal, Huntsville, Alabama, 35898
[3] Dept. of Physics, Duke University, Durham, NC 27708
[4] Laboratorium für Festkörperphysik, ETH-Zürich, 8093 Zürich, Switzerland

annikakriisa@gmail.com

Abstract. Remotely sensed microwave reflection was measured concurrently with standard magnetotransport in photo-excited high mobility GaAs/AlGaAs heterostructure. Experiments indicate strong reflection resonance on both sides of the magnetic field axis for linearly polarized microwave/terahertz photo-excitation over the examined frequency $30 < f < 330$ GHz band. In addition, there is evidence for electronic heating in the vicinity of cyclotron resonance (CR), which is indicated by reduced amplitude of the Shubnikov-de Haas oscillations. Effective mass extracted from the measurements was found to equal the CR mass within experimental error.

1. Introduction

The discovery and study of microwave-induced zero-resistance states and associated magnetoresistance oscillations have brought new interest in the experimental-[1-22] and theoretical-[23-36] study of magnetotransport at high filling factors or low magnetic fields in the GaAs/AlGaAs 2D electron system.[1-36] Along with this interest in magnetotransport at high filling factors comes new concern for the microwave reflection, absorption, and transmission properties of the GaAs/AlGaAs 2DES at low magnetic fields, including the remote detection of the cyclotron resonance (CR), and the correlation of CR with magnetotransport. Thus, we developed a measurement technique that allows measuring the power signal of the microwaves reflected back from the photo-excited high mobility GaAs/AlGaAs device, while measuring standard magnetotransport, i.e., the diagonal resistance, R_{xx}, and the Hall resistance, R_{xy}, over wide frequency range from 30 to 330 GHz. From the observations and results, we identify CR and find the electron effective mass to equal CR mass.

2. Experimental

Experiments were conducted on two different samples of 200 μm-wide two dimensional GaAs/AlGaAs heterostructure Hall bars with gold-germanium contacts. For the first sample the carrier density $n \approx 2.4 \times 10^{11}$ cm^{-2} and mobility $\mu \approx 10 \times 10^{6}$ cm^2/V s at temperature $T = 1.7$ K and for the second sample (data not shown in paper) $n \approx 3.3 \times 10^{11}$ cm^{-2} and $\mu \approx 6 \times 10^{6}$ cm^2/V s at $T = 1.7$ K. Samples were mounted at the end of a long waveguide which was then inserted into a liquid helium cryostat with a superconducting solenoid as shown in Figure 1(a). During the experiments, the magnetic field, B, was applied perpendicular to the 2DES and temperature was kept constant at 1.7 K. In the 30-50GHz band, linearly polarized microwaves were excited within the waveguide sample holder using an MW launcher coupled to a HP83650B MW source. A 6x-multiplier MM wave module provided radiation over the 65-110 GHz band, while 12x and 18x MW Virginia Diodes Multipliers generated microwave radiation from 140 to 330 GHz. Above 50 GHz, waveguides connected the source to the sample holder.

Standard four terminal low frequency (13 Hz) lock-in technique was used to collect the oscillatory magnetoresistive response, R_{xx}, of the high mobility GaAs/AlGaAs sample in sweeping B and at constant microwave frequency, f. Additionally to recording R_{xx}, the microwave signal, reflected back from irradiated sample, was simultaneously collected using a power detector and read out with a power meter. As shown in Figure 1(a), a waveguide coupler on top of the waveguide holding the sample was used to direct the reflected microwaves to the detector. The microwave power at the source was ~10^{-3} Watts up to 110 GHz, while the power of microwave signal reflected back from sample and collected with the power meter was ~ 10^{-12} Watts.

(a) (b)

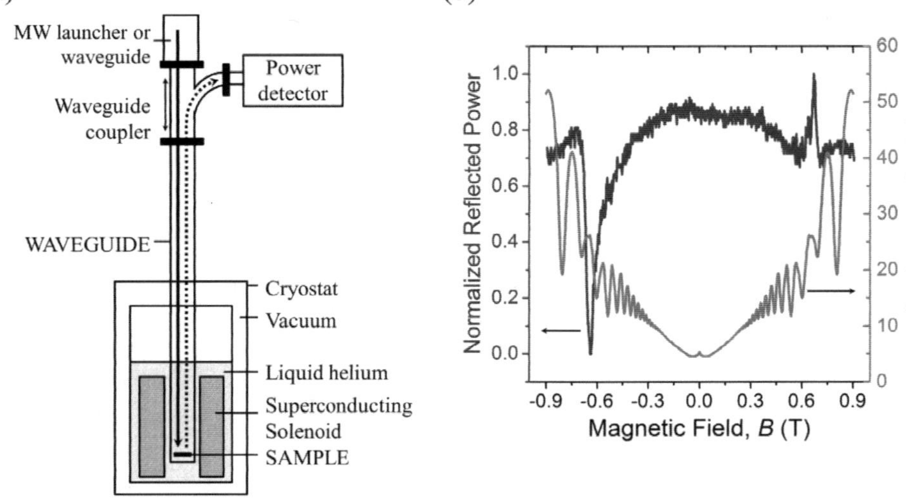

Figure 1 (a) A schematic of the experimental setup showing long waveguide, with GaAs/AlGaAs sample at bottom end, in the liquid helium cryostat within a superconducting solenoid. Sample was irradiated with microwaves (long solid line with arrow) and the microwave signal reflected back (long dotted line with arrow) was collected using a power detector on top of waveguide. (b) Diagonal resistance, $R_{xx,}$ (right ordinate) and the reflected microwave power (left ordinate) as a function of magnetic field for samples of high mobility 2D GaAs/AlGaAs heterostructures at temperature T = 1.7 K and frequency f = 267 GHz.

3. Results and Discussion

Figure 1(b) shows the normalized reflected power signal (left ordinate) along with the diagonal resistance, R_{xx}, (right ordinate) as a function of magnetic field measured in a GaAs/AlGaAs 2DES with microwave excitation at 267 GHz. During measurements, linearly polarized microwaves were used, thus resonances on both sides, positive and negative magnetic field, is seen. In Figure 1(b), reflected power signal shows easily distinguishable resonance peaks. Though the peaks are not identical in shape, they do occur at same place on both side of the magnetic field axis, at $|B|$ = 0.66 ± 0.02 T. These strong disturbances in reflected power signal were observed over examined frequency band of 30 to 330 GHz.

Above $|B|$ = 0.3 T, the diagonal resistance, R_{xx} in Figure 1(b) shows regular Shubnikov-de Haas effect (SdH) with apparent reduction in the oscillations amplitude between $|B|$ = 0.60 T and $|B|$ = 0.69 T. In 2DES, temperature damping of SdH amplitude is a well-known, although not very precise, method for locating cyclotron resonance (CR).[5] In the vicinity of CR, photo-excitation of electrons leads to the elevation of temperature or resonant electron heating. As SdH amplitude is particularly sensitive to heating, small increase in temperature near CR leads to the reduction of oscillations amplitude.[5,37] Thus, these strong resonance peaks in the reflected power signal and the resonant decrease in the SdH amplitude point to cyclotron resonance.

To better determine the physical origin of the resonance peaks in reflected power signal, their magnetic field positions were plotted as a function of the microwave frequency as shown in Figure 2. It is evident in Figure 2 that the resonance peak position in magnetic field, shown as filled circles, increase linearly with the MW frequency. Note here that each data point is an average over two peak locations in B, one at positive and one at negative B. The dashed line in Figure 2 is not a fit through the resonance positions, but a calculation of cyclotron frequency of electrons in an infinitely large GaAs/AlGaAs 2DES, that is $f_C = eB/2\pi m^*$,[37] where the effective mass $m^* = 0.067m_o$ (m_o is free electron mass). For the purpose of visual clarity, the linear fit through the data points is not shown in Figure 2, as it is almost indistinguishable from the dashed line shown. From the slope of linear fit (not shown), $dB/df_{MW} = 0.00242 \pm 0.00007$ T/GHz, we directly determined the effective mass ratio $m^*/m_o = (0.0678 \pm 0.0020)$. For the second sample measured (data not shown), the effective mass ratio $m^*/m_o = (0.0685 \pm 0.0022)$. Both mass ratios are within the experimental error of cyclotron effective mass, $m^* = 0.067m_o$ for GaAs/AlGaAs devices. These results imply that the position of resonance peaks in reflected power signal coincide with the cyclotron resonance of electrons in a bulk 2DES. This is explained with the idea that when the sample undergoes cyclotron resonance, there will be enhanced MW absorption by the sample. As a consequence, there will be a change in the reflected power, which will be the signature of CR.

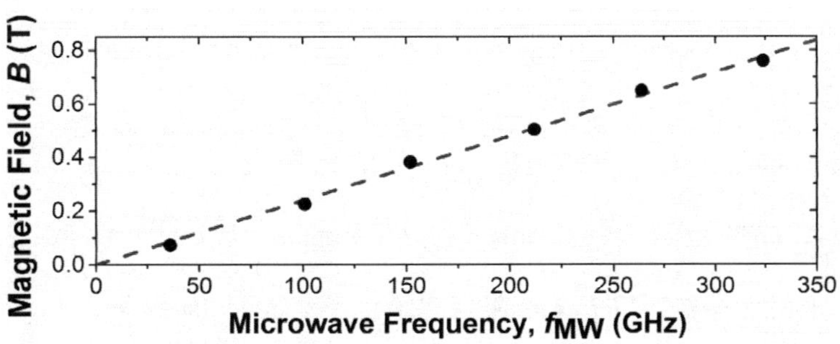

Figure 2 Filled circles show the location of resonance peaks in the reflected power signal in magnetic field as a function of incident microwave frequency, f_{MW}. Dashed line is not a fit through the data, and instead demonstrates the cyclotron frequency of electrons, $f_C = eB/2\pi m^*$, where the effective mass $m^* = 0.067m_o$.

4. Conclusions

We investigated microwave reflection in high mobility GaAs/AlGaAs over wide range of microwave frequencies. We found that the reflected power signal displays easily distinguishable resonance peaks that point to the cyclotron resonance. Since the radiation frequency, f, was well known in the experiment, and the magnetic field, B, was calibrated, the effective mass of the carriers was directly determined by following the positions of the peaks in B vs. f. The effective mass was found to equal the cyclotron resonance mass within experimental error. The results were reproducible in two separate samples of high mobility GaAs/AlGaAs heterostructures.

Acknowledgements
Magnetotransport measurements were supported by the U.S. Department of Energy, Office of Basic Energy Sciences, Material Science and Engineering Division under DE-SC0001762. Additional support was provided by the Army Research Office under W911NF-14-2-0076 and W911NF-15-1-0433.

References

[1] Mani R G, Smet J H, von Klitzing K, Narayanamurti V, Johnson W B and Umansky V 2002 *Nature (London)* **420** 646; 2004 *Phys. Rev. Lett.* **92** 146801; 2004 *Phys. Rev. B* **69** 193304
[2] Mani R G, Narayanamurti V, von Klitzing K, Smet J H, Johnson W B and Umansky V 2004 *Phys. Rev. B* **70** 155310; 2004 *Phys. Rev. B.* **69** 161306
[3] Zudov M A, Du R R, Pfeiffer L N and West K W 2003 *Phys. Rev. Lett.* **90** 046807
[4] Mani R G 2004 *Physica E* **22** 1; 2004 ibid. **25** 189; *Phys. Rev. B* **72** 075327; 2007 *Appl. Phys. Lett.* **91** 132103; 2008 *Physica E* **40** 1178; *Appl. Phys. Lett.* **92** 102107
[5] Kovalev A E, Zvyagin S A, Bowers C R, Reno J L and Simmons J A 2004 *Solid State Commun.* **130** 379
[6] Zudov M A 2004 *Phys. Rev. B* **69** 041304
[7] Simovic B, Ellenberger C, Ensslin K, Tranitz H P and Wegscheider W 2005 *Phys. Rev. B* **71** 233303
[8] Smet J H, Gorshunov B, Jiang C, Pfeiffer L, West K, Umansky V, Dressel M, Meisels R, Kuchar F and von Klitzing K 2005 *Phys. Rev. Lett.* **95** 116804
[9] Wiedmann S, Gusev G M, Raichev O E, Lamas T E, Bakarov A K and Portal J C 2008 *Phys. Rev. B.* **78** 121301
[10] Konstantinov D and Kono K 2009 *Phys. Rev. Lett.* **103** 266808
[11] Mani R G, Johnson W B, Umansky V, Narayanmurti V and Ploog K 2009 *Phys. Rev. B* **79** 205320
[12] Mani R G, Gerl C, Schmult S, Wegscheider W and Umansky V 2010 *Phys. Rev. B* **81** 125320
[13] Ińarrea J, Mani R G and Wegscheider W 2010 *Phys. Rev. B* **82** 205321
[14] Ramanayaka A N, Mani R G and Wegscheider W 2011 *Phys. Rev. B* **83** 162303
[15] Mani R G, Ramanayaka A N and Wegscheider W 2011 *Phys. Rev. B* **84** 085308
[16] Ramanayaka A N, Mani R G, Inarrea J and Wegscheider W 2012 *Phys. Rev. B.* **85** 205315
[17] Mani R G, Hankinson J, Berger C and de Heer W A 2012 *Nat. Commun.* **3** 996
[18] Mani R G, Ramanayaka A N, Ye T, Heimbeck M S, Everitt H O and Wegscheider W 2013 *Phys. Rev. B* **87** 245308
[19] Ye T, Mani R G and Wegscheider W 2013 *Appl. Phys. Lett.* **102** 242113; 2013 *Appl. Phys. Lett.* **103** 192106
[20] Mani R G and Kriisa A 2013 *Sci. Rep.* **3** 3478; Mani R G, Kriisa A and Wegscheider, W 2013 *Sci. Rep.* **3** 2747
[21] Ye T, Liu H C, Wegscheider W and Mani R G 2014 *Phys. Rev. B* **89** 155307
[22] Ye T, Liu H C, Wang Z, Wegscheider W and Mani R G 2015 *Sci. Rep.* **5** 14880
[23] Durst A C, Sachdev S, Read N and Girvin S M 2003 *Phys. Rev. Lett.* **91** 086803
[24] Ryzhii V 2003 *Phys. Rev. B* **68** 193402
[25] Ryzhii V and Suris R 2003 *J. Phys. Condens. Matter* 15, 6855
[26] Inarrea J and Platero G 2005 *Phys. Rev. Lett.* **94**, 016806; 2011 *Phys. Rev. B* **76** 073311
[27] Lei X L and Liu S Y 2005 *Phys. Rev. B* **72** 075345; 2006 *Appl. Phys. Lett.* **89** 182117
[28] Lei X L 2009 *Physica E* **42** 63
[29] Koulakov A A and Raikh M E 2003 *Phys. Rev. B* **68** 115324
[30] Lei X L and Liu S Y 2003 *Phys. Rev. Lett.* 91 226805
[31] Dmitriev I A, Vavilov M G, Aleiner I L, Merlin A D and Polyakov D G 2005 *Phys. Rev. B* **71** 115316
[32] Chepelianskii A D and Shepelyansky D L 2009 *Phys. Rev. B* **80** 241308
[33] Mikhailov S A 2011 *Phys. Rev. B* **83** 155303 (2011)
[34] Inarrea J 2011 *Appl. Phys. Lett.* **99** 232115
[35] Lei X L and Liu S L 2012 *Phys. Rev. B* **86** 205303
[36] Beltukov Y M and Dyakonov M I 2016 *Phys. Rev. Lett.* **116** 176801
[37] Miura N 2007 *Physics of Semiconductors in High Magnetic Fields* (New York: Oxford University Press)

IOP Conf. Series: Journal of Physics: Conf. Series **864** (2017) 012058 doi:10.1088/1742-6596/864/1/012058

Ab initio Calculation of The Magnetic Properties of Oxygen Impurity Complexes in CdSe

W Cheng[1], L Liu[2], P Y Yu[3]

[1] Key Laboratory of Beam Technology and Material Modification of Ministry of Education, College of Nuclear Science and Technology, Beijing Normal University, Beijing 100875, P.R. of China
[2] Changchun Institute of Optics, Fine Mechanics and Physics, Chinese Academy of Sciences, Changchun 130023, P.R. of China
[3] Physics Department, University of California and Materials Science Division, Lawrence Berkeley National Laboratory, Berkeley, CA 94720, USA

Email: chengwei@bnu.edu.cn

Abstract. We have investigated the magnetic dipole moments of interstitial oxygen molecules (O_2) and their interactions in CdSe by using spin-dependent first-principle calculations based on the local density functional (DFT) theory. We constructed supercells of $Cd_{64}Se_{64}$ by repeating the primitive cell Cd_2Se_2 4x4x2 times. Then an O_2 molecule aligned along c-axis was added to the center of the CdSe cage. The charge densities of the complexes were then computed for both spin-up and spin-down electrons. Their difference indicates that O_2 molecule in CdSe is *paramagnetic,* although its magnetic dipole moment is lower than that in free space. Two such O_2 molecules were then placed either (a) parallel to each other in side-by-side supercells or (b) in neighboring supercells on top of each other. The computed energies of the resultant magnetic structures suggest that the two O_2 magnetic moments interact anti-ferromagnetically in case (a) but do not interact in (b).

1. Introduction

Cadmium Selenide (CdSe) has many applications including the fabrication of solar cells[1], photo-chemical cells[2] and high energy radiation detectors[3]. Oxygen is often present in the growth or annealing of CdSe samples. The effects of oxygen on the properties of CdSe samples are well-documented already [4,5] . The presence of oxygen dopants in CdSe and CdTe can be detected with great sensitivity via far-infrared (FIR) absorption and these FIR spectra have revealed detailed microscopic structures of oxygen related defects [6]. In particular, experiment and first-principle calculations have suggested that some of the observed high frequency FIR modes may involve complexes containing O_2 [7] . Oxygen molecules are well-known to be paramagnetic in gaseous and liquid form. One interesting question is whether O_2 in CdSe is also paramagnetic? In this paper, we have applied first-principle spin-dependent density-functional (DFT) method to investigate the magnetic dipole moments of interstitial O_2 in CdSe for comparison with those of oxygen molecules in free space. After finding that oxygen molecules are paramagnetic in CdSe we have investigated the interaction between two adjacent oxygen molecules in CdSe. The interaction was found to be anti-ferromagnetic when the two molecules lie adjacent to each other within the same hexagonal channel of the wurtzite lattice.

2. Theory and Model

IOP Conf. Series: Journal of Physics: Conf. Series **864** (2017) 012058 doi:10.1088/1742-6596/864/1/012058

Our calculation has been performed using the projector-augmented wave (PAW) method as implemented by the commercial softwares VASP and MedeA.[8] The density functional chosen was based on the Perdew-Burke-Ernzerhof (PBE) functional within the Generalized Gradient Approximation (GGA) [9]. The energy cutoff was chosen to be 520eV and the spacing of K-points was set to be <2 nm^{-1}. The k-mesh was set to 5x5x5. The calculation explored the full Brillouin zone and included an interaction range of about 1 nm.
Supercells $Cd_{64}Se_{64}O_2$ were constructed with MedeA. This supercell cell was obtained by repeating the primitive cell Cd_2Se_2 4x4x2 times.O_2 molecule aligned along *c*-axis was added to the center of the CdSe cage. After relaxation of all atoms, the oxygen molecule is found to be at the center of a hexagonal cage in the wurtzite lattice. A schematic representation of the relaxed supercell is shown in Figure 1.

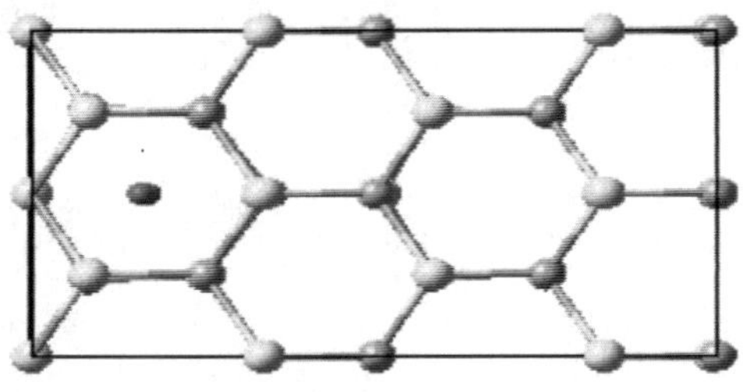

(a)

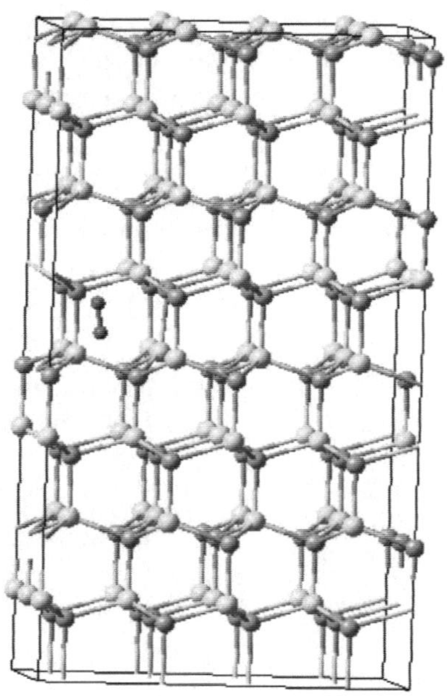

(b)

Figure 1 (a) Top view of the supercell showing the location of the oxygen molecule lying at the center of the hexagonal channel of the wurtzite CdSe lattice. (b) Side view of the CdSe lattice showing the orientation of the linear oxygen molecule as being slightly tilted away from the c-axis. The bond length of the oxygen molecule is extended from 0.1235nm in free space to 0.1307nm in CdSe

3. Results and Discussions

3.1 Electron Density Maps

In Figures 2 (a) and (b) we plot the computed densities for both spin-up and spin-down electrons. The magnetic moment of the O_2 molecule is proportional to the difference between these two spin dependent electron densities. This difference is plotted in Figure 3 (a). For comparison we plot in Figure 3 (b) the corresponding difference for a supercell consisting of a single O_2 molecule. Figure 3 (a) shows that the O_2 molecule in CdSe is paramagnetic like the free O_2 molecule. However, the magnitude of the magnetic moment of the O_2 molecule is reduced by about 5% when located inside CdSe. We note that the bond length of the O_2 molecule in CdSe is increased almost by the same amount. Both results can be attributed to screening of the oxygen-oxygen bond by the valence electrons of CdSe. This screening effect is relatively small because of the location of the O_2 molecule inside the lattice channels.

Figure 2 Density plots for (a) spin-up and (b) spin-down electrons.

(a) **(b)**

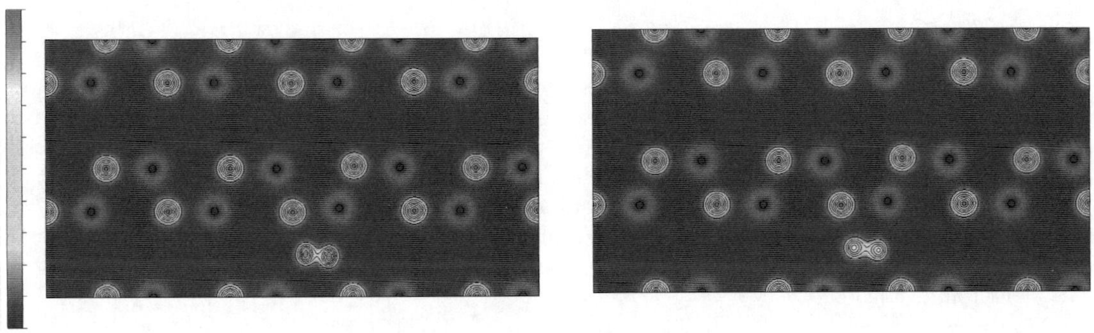

Figure 3 (a) Difference between electron densities with spin up and spin down in CdSe: O_2 ; (b) Difference between electron densities with spin up and spin down in an empty lattice containing only one O_2 molecule.

(a) **(b)**

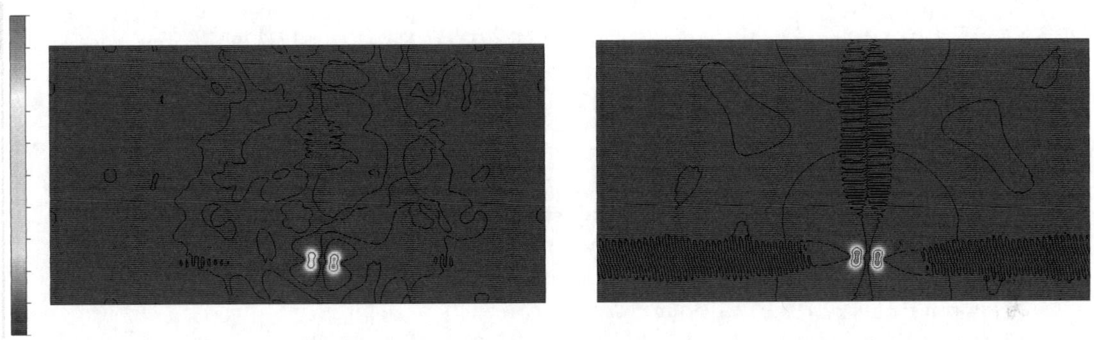

It is interesting to note that the spin density difference on the two oxygen atoms in CdSe is not the same. This is because the inversion symmetry between the two oxygen atoms in the O_2 molecule is broken when the molecule is located inside CdSe.

3.2 Interaction Between Two O_2 molecules in CdSe

We next investigate the magnetic interaction between the magnetic moments of two nearest neighbor interstitial O_2 molecules. Such interaction is expected to occur when the CdSe sample contains a high concentration of O_2 molecules. The fact that the magnetic moment of the O_2 molecule is oriented along the hexagonal channels of CdSe may lead to interesting 1D magnetic phases. We have considered two ways that the two O_2 molecules can be placed inside the CdSe lattice interstitially: (a) the two molecules are located parallel to each other in adjacent channels and (b) the two molecules are located in supercells on top of each other within the same hexagonal channel. We compute the VASP free energies of the supercell $Cd_{128}Se_{128}O_4$ containing two O_2 molecules. The supercell was not relaxed since we expect the interaction between the two O_2 molecules to be relatively weak as their unpaired charges are strongly localized around each molecule. The magnetic moments of the two O_2 molecules are allowed to be aligned either parallel (ferromagnetic) or anti-parallel (anti-ferromagnetic) to each other. We found that in case (a) there was negligible interaction between their magnetic moments because there was no difference between the VASP energies of the ferromagnetic and anti-ferromagnetic alignments. However, in case (b) the ferromagnetic state is higher in energy than the anti-ferromagnetic state by 77 meV indicating that the interaction is anti-ferromagnetic. This difference is mainly a result of the bigger distance between the two O_2 molecules in case (a).

4. Conclusions

Using first-principle spin dependent DFT technique we have computed the electron density distribution for spin-up and spin-down electrons in CdSe doped with interstitial O_2 molecules. We found that O_2 molecules remain paramagnetic in CdSe although its magnetic moment becomes reduced by about 5% as compared to those of free O_2 molecules. We found that when the two O_2 molecules are located in neighboring channels in the CdSe lattice, their magnetic interaction is negligible. However, *when they are located in the same channel in unit cells above each other, then their interaction is strongly anti-ferromagnetic.* The strong magnetic moment of interstitial O_2 molecules in CdSe can be exploited to determine their concentration and to study their local environment in CdSe.

Acknowledgments

The authors acknowledge Professor D. S. Wang at the Institute of Physics for helpful discussion and his support in making the computation arrangement through his grant (NSFC-10634070). Part of the computation described in this research is carried out on the ScGrid of Supercomputing Center, Computer Network Information Center of the Chinese Academy of Sciences. The work is also supported by the Scientific Research Foundation for the Returned Overseas Chinese Scholars, the National Basic Research Program of China under Grant No. 2010CB832903. This work was started when two of the authors (W.C. and L.L.) were visiting scientists at the Lawrence Berkeley National Laboratory. P.Y.Y. acknowledges helpful discussions with Prof. Steven G. Louie.

References

[1] Gruszecki T and Holmstrom B 1998 *Solar Energy Materials and Solar Cells* **51** 243
[2] Hodes G *et al* 1982 *Thin Solid Films* **90** 433
[3] Roth M 1989 *Nucl. Instr. Meth.* A **283**, 291
[4] Szabo J P and Cocivera M 1987 *J. Appl. Phys.* **61** 4820
[5] Fortin B H *et al* 1985 *Thin Solid Films* **131** 51
[6] Chen G *et al* 2008 *Phys. Rev. Lett.* **101** 195502 and references therein.
[7] Cheng W *et al* 2013 *J. Appl. Phys.* **114** 194901 and references therein
[8] Blochl P E 1994 *Phys. Rev.* B **50**, 17953 ; Kresse G and Joubert D 1999 *Phys. Rev.* B **59**, 1758. MedeA is a trade mark of Materials Design.
[9] Perdew J P *et al* 1996 *Phys. Rev. Lett.* **77**, 3865 ; *ibid* 1997 **78**, 1396

IOP Conf. Series: Journal of Physics: Conf. Series **864** (2017) 012059 doi:10.1088/1742-6596/864/1/012059

Spin-dependent transport in multi-terminal Aharonov-Casher ring with quantum dot

Taku Matsunaga[1], **Rui Sakano**[2], **and Mikio Eto**[1]

[1] Faculty of Science and Technology, Keio University, Hiyoshi, Yokohama 223-8522, Japan
[2] Institute for Solid State Physics, University of Tokyo, Kashiwa, Chiba 277-8581, Japan

E-mail: eto@rk.phys.keio.ac.jp

Abstract. The Aharonov-Casher (AC) effect is theoretically examined in a mesoscopic ring with an embedded quantum dot in a multi-terminal geometry. We examine a realistic model for such a system with spin-orbit interaction in the ring. A spin-polarized current can be generated in a three-terminal device even in the absence of magnetic field, whereas it cannot in a two-terminal device. The spin polarization is enhanced by the resonant tunneling around the Coulomb peak and Kondo effect in the Coulomb blockade regime. The efficiency for the spin polarization is discussed in realistic devices.

1. Introduction

A mesoscopic ring with an embedded quantum dot, so-called Aharonov-Bohm (AB) interferometer, is an important device to investigate the coexistence of the interference effect in a ring and single-electron transport through a quantum dot. It is known that the transport measurement of the phase shift through the quantum dot is not possible in a two-terminal setup because the conductance satisfies the relation of $G(\phi) = G(-\phi)$ by the Onsager's reciprocal theory, where ϕ is the AB phase of the magnetic flux penetrating the ring [1], whereas it is possible in a multi-terminal setup [2,3].

In the present paper, we theoretically study the Aharonov-Casher (AC) effect in such a device, which is caused by the spin-orbit interaction in the ring. In this system, various transport phenomena were reported, e.g., control of the spin-polarized current using a magnetic field in a two-terminal setup [4], modulation of the Kondo effect by the AB and AC effects [5,6], and generation of spin-polarized current in a three-terminal setup [7]. We examine the spin polarization of the current in another geometry with three terminals, depicted in Fig. 1. We show that (i) a spin-polarized current can be generated in a three-terminal device even in the absence of magnetic field, whereas it cannot in a two-terminal device, and (ii) the spin polarization is enhanced by the resonant tunneling around the Coulomb peak and Kondo effect in the Coulomb blockade regime.

Note that a mesoscopic ring with an embedded quantum dot is usually examined using a simple model in which the quantum dot is connected to leads L and R by the tunnel coupling and an arm of the ring is described by the direct coupling W between the leads [8]. We examine a general model with $W_{k',k}$ which depends on state k' in lead R and k in lead L. Then we have a relevant parameter, p_α ($\alpha = L, R$) defined in Eq. (4), which is the overlap integral between the conduction mode coupled to the quantum dot and that coupled to the upper arm of the ring

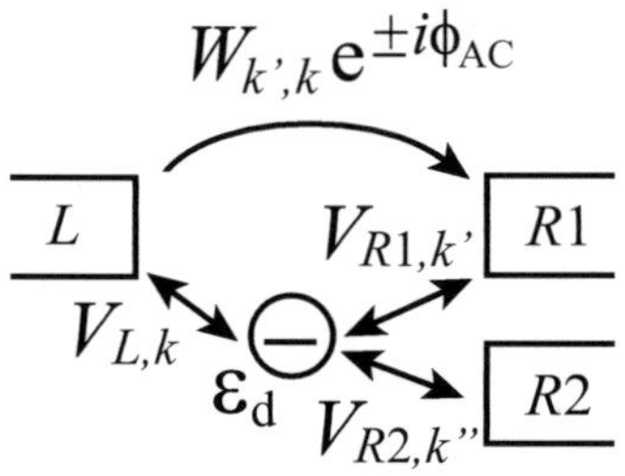

Figure 1. Model for a mesoscopic ring with an embedded quantum dot in a three-terminal geometry, in the absence of magnetic field. A quantum dot with single level ε_d is connected to leads L, $R1$, and $R2$ by tunnel couplings, $V_{L,k}$, $V_{R1,k'}$, and $V_{R2,k''}$, respectively. The upper arm of the ring is described by the coupling $W_{k',k}$ between leads L and $R1$, which depends on states k and k' in the leads. The spin-orbit interaction is taken into account by the AC phase $\pm\phi_{\mathrm{AC}}$ for spin $\sigma = \pm 1$. The electro-chemical potentials in the leads are set to be $\mu_{R1} = \mu_{R2}$ and $\mu_L - \mu_{R1} = eV$ with applied bias V ($\mu_\alpha \to \varepsilon_{\mathrm{F}}$ as $V \to 0$).

in lead α [9]. The model in Ref. [8] corresponds to our model with $p_\alpha = 1$ (when $V_{R2,k''} = 0$ and $\pm\phi_{\mathrm{AC}} \to \phi$ in Fig. 1), which is justified in the case of single channel in the leads. We evaluate the spin polarization with changing p_α to discuss the efficiency of realistic devices.

2. Model and calculation method

In the model depicted in Fig. 1, we assume a single level ε_d in the quantum dot and no magnetic field. The AC phase ϕ_{AC} by the spin-orbit interaction in the ring is attached as $W_{k',k}e^{\pm i\phi_{\mathrm{AC}}}$ for spin $\sigma = \pm 1$. The Hamiltonian is given by $H = H_{\mathrm{dot}} + H_{\mathrm{leads}} + H_T$, where

$$H_{\mathrm{dot}} = \sum_\sigma \varepsilon_d d_\sigma^\dagger d_\sigma + U d_\uparrow^\dagger d_\uparrow d_\downarrow^\dagger d_\downarrow \tag{1}$$

$$H_{\mathrm{leads}} = \sum_{\alpha=L,R1,R2} \sum_{k,\sigma} \varepsilon_k c_{\alpha,k\sigma}^\dagger c_{\alpha,k\sigma} \tag{2}$$

$$H_T = \sum_{\alpha=L,R1,R2} \sum_{k,\sigma} V_{\alpha,k}\left(c_{\alpha,k\sigma}^\dagger d_\sigma + \mathrm{h.c.}\right) + \sum_{k,k',\sigma}\left(W_{k',k}e^{\pm i\phi_{\mathrm{AC}}} c_{Rk'\sigma}^\dagger c_{Lk\sigma} + \mathrm{h.c.}\right), \tag{3}$$

where $d_\sigma^\dagger$ and d_σ ($c_{\alpha,k\sigma}^\dagger$ and $c_{\alpha,k\sigma}$) are creation and annihilation operators of an electron with spin σ in the quantum dot (with wavenumber k and spin σ in lead α), respectively. The Coulomb interaction in the quantum dot is $U = 0$ ($U \neq 0$) to examine the resonant tunneling (Kondo effect).

We have three parameters relevant to the transport, $\pi\sum_k |V_{\alpha,k}|^2 \delta(\varepsilon_k - \varepsilon_{\mathrm{F}}) = \Gamma_\alpha$, $\pi\sum_k w_{\alpha,k}\delta(\varepsilon_k - \varepsilon_{\mathrm{F}}) = x_\alpha$, and

$$\pi \sum_k V_{\alpha,k}\sqrt{w_{\alpha,k}}\delta(\varepsilon_k - \varepsilon_{\mathrm{F}}) = \sqrt{\Gamma_\alpha x_\alpha} p_\alpha \tag{4}$$

($|p_\alpha| \le 1$) in lead α, where ε_{F} is the Fermi level in the leads. $|p_\alpha| = 1$ for the case of single channel in the leads, whereas $|p_\alpha| < 1$ for the case of more than one channel.

We express the current $I_\pm$ for spin $\sigma = \pm 1$ from lead L to $R1$, where $\mu_{R1} = \mu_{R2}$ and $\mu_L - \mu_{R1} = eV$ ($\mu_\alpha \to \varepsilon_{\mathrm{F}}$ as $V \to 0$), in terms of retarded Green function of the quantum dot. The Green function is obtained analytically for $U = 0$ while it is evaluated using the Bethe ansatz exact solution [10,11] with Friedel sum rule for the Kondo effect with $U \neq 0$. The spin-dependent conductance, $G_\sigma = dI_\sigma/dV|_{V=0}$, is calculated at temperature $T = 0$.

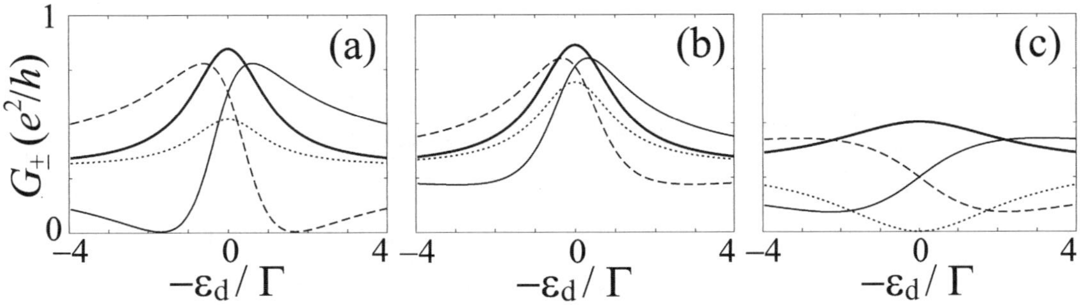

Figure 2. Conductance $G_\pm$ for spin $\sigma = \pm 1$ as a function of energy level ε_d in the quantum dot. The Fermi level in the leads is set to be $\varepsilon_F = 0$. $\Gamma_L = \Gamma_{R1} = \Gamma/2$, $x_L = x_{R1} = 0.3$, and $p_L = p_{R1} \equiv p$. (a) $\Gamma_{R2}/\Gamma = 1/5$ and $p = 1$. (b) $\Gamma_{R2}/\Gamma = 1/5$ and $p = 0.75$. (c) $\Gamma_{R2}/\Gamma = 5/3$ and $p = 1$. The AC phase is $\phi_{AC} = 0$ (solid line; $G_+ = G_-$), $\pi/2$ (G_+ by dotted line and G_- by bold solid line), and π (broken line; $G_+ = G_-$).

3. Calculated results

In the case of $U = 0$, we show the spin-dependent conductance G_σ as a function of energy level ε_d in the quantum dot (Fermi level in the leads is $\varepsilon_F = 0$). $p_L = p_{R1} \equiv p$. $p = 1$ in panel (a) and 0.75 in panel (b) when the tunnel coupling to lead $R2$ is small ($\Gamma_{R2}/\Gamma = 1/5$, where $\Gamma_L = \Gamma_{R1} = \Gamma/2$). $G_+ = G_-$ when $\phi_{AC} = 0$ (solid line) and π (broken line), whereas $G_+ \neq G_-$ when $\phi_{AC} = \pi/2$ (dotted and bold solid lines). Note that $G_+ = G_-$ always holds in the two-terminal geometry ($V_{R2,k''} = 0$). We find an asymmetric Fano resonance for $G_\pm$ when $\phi_{AC} = 0$ and π in panel (a) due to the large interference effect between the wavefunction through the upper arm of the ring and that through the quantum dot. The curve of $G_\pm$ becomes closer to the symmetric Breit-Wigner resonance in panel (b) since the interference is weaker with smaller p. For $U \neq 0$, the Fano-Kondo effect is seen when $p = 1$ [8], which changes to the conventional Kondo plateau as p decreases (not shown here).

With an increase in tunnel coupling to lead $R2$, the difference between G_+ and G_- becomes large until $\Gamma_{R2}/\Gamma = 5/3$. At this critical value of Γ_{R2}, the conductance is plotted in Fig. 2(c) when $p = 1$. $G_+ = 0$ if ε_d matches the Fermi level in the leads (resonant tunneling). This means that the perfect spin polarization can be realized by tuning the parameters to this condition in the case of $p = 1$.

To evaluate the spin-polarized current, we introduce the spin polarization ratio, $\eta = (G_+ - G_-)/(G_+ + G_-)$. For $U = 0$, we obtain the analytical expression

$$\eta = \frac{4\sqrt{x}\Gamma\Gamma_{R2}p^2 \sin\phi_{AC}}{2\left[2\sqrt{x}\Gamma\varepsilon_d\cos\phi_{AC} + 4x\Gamma\Gamma_{R2} + x(x+5)\Gamma^2\right]p^2 - \Gamma^2\left[x(x+4)p^4 + (x+1)^2\right] - 4x\left[\varepsilon_d^2 + (\Gamma+\Gamma_{R2})^2\right]}.$$

From this expression, we find $\eta = 0$ in the two-terminal set-up ($\Gamma_{R2} = 0$) or $\phi_{AC} = 0, \pi$. We also find that $\eta = \pm 1$ when $\Gamma_{R2}/\Gamma = 5/3$, $\phi_{AC} = \mp\pi/2$, and $p = 1$ only. We plot η in Fig. 3(a) when $\Gamma_{R2}/\Gamma = 5/3$ and $\phi_{AC} = \pm\pi/2$. $\eta = \pm 1$ at the resonance for $p = 1$ (curves a), but $|\eta| < 1$ in realistic situations with $|p| < 1$ (curves b and c).

Finally, we examine the spin polarization by the Kondo effect in the Coulomb blockade regime. We plot η when $U = 24\Gamma$ ($U = 9\tilde{\Gamma}$ with level broadening $\tilde{\Gamma} = \Gamma_L + \Gamma_{R1} + \Gamma_{R2}$) in Fig. 3(b). The other parameters are the same as in Fig. 3(a). We find a plateau of $\eta = \pm 1$ in the Coulomb blockade regime ($\varepsilon_d < \varepsilon_F < \varepsilon_d + U$) when $p = 1$. This is because the resonance level is fixed at the Fermi level by the Kondo effect. The absolute value of η at the plateau becomes smaller for smaller p.

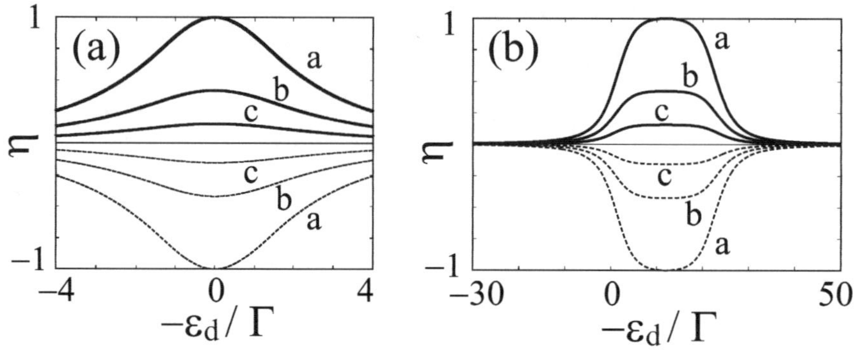

Figure 3. Spin polarization ratio $\eta = (G_+ - G_-)/(G_+ + G_-)$, as a function of energy level ε_d in the quantum dot. The Fermi level in the leads is set to be $\varepsilon_{\rm F} = 0$. $\Gamma_L = \Gamma_{R1} = \Gamma/2$, $\Gamma_{R2}/\Gamma = 5/3$, and $x_L = x_{R1} = 0.3$. (a) $U = 0$ and (b) $U = 24\Gamma$ (Kondo effect). Bold solid (dotted) lines correspond to the AC phase of $\phi_{\rm AC} = -\pi/2$ $(\pi/2)$, whereas solid lines of $\eta = 0$ to $\phi_{\rm AC} = 0$ and π. $p \equiv p_L = p_{R1} = 1$ (curve a), 0.75 (curve b), and 0.5 (curve c).

4. Conclusions

The generation of spin-polarized current has been studied theoretically using a mesoscopic ring with an embedded quantum dot in a multi-terminal geometry, considering the AC effect by the spin-orbit interaction in the ring. We have examined a realistic model with a parameter p for the overlap integral between the conduction mode coupled to the quantum dot and that coupled to the upper arm of the ring. We have shown that a spin-polarized current is generated in a three-terminal device even in the absence of magnetic field, whereas it cannot in a two-terminal device. The spin polarization is enhanced by the resonant tunneling around the Coulomb peak and Kondo effect in the Coulomb blockade regime. In an ideal situation with single conduction mode in the leads ($|p| = 1$), the perfect spin polarization can be realized by tuning the parameters in the device, whereas the spin polarization is less in realistic devices with $|p| < 1$.

Acknowledgments

This work is partly supported by a Grant-in-Aid for Scientific Research (S) (No. 26220711) from Japan Society for the Promotion of Science and MEXT Grant-in-Aid for Scientific Research on Innovative Areas "Science of hybrid quantum systems" (No. 15H05870).

References

[1] Yacoby A, Heiblum M, Mahalu D and Shtrikman H 1995 *Phys. Rev. Lett.* **74** 4047
[2] Schuster R, Buks E, Heiblum M, Mahalu D, Umansky V and Shtrikman H 1997 *Nature* **385** (London) 417
[3] Takada S *et al* 2014 *Phys. Rev. Lett.* **113** 126601
[4] Sun Q, Wang J and Guo H 2005 *Phys. Rev.* B **71** 165310
[5] Lim J S, Crisan M, Sánchez D, López R and Grosu I 2010 *Phys. Rev.* B **81** 235309
[6] Vernek E, Sandler N and Ulloa S E 2009 *Phys. Rev.* B **80** 041302(R)
[7] Chi F and Zheng J 2008 *Appl. Phys. Lett.* **92** 062106
[8] Hofstetter W, Konig J and Schoeller H 2001 *Phys. Rev. Lett.* **87** 156803
[9] Kubo T, Tokura Y, Hatano T and Tarucha S 2006 *Phys. Rev.* B **74** 205310
[10] Okiji A and Kawakami N 1982 *Phys. Soc. Jpn.* **51** 3192
[11] Wiegmann P B and Tsvelick A M 1983 *J. Phys.* C: *Solid State Phys.* **16** 2281

Macroscopic transport of a current-induced spin polarization

S Ullah[1], G J Ferreira[2], G M Gusev[1], A K Bakarov[3] and F G G Hernandez[1]

[1] Instituto de Física, Universidade de São Paulo, São Paulo 05508-090, SP, Brazil
[2] Instituto de Física, Universidade Federal de Uberlândia, Uberlândia 38400-902, MG, Brazil
[3] Institute of Semiconductor Physics and Novosibirsk State University, Novosibirsk 630090, Russia

E-mail: felixggh@if.usp.br

Abstract. Experimental studies of spin transport in a two-dimensional electron gas hosted by a triple GaAs/AlGaAs quantum well are reported. Using time-resolved Kerr rotation, we observed the precession of the spin polarization about a current-controlled spin-orbit magnetic field. Spatially-resolved imaging showed a large variation of the electron g-factor and the drift transport of coherent electron spins over distances exceeding half-millimetre in a direction transverse to the electric field.

1. Introduction

For the successful implementation of new functionalities using the spin degree of freedom in technological platforms, experimental techniques are required for the generation of spin polarization, transport across relevant length scales and polarization detection. For device applications of non-equilibrium electron spins, triggered by either optical [1] or electrical techniques [2], the generation and detection of spin currents should be accomplished without the use of extremely strong magnetic fields and desirably using only electrical voltages. The electrical generation of spin polarization have been widely studied in semiconductor structures of different dimensionality such as bulk samples [3, 4] and in n- and p-doped quantum wells (QWs) [5, 6]. The transport of spins polarized by an electrical current is of great interest not only due to the emergence of new phenomena but also due to possible applications in spin based electronics [7, 8]. Transverse transport of in-plane current-induced spin polarization (CISP) have been studied in L-shaped strained n-InGaAs channels [9]. Furthermore, it has been reported that the spins polarized in the out-of-plane direction by the spin Hall effect (SHE) can be drive long distances in n-doped GaAs epilayers [10]. In this paper, we report on the electrical generation of spin polarization in two-dimensional electron gases under the action of spin-orbit fields and the spin transport over distances exceeding half millimetre away from current path in a transverse direction.

2. Materials

The structure discussed in this manuscript is GaAs/AlGaAs triple quantum well (TQW), containg a dense two dimensional electron gas (2DEG) grown in the [001] direction. The structure is symmetrically δ-doped, with total density $n_s = 9.6 \times 10^{11} cm^{-2}$ and low temperature mobility $\mu = 5.5 \times 10^5 cm^2/Vs$. It contains a 26-nm-thick GaAs central well with electron density of about $1.4 \times 10^{11} cm^{-2}$ and two 12-nm-thick lateral wells with approximately equal electron density of $4.1 \times 10^{11} cm^{-2}$, each separated by 1.4-nm-thick $Al_{0.3}$Ga0.7As barriers. Fig. 1 (a) shows the calculated TQW band structure and subband charge density, where the black lines show the potential profile and the color lines show the first (dark yellow), second (orange) and third (violet) occupied subbands.

For the electrical generation of spin polarization the sample was patterned into a six-contact Hall bar geometry. The main channel have length L = 500 μm between the side probes (in the y-axis), width $W = 200$ μm and four transverse channels of equal width 15 μm that extend out from the main channel (in the x-axis) as shown schematically in Fig. 1 (b).

3. Experimental results

The current-induced spin polarization was measured by using the optical Kerr rotation (KR) technique. The spin polarization was generated by an electrical voltage modulated at f_2 = 1.1402 kHz for lock-in detection. A weak linearly polarized probe beam emitted from a pulsed Ti:sapphire laser was scanned on the structure for the spatial detection of the induced spin polarization. The polarization of the reflected beam was measured by a balanced photodetector and lock-in amplifier operating at f_2. The laser wavelength was tuned for maximum KR signal. The sample was immersed in the variable temperature insert of a split coil superconductor

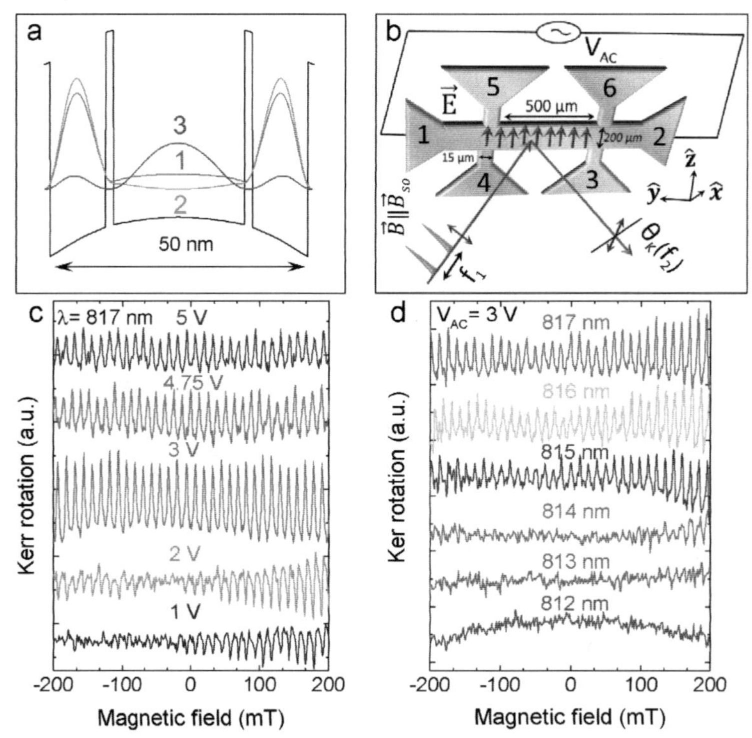

Figure 1. Longitudinal configuration: (a) TQW band structure and subband charge density. (b) Schematic for the optical amplification of the CISP. (c) KR versus B measured for several applied voltages. (d) B scan of KR signal at different pump-probe wavelengths. Measurements were taken at a temperature T = 1.2 K and probe power of 300 μW.

magnet in the Voigt geometry.

3.1. Longitudinal spin transport

In the longitudinal configuration, the KR was measured as a function of external magnetic fields while applying V_{AC} at the central channel (1-2) [see Fig. 1 (b)]. If the ensemble coherence time (T_2^*) is longer than the pulse repetition (with MHz frequency) the signal does not fully decay from the previous pulse and overlaps with the signal generated by the following pulses. Due to the constructive interference at certain resonant fields, the amplification of the precession (with GHz frequency) of those electrically polarized spin occurs and result in sharp resonance peaks. These resonance peaks with spacing ΔB obey the periodic condition $\Delta B = (hf_1/g\mu_B)$ where f_1 is the laser repetition rate [2]. Fig. 1 (c) displays the RSA pattern as a function of the external magnetic field for several applied voltages. The CISP amplitude increases with the increase of applied voltage and reaches its maximum value at 3 V. However, further increase of the applied voltage results in a strong reduction of the amplitude possibly due to heating effects [2]. Fig. 1 (d) displays the dependence of the CISP amplitude on the probe wavelength at a fixed voltage of 3 V. Furthermore, the hole spin coherence time involved in the generation of the electron spin coherence, causes the reinforcement of the resonant peak amplitude at higher magnetic fields [11]. The experimental conditions for maximum signal amplitude, namely $V_{AC} = 3$ V and $\lambda =$ 817 nm, were used in the following section.

3.2. Transverse spin transport

Now we switch from to the longitudinal transport to the transverse transport. In contrast to the longitudinal transport, where the spin polarization was measured in a region where the local current flows, we measured the spin polarization in a region with minimal electric field. For example, when V_{AC} is applied along the main channel (y-direction) the transverse transport will be into the side arms (x-direction) of the Hall bar. The reflectivity of the structure imaged as function of the position (x,y) is shown in Fig. 2 (b). The reflectivity map displays the central channel and the four transverse voltage probes. The solid lines highlight the possible current paths and the red regions separate the conducting areas where the white lines are the edges of those regions. In Fig. 2 (b), we imaged the KR signal as a function of B while applying V_{AC} in the x-axis using contacts 5-4 and scanning the probe spot at (0,y) approximately 250 μm away from the current path. We observed the transverse drift of the CISP in a scan of about $\Delta y = 0.4$ mm. We also observed that the spacing ΔB between the RSA peaks varies with the position. One can clearly see that the outer peaks are shifted towards higher fields for an increasing y-position indicating the variation of the g-factor as plotted (●) in Fig. 2 (b). Recently, the g-factor modification in a bulk InGaAs epilayer by an in-plane electric field was studied [12]. They found a dependence of Δg on the drift velocity: $\Delta g \propto v_d^2$. Furthermore, the dependence of the electrical g tensor variation on the magnetic field was also reported in quantum wells [13]. Our data gives a g-factor variation in agreement with those reports.

To extract the spin lifetime and polarization amplitude, the magnetic field scans were fitted using a Lorentzian function $\Theta_K = A/[(\omega_L T_2^*)^2 + 1]$ with $T_2^* = \hbar/(g\mu_B B_{1/2})$, where ω_L is the Larmor frequency, $\hbar$ is the reduced Planck's constant, μ_B is the Bohr magneton and $B_{1/2}$ is the peak half-width. The extracted parameters are plotted in Fig. 2 (c) and (d). Surprisingly we found that in our structure the CISP transverse drift can drive a constant T_2^* of about 6 ns by almost half millimeter. The transport length of $\ell_s = 0.625 \pm 0.058$ mm was extracted from fitting the data in Fig. 2 (d) by $exp(-y/\ell_s)$. Those results of T_2^* and ℓ_s are independent of the field chosen for analysis. More detailed analysis, theoretical calculations and transport along x-axis are published in [14].

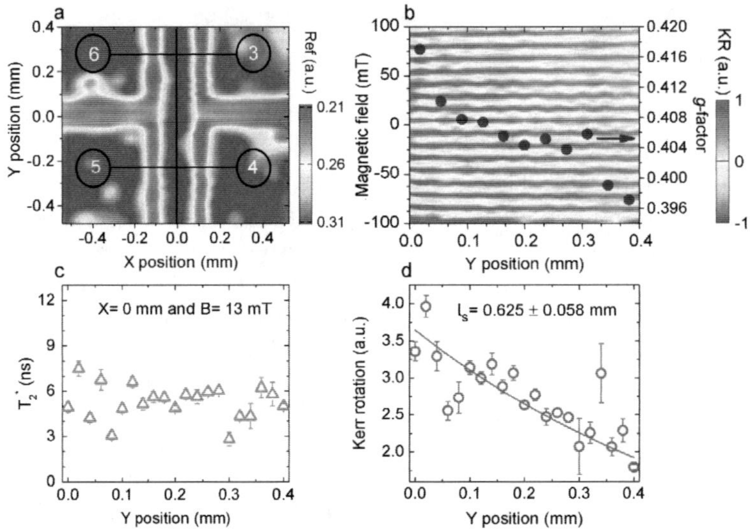

Figure 2. Transverse configuration: (a) Reflectivity map of the device. (b) KR as a function of B along (0,Y) mm axis with V_{AC}= 3 V. The symbol (●) shows the spatial variation of electron g-factor. The extracted values of T_2^* (△) and Amplitude (○) are plotted in (c) and (d) respectively, the solid line is an exponential fit to the data. T = 5 K.

4. Conclusion

We performed a detailed study on the CISP transport in a two-dimensional electron gas confined in GaAs/AlGaAs quantum wells. We found long T_2^* in the nanoseconds range and macroscopic drift distances in a direction transverse to the applied electric field. Additionally, a large spatial variation of the electron g-factor was observed.

5. Acknowledgments

F.G.G.H. acknowledges financial support from Grant No. 2009/15007-5, 2013/03450-7, 2014/25981-7 and 2015/16191-5 of the São Paulo Research Foundation (FAPESP). S.U. acknowledges TWAS/CNPq for financial support.

References

[1] Ullah S, Gusev G M, Bakarov A K and Hernandez F G G 2016 *J. Appl. Phys.* **119** 215701
[2] Hernandez F G G, Gusev G M and Bakarov A K 2014 *Phys. Rev.* B **90** 041302(R)
[3] Kato Y K, Myers R C, Gossard A C and Awschalom D D 2004 *Phys. Rev. Lett.* **93** 176601
[4] Ster N P, Ghosh S, Xiang G, Zhu M, Samarth N and Awschalom D D 2006 *Phys. Rev. Lett.* **97** 126603
[5] Sih V, Myers R C, Kato Y K, Lau W H, Gossard A C and Awschalom D D 2005 *Nat. Phys.* **1** 31
[6] Silov A, Blajnov P, Wolter J, Hey R, Ploog K and Averkiev N 2005 *Semiconductors* **39** 1323
[7] Aronov A G and Lyanda-Geller Yu B 1989 *JETP. Lett.* **50** 431
[8] Edelstein V M 1990 *Solid state commun.* **73** 233
[9] Kato Y K, Myers R C, Gossard A C and Awschalom D D 2005 *Appl. Phys. Lett.* **87** 022503
[10] Sih V, Lau W H, Myers R C, Horowitz V R, Gossard A C and Awschalom D D 2006 *Phys. Rev. Lett.* **97** 096605
[11] Yugova I A, Sokolova A A, Yakovlev D R, Greilich A, Reuter D, Wieck A D and Bayer M 2009 *Phys. Rev. Lett.* **102** 167402
[12] Luengo-Kovac, Macmahon M, Huang S, Goldman R S and Sih V 2015 *Phys. Rev.* B **91** 201110(R)
[13] Chen Y S, Fält S, Wegscheider and Salis G 2014 *Phys. Rev.* B **90** 121304(R)
[14] Hernandez F G G, Ullah S, Ferreira G J, Kawahala N M, Gusev G M and Bakarov A K 2016 *Phys. Rev.* B **94** 045305

Influence of Annealing Temperature on Mn-doped ZnO Diluted Magnetic Semiconductor Thin Films

Yu Zhang, Naho Nasu, and Xinwei Zhao

Department of Physics, Tokyo University of Science, Tokyo 1628601, Japan

xwzhao@rs.kagu.tus.ac.jp

Abstract. In this paper, we report the effects of Mn^{2+} and the defect on ferromagnetism in Mn-doped ZnO thin films prepared by radio frequency magnetron sputtering at different annealing temperature below 600 °C. The result of superconducting quantum interference device reveals that the sample which annealed at 500 °C realized ferromagnetism till room temperature. The X-ray diffraction results shows that the samples are wurtzite structure and the sample annealed at optimum temperature gets relatively better crystallinity. X-ray photoelectron spectroscopy results show clear evidence of oxygen vacancy increment for the samples which annealed at 500 °C. Our results indicate that Mn^{2+} ions replace the Zn sites and no evidence for either metallic Mn or Mn-related oxide in the samples. It was concluded that the existence of ferromagnetism was closely related to the concentration of Mn^{2+} and the defect of the ZnO matrix.

1. Introduction

Motivated by the theoretical prediction in 2000 that ZnO could be one of the most promising candidates for the ferromagnetic diluted magnetic semiconductors (DMSs) which were view as a future spintronics materials [1], various researches on Mn-doped ZnO (ZnO: Mn) DMSs have been carried out. Then the solubility of Mn into ZnO was found to exceed the equilibrium limit about 13% to 35% by pulsed laser deposition preparing method [2]. Although many investigations have been done, few of the experiments realized the ferromagnetic ZnO: Mn

IOP Conf. Series: Journal of Physics: Conf. Series **864** (2017) 012061 doi:10.1088/1742-6596/864/1/012061

DMS materials. Then challenges became how to improve the Curie temperature reach up to 300K and how to explain the origin of the ferromagnetism.

In this paper, we report the ferromagnetic ZnO: Mn thin films formed by radio frequency (rf) magnetron sputtering. The result shows that the ZnO: Mn thin film which annealed at 500 °C is ferromagnetic till 300 K.

2. Experiment

The Mn-doped ZnO thin films were grown on p-Si (100) substrate by rf magnetron sputtering, with the composite target of ZnO and Mn_2O_3 (10 wt %). The post annealing temperatures of 400 °C, 500 °C and 600 °C were used and the corresponding samples are herein after referred as A400, A500 and A600, respectively.

The crystal structures of the samples were analyzed by X-Ray diffraction (XRD RINT) with Cu Kα radiation (λ=1.5405 Å) in. Magnetic properties of the samples were studied with Quantum Design MPMS-XL, a superconducting quantum interference device (SQUID) in a temperature range from 5 K to 300 K. X-ray photoelectron spectroscopy (XPS) measurements were obtained by using a JPS-9010 spectrometer. The Raman spectra were excited with the 532 nm line by a Nikon Eclipse Ti-s Raman microscope.

3.Results and Discussion

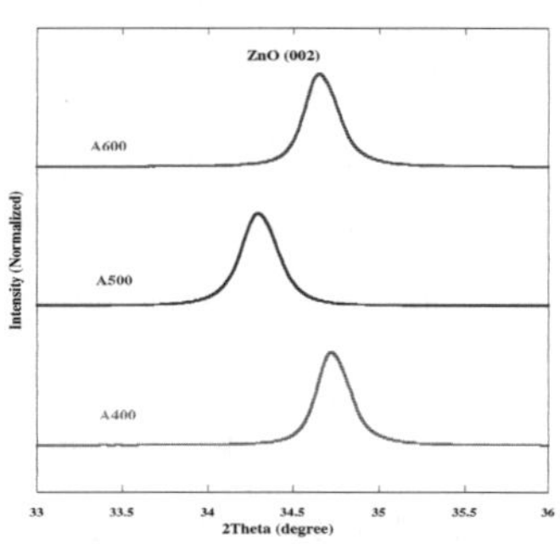

Fig. 1. X-ray diffractograms for A400, A500 and A600.

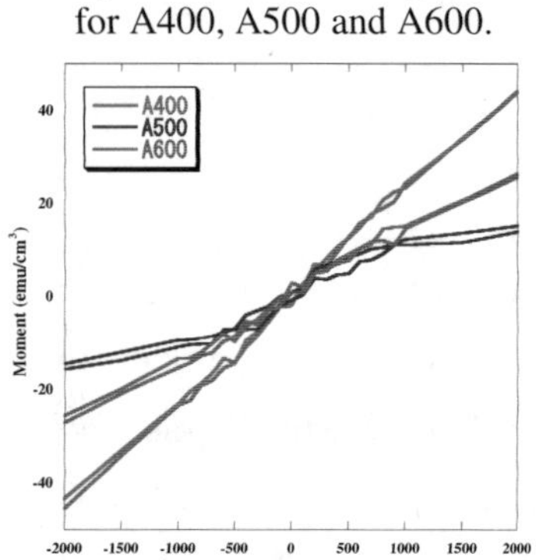

Fig. 2. M-H curves of A400, A500 and A600 at 300K.

Figure 1 shows the XRD patterns for the ZnO: Mn thin films. The XRD patterns, showing the single peak of ZnO (002), indicated that all the samples have c-axis preferred orientation with the wurtzite ZnO structures and no secondary phase exists. This suggests that Mn is substituting in ZnO and Mn doping did not alter the ZnO structure significantly. The c-axis lengths are 2.5810 Å, 2.610 Å, and 2.587 Å for A400, A500, and A600, respectively.

Figure 2 shows the M-H curves for the samples measured by SQUID at 300 K, indicating that the magnetism has changed with the annealing temperature. It is seen that the A400 sample shows no evidence of ferromagnetism. After annealing temperature rising to 500 °C, the thin films became ferromagnetism. The results present that A500 sample got room temperature ferromagnetic property and ruled out the possibility that the ferromagnetic property came from the Mn-related oxide whose Curie temperature or Neel temperature were below 120 K [3]. When the annealing temperature reach up to 600 °C, the coercive force of the sample downed from 80 Oe to 60 Oe. Meanwhile the remnant magnetization decreased from 0.8 emu/cm^3 to 0.6 emu/cm^3.

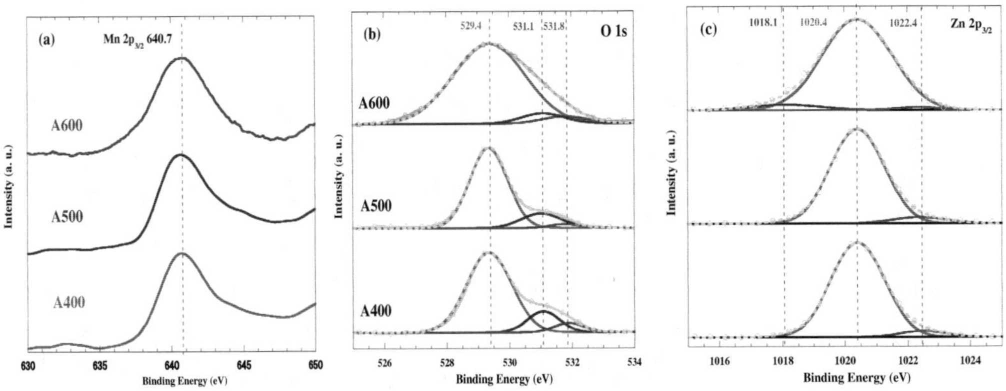

Fig. 3. Core level XPS spectra of (a) Mn $2p_{3/2}$, (b) O 1s, and (c) Zn $2p_{3/2}$.

Figure 3 shows the XPS core level spectra of Mn $2p_{3/2}$, O 1s and Zn $2p_{3/2}$ of all the samples. The binding energies of all the spectra were calibrated by the Au $4f_{7/2}$ peak (83.8 eV). The XPS core level spectra were deconvoluted by Gaussia peak fitting with similar spectral features. Figure 3 (a) reveals the Mn $2p_{3/2}$ core level XPS spectra are centered at 640.7 eV which correspond to Mn^{2+}. As the c-axis length increase with the concentration of the Mn^{2+}, because the radius of Mn^{2+} (0.066 nm) ion is longer than that of Zn^{2+} (0.060 nm), The conclusion that the A500 samples which showed stronger ferromagnetism get highest Mn^{2+} concentration and related-better structure can be indicated. The O 1s peak, shown in Fig. 3 (b), spectral contributions that were successfully disentangled by fitting analysis, resulting in two nearly Gaussian components centered at 529.35eV, 531.0 eV and 531.8 eV which are attributed to the Zn-O bands in ZnO lattice, oxygen vacancies and chemisorbed oxygen within the ZnO matrix respectively [4]. The ratios of the 531.0 eV/529.5 eV integrated intensities are 0.18, 0.19 and 0.07. It is reasonable to assume that A400 and A500 got almost the concentration of oxygen vacancy, while the A600 sample got the least. In our work, the main peak in Zn $2p_{3/2}$ is from the oxide form of Zn in ZnO as shown in Fig. 3 (c). The Zn $2p_{3/2}$ peaks of A400 and A500 are consisted of two peaks centered at 1020.4 eV and 1022.4 eV. The lower energy peak is attributed to the O^{2-} in the wurtzite structure of the hexagonal Zn^{2+} ion array, while the higher peak is attributed to the zinc vacancies [5].The ratios of the 1022.4 eV/1020.4 eV integrated intensities are 0.06, 0.07 and 0.02 for A400, A500 and A600, respectively. The peaks are found to shift toward the lower binding energy values as compared to the values reported in literature which can be attributed to the partial substitution of lattice by Mn^{2+} ions in ZnO and Zn-Mn bonding structure [6]. When the annealing temperature rise to 600 ^{o}C, the peak for Zn $2p_{3/2}$ is broaden. The phenomenon that one new peak appears at 1018.1 eV is related to the decrease of oxygen vacancy [5].

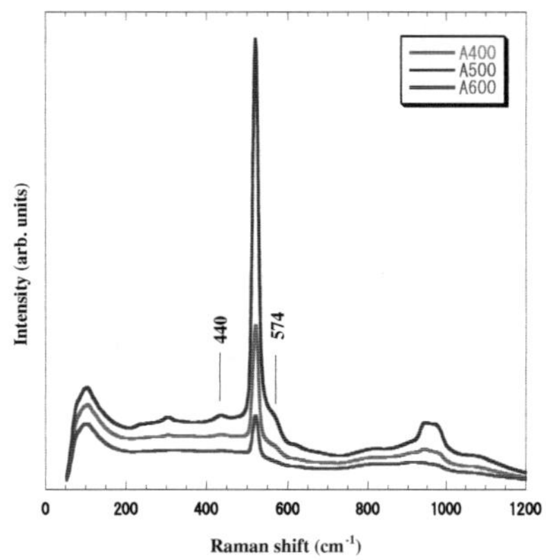

Fig. 4. Raman sprctra of A400, A500 and A600.

Figure 4 shows that the Raman spectra of the samples. The peak at 440 cm^{-1} can be assigned to the high frequency branch of the E_{2H} mode of ZnO [7]. The peak at 520 cm^{-1} is formed by the substrate (520 cm^{-1}) and the donor defects (524 cm^{-1}) which can be doubly occupied oxygen vacancies and zinc interstitials of ZnO matrix. The Raman bands at 574 cm^{-1} is related to the Zn vacancy [7]. As the figure showed, the A500 samples get more donor defects and Zn vacancy, while the A600 get the least of them. It supposed that some of the Mn doped in the ZnO matrix and occupied the site of Zn vacancy when the samples deposited. When the annealing temperature came to 600 °C, some of the Zn^{2+} ions replaced the Mn^{2+} and the oxygen vacancy decreased.

4. Conclusion

The XRD result shows that all the samples are wurtzite structure and the sample annealed at 500 °C gets relatively higher Mn^{2+} concentration. A clear evidence that a certain amount of Mn^{2+} is necessary to realize ferromagnetism. The XPS and Raman measurements indicate that the defect of the ZnO matrix decreases while the annealing temperature rises up and either metallic Mn or Mn-related oxides existed in the samples. Combining on the results of XRD, XPS and Raman, we draw the conclusion that the existence of ferromagnetism was closely related to the concentration of Mn^{2+} and the defect of the ZnO matrix.

5. Acknowledgements

The authors would like to express their thanks to the Green & Safety Project of Toyo University of Science.

References

[1] Dietl T, Ohno H, Matsukura F, Cibert J and Ferrand D 2000 *Sci.* **287** 1019

[2] Fukumura T, Jin Z and Ohtomo A 1999 *Appl. Phys. Lett.* **75** 3366

[3] Liu C, Yun F and Morkoc H 2005 *J. materials Science: Materials in electronics* **16** 555

[4] Ilyas U, Rawat R S, Tan T L, Lee P, Springhama S V , Zhang S, fengji L, Chend R and Sund H D 2011 *Appl. Surf. Sci.* 258 890

[5] Tay Y Y, Sun C Q and Chen P 2006 *Appl. Phys. Lett.* **88** 173118

[6] Saw K G , Ibrahim K, Lim Y T and Chai M K 2007 *Thin Solid Films* **515** 2879

[7] Yang L W, Wu X L, Huang G S, Qiu T and Yang Y M 2005 *J. Appl. Phys.* **97** 014308

IOP Conf. Series: Journal of Physics: Conf. Series **864** (2017) 012062 doi:10.1088/1742-6596/864/1/012062

Thin SOI lateral IGBT with band-to-band tunneling mechanism

Qiang Fu[1,*], Zhaohuan Tang[1], Kaizhou Tan[1], Zhikuan Wang[2], and Yong Mei[2]

[1] Science and Technology on Analog Integrated Circuit Laboratory, Chongqing 400060, China

[2] Sichuan Institute of Solid-State Circuits, China Electronics Technolgy Group Corporation, Chongqing 400060, China

E-mail: fredfq@163.com

Abstract. In this paper, a novel 200V lateral IGBT on thin SOI layer with a band-to-band tunneling junction near the anode is proposed. The structure and the operating mechanism of the proposed IGBT are described and discussed. Its main feature is that the novel IGBT structure has a unique abrupt doped p++/n++ tunneling junction in the side of the anode. By utilizing the reverse bias characteristics of the tunneling junction, the proposed IGBT can achieve excellent reverse conducting performance. Numerical simulations suggest that a low reverse conduction voltage drop V_R=−1.6V at a current density of 100A/cm^2 and a soft factor S=0.63 of the build-in diode are achieved.

1. Introduction

The lateral insulated gate bipolar transistor (LIGBT) is a most promising power device for integrated power ICs, smart power circuits and RF applications [1-3]. SOI (Silicon-On-Insulator) has developed as the major candidate to improve many performance of power devices and power ICs, such as high speed, superior isolation and strong immunity to latch-up [4]. Thin SOI LIGBTs synthesize the advantages of the both. However, in various power circuit applications, IGBT often needs an anti-parallel diode to conduct current in a reverse path because of its uni-directional conducting characteristic. It is conventional that the anti-parallel diode is externally co-packed with the IGBT, and it leads to additional testing time, additional cost and area consumption to accommodate the on-chip diodes. The reverse conducting IGBT (RC IGBT) is thus appeared for this need, and many studies and explorations have been done for the RC IGBT [5-7].

In recent years, more and more scientists and researchers have given a lot of efforts to integrate quantum tunneling effect into CMOS devices and power devices [8,9], such as in tunnel field effect transistors (TFETs) and metal-oxide-semiconductor field effect transistors (MOSFETs). These devices also show very good performance characteristics.

In this paper, a novel 200V lateral IGBT on thin SOI layer with a band-to-band tunneling junction near the anode (named BTB-LIGBT) is proposed. The structure and the operating mechanism of the proposed IGBT are described and discussed.

2. Device structure and operation mechanism

Figure 1 shows the schematic cross-sectional view of the novel BTB-LIGBT, which features a unique abrupt highly doped p++/n++ tunneling junction in the side of the anode. As shown in Fig. 1, the difference is at the anode side of the device compared to the conventional LIGBT on thin SOI. For convenience, we set the n++ pocket doping and the p++ pocket doping equal ($N_{NT}=N_{PT}=N_T$), and set the n++ pocket width and the p++ pocket width equal ($L_{NT}=L_{PT}=L_T$). L_{Cell} is the length of a LIGBT cell pitch. B_{OX} and T_{SOI} represent the thickness of the buried oxide and SOI layer, respectively. L_{drift} and N_{drift} are the length and the doping of n-drift region, respectively.

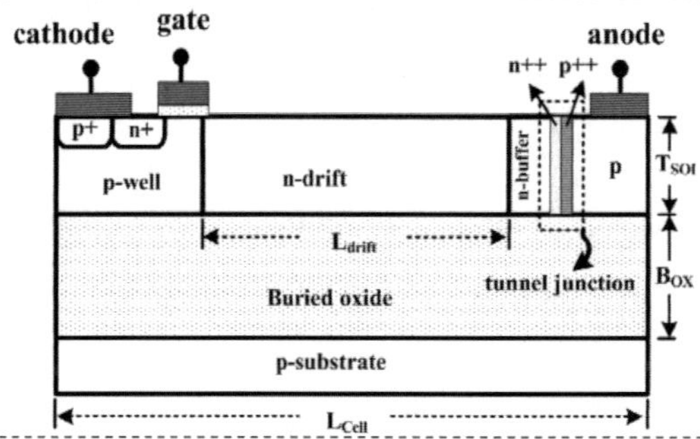

Figure 1. The schematic cross-section of the proposed BTB-LIGBT.

The geometry and doping concentration profile near the anode of the simulated BTB-LIGBT are shown in Figure 2. Because of the exist of the tunneling junction the BTB-LIGBT also exhibits the distinct tunneling current, which is the same as other tunnel devices. Figure 3 shows the typical static current-voltage (I–V) characteristics of the BTB-LIGBT. Compared to the conventional IGBT, the BTB-LIGBT shows a visible negative differential resistance (NDR) phenomenon, but it can achieve excellent reverse conducting performance.

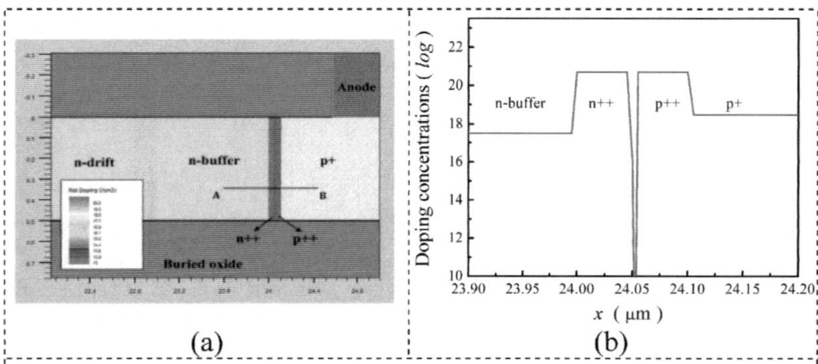

Figure 2. (a) Net doping profile schematic near the anode of the simulated BTB-LIGBT. (b) Doping concentrations near the tunneling junction along line AB.

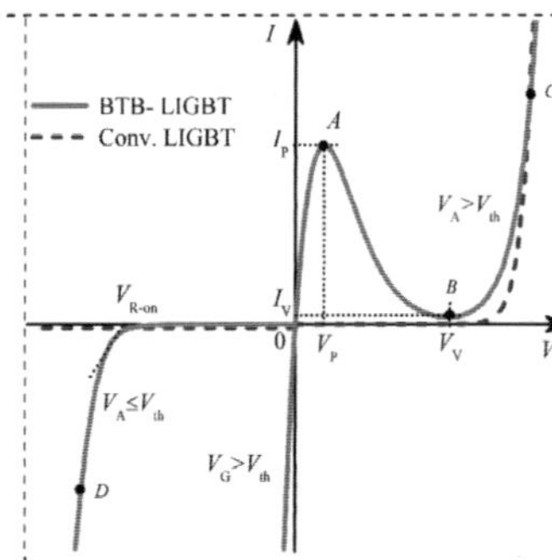

Figure 3. The typical I–V characteristics of the BTB-LIGBT.

The schematic energy band profile at the anode end in BTB-LIGBT, along line AB, is shown in Figure 4. In the first of forward conduction, the electrons can tunnel from the E_C band to the E_V band (from n++ side to p++ side). Thus, the current first increases to a maximum value, then decreases to a minimum value. When the anode voltage V_A is larger than 0.7V, the current increases exponentially with V_A. When a forward bias is applied, the tunnel junction is also reverse biased. A mass of electrons can tunnel from the valence band into the conduction band, and there is no NDR.

3. Numerical simulations and discussion

The results in this paper are obtained by Silvaco atlas device simulator. In numerical simulations for the 200V BTB-LIGBT on thin SOI layer, we consider the proposed device with following parameters values: L_{Cell}=25μm, B_{OX}=5μm, T_{SOI}=0.5μm, L_{drift} =20μm, N_{drift} =1×10^{16}cm^{-3}. In order to have a fair baseline, the gate voltage V_G is 15V for forward conducting mode, and is 0V for reverse conducting mode.

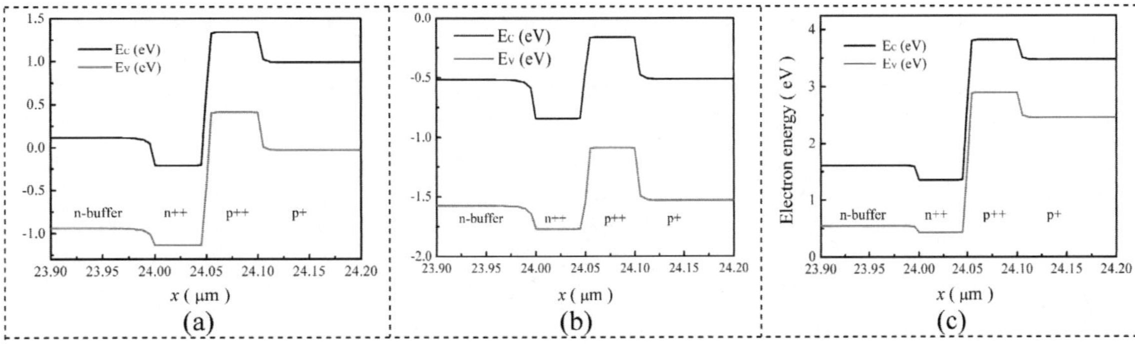

Figure 4. Schematic energy band profile at anode end in BTB-LIGBT. (a) Thermal equilibrium, V_G=0V, V_A=0V; (b) forward bias, V_G=15V, V_A=1.5V; (c) reverse bias, V_G=0V, V_A=–2.5V.

Figure 5 shows the forward I-V characteristics of the BTB-LIGBT with different N_T and different L_T. The tunneling current peak value I_P increases with an increasing N_T and a decreasing L_T. When $N_T \leq 1\times10^{20}\text{cm}^{-3}$ and $L_T \geq 80$nm, the NDR disappears basically. There is another interesting phenomenon, when $V_A \geq 0.6$V, the tunneling current falls to zero and the normal diffusion current starts to dominate. And the diffusion current decreases with a increasing N_T and an increasing L_T because of the injection efficiency limit.

The main advantage of the BTB-LIGBT is that it can achieve excellent reverse conducting performance. Figure 6 shows the reverse I-V characteristics of the BTB-LIGBT with different N_T and different L_T. The reverse tunneling current increases with an increasing N_T, but the L_T has little influence on it. When L_T=50nm, $N_T=5\times10^{20}\text{cm}^{-3}$, the reverse conduction voltage drop V_R is –1.6V at a current density of 100A/cm^2.

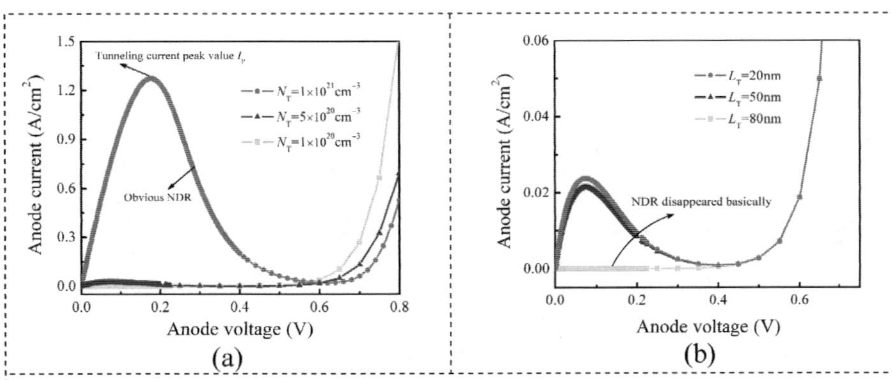

Figure 5. Forward I-V characteristics of the BTB-LIGBT with different parameters: (a) with N_T (fixed L_T=50nm); (b) with L_T (fixed $N_T=5\times10^{20}\text{cm}^{-3}$).

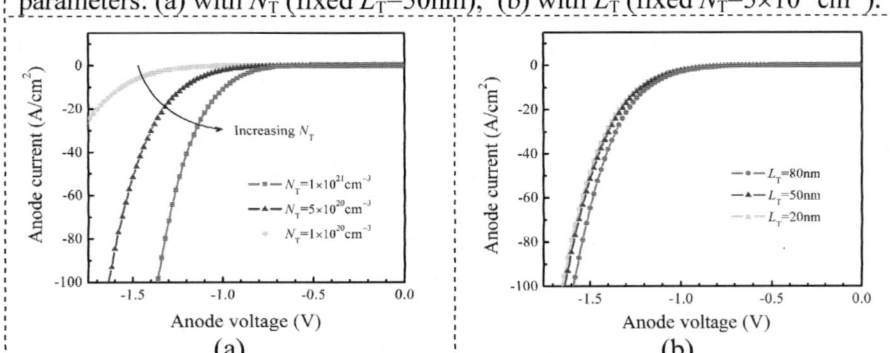

Figure 6. Reverse I-V characteristics of the BTB-LIGBT with different parameters: (a) with N_T (fixed L_T=50nm); (b) with L_T (fixed $N_T=5\times10^{20}\text{cm}^{-3}$).

Reverse recovery waveforms of the BTB-LIGBT during the reverse recovery process are shown in Figure 7. In these waveforms, we set L_T=50nm, $N_T=5\times10^{20}\text{cm}^{-3}$. The load inductance is 10nH. From the current waveform we can see that the reverse recovery time t_{rr} of the built-in diode in BTB-LIGBT is 10ns and the soft factor S of it is 0.63.

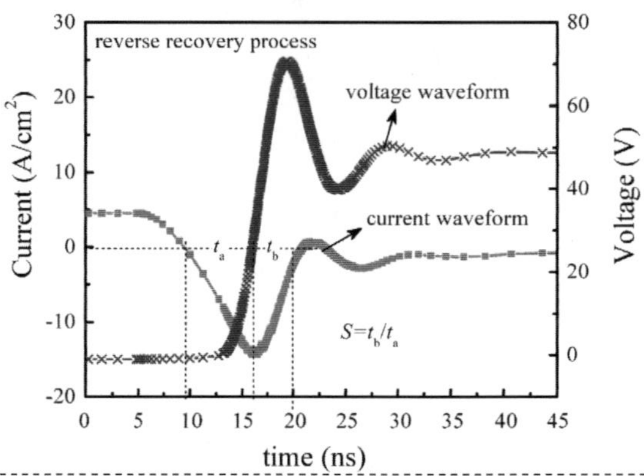

Figure 7. Reverse recovery waveforms of the BTB-LIGBT.

4. Conclusion

A novel 200V lateral IGBT on thin SOI layer with band-to-band tunneling mechanism is proposed in this paper, which introduces an abrupt doped p++/n++ tunneling junction in the side of the anode. Because of the unique reverse conducting characteristics of the tunneling junction, the proposed BTB-LIGBT achieves excellent reverse conducting performance which the conventional LIGBT has not. And some numerical simulation results are presented and discussed.

References

[1] Pathirana V, Napoli E, Udrea F, et al. 2006 *IEEE Transactions on Power Electronics* **21** 1521
[2] Liu S, Sun W, Qian Q, et al. 2010 *Solid State Electronics* **54** 1598
[3] Fu Q, Zhang B, Luo X, Wang Z, and Li Z J 2013 *Micro & Nano Letters* **8** 386
[4] Udrea F 2007 *Proc. in IEEE Bipolar/bicmos Circuits & Technology Meeting* 74
[5] Jiang H, Zhang B, Chen W, et al. 2012 *IEEE Electron Device Letters*33 417
[6] Takahashi H, Yamamoto A, Aono S, et al. 2004 *Proc. in IEEE ISPSD* 133
[7] Napoli E, Spirito P, Strollo A G M, et al. 2002 *IEEE Electron Device Letters* **23** 532
[8] Ye H, Haldar P 2008 *IEEE Transactions on Electron Devices* **55** 1524
[9] Ganapathi K, Yoon Y, Salahuddin S 2010 *Applied Physics Letters* **97** 033504

IOP Conf. Series: Journal of Physics: Conf. Series **864** (2017) 012063 doi:10.1088/1742-6596/864/1/012063

Effect of molecular layers on charge transport in nanowires

V. Handziuk[1], S. Pud[1], M. Coppola[1], A. Kisner[2] and S. Vitusevich[1]

[1]Forschungszentrum Jülich, Peter Grünberg Institute (PGI-8), 52425 Jülich, Germany
[2]Department of Cell Biology and Neuroscience, Rutgers University, 604 Allison Road Piscataway, NJ, USA

E-mail: v.handziuk@fz-juelich.de

Abstract. Stable ultrathin gold nanowires (Au NWs, 2nm diameter) were produced using wet chemical synthesis, which implied oleylamine molecules, and subsequently assembled onto gold electrodes. Transport properties of fabricated structures were studied using the noise spectroscopy technique. Lorentzian-shaped components were revealed in the noise spectra. The characteristic frequencies of these components were independent of the bias applied, which thus excludes self-heating of the nanowires. Noise peculiarities were shown to be the result of the presence of oleylamine molecules on the NWs. The evidence of oleylamine on the surface of the fabricated structures was confirmed by high-resolution transmission electron microscopy. Organic monolayers can strongly influence charge properties in nanosystems. These results should be taken into account in designing nanowire-based devices for molecular electronics.

1. Introduction

Ultrathin metal and semiconductor nanowires are attractive objects for investigations due to their unique properties[1] and wide range of possible applications[2,3]. Nanowires possess high surface-to-volume ratio, which makes them suitable for (bio)sensing applications. Metal NWs may be used as quasi one dimensional conductors for applications in nanoscale electronics. Studies of charge transport in such nanowires are therefore very important. We produced bundles of stable ultrathin Au NWs with diameters in the range of 2÷5nm and lengths of several microns[1] using wet chemical synthesis. After synthesis, the nanowires were assembled onto large gold electrodes using a microfluidic channel. The detailed sample fabrication procedure is described in Ref. [1]. Due to specific synthesis conditions, oleylamine residuals may remain on the surface of the nanowires.

In this contribution, we investigate the transport properties of ultrathin Au NWs using a low-noise spectroscopy technique. We show that the noise peculiarities revealed are the result of oleylamine influence on charge transport in nanowires.

2. Results and discussion

The normalized current noise spectrum, measured for synthesized ultrathin Au NW, is shown in Fig. 1a. At least two well-resolved Lorentzian-shaped components (at 3 Hz and 58 Hz) were revealed in the noise spectra. Further studies were performed on 2÷5nm Au NW structures without oleylamine. After measurements of bare Au NWs oleylamine molecules were deposited onto the nanowires. The current noise spectral density graphs are shown in Fig. 1b. The noise follows 1/f behavior in the bare metal NWs while several Lorentzian-shaped components can be recognized when oleylamine was deposited on the structure surface. Thus the appearance of Lorentzian-shaped noise components has to be addressed in fluctuation processes due to molecules covering the nanowire surface.

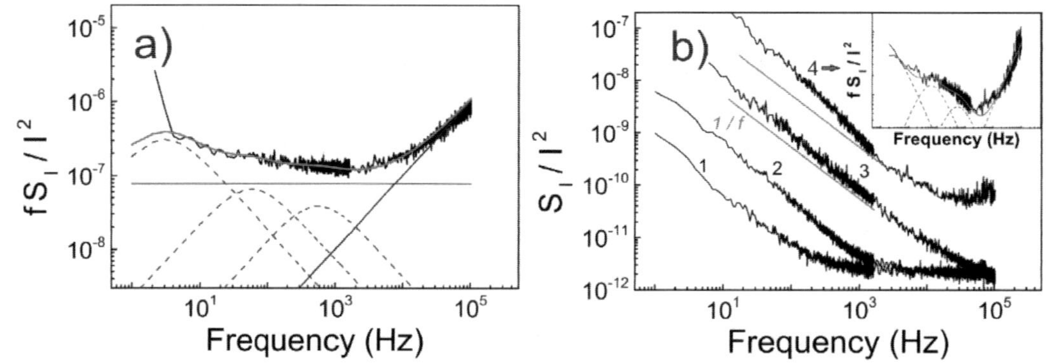

Fig. 1 a) Normalized current noise spectral density, measured for Au NWs with oleylamine molecules; b) Noise spectral density measured for bare Au NWs with diameters in the range of 2 nm to 5 nm (curves 1–3) and for 5 nm diameter Au NW with oleylamine deposited on the nanowire surface (curve 4). For clarity, curve 4 is replotted in the inset. Two well-resolved Lorentzian-shaped components at characteristic frequencies of 3 Hz and 58 Hz are related to molecule-nanowire interactions.

It should be noted that oleylamine was used to assist the growth of Au NWs. After synthesis these molecules may remain on the surface of the nanowires forming an insulating layer between NWs and contacts. Results of transmission electron microscopy confirm the formation of a uniform molecular layer of oleylamine molecules (see Fig. 2).

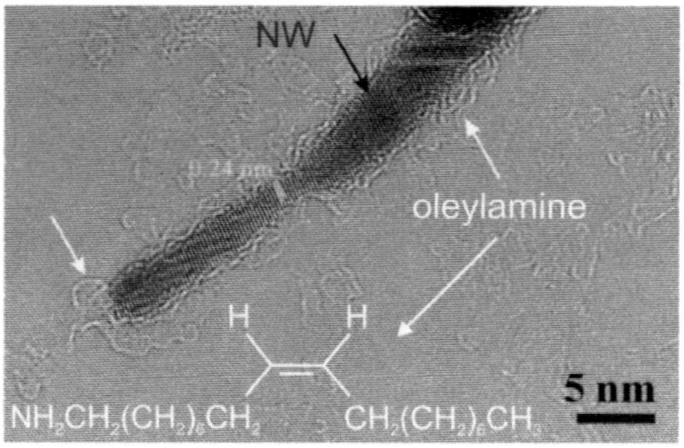

Fig. 2 High-resolution transmission electron micrograph of Au NW on a graphene sheet

The presence of such molecular interfaces influences the dynamics of charge transport in the structures.

3. Summary

To conclude, the presence of monolayers of organic molecules on the surface of nanowire structures influences their transport properties. Noise spectroscopy is a powerful method for studying these layers. The revealed peculiarities should be taken into account in designing NW-based devices for molecular electronics.

References

[1] Pud, S. et al. *Small* **9**, 846–852 (2013).
[2] Khalil, A., Singh Lalia, B., Hashaikeh, R. & Khraisheh, M. *J. Appl. Phys.* **114**, (2013)
[3] Ye, S., Rathmell, A. R., Chen, Z., Stewart, I. E. & Wiley, B. J. *Adv. Mater.* **26,** 6670–6687 (2014)

IOP Conf. Series: Journal of Physics: Conf. Series **864** (2017) 012064 doi:10.1088/1742-6596/864/1/012064

Si Interface Barrier Modification on Memristor for Brain-Inspired Computing

Wei Wu, Huaqiang Wu, Bin Gao and He Qian

Institute of Microelectronics, Tsinghua University, Beijing 100084, China

wuhq@tsinghua.edu.cn

Abstract. Memristor is an emerging technology aimed at implementing neuromorphic computing in hardware system. Resistive random access memory (RRAM) is a kind of memristor with excellent performance, but abrupt switching in the set process influences the efficiency of neuromorphic system. In this study, we present an interface switching memristor device based on TiN/Si/TaOx/TiN stack and CMOS compatible fabrication process to achieve gradually resistive switching both in set and reset processes. The devices show a more than 10 switching window. The related switching mechanism is discussed.

1. Introduction

Von Neumann architecture computer is inefficiency in parallel computing, and consumes large amount of energy in transferring data between memory and arithmetic logic unit [1]. Brain-inspired neuromorphic computing is an emerging computing paradigm proposed to solve these problems. Memristor is an emerging device aimed at implementing neuromorphic computing in hardware system. The main performance metrics of memristor are device dimensions, energy consumption, operating speed, multi-level states, dynamic range, retention, and endurance. Resistive random access memory (RRAM) is a kind of memristor which exhibits excellent performance, such as low energy consumption, ultrafast operating speed, etc. Multi-level states can be achieved in reset process. However, most of RRAM based on local filamentary switching mechanism is hard to realize gradual SET characteristics [2]. It will influence the efficiency of neuromorphic computing hardware system.

In this work, to implement gradual set in RRAM based memristor, we demonstrate an analog switching memristor with TiN/Si/TaO_x/TiN structure based on interfacial switching mechanism. It exhibits gradual switching in both set and reset process. Amorphous silicon layer and TaO_x layer with different thicknesses were fabricated by sputtering to research the structure effect on the device performance. We find 6nm Si/5nm TaO_x device presents the best performance. The device can be switched gradually between several million ohm and tens million ohm, and the dynamic resistance range is larger than 10. The 6nm Si barrier layer is used to suppress the device to device variability and improve the reliability for device operation [3]. In addition, the fabrication process of the device is CMOS compatible at room temperature. The switching mechanism is also discussed. This work will help to boost the application of RRAM based neuromorphic device for the future high efficiency brain-inspired computing.

2. Experiment details

2.1. Device fabrication

In this study, memristors with TiN/Si/TaOx/TiN stacks was fabricated to implement gradual set process in RRAM. TiN is used as both top electrode and bottom electrode. The active layer consists of Si and TaOx. The process scheme of TiN/Si/TaOx/TiN memristor is shown in Fig. 1. The TiN (40nm) bottom electrode was deposited by magnetron sputtering on 200nm SiO_2, followed by a lift-off process to pattern bottom electrode. TiN resistivity is less than 120 μΩ*cm. Then Si barrier layer and TaOx active layer were deposited by magnetron sputtering. Next, TiN (40nm) top electrode was deposited and patterned. Final, a reactive ion etching process to expose the bottom electrode. The device size includes 2 μm x 2 μm , 3 μm x 3 μm , 5 μm x 5 μm , 10 μm x 10 μm , 50 μm x 50 μm ,100 μm x 100 μm and 500 μm x 500 μm .

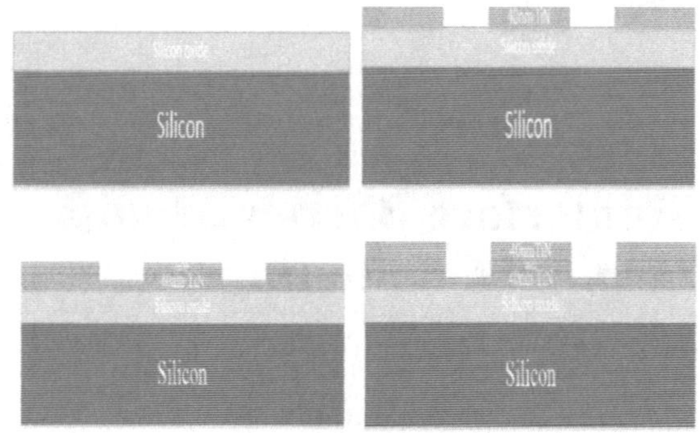

(a)

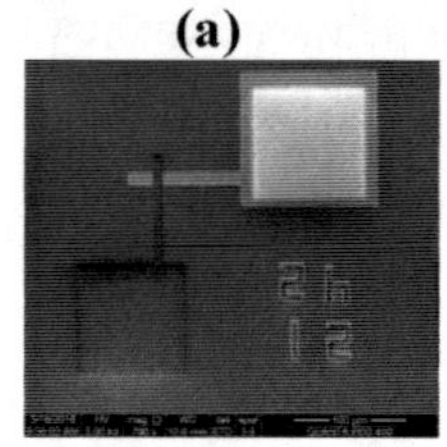

(b)

Figure 1. (a) Fabrication process of the memristor. (b) SEM photograph of the device.

2.2. Electrical characteristics

Cascade probe station and Keithley B1500 semiconductor analyzer was used to measure the device characteristics. The bottom electrode is grounded and the operation voltage is applied to top electrode. Fig. 2 shows resistive-switching I-V curves of TiN/Si/TaOx/TiN memristor measured by DC sweep. The memristor is initial high resistance state (HRS). The negative bias can gradually set device resistance from the HRS to the low resistance state (LRS). During the positive bias, the device resistance can be gradually reset form the LRS to the HRS. The on-off ratio is larger than 10x. Bipolar switching behavior is observed on this device and there is no abrupt resistance change both in set and reset process, which means this analog switching RRAM is more suitable for brain-inspired computing than traditional filamentary switching RRAM. As shown in Fig. 3, continues negative bias sweep is applied to the top electrode and the device resistance can be set to lower resistance state after each cycle, which is attributed to interface barrier modification effect[4].

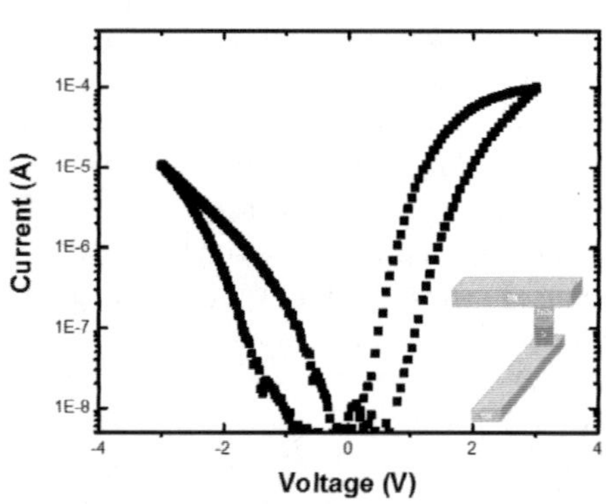

Figure 2. Typical bipolar Resistive switching I-V curves of the memristor.

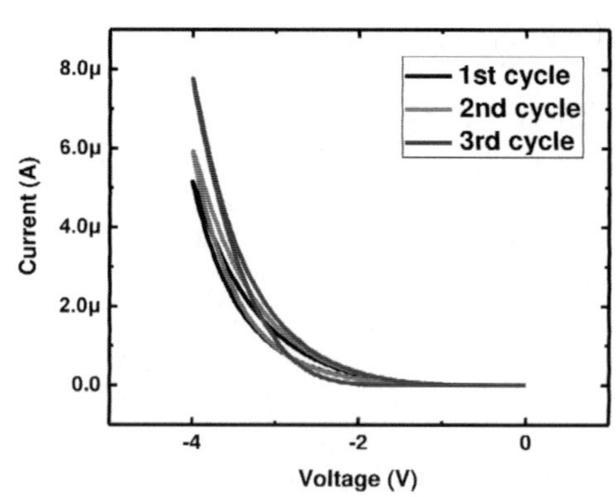

Figure 3. Resistance gradually switching characteristic in set process.

IOP Conf. Series: Journal of Physics: Conf. Series **864** (2017) 012064 doi:10.1088/1742-6596/864/1/012064

3. Results and discussion

Amorphous silicon and TaOx with different process parameters and different thicknesses were sputtered to verify the effect of structure on memristor performance. As shown in Fig. 4, the device with 8nm Si and 10nm TaOx presents analog switching but too high operation voltage. To decrease the operation voltage, a thinner Si layer is applied to the device. The operation voltage can decrease to 10V but presenting altra small on-off ratio at 1V. In order to increase the on-off ratio of the device, TaOx layer decrease to 7nm and a 2x window is observed at 1V during the reset process. Continuely dereasing the TaOx layer, the device with 6nm Si/5nm TaOx presents more than 10x on-off ratio at 1V and less than 3V operation voltage. So decreaseing the silicon layer and TaOx can both decrease operation voltage and decreasing TaOx can increase on-off ratio obviously. Due to the limitation of sputter equipment, it is hard to control the thinkness of material under 5nm. We find the device with 6nm Si/5nm TaOx presents the best performance.

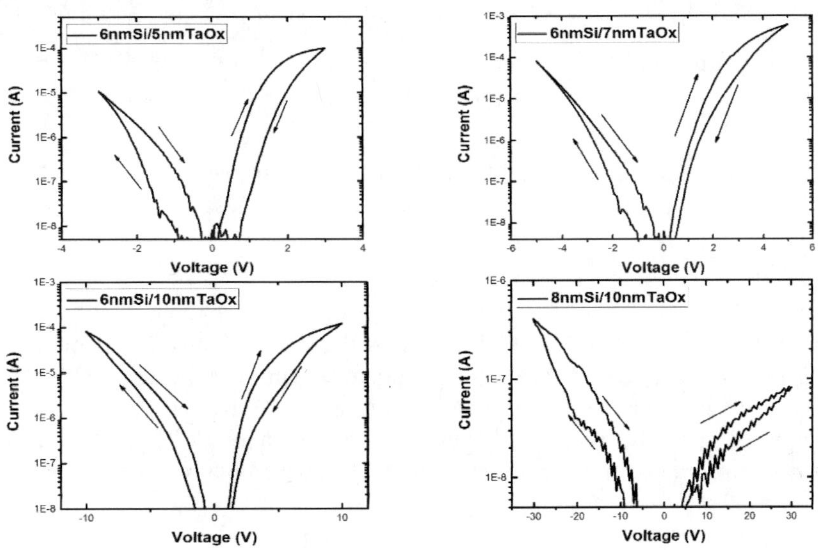

Figure 4. I-V curves of the memristors with different structures.

To elucidate the switching mechanism, devices with different sizes were measured under the same condition (Fig. 5). 100x100um^2 and 50x50um^2 were operated with self-compliance, and the self-limited current increases with device size, which proves the switching mechanism takes place over all the active area. It is different form the filamentary switching machanism, which just takes place over the filament region. Compared with the filamentary based RRAM, device resistance in set process is more controllable. Another advantage of the analog switching RRAM is that the device resistance will scalling with area, which is extremely important for lower power comsumption application in neuromorphic computing.

All TaOx-based RRAM are related to filamentary formation[4] or schottky barrier modification[5]. In this device, it is believed that the device formed a interface layer between Si and TaOx during the process[3]. During the set process, a negtive voltage is applied to the top electrode and oxygen vacancy (Vo) move to the interface layer. The interface barrier will be lower because of a large amount of Vo. During the reset process, a positive voltage is applied to top electrode and the oxygen vacancy will move back to Si. So the interface barrier will be higher and the device resistance can be reset from LRS to HRS (shown in fig. 6).

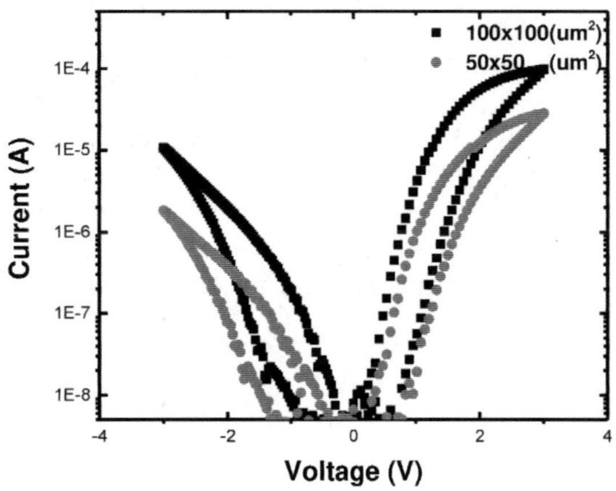

Figure 5. I-V curves of the memristor with different size.

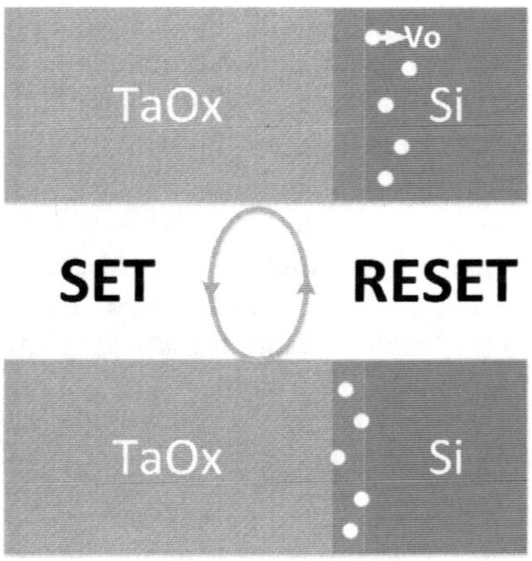

Figure 6. Schematics of switching process

4. Conclusion

A TiN/Si/TaOx/TiN memristor has been fabricated and characterized. We demonstrate a interface switching memristor, with simple structure, shows gradual set and reset process. The device can be used for implementing brain-inspired neuromorphic computing in hardware system. It provides superior performance compared with the typical filamentary RRAM in gradully switching and low power application. It will help to boost the application of RRAM based neuromorphic device for the future high efficiency brain-inspired computing.

References

[1] Burr G. W., Narayanan P., Shelby R. M., Sidler S. 2015 Large-scale neural networks implemented with non-volatile memory as the synaptic weight element: Comparative performance analysis (accuracy, speed, and power). *IEEE International Electron Devices Meeting. IEEE.*

[2] Yu S., Gao B., Fang Z., Yu H., Kang J., Wong, H. - . P. 2013 A low energy oxide-based electronic synaptic device for neuromorphic visual systems with tolerance to device variation. *Advanced Materials,* **25**(12), 1774-9.

[3] Govoreanu B., Crotti D., Subhechha S., Zhang L. 2015 A-vmco: a novel forming-free, self-rectifying, analog memory cell with low-current operation, nonfilamentary switching and excellent variability.

[4] Huang X., Wu H., Sekar D. C., Nguyen S. N. 2015 Optimization of TiN/TaOx/HfO2/TiN RRAM Arrays for Improved Switching and Data Retention. *Memory Workshop. IEEE.*

[5] Wang Y. F., Lin Y. C., Wang I. Lin, T. P., Hou T. H. 2015 Characterization and modeling of nonfilamentary ta/taox/tio2/ti analog synaptic device. *Scientific Reports,* 5.

IOP Conf. Series: Journal of Physics: Conf. Series **864** (2017) 012065 doi:10.1088/1742-6596/864/1/012065

Efficient hybrid solar cell with P3HT:PCBM and Cu_2ZnSnS_4 nanocrystals

Se-Jeong Jang, Nhu Thuy Ho and Yong Soo Kim*

Department of Physics and Energy Harvest-Storage Research Center, University of Ulsan, Ulsan 44610, South Korea

*Email: yskim2@ulsan.ac.kr

Abstract. Recently, Cu_2ZnSnS_4 (CZTS) with band gap about 1.50 eV is predicted to become an ideal light absorption material due to the abundant component elements in the crust being nontoxic and environmentally friendly. In this study, we propose a hybrid solar configurations with solution-processed CZTS nanocrystals and P3HT:PCBM bulk heterojunction. The forming double heterojunction, as charge can be separated at both the P3HT:PCBM and CZTS:PCBM interface is attributed to enhance the light harvesting efficiency. As a result, organic solar cells with CZTS nanocrystals show the higher efficiency 3.32 % compare to 2.65 % of reference organic solar cells. Improvement of the power conversion efficiency of 25% is obtained by the increase of the short-circuit current and fill factor.

1. Introduction

Recently, the quaternary semiconductor Cu_2ZnSnS_4 (CZTS) has received considerable attention as a material capable of driving the development of low-cost and high performance photovoltaics [1]. Typically, CZTS is used as active material of CZTS/CdS based thin film solar cell (SC). Although the high-efficiency ($\eta \approx 10$ %) is obtained, the containing high toxic cadmium (Cd) and using vacuum deposition method leads to an increase in the prices of products [2,3]. Hence, many efforts have been made to develop new CZTS based SCs with non-toxic materials and cost competitive. Alternative processing strategies including electrochemical and chemical-solution synthesis and deposition have got some reasonable success.[4,5] A heterojunction with CZTS Nanostructures (NSs) and [6,6]-phenyl-C61-butyric-acid-methyl-ester (PCBM) of efficiency of 0.90 % has been reported by Saha *et. al* [2]. Even the η of SCs using CZTS NSs are lower than expectation, the CZTS NSs with controlled phase formation still reveals a great potential for devising low-cost and green SCs.

In this paper, the hybrid SC with CZTS NSs and the blend of poly(3-hexylthiophene) (P3HT) and PCBM is reported. CZTS NSs were synthesized by solution-processed and simply deposited on substrate by spin coating method. The interpenetrating networks where the CZTS nanocrystals (NCs) as an additional absorber within the blended P3HT:PCBM is expected to improve the light harvesting and thus enhance the efficiency of fabricated SCs.

2. Experiment

2.1. Synthesis of CZTS nanocrystals and Device Fabrication

IOP Conf. Series: Journal of Physics: Conf. Series **864** (2017) 012065 doi:10.1088/1742-6596/864/1/012065

The CZTS NCs were made by sol-gel method [4]. The precursors for CZTS including oleylamine (OLA), copper (II) acetylacetonate ($Cu(acac)_2$), zinc(II) acetate ($Zn(CH_3COO)_2$), tin(II) chloride dehydrate ($SnCl_2.2H_2O$), sulfur (S) were used as received from Sigma Aldrich. The CZTS NCs are dispersed in methanol (10 wt%). The ZnO sol-gel solution was prepared by dissolving zinc acetate [$Zn(CH_3COO)_2 \cdot 2H_2O$] in 2-methoxyethanol (Sigma-Aldrich Co.) and ethanolamine. P3HT, PCBM (OSM company) were used as received. The inverted BHJ SCs were fabricated with a structure of ITO/ ZnO/ CZTS NCs/ P3HT:PCBM/ MoO_3/ Ag. The sol-gel solution of 0.75 M ZnO was spin coated on ITO surface. CZTS nanocrystals were deposited on ZnO layer by spin-coating method with different spin speed and annealed at 100°C for 10 min. Blended solutions of P3HT:PC61BM was then spin-coated. Finally, MoO_3 (5 nm) and Ag (100 nm) was thermally evaporated through shadow mask with defined area of 0.13 cm^2.

2.2. Characterization and measurement

The crystalline structure of CZTS NSs was analysed by the X-ray diffraction (XRD) technique. Atomic force microscopy (AFM) image was obtained using a Seiko E-Sweep atomic microscope in tapping mode. Keithley 2401 source meter was used to measure the current density-voltage (*J-V*) curves of the SCs under AM 1.5 simulated illumination. Impedance measurement was carried out at room temperature using impedance spectroscopy (IviumStat, Ivium Tech) in the frequency range from 1 MHz down to 1 Hz, and its results were fitted by the Z-view program. The external quantum efficiency (EQE) spectra of devices were obtained using a solar cell spectral response/QE/IPCE measurement system (Newport Co., Oriel IQE-200).

3. Discussion

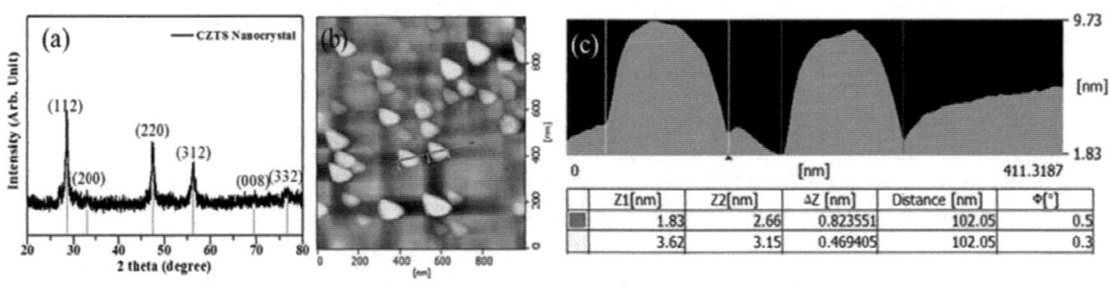

Figure 1. (a) X-ray diffraction, (b) AFM image of CZTS NPs (scan area: 1 μm × 1 μm), and (c) height profile along the white line in the AFM image.

Fig. 1(a) showed an XRD spectrum of the as-synthesized CZTS NCs. The diffraction peaks presented kesterite-type tetragonal phase (JCPDS No. 26–0575) [2, 4]. It reveals that all the CZTS has good crystallinity. AFM image of CZTS NCs on Si wafer in Fig. 1(b) show that those were fully diluted in methanol and retained its monodisperse morphology during the mixing and spin coating process. The density of NCs on substrate could be easily controlled by spin speed during the deposition process. The cross-section AFM image verified the flake-like structure of CZTS NCs that have diameter of 100 nm and thickness of 10 nm (Fig. 1 (c)).

Fig. 2(a) and (c) present schematic and band alignment of the reported device. The flake-like CZTS NCs were first spin coated on the electron transport layer (ZnO). Then the blended of P3HT:PCBM solution was following dropped and spin coated on the top of it to form an interpenetrating network. Therefore, the active layer composed of a CZTS NCs and bulk-heterojunction polymer that filled the vacancy or pores among the CZTS NCs. This structure was attributed to form double heterojunction,

IOP Conf. Series: Journal of Physics: Conf. Series **864** (2017) 012065 doi:10.1088/1742-6596/864/1/012065

as charge can be separated at both the P3HT:PCBM and CZTS:PCBM interface. A similar structure of ZnO/CZTS/P3HT:PCBM was fabricated on SiO_2 substrate for FE-SEM measurement. Fig. 2(b) shows the cross-section image of reported device with the presents of CZTS NCs in the active layer. The P3HT:PCBM polymer covered on the top and filled vacancies among CZTS NCs.

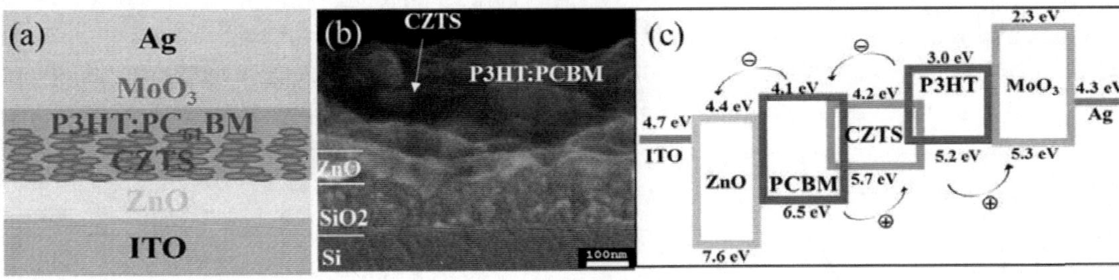

Figure 2. (a) Configuration of hybrid solar cell, (b) cross section FE-SEM image of device, and (c) systematical energy level diagram.

The *J–V* characterizations of the devices with and without CZTS NCs were shown in Fig. 3(a). The density of the CZTS NCs in the devices was varied by speed of spin coating process. The parameters of reported solar cells were shown in Table 1. Without using CZTS NCs, the reference sample exhibited the current density (J_{sc}) of 7.21 mA/cm^2, open-circuit voltage (V_{oc}) of 0.59 V, fill factor (*FF*) of 0.62 and the efficiency (η) of 2.64 %. With the present of the CZTS NCs at the spin speed of 400 rpm, the η of device was decreased. Poor homogeneous packing of CZT NCs with high density could restrict the polymer covering. When the spin speed of 800 rpm was applied, the J_{sc} of sample is dramatically increased leading to improvement in efficiency.

Table 1. Hybrid SC performance parameters according to CZTS NCs spin coating speed

CZTS Spin speed	J_{sc} (mA/cm^2)	V_{oc} (V)	*FF* (%)	η (%)	R_s (Ω)	R_{sh} (Ω)
Reference	7.21	0.59	62	2.64	8.32	4.37
400 rpm	6.52	0.59	64	2.46	7.19	5.11
600 rpm	7.62	0.59	65	2.92	7.87	5.97
800 rpm	8.28	0.59	68	3.32	7.36	7.49
1000 rpm	8.06	0.59	61	2.90	8.32	5.77
1200 rpm	7.61	0.59	61	2.73	8.24	3.97

The best performance was archived with J_{sc} of 8.28 mA/cm^2, V_{oc} of 0.59 V, *FF* of 0.68 and η of 3.32 %. The increase of J_{sc} and *FF* boost the efficiency rising nearly 25 % in compare to that of reference sample. While the J_{sc} was based on the generation and collection of light-generated carriers, deviations of *FF* could be explained by the parasitic loss mechanisms of series (R_s) and shunt resistance (R_{sh}) in Table 1. Then the device performance became worse when the spin coating speed was increased over 800 rpm. It is notable that EQE spectra shown in Fig. 3 (b) was improved in long wavelength (>500 nm) region. This was attributed to the enhancement of light absorption by CZTS NCs (E_g of 1.5 eV) and generation of electron-hole pairs at CZTS:PCBM junction.

Nyquist plots of impedance data and simulated spectra are shown in Fig. 3(c). Impedance responses of SCs were modelled by an equivalent circuit consisting of a series resistance R_c accounted for connection resistances, *R* elements account for the junction resistance and *C* presents for cell capacitance. From the impedance spectra, it clearly depicts that the semicircle's diameter for reference device is much larger than that of the CZTS device. In sample without CZTS, the simulated spectra showed R_c of 20 Ω, R of 613.6 Ω, C of 13.1 nF. When device was fabricated with CZTS nanocrystals, the simulated data of device showed R_c of 15 Ω, R of 64.9 Ω, and C of 15.1 nF. It is interesting to find that with the CZTS NCs the junction R of device in complex impedance plane is apparently decreased than that of reference sample [6]. Furthermore, the excessed electrons and holes in conduction and valance bands separation might cause slightly increased capacitance [7]. In other words, the CZTS NC not only enhance the light harvesting but also serve as extra charge transport path for photo-generated carriers, thus leading to improve the charge transfer and decreased energy loss.

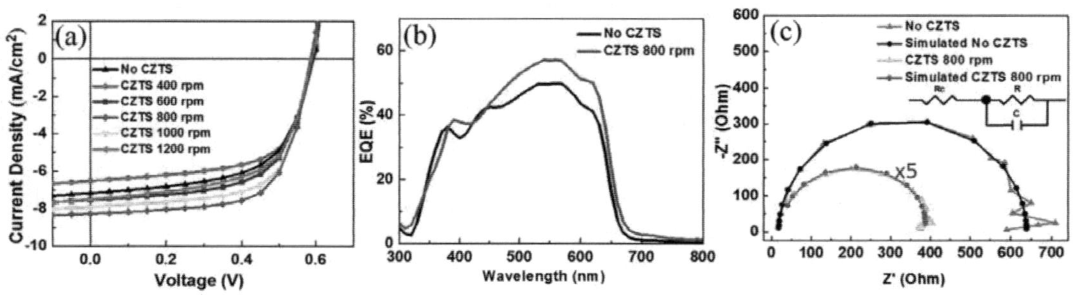

Figure 3. (a) *J-V* characteristics of reported devices, (b) external quantum efficiency spectra of solar cells, and (c) Nyquist presentation of impedance data and equivalent circuit for curve simulation.

4. Conclusion

In this report, we demonstrated a novel hybrid solar cell model using the earth-abundant, non-toxic materials and solution-processed CZTS/P3HT:PCBM. The optical and electrical effect of CZTS nanocrystals on performance of devices have been carefully demonstrate through *J-V*, EQE and impedance measurement. The present of CZTS nanocrystals was attributed to serve as an additional absorber and form extra charge transport path within the photoactive layer. Consequently, optimum device shows η of 3.32 %, 25 % higher than efficiency of reference device without CZTS nanocrystals. To the best of our knowledge, this is one of the highest efficiency of hybrid solar cells based on CZTS material

Acknowledgements

This research was supported by the "Leaders INdustry-university Cooperation" Project, supported by the Ministry of Education (MOE).

References

[1] Lai Y, Liu F and Su Z 2013 *Int. Photonics and Optoelectronics Meetings* (Wuhan: China) p ASu3A.2
[2] Saha S K, Guchhait A and Pal A J 2012 *Phys. Chem. Chem. Phy.* **14** 8090
[3] Barkhouse D A R, Gunawan O, Gokmen T, Todorov T K and Mitzi D B 2012 *Prog. Photovoltaics* **20** 6
[4] Steinhagen C, Panthani M G, Akhavan V, Goodfellow B, Koo B and Korgel B A 2009 *J.Am. Chem. Soc.* **131** 12554
[5] Wang Y, Li C, Yin X, Wang H and Gong H 2013 *ECS J. Sol. State Sc.* **2** Q95
[6] Li Z, Li S, Zhang Z, Zhang X, Li J, Liu C, Shen L, Guo W and Ruan S 2016 *Phys. Chem. Chem. Phy.* **18** 11435
[7] Suman C K, Seunguk N, Seohee K, Sin-Doo L, Changhee L, Donggu L and Jeongho P 2008 *J. Korean Phys. Soc.* **53** 3278

Effect of pentacene/Ag anode buffer and UV-ozone treatment on durability of small-molecule organic solar cells

S Inagaki, S Sueoka and K Harafuji

Department of Electrical and Electronic Engineering, Ritsumeikan University, Kusatsu, Shiga 525-8577, Japan

Abstract. Three surface modifications of indium tin oxide (ITO) are experimentally investigated to improve the performance of small-molecule organic solar cells (OSCs) with an ITO/anode buffer layer (ABL)/copper phthalocyanine (CuPc)/fullerene/bathocuproine/Ag structure. An ultrathin Ag ABL and ultraviolet (UV)-ozone treatment of ITO independently improve the durability of OSCs against illumination stress. The thin pentacene ABL provides good ohmic contact between the ITO and the CuPc layer, thereby producing a large short-circuit current. The combined use of the abovementioned three modifications collectively achieves both better initial performance and durability against illumination stress.

1. Introduction

Organic solar cells (OSCs) are considered to be promising candidates for renewable energy applications because they possess several desirable properties such as flexibility, low cost, large size, and ease of fabrication. OSCs, however, still have two major disadvantages: low power-conversion efficiency (η_p) and a short lifetime. Several degradation mechanisms that contribute to the short lifetime of OSCs have been reported. In the course of OSC degradation, the current–voltage (*J–V*) characteristics frequently become distorted from an exponential diode curve to a curve with an S-shaped kink in the positive-bias region. This leads to decreases in the fill factor (*FF*) and in η_p [1]. Several theoretical models have been proposed to explain the mechanisms responsible for the S-shaped kink in the curve, e.g., charge accumulation induced by the cathode buffer layer [2] and charge blocking originating from the presence of strong interface dipoles [3].

The effects of three types of stress, i.e., illumination, air exposure, and electrical stress, on device performances have been systematically examined in CuPc/fullerene (C_{60}) OSCs [4]. Understanding the degradation mechanisms resulting from illumination stress is of primary importance because OSCs generate power by receiving light energy. In pentacene/C_{60} OSCs exposed to constant solar illumination, differences in the temporal changes in the *J–V* characteristics are clearly present between devices kept in air, under vacuum, and under vacuum with an ultraviolet (UV) filter [5]. Under UV radiation, reorganization of pentacene near the indium tin oxide (ITO) substrate is considered to generate a gap that leads to a reduction in contact quality between ITO and pentacene. The interface conditions between ITO and the donor layer certainly play a decisive role in achieving durability under illumination.

One promising method for improving durability is the introduction of a thin layer of inorganic and organic materials between the ITO anode and the donor layer, called an anode buffer layer (ABL). The stability of OSCs under irradiation by light is reported to be markedly enhanced by insertion of a

IOP Conf. Series: Journal of Physics: Conf. Series **864** (2017) 012066 doi:10.1088/1742-6596/864/1/012066

MoO_3 ABL [6]. Organic ABLs also play various roles in OSC improvement, such as modification of the molecular orientation of donor, which enhances charge transport, adjustment of the work function (WF), and reduction of the potential-barrier height of hole transport. The S-shaped kink is suppressed under illumination upon insertion of a set of ultrathin Ag and thin pentacene layers [1]. Furthermore, UV-ozone treatment of the ITO surface is a useful technique for optimizing the surface conditions. This treatment eliminates organic contaminants, tunes the anode WF, and increases the hole-transport efficiency [7].

Although several surface modifications of ITO have been proposed to improve OSC performance, only few systematic studies have investigated the durability of OSCs against illumination. This study investigates the effects of two types of ABLs: ultrathin Ag and thin pentacene layers. Moreover, against illumination, the effect of UV-ozone treatment on the durability of small-molecule OSCs with an ITO/anode buffer/CuPc/C_{60}/bathocuproine (BCP)/Ag structure is investigated.

2. Experimental method

Figure 1 shows the schematic structures of three types of organic solar cells: Devices A–C. All three devices comprise a 20-nm-thick CuPc layer as the donor, a 40-nm-thick C_{60} layer as the acceptor, and a 10-nm-thick BCP layer as the cathode buffer layer. Device A is a standard device. Device B has an additional ABL of a 3-nm-thick pentacene layer. Device C has an ABL composed of a 0.5-nm-thick Ag layer and 3-nm-thick pentacene layer. CuPc is purified three times via thermal gradient sublimation before use. The other organic materials are used as-purchased.

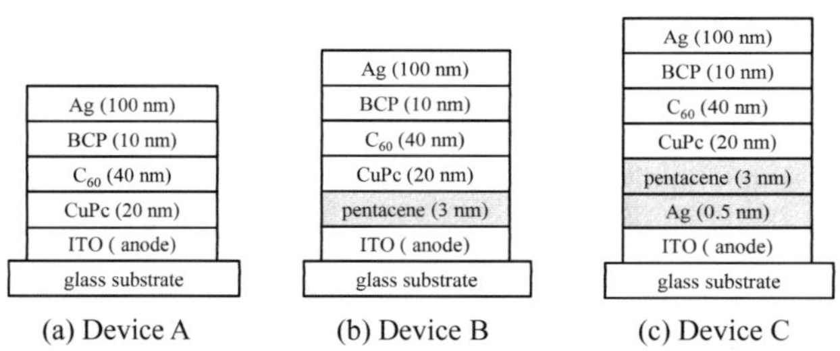

Figure 1. Schematic structures of three types of organic solar cells.

An ITO substrate with a thickness of 1600 Å is cleaned three times using ultrasonic treatment in detergent, deionized water, acetone, and ethanol, in this order, for 5 min each. Subsequently, a 10-min UV-ozone treatment is performed on the ITO surface with an intensity of 13.5 mW/cm^2. Organic materials comprising CuPc, C_{60}, and BCP are evaporated onto the ITO substrate at a rate of 0.5 Å/s under a pressure of less than 1.0×10^{-3} Pa. Pentacene is evaporated at a rate of 0.2 Å/s, the Ag cathode material is evaporated at a rate of 1.0 Å/s, and the ultrathin Ag anode buffer is evaporated onto the ITO substrate at a rate of 0.1 Å/s. The top Ag electrode is thermally evaporated via a 6 mm^2 stainless-steel mask with an active area of 2.4×2.5 mm^2.

The *J–V* measurements are performed using xenon lamp illumination at an intensity of 100 mW/cm^2 and an air mass 1.5 global spectrum (AM1.5G). All measurements are performed in air at room temperature. Repetitive illumination through the bottom ITO electrode is imposed on the device. Repetitive illumination refers to 100 cycles, each comprising 3 s of illumination and subsequent darkness maintained for 12 s. The repetition period, T, is 15 s. Device parameters are monitored as a function of the number of illumination cycles.

3. Experimental results and discussion

Table 1 summarizes the experimental results for the initial device parameters. Here, J_{sc} is the short-circuit current density and V_{oc} is the open-circuit voltage. Devices B and C with pentacene as an anode buffer have larger values of J_{sc} compared with Device A with no pentacene; this results in an increase in η_p. The effect of UV-ozone treatment on the initial performance is unclear.

Table 1. Initial device parameters for Devices A–C with and without UV-ozone treatment.

Device	UV exposure time (min)	J_{sc} (mA/cm^2)	V_{oc} (V)	*FF*	η_p (%)
A	0	4.2	0.44	0.59	1.1
	10	4.2	0.44	0.57	1.0
B	0	4.8	0.46	0.58	1.3
	10	4.5	0.45	0.59	1.2
C	0	4.4	0.46	0.56	1.2
	10	4.4	0.43	0.54	1.0

Figures 2(a) and 2(b) show the *J–V* dynamics for the 1st, 50th, and 100th cycles and changes in J_{sc}, V_{oc}, *FF*, and η_p for Device A without UV-ozone treatment, respectively. The *J–V* curve becomes sharply S-shaped during the 100th cycle. In addition, the decrease in the gradient of the *J–V* curve in the first and fourth quadrants is substantial. This decreasing gradient indicates an increase in series resistance, R_s, at the ITO/CuPc interface. With increasing number of cycles, η_p decreases significantly, accompanied by a large decrease in *FF*. After the 100th cycle, *FF* and η_p decrease by 40% and 50%, respectively, relative to their initial values, while J_{sc} and V_{oc} decrease by only 8% and 9%, respectively.

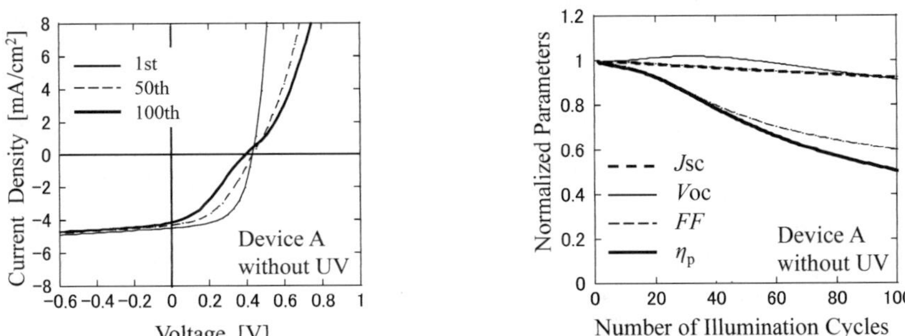

Figure 2. Experimental results for Device A without UV-ozone treatment. (a) *J–V* dynamics and (b) changes in J_{sc}, V_{oc}, *FF*, and η_p as a function of the number of illumination cycles.

Figures 3(a)–(c) depict the *J–V* dynamics for Device B without and with UV-ozone treatment, and Device C without UV-ozone treatment, respectively. The *J–V* curve for Device B without UV-ozone treatment becomes sharply S-shaped during the 100th cycle. Conversely, the *J–V* curve for Device B with UV-ozone treatment does not show the S-shaped kink; however, the gradient of the *J–V* curve in the first and fourth quadrants decreases slightly with increasing number of cycles. The *J–V* curve for the first cycle for Device C without UV treatment is almost coincident with that for the 100th cycle; i.e., only slight degradation occurs after 100 cycles.

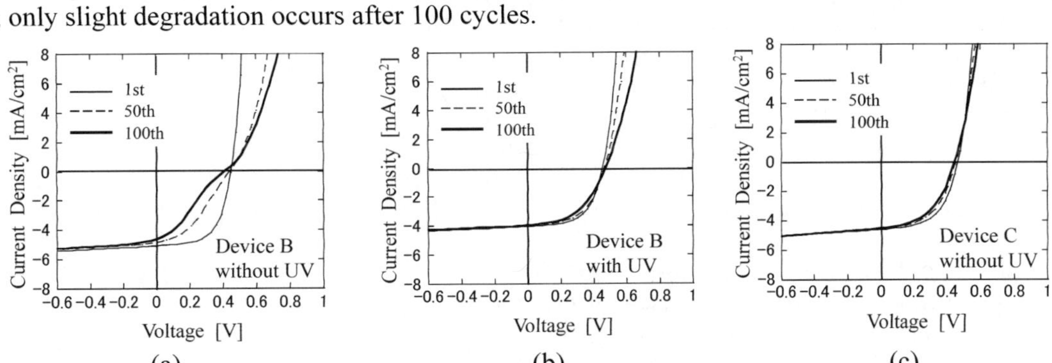

Figure 3. Experimental results of *J–V* dynamics. (a) Device B without UV, (b) Device B with UV, and (c) Device C without UV.

Figure 4 shows the experimental results for Device C with UV-ozone treatment, with the *J–V* dynamics depicted in Figure 4(a). The *J–V* curve for the first cycle is almost completely coincident with that for the 100th cycle. Figure 4(b) shows the changes in J_{sc}, V_{oc}, *FF*, and η_p. The graph demonstrates that η_p decreases by only 7% at the 100th cycle, accompanied by a slight decrease in V_{oc} and *FF*. Moreover, J_{sc} remains very close to its initial value even during the 100th cycle.

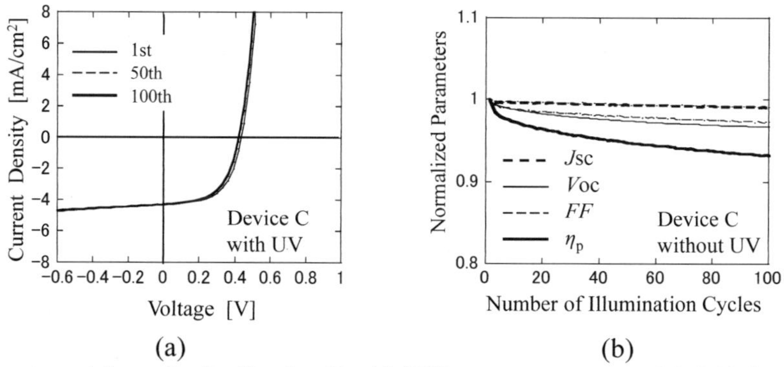

Figure 4. Experimental results for Device C with UV-ozone treatment. (a) *J–V* dynamics, and (b) changes in J_{sc}, V_{oc}, *FF*, and η_p as a function of the number of cycles.

A separate atomic force microscopy analysis of a 2-nm-thick CuPc on ITO indicates that the heat generated by the UV component of illumination possibly causes a change in the CuPc crystal structure near the ITO/CuPc interface; this deteriorates the quality of the electrical contact. The present study demonstrates that two types of surface modifications on ITO, i.e., an ultrathin Ag layer and UV-ozone treatment, independently and collectively improve the durability of OSCs against illumination stress. The thin pentacene layer itself does not lead to any durability improvement but its combined use with the Ag layer and UV-ozone treatment yields better initial device performance. Using scanning electron microscopy, we separately observed that the Ag layer was discretely deposited onto ITO. Resonant light absorption in Ag clusters may accumulate trapped electrons near the interface, reduce the barrier height of hole transport, and suppress the S-shaped kink. The UV-ozone treatment creates hydrophilic conditions with the OH functional groups on ITO [8]. The attraction between H atoms from the OH functional groups and π electrons from the CuPc molecules might prevent disorder in the CuPc crystals during illumination.

4. Conclusion

The thin pentacene layer on ITO provides good ohmic contact at the ITO/CuPc interface, thereby resulting in a large short-circuit current. An ultrathin Ag layer and UV-ozone treatment of ITO independently and collectively improve the durability against illumination stress.

References

[1] Oida T and Harafuji K 2013 *Japan. J. Appl. Phys.* **52** 011601
[2] Wang J C, Ren X C, Shi S Q, Leung C W, Chan P K L 2011 *Org. Electron.* **12** 880
[3] Kumar A, Sista S, Yang Y 2009 *J. Appl. Phys.* **105** 094512
[4] Harafuji K, Sato H, Matsuura T, Omoto Y, Kaji T, Hiramoto M 2014 *Japan. J. Appl. Phys.* **53** 122303
[5] Sullivan P, Jones T S 2008 *Org. Electron.* **9** 656
[6] Kanai Y, Matsushima T, Murata H 2009 *Thin Solid Films* **518** 537
[7] Sugiyama K, Ishii H, Ouchi Y, Seki K 2000 *J. Appl. Phys.* **87** 295
[8] Farina F E and Harafuji K 2015 *KJF Int. Conf. on Organic Materials for Electronics and Photonics (Jeju)*.

Energy transfer to enhance the characteristics emission from lanthanide-doped nanocrystals

Yangqing Wu, Shaobing Lin, Xiaowei Zhang, Jun Xu*, Kunji Chen
School of Electronic Science and Engineering, National Laboratory of Solid State Microstructures, Collaborative Innovation Centre of Advanced Microstructures, Nanjing University, Nanjing, 210093, China.

*E-mail: junxu@nju.edu.cn

Abstract. We report the improved characteristic emission of Er^{3+} ions by co-doping wide-band-gap SnO_2 nanocrystals (NCs), which as sensitizers of Er^{3+} ions. It is found that the Er^{3+}-related near infrared emission is enhanced by three orders of magnitude, which can be attributed to the effective energy transfer process occurs between SnO_2 nanocrystals and Er^{3+} ions. Meanwhile, we also prepare the Er^{3+}-Yb^{3+} co-doped $NaYF_4$ NCs with Gd^{3+} dopant ions to obtain high-efficiency up-conversion emission by energy transfer from Yb^{3+} to Er^{3+}. We demonstrate experimentally that Gd^{3+} dopant ions play key roles in the formation of β-$NaYF_4$ NCs which can improve the energy transfer efficiency from Yb^{3+} to Er^{3+} obviously. Consequently, the up-conversion emission is enhanced by more than five times.

1. Introduction

Recently, the studies on down- and up-conversion process in Lanthanide-doped nanomaterials have attracted much attention since it can be used to improve the performance of many kinds of optoelectronic devices [1,2]. However, due to small absorption cross sections and small radiative decay rates of trivalent lanthanide ions, the emission efficiency in lanthanide-doped nanomaterials is still low, which prevents their practical applications. To increase the absorption of Lanthanide-doped doped nanocrystals (NCs), the host materials are often additionally doped with strongly absorbing ions (Yb^{3+}) or wider band gap materials (ZnO NCs, In_2O_3 NCs) called sensitizers, which should also ensure efficient energy transfer to the activator.

In our previous work, the near-infrared (1.54 um) emission intensity of Er^{3+} ions under 325nm excitation was enhanced by 100 times by co-doping 3 nm In_2O_3 NCs as sensitizers[3]. Here, we report the improved down-shift emission of Er^{3+} ions by co-doping wide-band-gap SnO_2 NCs, which as sensitizers of Er^{3+} ions[4]. Meanwhile, we also prepare the Er^{3+}-Yb^{3+} co-doped $NaYF_4$ NCs with Gd^{3+} dopant ions to obtain high-efficiency up-conversion emission by energy transfer from Yb^{3+} to Er^{3+}. We demonstrate experimentally that Gd^{3+} dopant ions play key roles in the formation of β-$NaYF_4$ NCs which can improve the energy transfer efficiency from Yb^{3+} to Er^{3+} obviously.

2. Experimental

A sol-gel process was used to fabricate the SiO_2 thin films co-doped with Er^{3+} ions and SnO_2 NCs, as we previously reported[4].The $NaYF_4$:Yb,Er NCs were synthesized by a hydrothermal method. Briefly, 1.5 g NaOH was dis-solved in 7.5 mL deionized water, then mixed with 25 mL of ethanol and 25 mL of oleic acid under stirring. 10 mL 0.2 M $RE(NO_3)_3$ and 5 mL 2 M NH_4F were added to the resulting mixture. The solution was transferred into a 100 mL of Teflon-lined autoclave and heated at 180℃ for 2 h. The NCs were washed with water and ethanol several times and collected by centrifugation.

Finally, the NCs were dry at 60℃ for 12 h and re-dispersed into ethanol to form an aqueous dispersion and were ready for use.

3. Results and discussion

The amount of Er^{3+} ions added into SiO_2 thin films was fixed at 5% (molar ratio). The SnO_2 NCs are about 4-6 nm as shown in inset of figure 1. Figure 1 shows the photoluminescence (PL) spectra of the SiO_2 thin films doped with SnO_2 NCs and Er^{3+}ions. Sn-free and 20% Sn films show the same peak at 1540 nm under laser excitation at 325 nm. After co-doping of SnO_2 NCs, the down-shift emission of 20% Sn sample is more than three orders of magnitude to the Sn-free sample due to the energy transfer process occurs between SnO_2 NCs and Er^{3+} ions. The wider band gap SnO_2 NCs have large absorption cross-sections to incident UV-light and can effectively improve the density of sensitizing Er^{3+} ions, then improve the characteristic emission of Er^{3+} ions.

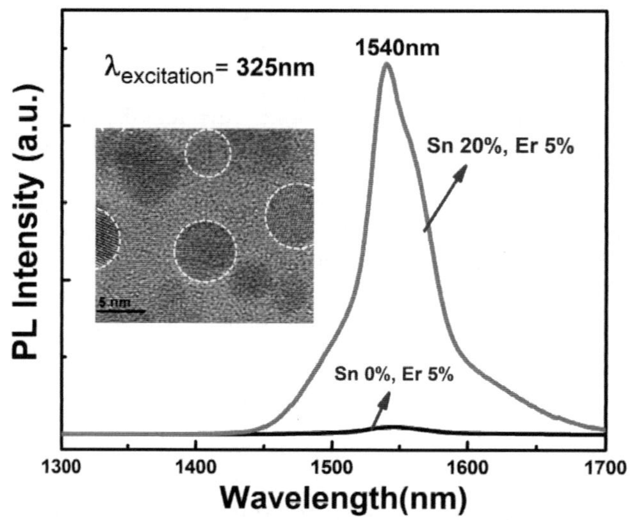

Figure 1. PL intensities at 1540 nm for 5 mol% Er^{3+} ions and 20 mol% Sn co-doped silica films excited at 325 nm (black line, multiplied by a factor of 1000 to allow the comparison); inset is high-resolution TEM image of SnO_2 NCs doped in SiO_2 thin films.

Figure 2(a) shows up-conversion PL intensities of $NaYF_4$:Yb,Er NCs for Gd0 (0 mol% Gd^{3+} doped ions) and Gd15 (15 mol% Gd^{3+} doped ions) samples excited at 980 nm with a laser power of 656 mW. The electron of Yb^{3+} is excited at 980 nm in $NaYF_4$:Yb,Er NCs and then it can transfer the energy to the Er^{3+}. As shown in Figure 2(a), there are three major emission bands at 521nm (green light), 539 nm (green light), and 654 nm (red light), which are assigned to the $^2H_{11/2}$ to $^4I1_{5/2}$, $^4S_{3/2}$ to $^4I_{15/2}$, and $^4F_{9/2}$ to $^4I1_{5/2}$ transitions of Er^{3+} ions, respectively. Figure 2(a) shows the up-conversion luminescence intensities of Gd15 sample clearly stronger than Gd0 sample due to a small residual amount of β-$NaYF_4$ phase appearing, as a well consistent with the presence of two phases observed by XRD in Figure 2(b). As shown in Figure 2(b), the diffraction peaks of the Gd0 sample agree well with those calculated for the α-$NaYF_4$ phase (JCPDS card, No.77-2042). However, there was a small residual amount of β-$NaYF_4$ phase (JCPDS card, No.16-0334) of Gd15 sample. In Figure 2(b) inset TEM images, we can see a few larger $NaYF_4$ rods of Gd15 sample, which confirm β-$NaYF_4$ phase is formed. And we see no rods in Gd0 sample. We conclude that Gd^{3+} dopant ions play key roles in the formation of β-$NaYF_4$ NCs. It can be partly attributed to the crystal structure of β-$NaYF_4$ phase consists of an ordered array of F^{2-} ions with two types of relatively low-symmetry cation sites selectively occupied by Na^+ and RE^{3+} ions, resulting in significant electron cloud distortion of the cations to accommodate the structural change. Importantly, light lanthanides with large ionic radii (Gd^{3+}) exhibit a high tendency towards electron cloud distortion owing to increased dipole polarizability, and thus favour the hexagonal structures[5]. The Gd^{3+} ions induce phase changed which may improve the energy transfer efficiency from Yb^{3+} to Er^{3+}, then improve the up-conversion emission.

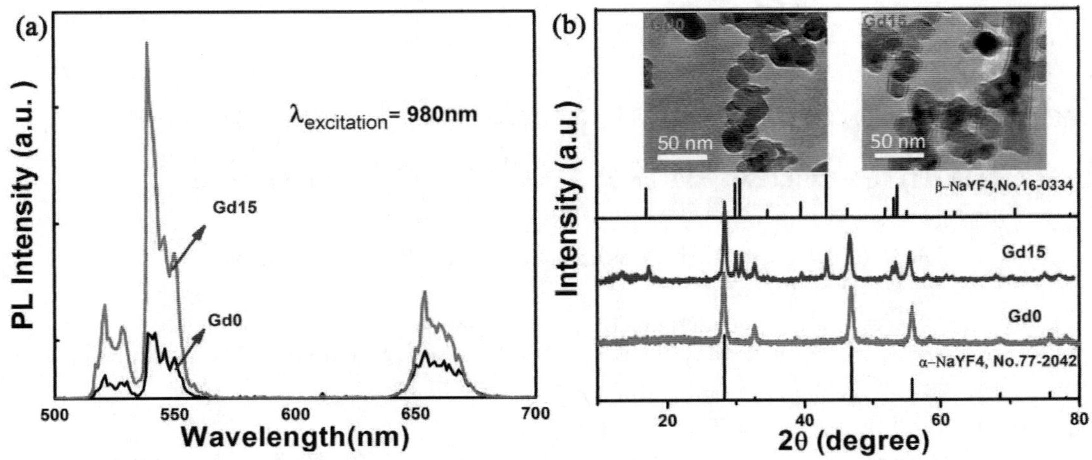

Figure 2. (a) up-conversion PL intensities of $NaYF_4$:Yb,Er NCs for Gd0 and Gd15 samples excited at 980 nm with a laser power of 656 mW; (b) XRD patterns of $NaYF_4$:Yb,Er NCs for Gd0 and Gd15 samples, insets are TEM images of Gd0 and Gd15 samples.

4. Conclusions

We have fabricated SiO_2 thin films co-doped with Er^{3+} ions and SnO_2 NCs. We find that the Er^{3+}-related down-shift emission is enhanced by three orders of magnitude due to the effective energy transfer process occurs between SnO_2 NCs and Er^{3+} ions. Furthermore, we also have prepared the Er^{3+}-Yb^{3+} co-doped $NaYF_4$ NCs with Gd^{3+} dopant ions to obtain high-efficiency up-conversion emission by energy transfer from Yb^{3+} to Er^{3+}. We demonstrate experimentally the up-conversion emission of $NaYF_4$:Yb,Er NCs with Gd^{3+} dopant ions is enhanced by more than five times. We conclude that Gd^{3+} ions play key roles in the formation of β-$NaYF_4$ NCs which can improve the energy transfer efficiency from Yb^{3+} to Er^{3+} obviously. The present work provides a novel approach to enhance the characteristics emission from lanthanide-doped nanomaterials, which indicates the possibility to improve the down- and up-conversion efficiency for device applications.

5. Acknowledgments

This work was supported by "973 program" (2013CB632101) and NSFC (No.11274155) and "333 project" of Jiangsu Province (BRA2015284).

6. References

[1] Galli M, Politi A, Belotti M, Gerace D, Liscidini M, Patrini M and Andreani L C 2006 *Appl. Phys. Lett.* **88** 251114
[2] Deng R R, Qin F, Chen R F, Huang W, Hong M H and Liu X G 2015 *Nature nanotech.* **10** 237-242
[3] Lin T, Zhang X W, Xu J, Swihart M, Xu L, Chen K J 2013 *Appl. Phys. Lett.* **103** 181906
[4] Zhang X X, Lin T, Zhang P, Xu J, Lin S B, Xu J and Chen K J 2014 *Opt. Express.* **22**, 369-376
[5] Wang F, Han Y, Lim C S, Lu Y H, Wang J, Xu J, Chen H Y, Zhang C, Hong M H and Liu X G, 2010 *Nature* **25** 1061-1065

IOP Conf. Series: Journal of Physics: Conf. Series **864** (2017) 012068 doi:10.1088/1742-6596/864/1/012068

Effect of doping concentration on Nd-related photoluminescence in TiO_2 with Al co-doping

Mariko Murayama[1], Yushi Yanagida[1], Shuji Komuro[2] and Xinwei Zhao[1]

[1]Department of Physics, Tokyo University of Science, Tokyo 162-8601, Japan
[2]Faculty of Science and Engineering, Toyo University, Saitama 350-8585, Japan

E-mail: xwzhao@rs.kagu.tus.ac.jp

Abstract. Nd doped, Nd-Al co-doped anatase phase TiO_2 thin films were fabricated on silicon substrates by laser ablation and post annealing. The result of measuring PL spectra showed intense emissions from a luminescent centre of Nd^{3+} ions in these thin films. The luminescence of Nd-Al co-doped samples were stronger and broader than only Nd doped samples. Doping concentration of Nd determined suitable co-doping concentration of Al and post annealing temperature. Also there is the difference of shape of the PL spectra which indicates that the difference of local fine structure around Nd. Based on these facts, XAFS spectra was measured and suggested that the coordination around Nd was changed by this Al-co-doping.

1. Introduction

Rare earth element doped semiconductors and oxide materials have been studied for long years because of their possibilities of being applicable materials for visible and infrared light emitting diodes and other optoelectronic devices[1-4]. Trivalent neodymium (Nd^{3+}) exhibits sharp luminescence in the infrared and near-infrared wavelength regions. Titanium dioxide (TiO_2)has been received attention as an optimal candidate for host materials to activate doped rare earth ions, because of its large band gap energy works as a window[5, 6]. One of keys of enhancing PL emission, which is distorted coordinations around rare earth dopant, was reported in our previous work[7, 8]. Distortion has been caused by growing annealing temperature, and the luminescent centre was optimally activated without co-doping. On this paper, enhancing PL emission with co-doping is focused. On our previous work[9], Nd-doped anatase phase TiO_2 (A-TiO_2:Nd) has been confirmed emitting intense luminescence and these emissions were enhanced with co-doping aluminium (Al). Other effects of sensitisation are improving the Nd^{3+}-emission lifetime and broadening shapes of PL spectra. These facts suggest that different concentration of dopants would change the bond length or coordination around Nd, and that enhance Nd^{3+}-related emissions. In this paper, we report PL and XAFS analysis for different concentrations of Nd and Al in TiO_2.

2. Experimental procedure

The Nd doped and Nd-Al co-doped TiO_2 (TiO_2:Nd, TiO2:Nd-Al) thin films were deposited by laser ablation using a YAG laser ($4_{\omega 0}$ = 266 nm, 1 J/cm^2). A TiO_2 ceramic target, including 0.2 or 1.0 wt% Nd_2O_3 and 0.0 or 0.5 wt% Al_2O_3, was used as the deposition source. A-TiO_2

thin films were fabricated in a vacuum chamber with an O_2 pressure of 1.5×10^{-2} Torr. As has reported in our previous work, selecting an appropriate atmosphere enables the production of the suitable crystalline phase A-TiO_2[10]. A-TiO_2:Nd/ Nd-Al thin films of 300 nm thick were formed on the Si (100) surfaces at room temperature. After the deposition, the samples were annealed at a temperature of 700 °C for Nd (0.2 wt%)-Al (0.0 or 0.5 wt%) and of 500 °C for Nd (1.0 wt%)-Al (0.0 or 0.5 wt%), for 3 min in the O_2 atmosphere for optical activation of Nd^{3+} ions and formation of the nanometer-sized TiO_2 crystals. The annealing temperatures were settled as the optimal conditions to enhance PL emissions at intra-shell $^4F_{3/2}{\rightarrow}^4 I_{11/2}$ transitions in Nd^{3+} ions (1080-1096 nm) with Al co-doping. Based on these recognitions, selected annealing temperature of Nd (0.2 wt%) and Nd (1.0 wt%) was different. Besides these samples, the samples for XAFS measurements were prepared, because Nd (0.2 wt%) is too dilute to measure XAFS spectra. Deposition time of Nd (0.2 wt%) samples were 3 times longer than the other samples. All samples were prepared with an O_2 pressure of 2.0×10^{-2} Torr, and post annealed at 700 °C for 3 min. The A-TiO_2:Nd/ Nd-Al thin films of appriximately 900 nm thick were formed. Other conditions of these samples are the same with the measuring PL sampels.

A He-Cd laser (λ = 325 nm) was used to excite the Nd^{3+} and to measure PL spectra from the TiO_2:Nd thin films. The X-ray absorption fine structure (XAFS) measurements using synchrotron radiation (SR) were performed at the High Energy Accelerator Research Organization BL 27B Photon Factory in Tsukuba prefecture, Japan. A Si double crystal set was used to monochromatise the SR beam. A seven-element pure Ge solid-state detector was used for counting the fluorescence X-rays proportional to the X-ray absorption of the TiO_2:Nd thin films. The X-ray incidence angle was fixed at 45°. The Athena of Demeter, which is a free programme for XAFS analysis created by Ravel B, was used for an ordinary process of XAFS analysis, that is, extracting the oscillations and fitting the Fourier spectra[11, 12].

3. Results and discussion

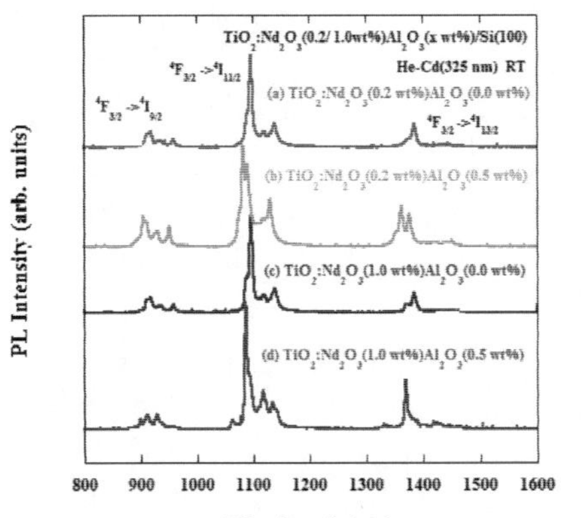

Figure 1: PL spectra of A-TiO_2:Nd_2O_3 (x wt%)-Al_2O_3 (y wt%):(a)x=0.2, y=0.0, (b)x=0.2, y=0.5,(c)x=1.0, y=0.0,(d)x=1.0, y=0.5.

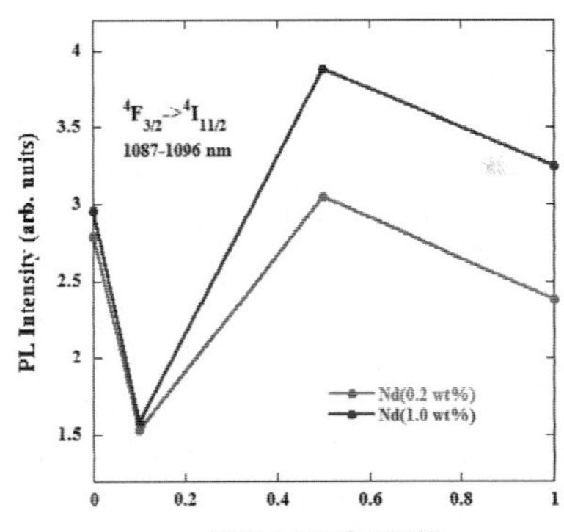

Figure 2: PL intensity ratio at $^4F_{3/2}{\rightarrow}^4 I_{11/2}$ transitions of A-TiO_2:Nd (0.2/ 1.0 wt%) with different Al co-doing concentrations.

Figure 1 shows PL spectra of A-TiO_2:Nd_2O_3 (0.2 wt%)-Al_2O_3 (0.0, 0.5 wt%) post annealed at 700 °C, with A-TiO_2:Nd_2O_3 (1.0 wt%)-Al_2O_3 (0.0, 0.5 wt%) post annealed at 500 °C. The PL emissions corresponding to $^4F_{3/2}{\rightarrow}^4I_J$ (J = 9/2, 11/2, 13/2) intra-4f transitions in Nd^{3+} ions can be observed. Every sample showed strong luminescent emissions.

This result demonstrates that Nd^{3+} with a crystal field allowing intra-$4f$ transitions is formed in crystal matrix by annealing. These facts indicate that the fabricated samples annealed at an appropriate temperature and concentration of co-doing results in a window for making the luminescence centre. Increasing of concentration of Al co-dopant made stronger PL intensity, sifted from the peak of non co-doping samples, and changed the shape of PL spectra with remaining similar tendency of non co-doping samples. Different PL intensities suggest difference of the transition probability, and different fine structures of PL spectra suggest difference of coordination around the Nd^{3+} ions, which is the luminescent centre.

Figure 2 shows PL intensity ratio at ${}^4F_{3/2}\rightarrow{}^4I_{11/2}$ transition of A-TiO_2:Nd (0.2/ 1.0 wt%) with different Al co-doping concentrations. As shown in Fig. 1, the shape of spectra changed with co-doping Al (0.5 wt%). Because of this reason, the most intensive peak of the PL emission of these samples was not the same wavelength, but the same transitions. The samples co-doped with 1.0 wt% Al were also shown for comparrison. The influence of Al co-doping on the Nd^{3+} with different Nd concentrations will be discussed by analysing the local microstructure around the Nd^{3+} ions.

The XAFS spectra at the Nd L_{III}-edge (6713 eV) of A-TiO_2:Nd with/ without co-doping Al are shown in Fig. 3. The lowest spectrum is from Nd_2O_3 powder of the standard sample. These series of samples are the same concentrations as showed in Fig. 1. The most intense peak means absorption of Nd dopant and Nd oxides. Every spectrum appeared at the same peak position as the Nd_2O_3 powder, which suggests the Nd^{3+} state was mainly formed in Nd_2O_3 thin films without any out diffusions. The spectra obtaining from the samples of doping Nd (0.2 wt%) were very noisy because of dilute doping.

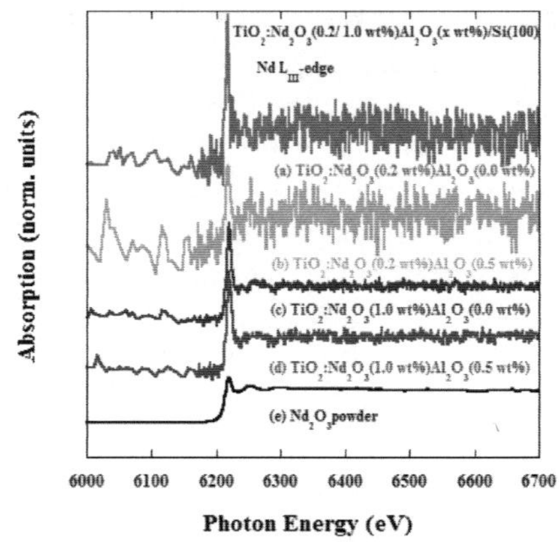

Figure 3: XAFS spectra at the Nd L_{III}-edge of A-TiO_2:Nd-Al thin films obtained at different concentrations. The lowest spectrum is that from the Nd_2O_3 powder of the standard sample.

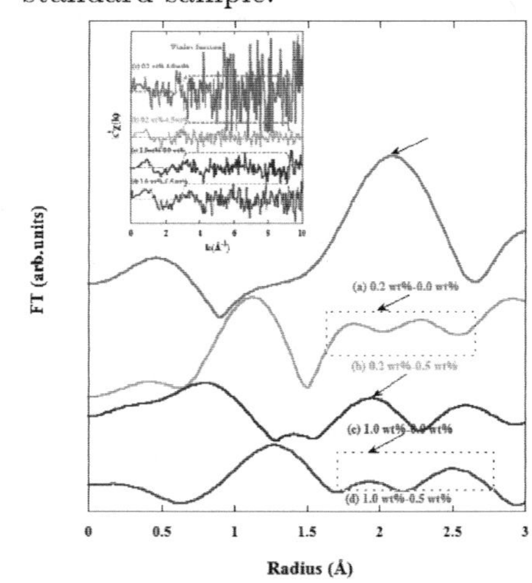

Figure 4: RSF around Nd delivered from FT of EXAFS oscillation, $k^2\chi$, of A-TiO_2:Nd-Al samples shown in Fig. 3 . The inset is k^2-weighted EXAFS oscillations (solid line) and window functions (dashed line) of A-TiO_2:Nd/ Nd-Al.

The inset of Figure 4 shows the k^2-weighted extended X-ray absorption fine structure (EXAFS), $\chi(k)$ oscillations of Nd in TiO_2:Nd/ Nd-Al. k is the photoelectron vector. A bigger weighting k^n means emphasising higher oscillation. k^2 were chosen because higher range of oscillations (after 9.3 Å) were noisy. The window function was performed in the range of 3.0-9.3 $Å^{-1}$, dk = 0.5. The chosen ranges in the XAFS analysis of this experiment are all the same because of comparison.

Figure 4 shows the radial structural function (RSF) around Nd obtained by the Fourier transform (FT) of k^2-weighted XAFS spectra in inset, $k^2\chi$ of A-TiO_2:Nd/ Nd-Al. These values of bonding length are shorter than the actual distance, because the result does not account for the phase shift of the photoelectrons by scattering of atoms in the FT. There were peaks located at an average distance of less than 1.3 Å owing to the low frequency waves in the background. This range was not taken into consideration owing to this reason. The first nearest neibougher (NN) of Nd in the samples without co-doping showed the similar spectra as a case of TiO_2:Sm in our previous work[8]. On the other hand, the first NN of Nd in the samples with Al co-doping were broadened and splitted. This result indicates that local fine structure of Nd was changed by co-doping Al as suggestion from the result of PL measurement.

The optimal annealing temperature and concentration of co-doping was selected to prepare the most effective condition as sensitiser. In our previous work, enhancing PL emission, which is distorted coordinations around rare earth dopant, was reported[7, 8]. The distortion has been caused by growing annealing temperature without co-doping, and the luminescent centre was optimally activated. The way of approaching to enhance PL emission is different. One is optimally annealed samples without co-doped and the other is samples with co-doping. Both samples showed intense PL spectra. As shown in Fig. 4, changing local fine structure of Nd^{3+} was confirmed. These results predicted that not the same way, but the same tendency, the coordinations around Nd ions were distorted both in concentrations controlled and annealing temperature controlled samples. These XAFS spectra of samples have not been completely fitted because of noise. Hence this is still in hypothetical process, but further experiments have been already planned to gain the XAFS spectra to analyse the second NN.

4. Conclusion

Nd^{3+} ions doped, Nd^{3+}-Al co-doped anatase (A-)TiO_2 thin films were deposited by laser ablation technique. The crystal phase was properly selected by under controlling of the O_2 pressure. After ablating, the samples were annealed. The samples were estimated by PL and XAFS. The samples of co-doping Al (0.5 wt%) of A-TiO_2:Nd exhibited stronger PL emission than non co-doping samples. This result suggests co-doped Al sensitises PL emission, and this effect appeared when the optimal concentration and annealing temperature were selected. The result of PL measurement revealed different fine structure of the PL spectra which suggests the difference of coordination around the Nd^{3+} ions of the luminescent centre. Based on these results, XAFS spectra were measured. The result of the XAFS analysis revealed the local structures of A-TiO_2:Nd and A-TiO_2:Nd-Al. The coordination around Nd^{3+} of non co-doped and co-doped samples was different. The first NN of Nd in co-doping samples was splitted and broadened. This result suggests the coordination around Nd^{3+} was distorted as the without co-doping samples optimally annealed, and showed the intense PL emission.

5. Acknowledgement

The authors would like to thank Mr Sakurai and Mr Aizawa of Tokyo University of Science for the stimulative discussion on previous work. Additionally, we thank the High Energy Accelerator Research Organization for experimental support and technical advice.

References

[1] Kenyon A J 2002 *Progress in Quantum Electronics* **26** 22584
[2] Nazarov A N, Tiagulskyi S I, Tyagulskyy I P, Lysenko V S, Rebohle L, Lehmann J, Prucnal S, Voelskow M and Skorupa W 2010 *J. Appl. Phys.* **107** 123112
[3] Ishii M, Towlson I B, Poolton N, Harako S, Zhao X, Komuro S and Hamilton B 2012 *J. Appl. Phys.* **111** 053514
[4] Zhu C, Lv C, Gao Z, Wang C, Li D, Ma X and Yang D 2015 *Appl. Phys. Lett.* **107** 131103
[5] Yang Y, Lv C, Zhu C, Li S, Ma X and Yang D 2014 *Appl. Phys. Lett.* **104** 201109
[6] Kima J H and Holloway P H 2004 *Appl. Phys. Lett.* **85** 1689
[7] Ishii M, Komuro S and Morikawa T 2003 *J. Appl. Phys.* bf 94 3823
[8] Sakurai J, Harako S, Ohtsuki T, S. Komuro, Hirao N, Kasahara R and Zhao X 2012 *Japan. J. Appl. Phys.* **51** 06FG03
[9] Aizawa Y, Ohtsuki T, Harako S, Komuro S, Hirao N and Zhao X 2014 *Japan. J. Appl. Phys.* **52** 06JG06
[10] Jing F, Harako S, Komuro S and Zhao X 2009 *J. Phys.* D **42** 085109
[11] Ankudinov A L, Bouldin C E, Rehr J J, Sims J and Hung H 2002 *Phys. Rev.* B **65** 104107
[12] Ankudinov A L, Ravel B, Rehr J J and Conradson S D 1998 *Phys. Rev.* B **58** 7565

IOP Conf. Series: Journal of Physics: Conf. Series **864** (2017) 012069 doi:10.1088/1742-6596/864/1/012069

Charge carrier kinetics in GeSi/Si quantum dots probed by mid-infrared radiation

A N Sofronov[1], L E Vorobjev[1], D A Firsov[1], R M Balagula[1], and A A Tonkikh[2,3]

[1]Peter the Great St. Petersburg Polytechnic University, 195251, Polytechnicheskaya 29, St. Petersburg, Russia
[2]Scia-Systems GmbH, Annaberger-Str. 240, Chemnitz, Germany
[3]Institute for Physics of Microstructures RAS, 603950, GSP-105, Nizhnii Novgorod, Russia

e-mail: sofronov@rphf.spbstu.ru

Abstract. The mid-infrared photo-induced absorption relaxation kinetics of the self-assembled array of GeSi quantum dots in Si matrix was studied in the conditions of short pulsed interband optical excitation. The measured absorption decay curves directly show the temporal evolution of the population of the QD ground states. The analysis of the experimental data allowed us to estimate characteristic recombination and capture times, and the electron localization energy in the vicinity of QD.

1. Introduction

GeSi/Si quantum dot structures are promising objects for the development of the quantum dot infrared photodetectors (QDIP) [1,2]. Germanium quantum dots (QDs) in silicon matrix self-assemble during the strain-driven epitaxy. The built-in strain affects the conduction and valence band profiles in the vicinity of the quantum dot interface. The details of the band alignment are important for the performance of the QDIPs. The mid-infrared optical spectroscopy is a powerful instrument for study of the energy structure of the charge carriers in QD structures, while kinetics of the absorption decay allows one to resolve the information about characteristic relaxation times. In the present work, we study the decay curves of the photo-induced absorption after a short excitation of the electron-hole pairs in GeSi/Si QDs in a wide temperature range.

2. Samples and methods

The charge carrier kinetics studies were carried out in a dense array of the surfactant-mediated GeSi/Si (001) undoped quantum dot structures. The details of growth process are described in Ref. [3].

Non-equilibrium charge carriers were created in structure by means of short 10 ns pulse of interband optical excitation from the solid state YAG:Nd laser with frequency multiplication with pulse energy up to 9 mJ. The temporal evolution of the population of the hole ground state inside the quantum dot was monitored measuring a decay of the intraband mid-infrared (mid-IR) optical absorption of the sample after the pulsed interband excitation. The wide-band globar coupled with prismatic monochromator was used as a source of tuneable mid-IR probe radiation. The intensity of the mid-IR radiation transmitted through the sample was measured with liquid-nitrogen-cooled MCT detector with about 100 ns rise time.

The photon energy of the mid-IR probe light was selected to match the energy of the transition between dot ground state and the continuum, as it is known from the previous steady-state photo-induced absorption and photoconductivity spectroscopy [4]. So the mid-IR optical absorption change under optical excitation is directly proportional to the number of holes in the ground state of QD.

3. Experimental results and discussion

The photo-induced absorption decay curves were measured in a wide range of the excitation levels at lattice temperatures T_L from 80 K up to 300 K with a 10 K interval. Typical decay curves for high and low temperatures are plotted in figure 1 at probe photon energy 0.3 eV as a time dependencies of the optical transmission change $|\Delta T|$.

The low-temperature (below 200 K) relaxation curves demonstrate two-stage behavior with a fast (few μs) and slow (tens of μs) characteristic decay times. A slow relaxation time tends to decrease with an increase in temperature, while the fast relaxation time slightly grows up. Inset in figure 1 shows the temperature dependencies of the photo-induced absorption at two moments of time inside the fast (3 μs after the excitation) and slow (25 μs after the excitation) decay stages.

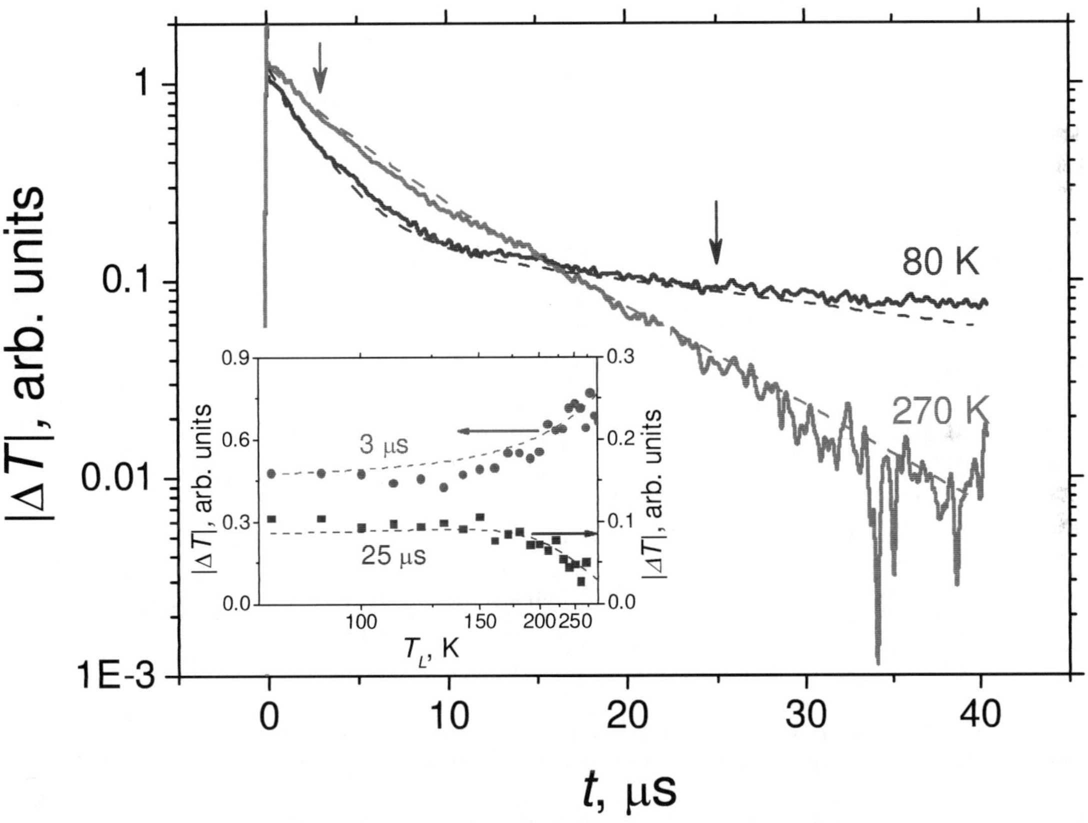

Figure 1. Typical decay curves of photo-induced absorption at 0.3 eV for high and low temperatures after short pulsed excitation. Inset shows the temperature dependence of the optical transmission change at the moments of 3 μs and 25 μs after the excitation. Dashed curves show the biexponential fit.

IOP Conf. Series: Journal of Physics: Conf. Series **864** (2017) 012069 doi:10.1088/1742-6596/864/1/012069

We attribute the fast decay component to the direct and/or indirect in real space recombination [5] of the electrons and holes in the quantum dots. Then, the inverse fast time shows the probability of recombination proportional to the number of electrons localized in the vicinity of the quantum dot. It decreases with temperature due to the thermal electron delocalization.

In order to explain the experimentally observed biexponential decay we suggest to introduce the positive contribution to the QD depopulation rate that exponentially decays itself. This contribution can stand for the capture of the bulk holes in silicon matrix to the quantum dots. Then, the inverse slow time shows the probability of capture of the holes to the dot ground state.

The dashed curves in figure 1 show the result of biexponential fit of the whole set of experimental $|\Delta T|$ curves with additional conditions on the inverse fast and slow times that were assumed to be activation-like dependent on temperature with activation energies E_e and E_h, respectively. The resulting temperature dependencies of the fast and slow times are plotted in figure 2. The corresponding activation energies E_e = 25 meV and E_h = 70 meV were derived.

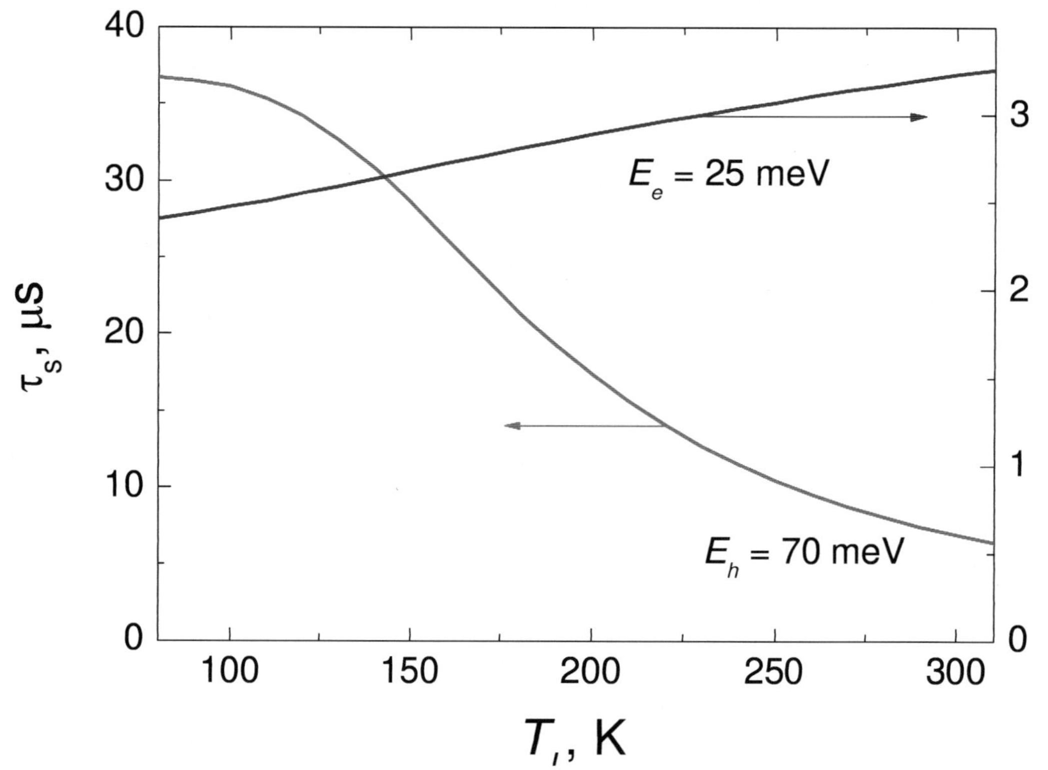

Figure 2. Temperature dependencies of the fast (τ_f, right axis) and slow (τ_s, left axis) decay times derived from the biexponential fit of the experimental absorption decay curves.

The localizations of the electrons in the vicinity of the QDs with the binding energy E_e arises due to the strain-induced band bending at the quantum dot interface. The 25 meV binding energy derived from the absorption decay curves analysis is comparable with the calculated energy of electron strain-induced localization [6].

The presented experimental results show the significant temperature dependence of the hole capture time. The capture probability rises with temperature with relatively large 70 meV activation energy. Again, this energy can be treated as a strain-induced valence band barrier at the QD interface for the bulk holes in silicon matrix. However, the QD capture time can be significantly dependent on such factors as initial population and charge state of the QD, and it is necessary to exploit a set of master equations for detailed quantitative analysis of the QD state population kinetics.

4. Conclusion

In this work, we present the temperature evolution of the ground state population kinetics in the GeSi/Si quantum dot structures. The characteristic recombination and capture times and the electron binding energy were estimated. It was shown that non-equilibrium population of the hole ground state in the GeSi/Si quantum dots is highly temperature dependent.

Acknowledgements

This work was supported by the Ministry of Education and Science of the Russian Federation (state assignment) and the Russian Federation President grant for young scientists (MK-6064.2016.2).

References

[1] Yakimov A I, Kirienko V V, Armbrister V A, Bloshkin A A, and Dvurechenskii A V 2015 *Appl. Phys. Lett.* **107** 213502
[2] Yakimov A I, Kirienko V V, Bloshkin A A, Armbrister V A, Kuchinskaya P A, and Dvurechenskii A V 2015 *Appl. Phys. Lett.* **106** 032104
[3] Tonkikh A and Werner P 2013 *Physica Status Solidi B* **250** 1795
[4] Sofronov A N, Vorobjev L E, Firsov D A, Panevin V Y, Balagula R M, Werner P, and Tonkikh A A 2015 *Superlattices and Microstructures* **87** 53
[5] Kamenev B V, Tsybeskov L, Baribeau J-M, and Lockwood D J 2005 *Phys. Rev. B* **72** 193306
[6] Yakimov A I, Dvurechenskii A V, Nikiforov A I, Bloshkin A A, Nenashev A V, and Volodin V A 2006 *Phys. Rev. B* **73** 115333

IOP Conf. Series: Journal of Physics: Conf. Series **864** (2017) 012070 doi:10.1088/1742-6596/864/1/012070

Electric field influence on mid-infrared absorption and interband photoluminescence in tunnel-coupled GaAs/AlGaAs quantum wells

M Ya Vinnichenko, A N Sofronov, R M Balagula, D A Firsov, L E Vorobjev

Peter the Great St. Petersburg Polytechnic University, 29 Polytechnicheskaya str., St. Petersburg 195251, Russia

e-mail: mvin@spbstu.ru

Abstract. The influence of transverse and lateral electric field on the mid-infrared intersubband light absorption is experimentally investigated in tunnel-coupled GaAs/AlGaAs quantum wells. Observed absorption modulation in electric field is related to the electron redistribution between the quantum well states resulting in variation of space charge in the structure. This phenomenon in transverse electric field is connected with a change of potential profile, whereas in lateral field effect is caused by electron heating. Electron heating also leads to a modulation of near-infrared interband photoluminescence under lateral electric field.

1. Introduction

Charge carrier heating in electric field in quantum wells can affect the characteristics of various optoelectronic devices (lasers, modulators etc.). Lateral and transverse electric fields have different impact on mid-infrared absorption and interband photoluminescence in quantum wells (QW). It is known that radiation frequency modulation is more efficient than intensity modulation in communication systems [1]. One of the techniques that can rapidly change the frequency of quantum cascade laser radiation is external electric field modulation of the refractive index of quantum well layers, introduced into the quantum cascade laser waveguide. A possible way to create such a system is an introduction of the layers with tunnel-coupled quantum wells into the laser waveguide. Complicated band structure of tunnel-coupled quantum wells facilitates the observation of the electric field influence on the QW optical properties. The first step toward achieving this objective is to study the refractive index and absorbance variation of the layers with tunnel-coupled quantum wells in the mid-IR region under electric field.

The investigation of these phenomena in double tunnel-coupled QW is still ongoing. Electric field causes the redistribution of electrons between size quantization levels due to carrier heating. The photoluminescence (PL) study provides an easy way to analyze heating. Earlier the influence of lateral electric field on the photoluminescence spectrum was investigated in [2]. In contrast to our work, the modulation of PL in [2] was not directly connected with electron heating. Carrier heating effect on the intersubband absorption in tunnel-coupled quantum wells was studied in [3] at a fixed radiation wavelength. Comparing the effect of an electric field and lattice heating on the spectra of intersubband absorption and interband PL, the temperature of the hot electrons was estimated [3].

This work is devoted to the complex investigation of the influence of different configurations of electric field on optical properties of tunnel-coupled GaAs/AlGaAs quantum wells.

IOP Conf. Series: Journal of Physics: Conf. Series **864** (2017) 012070 doi:10.1088/1742-6596/864/1/012070

2. Samples and experiment

Structures with 100 pairs of doped GaAs/AlGaAs QWs with different width were studied. The structures were grown by molecular-beam epitaxy on a semi-insulating GaAs substrate. In order to characterize the structure, interband photoluminescence spectra without electric field were measured with Horiba FHR-640 grating monochromator with a Symphony II CCD-camera and YAG laser as a source of interband optical pumping. Spectra of PL modulation in electric field were studied using a Si photodetector. In order to apply of transverse electric field, mesa was etched from one of the structures and Cr/Au contacts were deposited on its surface in vacuum. Indium contacts to the structure intended for the lateral electric field application were manually deposited and annealed at the nitrogen atmosphere. The facets of the samples were mechanically grounded and polished at 45° angle to create a multipass sample geometry for intersubband optical absorption studies. Such geometry is based on total internal reflection of radiation in the sample and increases the absorption in the sample by increasing effective optical path. Also this method allows to investigate absorption for two polarisations independently. The intersubband absorption variation spectra under lateral and transverse electric field were studied using a Bruker Vertex 80v vacuum Fourier-transform spectrometer in step-scan mode. In order to avoid sample overheating, the lateral electric field was applied to the sample in pulsed mode. Contribution of intersubband transitions to light absorption by the quantum well structures without electric field was determined from the ratio of transmission spectra for two light polarizations.

3. Results and discussions

Calculation showed the presence of 3 electron energy levels in paired QWs. The equilibrium interband photoluminescence spectra were measured at different lattice temperatures for both structures. Spectral peaks demonstrate a temperature shift corresponding to the bandgap change. At low temperatures, there is only one peak associated with the optical *e*1-*e*3 transitions, which means that the carriers are localized mainly at the *e*1 level. Carriers begin to fill the overlying energy level with the increase in temperature.

The spectra of intersubband absorption modulation demonstrate a decrease in *e*1-*e*3 absorption under transverse electric field (see figure 1). This phenomenon can be attributed to electron redistribution between the *e*1 and *e*2 levels in electric field induced by the modification of the QW potential profile. Saturation of modulation at particular transverse electric field can be caused by total depletion of *e*1 level. An electric field of opposite polarity does not induce a significant variation of the electron concentration on *e*1 level, therefore, no change in intersubband absorption was detected. Spectra of refraction index change in electric field were calculated from the obtained intersubband absorption modulation spectra using Kramers-Kronig relation. Results of the calculation show that maximal change of refraction index under transverse electric field is approximately 10% of the bulk GaAs refraction index. This result suggests that tunnel-coupled quantum well structures can be implemented in the development of frequency electro-optical modulators for the mid-IR region radiation emitters.

In lateral electric field (see figure 2), we found a decrease in *e*1-*e*3 absorption, an increase in *e*2-*e*3 absorption and a slight red shift of *e*1-*e*3 absorption peak. These phenomena can be related to the hot electron redistribution between the *e*1 and *e*2 levels and subsequent variation of space charge in the structure since the wave functions of e1 and e2 electron levels are localized at wide and narrow QW of the pair, respectively.

The electron temperature T_e was estimated from the comparison of the absorption change related to the lateral electric field and to the lattice temperature. The experimentally determined electron temperature was $T_e = 97$ K at $E = 695$ V/cm and $T_e = 103$ K at $E = 1045$ V/cm at lattice temperature $T = 77$ K.

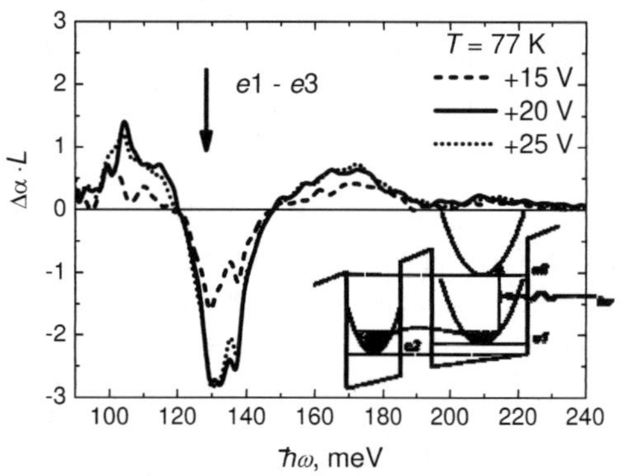

Figure 1. Absorption modulation spectra for *p*-polarized light under different transverse electric fields. The arrow shows the calculated electron transition *e*1-*e*3 energy. Electron redistribution under transverse electric field is schematically shown in inset. *L* is the optical path.

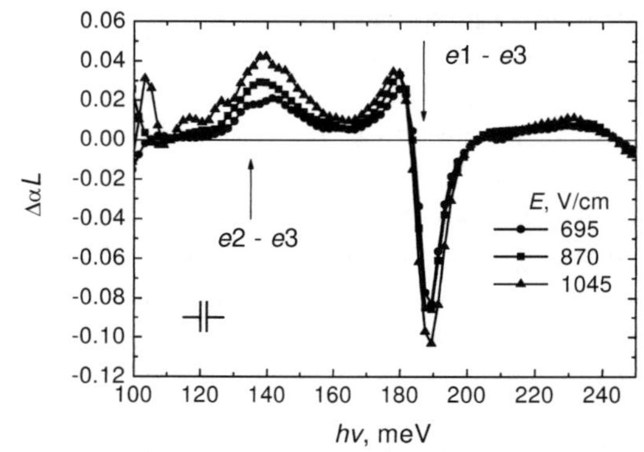

Figure 2. Spectra of intersubband absorption modulation under different lateral electric fields *E* at $T = 77$ K. The arrows indicate the calculated electron transition energies.

The interband photoluminescence variation spectra were measured at different lateral electric fields at $T = 77$ K. Under applied electric field, the hot carrier temperature was not high enough to significantly change the concentration of carriers at hole and electron levels *lh*1, *hh*2 and *e*2. Only modulation of interband photoluminescence associated with the *e*1-*hh*1 optical transitions was observed (see figure 3). It should be noted that interband PL modulation was observed at a high level of optical pumping. In such conditions, rather high nonequilibrium carrier concentration is created, therefore we can consider the equality of electron and hole temperatures due to the strong electron-hole scattering. The electron temperature T_e was estimated by means of the shortwave spectra shoulder analysis. To derive T_e value one should obtain the energy value $h\nu_0$ which corresponds to the zero photoluminescense modulation $\Delta I_{PL} = 0$ or, in other terms, $dI_{PL}/dT_e = 0$. Photoluminescence intensity I_{PL} is proportional to the number of spontaneous photons of interband radiation in elementary volume during elementary time period at the frequency range from ν to $\nu + \Delta\nu$, which, in turn, is proportional to the carrier distribution function. Taking a derivative one can obtain nonequilibrium charge carrier temperature using Boltzmann distribution function: $T_e = (h\nu_0 - \varepsilon_g)/2k$, were k is Boltzmann constant, ε_g is effective band gap. At $T = 77$ K, the electron temperature rise due to the heating was determined to be about 30 K. This temperature was close to the temperature obtained from the analysis of intersubband absorption modulation spectra.

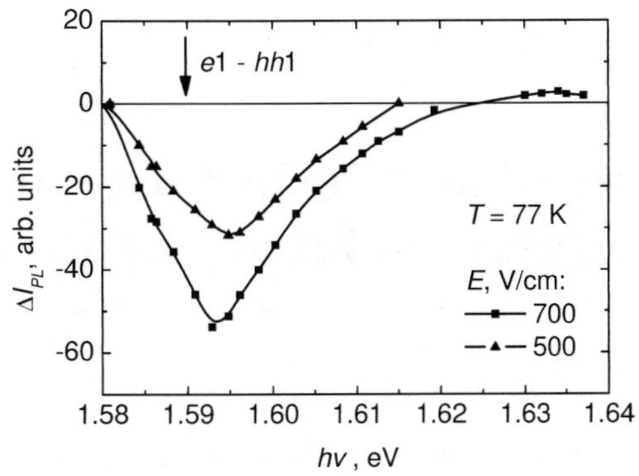

Figure 3. Modulation of interband photoluminescence at $T = 77$ K under lateral electric field E. Arrow shows the calculated energy of interband transitions $e1$-$hh1$.

4. Conclusion

The influence of the lateral and transverse electric fields on the mid-infrared intersubband light absorption and near-infrared interband photoluminescence in tunnel-coupled GaAs/AlGaAs quantum wells is experimentally observed. The temperature modification of the intersubband absorption spectra was studied as well. The effect of a transverse electric field on the intersubband absorption is explained by the redistribution of electrons between the first quantum levels and a change in the energy spectrum of the structure. Modulation of intersubband absorption and interband photoluminescence under lateral electric field is explained by the redistribution of hot charge carriers between the two lowest electronic states belonging to different quantum wells off the pair and by subsequent variation of space charge in the structure. The electron temperature T_e was estimated from the comparison of equilibrium intersubband absorption spectra and modulation of absorption spectra under lateral electric field. The analysis of interband photoluminescence modulation spectra allowed us to derive electron temperature and compare it with temperature obtained from the absorption spectra analysis. The temperature values obtained by two methods were close. The value of the observed absorption variation proves the application prospects of tunnel-coupled quantum wells as a basis for fast infrared light modulators.

Acknowledgments

This work has been supported by the Russian Foundation for Basic Research (Projects №№ 14-02-00336, 16-32-60085), the Ministry of Education and Science of the Russian Federation (state assignment) and the Russian Federation President grant for young scientists MK-4616.2016.2.

References

[1] R. Martini, C. Bethea, F. Capasso, C. Gmachl, R. Paiella, E. A. Whittaker, H. Y. Hwang, D. L. Sivco, J. N. Baillargeon, and A. Y. Cho, *Electron. Lett.* **38**, 181 (2002)
[2] Vorobjev L E, Vinnichenko M Ya, Firsov D A, Zerova V L, Panevin V Yu, Sofronov A N, Thumrongsilapa P, Ustinov V M, Zhukov A E, Vasil'iev A P, Shterengas L, Kipshidze G, Hosoda T and Belenky G 2010 *Semiconductors* **44(11)**, 1402-1405
[3] Zerova V L, Vorob'ev L E, Firsov D A, Towe E 2007 *Semiconductors* **41(5)** 596

IOP Conf. Series: Journal of Physics: Conf. Series **864** (2017) 012071 doi:10.1088/1742-6596/864/1/012071

Properties of intensive defect-related band in photoluminescence spectra of heavily doped $Al_xGa_{1-x}N$:Si layers

I V Osinnykh[1,3], T V Malin[1], V F Plyusnin[2,3], and K S Zhuravlev[1,3]

[1]Rzhanov Institute of Semiconductor Physics, Siberian Branch of the Russian Academy of Sciences, Lavrentieva av. 13, Novosibirsk 630090, Russia.

[2]Voevodsky Institute of Chemical Kinetics and Combustion of the Siberian Branch of the Russian Academy of Sciences, Institutskaya 3, Novosibirsk, 630090, Russia.

[3]Novosibirsk State University, Pirogova 2, Novosibirsk 630090, Russia.

igor-osinnykh@isp.nsc.ru

Abstract. We report photoluminescence investigations of heavily doped $Al_xGa_{1-x}N$:Si films grown by molecular beam epitaxy on sapphire substrates. The wide intensive defect-related band dominates in the photoluminescence spectra of $Al_xGa_{1-x}N$:Si films with the Al content higher than 0.38 covering the whole visible spectral range. This band is attributed to donor-acceptor and free electron-acceptor transitions involving the same acceptor. The acceptor ionization energy of about 1.87 eV for heavily doped AlN:Si was obtained, decrease of Al content leads to decrease of the acceptor ionization energy. The donor was assigned to the Si atom on the Ga/Al site; the acceptor might be (0/-) transition level of the C_N or (2-/3-) transition level of the V_{Al}.

1. Introduction

GaN and its alloys with AlN have emerged as perspective semiconductor materials with applications to laser sources for a blue-green spectral range. It was recently reported the finding of a blue-green superluminescence related to optical transitions through levels of defects in heavily doped $Al_xGa_{1-x}N$:Si layers with high Al content covering the whole visible and near-infrared region of the spectrum. [1]. This luminescence band was attributed to superposition of donor-acceptor and free electron-acceptor transitions involving the same acceptor [2]. Super-radiance within the broad band was obtained by pumping these samples up with an electron beam with power more than 200 kW. Overall, however, the study of luminescent properties and obtaining stimulated emission in the optical transitions via defect levels in wide band-gap solid alloys $Al_xGa_{1-x}N$:Si has not been given attention.

The aim of this work was further study of the characteristics of photoluminescence of heavily doped $Al_xGa_{1-x}N$:Si layers.

2. Experimental details

We investigated 1.2 μm-thick $Al_xGa_{1-x}N$:Si layers grown by molecular beam epitaxy techniques using ammonia as a source of active nitrogen on nitridated (0001)-oriented 400-μm-thick sapphire substrates

IOP Conf. Series: Journal of Physics: Conf. Series **864** (2017) 012071 doi:10.1088/1742-6596/864/1/012071

with an about 300-nm-thick AlN nucleation layer. The Al content (x_{Al}) in the layers was set by the Al-to-Ga flux ratio during the growth. The mix of 0.7% silane (SiH_4) with nitrogen (N_2) was used as the source of Si. The first series of studied layers was grown at the constant silane flux F_{SiH4} = 3 sccm, x_{Al} was varied from 0 to 1. The second series of studied layers was grown at the constant x_{Al} = 0.62, F_{SiH4} was varied from 0 to 6 sccm. The silicon concentration in the layers was estimated about 5×10^{19} cm^{-3} at F_{SiH4}=3 sccm. The concentration of free electrons in the samples under study was determined by measuring the Hall effect in Van der Pauw geometry in a constant magnetic field of 0.5 T at room temperature. Measurements of photoluminescence (PL) of the studied samples were conducted at the room temperature using a contunouos He-Cd laser with a photon energy E_{ph}= 3.81 eV and a pulsed 4th harmonic of Nd:YLF -laser (E_{ph}= 4.71 eV) for excitation. Measurements of photoluminescence excitation (PLE) spectra were carried out at room temperature on a FLS920 spectrofluorimeter with a 450 W xenon lamp as the excitation source. This setup allowed the measurement of both PLE and PL spectra.

3. Results

The PL spectra of heavily silicon doped $Al_xGa_{1-x}N$ layers belong to the first series at room temperature are shown in Figure 1. PL spectra of heavily doped GaN structure (x_{Al}=0) comprise only the band-edge ultraviolet band ($E_{em}\approx$ 3.6 eV) related to band-to-band transitions, the defect-related "yellow" band typical for undoped GaN is reduced likely because of Si incorporation in Ga vacancies that are considered the origin of the "yellow" band. The free-electron concentration in GaN structure was about 1.1×10^{20} cm^{-3}. The rise of x_{Al} leads to decrease of the intensity of band-edge PL and free-electron concentration (see the insert in Figure1) both whereas the very weak defect-related PL appears in the visible spectral range. The wide defect-related PL band dominates in PL spectra of the $Al_xGa_{1-x}N$ layers with x_{Al}>0.38. The increase of Al content from x_{Al}=0.47 to x_{Al}=1 leads to the blue shift of the energy position of the PL band from red-orange spectral range ($E_{em}\approx$ 2.0 eV) to violet spectral range ($E_{em}\approx$ 3.1 nm).

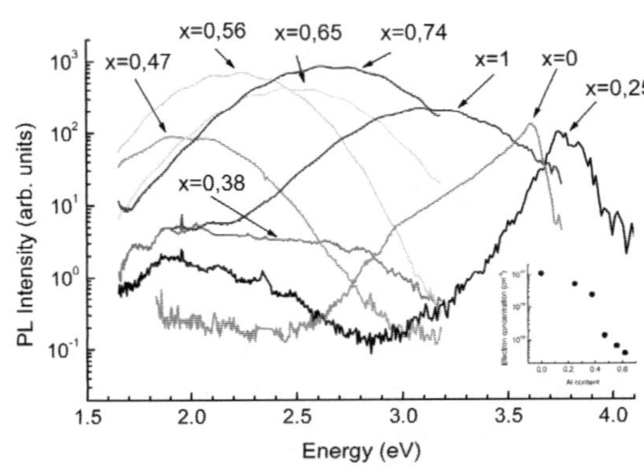

Figure 1. PL spectra of heavily doped $Al_xGa_{1-x}N$:Si layers grown at the constant silane flux F_{SiH4} = 3 sccm. Spectra were measured at room temperature. The insert shows dependence of the electron concentration on Al content.

Figure 2 shows the energy position of the defect PL band (E_{em}) and the band gap of the $Al_xGa_{1-x}N$ layers (E_g) estimated using the Vegard law:

$$E_{gAlGaN} = x_{Al}E_{gAlN} + (1 - x_{Al})E_{gGaN} - x_{Al}(1 - x_{Al})b = 1\cdot x^2 + 1.59\cdot x + 3.43 \quad (1)$$

here E_{gGaN}=3.43 eV [3] and E_{gAlN}=6.015 eV [4] are the GaN and AlN band gaps, respectively, and b=1.0 eV [5] is the bowing parameter. E_{em} rises weaker with x_{Al} increase than E_g. The intensity of the green band in PL spectra of the second series of samples ($E_m\approx$ 2.4 eV) increases with the rise of silicon concentration up to F_{SiH4} = 4.5 sccm then the PL intensity decrease with the further the rise of silicon concentration (see Figure 3). The electron concentration was about 1.6×10^{19} cm^{-3} and 4.1×10^{17} cm^{-3} for the samples grown at F_{SiH4}=1.5 sccm and F_{SiH4}=3.0 sccm respectively, the undoped sample and samples grown in highest silane flux were very weak conducting. That is an agreement with our previous results [6].

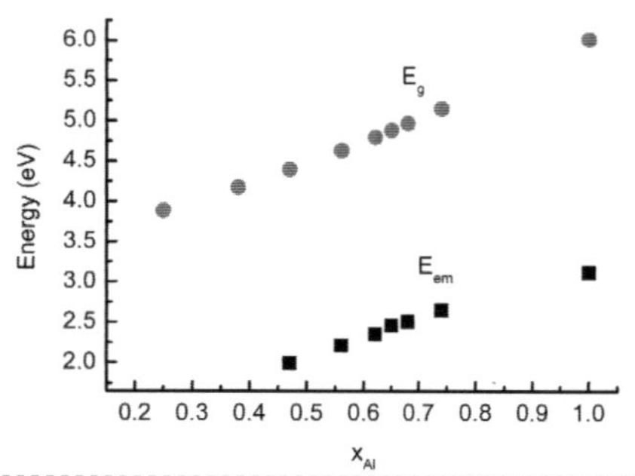

Figure 2. Dependences of band-gap (E_g) and the defect PL band energy position (E_m) on Al content.

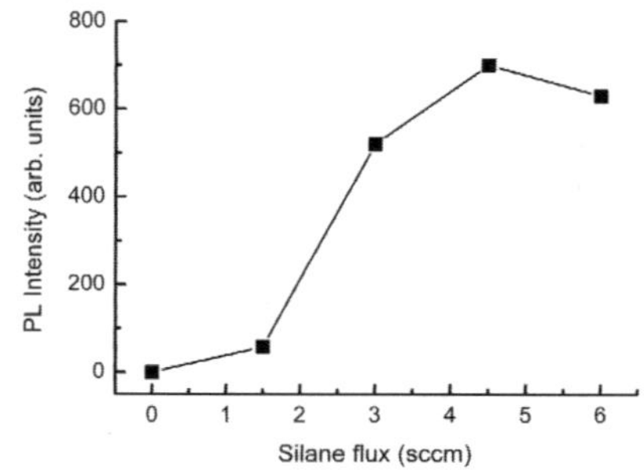

Figure 3. Dependences of the intensity of the defect PL band in $Al_{0.62}Ga_{0.38}N$ layers on silane concentration.

Figure 4 shows the PLE spectra of the $Al_{0.74}Ga_{0.26}N$ and AlN (x_{Al}=1) samples at detection energy of 3.08 eV and 2.7 eV respectively and PL spectra measured at the excitation energy 4.71 eV at room temperature. The peak in the PLE spectra marked E_1 at 5.2 eV and 6.0 eV for the $Al_{0.74}Ga_{0.26}N$ and AlN layers respectively corresponds to band-to-band transitions in heavily doped $Al_xGa_{1-x}N$ layers because its energy position matches with the E_g from expression (1). The second peak at E_2 eV at 4.7 eV and 4.94 eV for the $Al_{0.74}Ga_{0.26}N$ and AlN layers respectively was attributed to acceptor-to-band transitions in the heavily doped AlN layer. The large energy difference between the PL and PLE band E_2 peaks indicates a large Stock shift due to the strong electron-phonon coupling in a deep center. This results in a good agreement with our previous PL and PLE investigations of heavily doped $Al_xGa_{1-x}N$ layers with high x_{Al} [2].

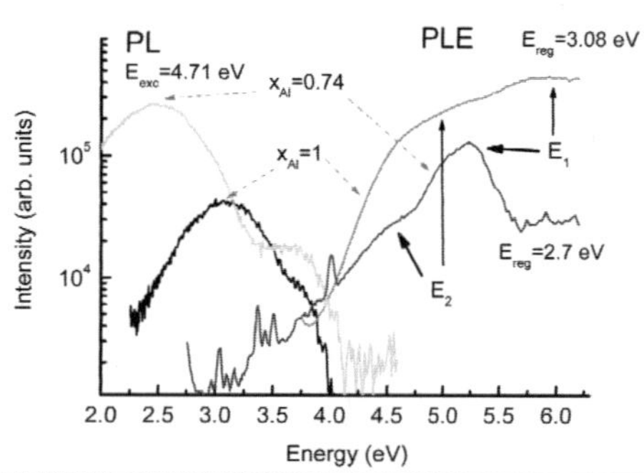

Figure 4. PLE and PL spectra of the $Al_{0.74}Ga_{0.26}N$ and AlN (x_{Al}=1) samples at room temperature.

4. Discussion

The large Stock shift is typical for the centers with the strong electron-phonon coupling. The configuration coordinate model describes such systems well [7, 8]. The configuration coordinate diagram describing optical absorption and emission at defects with the strong electron-phonon coupling is shown in Figure 5. S_{em} and S_{ab} are the numbers of photon with energies $h\nu_g$ и $h\nu_e$ involved in the transitions whole optical energies are E_m and E_{ab}, respectively. We assume that $S_{em} = S_{ab} = S$ and $h\nu_g = h\nu_e = h\nu_0$ .The emission peak energy (E_m) is given by

$$E_{em} = E_{gAlGaN} - E_D - E_A - S_{em} h\upsilon_0 \qquad (2)$$

here S_{em} is the number of photons with the energy $h\nu_g$ involved in the transitions. The Stock shift in this model should be

IOP Conf. Series: Journal of Physics: Conf. Series **864** (2017) 012071 doi:10.1088/1742-6596/864/1/012071

$$E_2 - E_{em} = 2S_{em}h\upsilon_0 + E_D \tag{3}$$

Since the intensity of green PL increases with increasing Si concentration, the observed PL band is related to the silicon impurity in $Al_xGa_{1-x}N$. The Si atom on the aluminum site Si_{Al} is considered as a deep donor with a binding energy of 250 meV in AlN [9-11]. This allows estimating $S_{em}h\nu_o$ value.

$$S_{em}h\upsilon_0 = 0.5(E_2 - E_{em} - E_D) \approx 0.8eV \tag{4}$$

After that we obtain the acceptor ionization energy in AlN

$$E_A = E_{gAlN} - E_{em} - E_D - S_{em}h\upsilon_0 \approx 1.87eV \tag{5}$$

The nature of a deep acceptor with a binding energy of about 1.87 eV is not yet completely clear. Possible candidates are (0/-) transition level of the C_N situated at about 1.88 eV [12] or (2-/3-) transition level of the V_{Al} [13] above the valence-band maximum. The possibility that Si_N act as a acceptor; however, the formation energy of Si_N is so high that its formation is highly unlikely [14].

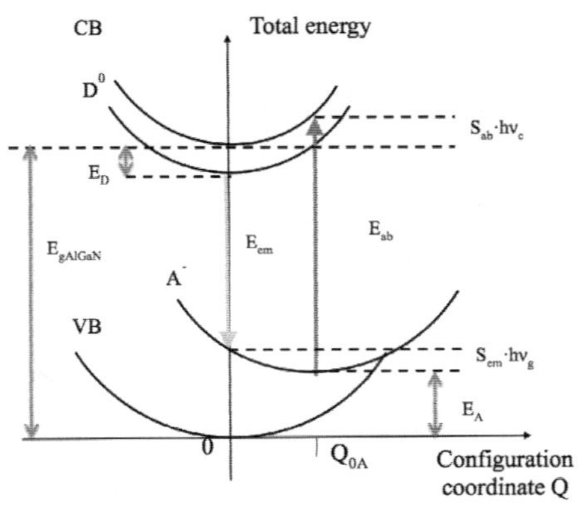

Figure 5. Configuration coordinate diagram describing optical band-to-band absorption and donor-acceptor pair emission at acceptor-like defects with the strong electron-phonon coupling. The A^- term corresponds to an occupied state of the defect, and the VB, CB, and D^0 terms correspond to unfilled excited states of the defect where an electron is in the valence band, the conductivity band, and occupies the shallow donor state, respectively. The blue and green arrows indicate optical transitions during excitation and recombination, respectively.

The decrease of x_{Al} from 1 to 0.74 leads to decrease difference in energy position of E_1 and E_2 peaks from 1.06 eV to 0.5 eV. It's known that Si on cation site is considered as a shallow donor with a binding energy of 28-30 meV in GaN [15-17], and remain as shallow donor with the silumar energy up to x_{Al}~0.85 with further deepening to 250 meV in AlN [10]. Assuming $E_D \approx 0.03$ eV we obtain $S_{em}h\nu_0 \approx 1.1$ eV and $E_A \approx 1.6$ eV for $Al_{0.74}Ga_{0.26}N$. So the decrease of Al content leads to decrease of E_A and increase of S_{em} or $h\nu_0$. From expression (2) and weaker dependence of E_{em} on x_{Al} than E_g (see figure 2) we obtain

$$\Delta(E_A + S_{em}h\upsilon_0 + E_D) = \Delta(E_{gAlGaN} - E_{em}) \approx 1.1 \cdot \Delta x_{Al} \tag{6}$$

5. Conclusions

The photoluminescence properties of the defect related emission attributed to the superposition of donor-acceptor and free electron-acceptor transitions in heavily doped $Al_xGa_{1-x}N$:Si layers with $x_{Al}>0.38$ have been investigated. The increase of Al content from $x_{Al}=0.47$ to $x_{Al}=1$ leads to the blue shift of the energy position of the PL band from red-orange spectral range to violet spectral range. The acceptor ionization energy was estimated to be 1.87 eV at $x_{Al}=1$, decrease of to x_{Al} leads to decrease of this value. The donor is supposed to be Si_{III}, while the origin of the acceptor is not completely clear. Possible candidates are (0/-) transition level of the C_N or (2-/3-) transition level of the V_{Al}.

Acknowledgments
This work was supported by Russian Foundation for Basic Research (grants no. 16-32-00773, 16-32-00765 and 16-02-00018).

IOP Conf. Series: Journal of Physics: Conf. Series **864** (2017) 012071 doi:10.1088/1742-6596/864/1/012071

References

[1] Bokhan P A, Gugin P P, Zakrevsky Dm E, Zhuravlev K S, Malin T V, Osinnykh I V, Solomonov V I and Spirina A V 2014 J. Appl. Phys. **116** 113103.

[2] Osinnykh I V, Malin T V, Plyusnin V F, Suranov A S, Gilinsky A M, and Zhuravlev K S 2016 Jpn. J. Appl. Phys. **55** 05FG09.

[3] Muth J F, Brown J D, Johnson M A L, Yu Zhonghai, Kolbas R M, Cook Jr J W and Schetzin J F 1999 MRS Internet J. Nitride Semicond. Res. **4S1** G5.2.

[4] Feneberg M, Leute R A R, Neusch B, Thonke K, and Bickermann M 2010 Phys. Rev. B **82** 075208.

[5] Koide Y, Itoh H, Khan M R H, Hiramatsu K, Sawaki N, and Akasaki I 1987 J. Appl. Phys. **61** 4540.

[6] Zhuravlev K S , Osinnykh I V, Protasov D Yu, Malin T V, Davydov V Yu, Smirnov A N, Kyutt R N, Spirina A V, Solomonov V I 2013 Phys. Status Solidi C **10** 315.

[7] Huang K and Rhys A 1950 Proc. R. Soc. A **204** 406.

[8] Henry C H and Lang D V 1977 Phys. Rev. B. **15** 989.

[9] Silvestri L, Dunn K, Prawer S, and Ladouceur F 2011 Appl. Phys. Lett. **99** 122109.

[10] Borisov B, Kuryatkov V, Kudryavtsev Yu, Asomoza R, Nikishin S, Song D Y, Holtz M, Temkin H 2005 Appl. Phys. Lett. **87** 132106.

[11] Taniyasu Y, Kasu M, and Makimoto T 2004 Appl. Phys. Lett. **85** 4672.

[12] Limpijumnong S and Van de Walle C G 2004 Phys. Rev. B **69** 035207.

[13] Zhang Y, Liu W, Niu H 2008 Phys. Rev. B **77** 035201.

[14] Bogusławski P and Bernholc J 1997 Phys. Rev. B **56** 9496.

[15] Wang H and Chen A-B 2000 J. Appl. Phys. **87** 7859.

[16] Martı ́nez-Criado G, Miskys C R, Cros A, Ambacher O, Cantarero A, Stutzmann M 2001 J. Appl. Phys. **90** 5627.

[17] Moore W J, Freitas J A, Lee S K, Park S S, Han J Y 2002 Phys. Rev. B **65** R 081201.

Circular photon drag effect in bulk semiconductors

V A Shalygin[1], M D Moldavskaya[1], S N Danilov[2], I I Farbshtein[3] and L E Golub[3]

[1] Department of Physics of Semiconductors and Nanoelectronics, Peter the Great St. Petersburg Polytechnic University, St. Petersburg 195251, Russia
[2] Terahertz Center, University of Regensburg, Regensburg 93040, Germany
[3] Ioffe Institute, St. Petersburg 194021, Russia

E-mail: shalygin@rphf.spbstu.ru

Abstract. We report on the observation of the circular photon drag effect in a bulk semiconductor. The photocurrent caused by a transfer of both linear and angular momenta of photons to charge carriers is detected in tellurium. Dependencies of the photocurrent on the light polarization and on the incidence angle agree with the symmetry analysis of the circular photon drag effect. Experimental spectral data on the photocurrent in mid-infrared range qualitatively agree with a microscopic model of the circular photon drag effect considering intersubband optical transitions of holes in tellurium.

1. Introduction

Circular photon drag effect (CPDE) represents an electrical current arising in medium under optical excitation due to simultaneous transfer of both linear and angular momenta from photons to free charge carriers. This photocurrent is sensitive to the light helicity and reverses its direction on switching from right- to left-circular polarization of light. In contrast to the intensively studied photon drag currents insensitive to the circular polarization (for review see [1]), the history of CPDE is rather short. While CPDE was predicted theoretically in 1980s [2, 3], it was demonstrated experimentally only in 2006 by studying quantum wells [4]. Further experiments also dealt exclusively with 2D systems such as photonic crystal slabs [5], grapheme [6–8], metamaterials [9] and quantum well structures [10]. However, CPDE has not been observed so far in any bulk (3D) material. The reason is that CPDE is forbidden by symmetry in cubic crystals like III–V semiconductors, or it is masked by other effects in those media where it is allowed.

In this work, we report on the observation of CPDE in a bulk semiconductor demonstrating that the photon drag current sensitive to the light helicity is possible in 3D crystals. For investigations we choose tellurium, because it demonstrates a few related phenomena, namely, electric-current-induced optical activity, the linear photon drag effect, and linear and circular photogalvanic effects [1, 11]. In contrast to 2D systems [4, 6–8], for tellurium we can choose a particular geometry where CPDE is not hidden by any other effect. We show experimentally that the values of CPDE current in tellurium are two orders of magnitude higher than in quantum well structures. The phenomenological model of CPDE is developed on the base of symmetry arguments accounting for birefringence of tellurium. Microscopic mechanism of the observed effect is discussed.

2. Phenomenological description of CPDE

In order to choose a proper geometry for observation of the CPDE current we perform the symmetry analysis of helicity-dependent photocurrents in tellurium. The point symmetry group of tellurium is D_3. In the plane perpendicular to the optical axis z, there are three rotation axes C_2. We denote one of them as x and also denote a perpendicular axis in the same plane as y (see inset in figure 1). We consider a radiation incident in the plane xz. Performing a symmetry analysis, we obtain the photocurrent which reverses its direction under switching from right-hand- to left-hand-polarized radiation. The density of this photocurrent j_{circ} proportional to the circular polarization degree of light P_{circ} is given by

$$j_z^{circ} = \gamma P_{circ} \frac{q_z}{q} E^2 , \quad (1)$$

$$j_x^{circ} = \tilde{\gamma} P_{circ} \frac{q_x}{q} E^2 + \tilde{T} P_{circ} \frac{q_x^2}{q} E^2 , \quad (2)$$

$$j_y^{circ} = T P_{circ} \frac{q_x q_z}{q} E^2 . \quad (3)$$

Here $\boldsymbol{q}$ and $\boldsymbol{E}$ are the radiation wave vector and electric field, respectively. The constants γ and $\tilde{\gamma}$ describe the circular photogalvanic effect caused solely by transfer of an angular momentum of photons to free carriers but not accompanied by a linear momentum transfer. The longitudinal CPDE current described by the constant $\tilde{T}$ is present due to a trigonal symmetry of tellurium. The transverse CPDE described by the constant T is caused by a nonequivalence of the z direction and the directions in the perpendicular plane xy, i.e., due to uniaxiality of tellurium. This CPDE current is odd in the incidence angle θ_0. Equations (1)–(3) demonstrate that the current j_y^{circ} transverse to the incidence plane xz is caused solely by the CPDE, in contrast to two other photocurrent components. Therefore in the experiments aimed at observation of CPDE we focus on the photocurrent j_y^{circ}.

To derive the dependence of j_y^{circ} on the incidence angle it is necessary to take into account that tellurium is a birefringent crystal. Since the ordinary and extraordinary beams propagate with different velocities, the CPDE current oscillates in space along the propagation direction. To obtain the resultant CPDE current density j_L^{CPDE} in a sample of thickness L, one can integrate j_y^{circ} over the L. As it was shown in our paper [12], the resultant CPDE current density depends on θ_0 as follows

$$j_L^{CPDE}(\theta_0) = T \frac{\omega}{c} P_{circ}^0 E_0^2 \tau_{ps}(\theta_0) \sin\theta_0 \frac{\sin[2\pi L / d(\theta_0)]}{2\pi L / d(\theta_0)} , \quad (4)$$

where the transmission coefficient $\tau_{ps}(\theta_0)$ and oscillation period $d(\theta_0)$ are determined in [12].

3. Experimental observation of CPDE and discussion of its microscopic mechanism

Experiments were performed on a p-Te single crystal (hole concentration $7 \cdot 10^{16}$ cm^{-3}) at room temperature. The lateral surface of the samples was a natural facet of the crystal, and the end faces were subjected to optical polishing. The crystal under investigation exhibited a natural optical activity and was levorotatory. The thickness of the sample in z-direction was L = 0.8 mm. To measure the photocurrent Jy, two contacts were located at the lateral surface of the sample (see inset in figure 1). We applied pulsed radiation of a Q-switched CO_2 laser operating in the spectral range from 114 to 135 meV. The radiation was focused on a spot of 0.5 mm diameter. The photocurrent was measured by means of a storage oscilloscope. The sample was illuminated by a laser beam under incidence angle θ_0. The laser radiation was linearly polarized. By applying a Fresnel $\lambda/4$ rhomb we modified the radiation polarization from linear to elliptical. The circular polarization degree of the light at the Fresnel rhomb output P_{circ}^0 is changed according to $P_{circ}^0 = \sin 2\varphi$ where φ is the azimuth of the Fresnel rhomb.

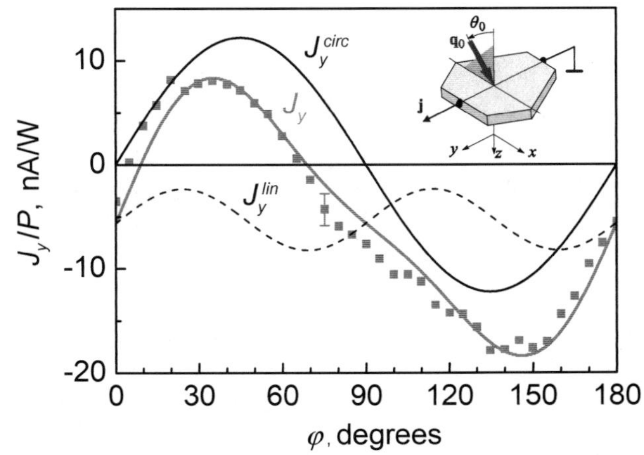

Figure 1. Dependence of the transverse photocurrent on the Fresnel rhomb azimuth at the photon energy $\hbar\omega = 134$ meV. Curves $J_y^{circ} \propto \sin 2\varphi$ and J_y^{lin} represent the circular and linear photocurrents, respectively. Inset shows the experimental geometry.

The photocurrent at each azimuthal angle φ depends linearly on the laser power P. Figure 1 shows the typical dependence of the transverse photocurrent normalized to the laser power (J_y/P) on the Fresnel rhomb azimuth (φ) under oblique incidence of the laser beam (at $\theta_0 = -15°$). The experimental data are well described by the following phenomenological expression:

$$J_y / P = C \sin 2\varphi + L_1 \sin 4\varphi + L_2, \tag{5}$$

where the first term is proportional to P_{circ}^0 and corresponds to the "circular" photocurrent $J_y^{circ} \propto \sin 2\varphi$ which we are interested in. Two other terms represent the "linear" photocurrent J_y^{lin} which appears under elliptically polarized excitation. The linear photocurrent is insensitive to the radiation helicity.

We analyzed the φ-dependencies of the transverse photocurrent J_y at various incidence angles θ_0 and revealed the dependence of the circular photocurrent amplitude C on the incidence angle. It is presented in figure 2(a) by circles. One can see that the circular photocurrent is mainly an odd function of θ_0 with an admixture of a small even contribution. According to the phenomenological arguments (see section 2), the CPDE current is an odd function of the incidence angle. Therefore we continued our analysis studying the odd in θ_0 part of C. It is defined as follows: $C_{odd}(\theta_0) = (C(\theta_0) - C(-\theta_0))/2$. Experimental data on $C_{odd}(\theta_0)$ are shown in figure 2(a) by squares. They are fitted well by the average CPDE current density $j_L^{CPDE}(\theta_0)$ given by equation (4). This finding confirms that the dominating odd in θ_0 contribution to the circular photocurrent is due to the CPDE. We note that the CPDE current detected in tellurium exceeds by two orders of magnitude the current observed in quantum-well structures [4].

We performed the same measurements and analysis at three other photon energies. The obtained dependence of $|C_{odd}|$ on the photon energy is shown in figure 2(b). In the spectral range of 116–134 meV, absolute value of the observed CPDE current decreases with the photon energy. Such a behavior qualitatively agree with a microscopic model considering intersubband optical transitions of holes in tellurium for the photon energies exceeding the intersubband gap [12]. In the framework of this model, CPDE current is estimated simulating shifts of carriers in real space occurring at photon absorption. Simulation gives $j_L^{CPDE} / I \approx 50$ nA/W at $\hbar\omega = 130$ meV and $\theta_0 = 10°$ (I is the radiation intensity). This is the same order of magnitude as that of the experimental data (see figure 2). Slightly smaller values of the current (~10 nA/W) are detected in the experiment because not all current generated in the laser spot area reaches the contacts. Part of the current is closed in the nonilluminated part of the sample.

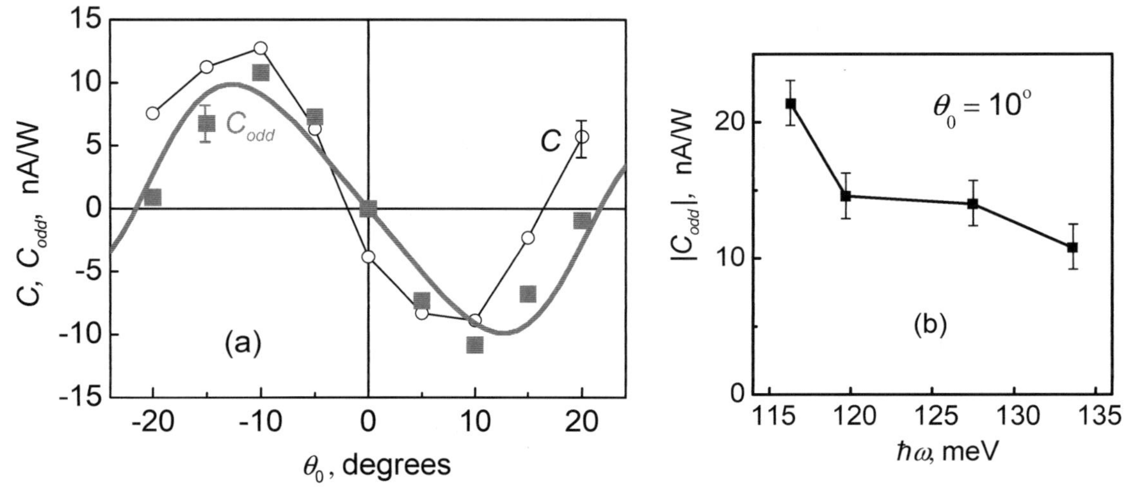

Figure 2. (a) Incidence angle dependence of the circular photocurrent amplitude C (circles) and its odd in θ_0 component C_{odd} (squares) at $\hbar\omega = 134$ meV. Thick line is a fit of C_{odd} by $j_L^{CPDE}(\theta_0)$ given by equation (4). (b) Dependence of the CPDE amplitude $|C_{odd}|$ on the photon energy.

4. Conclusions

The helicity-dependent photocurrent transverse to the light incidence plane was detected in bulk tellurium. The above analysis of the polarization state, incidence angle, and photon energy dependencies of the photocurrent confirms the observation of the CPDE. The CPDE current is shown to be an odd function of the incidence angle. The phenomenological model of CPDE was developed based on symmetry arguments accounting for birefringence of tellurium. Experimental spectral data on the photocurrent qualitatively agree with a microscopic model of the circular photon drag effect considering intersubband optical transitions of holes in tellurium. Due to the high sensitivity of CPDE, tellurium can be used for helicity-dependent photodetectors in mid-infrared range.

Acknowledgements

This work was supported by the Ministry of Education and Science of the Russian Federation (state assignment). The work of L.E.G. was supported by the Russian Foundation for Basic Research.

References

[1] Ganichev S D and Prettl W 2006 *Intense terahertz excitation of semiconductors* (Oxford (USA): Oxford University Press)
[2] Ivchenko E L and Pikus G E 1986 *Semiconductor Physics* ed V M Tuchkevich and V Ya Frenkel (New York: Consultant Bureau) p 427
[3] Belinicher V I 1981 *Sov. Phys. Solid State* **23** 2012
[4] Shalygin V A *et al* 2007 *JETP Letters* **84** 570
[5] Hatano T *et al* 2009 *Phys. Rev. Lett.* **103** 103906
[6] Karch J *et al* 2010 *Phys. Rev. Lett.* **105** 227402
[7] Jiang Ch *et al* 2011 *Phys. Rev.* B **84** 125429
[8] Olbrich P *et al* 2013 *Phys. Rev.* B **88** 245425
[9] Kang L *et al* 2015 *Adv. Mater.* **27** 4377
[10] Stachel S *et al* 2014 *Phys. Rev.* B **89** 1154354
[11] Shalygin V A *et al* 2012 *Physics of the Solid State* **54** 2362.
[12] Shalygin V A *et al* 2016 *Phys. Rev.* B **93** 045207

Excitation and decay of surface plasmon polaritons in *n*-GaN

V A Shalygin[1], G A Melentev [1], L E Vorobjev[1], V Yu Panevin[1], D A Firsov[1], V V Korotyeyev[2], Yu M Lyaschuk[2], V A Kochelap[2] and S Suihkonen[3]

[1] Department of Physics of Semiconductors and Nanoelectronics, Peter the Great St. Petersburg Polytechnic University, St. Petersburg 195251, Russia
[2] Department of Theoretical Physics, Institute of Semiconductor Physics NASU, Kyiv 03028, Ukraine
[3] School of Electrical Engineering, Aalto University, Espoo 02150, Finland

E-mail: shalygin@rphf.spbstu.ru

Abstract. We report on the studies of the surface plasmon polaritons in *n*-GaN epitaxial layers. The grating etched on the surface of the epitaxial layer is used to excite surface plasmon polaritons by means of terahertz photons. The experimental reflectivity spectrum for *p*-polarized radiation demonstrates a set of resonances associated with excitation of different surface plasmon polariton modes. Spectral peculiarities due to the diffraction effect have been also revealed. Emission of terahertz radiation is investigated under epilayer temperature modulation by electric current. The emissivity spectrum of the epitaxial layer with surface-relief grating shows peaks in the frequency ranges corresponding to the decay of the surface plasmon polariton modes. The characteristic features of the reflectivity and emissivity spectra are well described theoretically by a differential method with explicit integration scheme.

1. Introduction

During recent years, there has been a growing interest in the study of surface plasmon polaritons (SPPs) which represent electromagnetic (EM) waves propagating at the interface between a dielectric and a conductor, evanescently confined in the perpendicular direction. The remarkable properties of the SPPs are widely utilized in science and technology [1]. The modern trend in the field of plasmonics is terahertz (THz) plasmonics which deals with heavily doped semiconductors [2]. The frequency selective properties of semiconductor plasmonic structures in the THz frequency range may find a number of applications, particularly, in the fields of THz biosensing and THz bioimaging. Among different kinds of the semiconductor compounds that are considered as a platform for sensor devices, gallium nitride is assumed to be one of the most favourable.

In the present paper we report on the experimental and theoretical studies of the SPPs in heavily doped GaN epitaxial layers. Reflection and emission of radiation in the frequency range of 2–20 THz including the Reststrahlen band are investigated for samples with grating etched on the sample surface, as well as for samples with planar surface. A differential method with explicit integration scheme [3] is used for theoretical interpretation of the reflectivity and emissivity spectra of the samples with profiled surface. We treat these spectra in the terms of excitation and decay of SPPs.

2. Results and discussion

Wurtzite GaN epitaxial layers of 6.2 μm thickness were grown on sapphire substrate by MOVPE. The epilayers had the electron concentration of $3.6 \cdot 10^{19}$ cm^{-3} and the mobility of 122 cm^2/V·s at room temperature. A grating with a period of a = 86 μm and ridge width $w = a/2$ was etched on the surface of epilayer. The etch depth was 4.5 μm. The reference samples without grating were studied as well. In some experiments, the epilayer temperature was modulated by pulsed electric current.

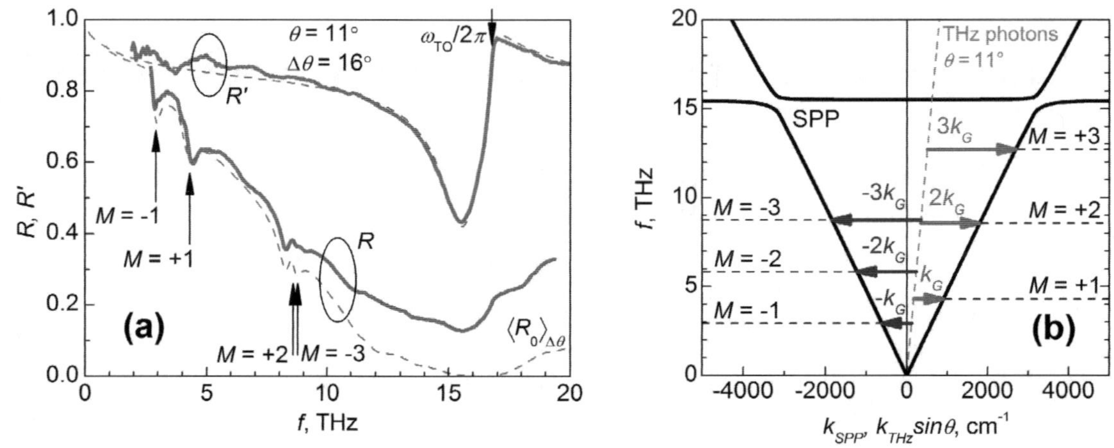

Figure 1. (a) Reflectivity spectra for p-polarized radiation for the samples with a surface-relief grating (R) and with a planar surface (R'). Experimental data are shown by solid lines, the results of theoretical simulation are depicted by dashed lines. (b) Dispersion dependences for surface plasmon polaritons (solid line) and THz photons at θ = 11° (dashed line). Horizontal dashed lines show spectral positions of the SPP resonances of different orders M.

The THz reflectivity spectra were investigated at room temperature using a Fourier spectrometer. The incident radiation beam had an angular aperture of $\Delta\theta$ = 16°. The axis of the beam formed the angle θ = 11° with the normal to the sample surface. The plane of incidence was perpendicular to the ridges of grating. Experimental reflectivity spectra for p-polarized radiation are plotted in figure 1(a). It should be emphasized that spectra for the grating and planar samples differ drastically.

For the planar sample, the recorded reflectivity spectrum (solid curve R') has two distinctive behaviors depending on the ratio between the contribution of electron and phonon subsystems to the complex permittivity. Theoretical simulation of the reflectivity taking into account the both contributions (dashed curve R') adequately describes the experimental data. In the lower frequency range (f<10 THz), the electron contribution dominates resulting in weak frequency dependence of the reflectivity. The strong frequency dependence of the reflectivity is observed in the spectral range of 14–20 THz where the contribution of the phonon subsystem dominates over electrons (Reststrahlen band). The significant decrease of the reflectivity corresponds to the low-frequency plasmon-phonon mode located closely to $\omega_{\mathrm{TO}}/2\pi$. Note that the real part of the permittivity is negative at f< 15 THz, thus in this spectral range the necessary condition for SPP propagation is satisfied.

For the grating sample, experiment demonstrates a significant decrease of the reflectivity (solid curve R in figure 1(a)) in the whole examined spectral range (2–20 THz). Firstly, one can see a sequence of sharp dips. They arises for p-polarized radiation only and we attribute them to an excitation of SPPs. Actually, it is known that only radiation having a component of electric field perpendicular to the grating ridges can excite surface waves. This causes dissipation of the radiation energy and is seen as a decrease of reflectivity. The excitation of SPPs has a resonant behavior, and the resonant frequencies can be found using the phase-matching condition [1]

$$k_{\mathrm{SPP}} = k_{\mathrm{THz}} \sin\theta + Mk_G, \qquad M = \pm 1, \pm 2, \pm 3, \ldots \tag{1}$$

and the SPP dispersion law for the epilayer/vacuum interface [1]

$$k_{\mathrm{SPP}}^2 = \left(\frac{\omega}{c}\right)^2 \frac{\varepsilon_{\mathrm{c}}(\omega)}{\varepsilon_{\mathrm{c}}(\omega)+1}, \tag{2}$$

where k_{SPP} and k_{THz} are the wave vectors of SPP and THz photon, ω is angular frequency, $k_{\mathrm{G}} = 2\pi/a$, $\varepsilon_{\mathrm{c}}(\omega)$ is the permittivity of the conducting epilayer. The integer numbers M correspond to different modes of the SPP resonances (figure 1(b)). The second feature of the grating sample reflection (R) is the collapse of the Reststrahlen band that is well pronounced for the planar sample (R').

We performed theoretical calculations of the grating sample reflectivity using the model and differential method with explicit integration scheme described in [3]. Simulated reflectivity of the zero diffraction order ($m = 0$) is shown in figure 1(a) by dashed curve $\langle R_0 \rangle_{\Delta\theta}$ (angle brackets denote averaging over $\Delta\theta$). Simulated spectrum matches well with the experimental spectrum in the spectral range of 0–10 THz. The frequencies, amplitudes, and widths of the resonance dips in the simulated and experimental spectra are close to each other. The SPP resonances are clearly seen at frequencies of 2.9, 4.4, 8.3 and 8.7 THz. According figure 1(b), we identified that these frequencies correspond to the SPP resonances of orders $M = -1, +1, +2, -3$, respectively.

A deviation between the simulated and observed reflectivity spectra arises at higher frequencies ($f > 10$ THz). This discrepancy is due to the diffraction phenomenon. Indeed, together with the excitation of SPP modes, the surface grating splits the incident beam into several reflected beams propagating within different solid angles. The angles between the zero order (R_0) and higher order (R_m, $m \neq 0$) diffraction beams decrease with increasing frequency (see figure 2). Consequently, in the high frequency range, the angular aperture of detection system (in our case $\Delta\theta = 16°$) can partially catch the beams of higher diffraction orders. The estimates show that the beams of the ±1st diffraction orders are partially overlap with the zero-order reflected beam at $f > 10$ THz (see figure 2(b)). That is why the observed reflectivity exceeds the simulated reflectivity of the zero diffraction order $\langle R_0 \rangle_{\Delta\theta}$ in this spectral range.

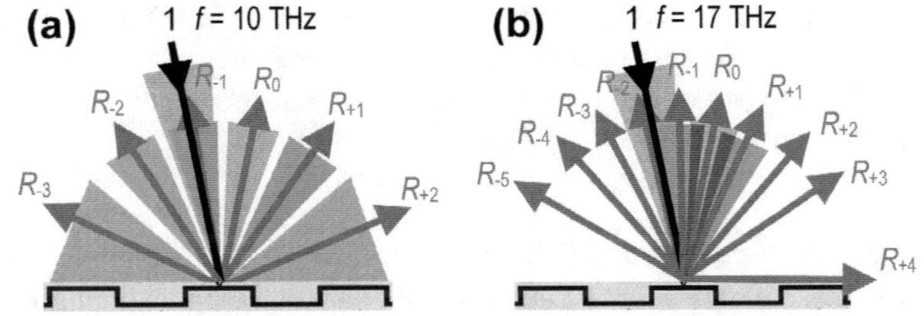

Figure 2. Reflected beams R_m of the different diffraction orders m for two frequencies of the incident radiation: $f = 10$ THz (a) and 17 THz (b). Arrow "1" denotes the direction of the incident beam ($\theta = 11°$). Colored sectors show the angular apertures of the reflected beams corresponding to the angular aperture of incident beam $\Delta\theta = 16°$.

Emission of THz radiation from the samples was measured under epilayer temperature modulation by the pulsed electric current. The spectral studies of the emitted radiation were carried out under sample cooled down to 9 K using a Fourier spectrometer operating in a step-scan mode. The THz radiation was collected in the direction perpendicular to the sample surface, the angular aperture was $\Delta\theta \approx 16°$. A silicon bolometer was used as a detector.

To clarify the distinctive emission properties related to SPPs, we measured spectral dependencies of the detector photoresponse for the grating and planar sample, denoted as U and U', respectively. The same electric power of 1500 W was applied to the sample in both cases providing Joule heating of the epitaxial layer up to ~100 K. The experimental results on THz emission from the grating and planar samples are presented in figure 3(a) by thick line which shows the ratio $U(f)/U'(f)$. Three peaks of THz emission in spectral intervals corresponding to the phase matching condition for SPPs and THz photons for $M = \pm1$, ±2 and ±3 (see figure 3(b)) were detected in the grating sample. Therefore, one

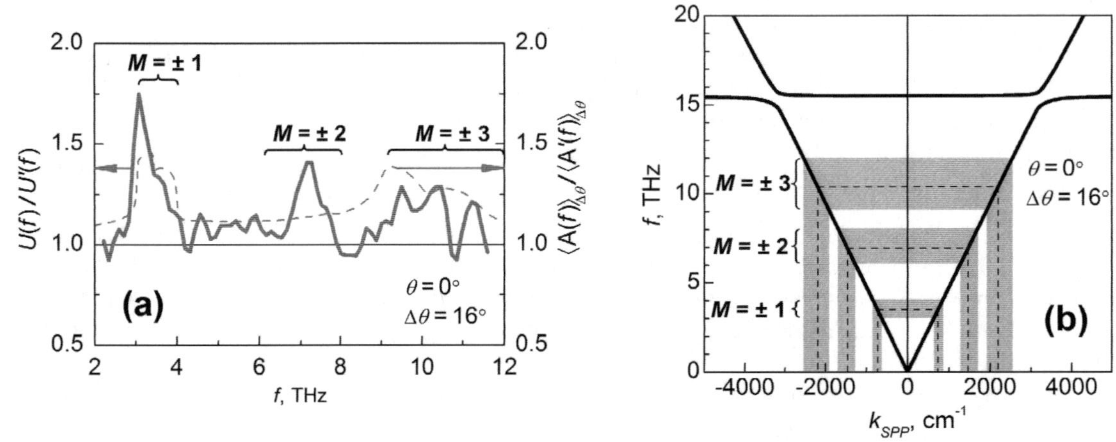

Figure 3. (a) Emissivity spectra of the grating sample normalized to the planar epitaxial layer. Solid line represents experimental data on the photoresponse ratio at applied electric power of 1500 W, dashed line shows the results of theoretical simulation. (b) Spectral intervals corresponding to the plasmonic resonances of different orders M (simulation at $\theta = 0°$, $\Delta\theta = 16°$).

can conclude that the characteristic emission peaks originate from the radiative decay of SPP excitations. In the framework of the model of blackbody-like radiation [3], this ratio is expected to be equal to the ratio $\langle A(f)\rangle_{\Delta\theta} / \langle A'(f)\rangle_{\Delta\theta}$, where A and A' are the simulated absorptivities of the grating and planar samples, respectively. Indeed, the calculated curve (dashed line in figure 3(a)) describes qualitatively the major features of the experimental spectrum. In particular, the calculated positions and widths of the SPP peaks for $M = \pm1$ and ±3 are very close to the experimentally measured. We associate a mismatch between theory and experiment for the second-order SPPs ($M = \pm2$) with the certain restrictions of the considered theoretical model.

3. Conclusions

The experimental and theoretical results of this work provide insights for the development of GaN-based devices aimed to absorb/emit terahertz radiation selectively. In particular, our studies can be applied for the development of portable sources of THz radiation operating under electric pumping. The effects of essential light absorption in a narrow spectral and angular intervals analyzed above can find applications in chemical and bio-chemical surface plasmon grating detectors working in THz frequency range.

Acknowledgements

This work was supported by the RFBR (projects Nos. 14-02-90444 and 16-32-60085), the Ministry of Education and Science of the Russian Federation (state assignment), the National Academy of Sciences of Ukraine (project 25-02-14), and the Academy of Finland (project No. 13251864).

References

[1] Maier S A 2007 *Plasmonics: Fundamentals and Applications* (New York: Springer)
[2] Yu N *et al* 2010 *Electron. Lett.* **46**(26) 52
[3] Melentev G A *et al* 2016 *J. Appl. Phys.* **119** 093104

IOP Conf. Series: Journal of Physics: Conf. Series **864** (2017) 012074 doi:10.1088/1742-6596/864/1/012074

Photoluminescence of the PbS quantum dots fabricated by the Langmuir-Blodgett technique

D.V. Gulyaev[1], L.L. Sveshnikova[1], S.A. Bacanov [1], A.K. Gutakovskii[1,2] and K.S. Zhuravlev[1,2]

[1] Rzhanov Institute of Semiconductor Physics, Siberian Branch of Russian Academy of Sciences, 13, Lavrentiev avenue, Novosibirsk, 630090, Russia,
[2] Novosibirsk State University, 2, Pirogov Street, Novosibirsk, 630090, Russia

E-mail: gulyaev@isp.nsc.ru

Abstract. The optical and structure properties of the PbS nanoclusters in the Langmuir-Blodgett matrix as well as their properties after the matrix removal in the ammonium atmospheres have been studied. According to the electron microscopy data, it has been established that there are PbS nanoclusters formed on the surface of the samples. It has been found that the matrix removal results in a significant enlargement of the PbS nanocrystals accompanied by the shift of the luminescence into the infra-red range. It has been shown that the PL intensity of PbS quantum dots (QDs) in the matrix decreases with the temperature decrease due to the carrier transfer into defect QDs whereas the the luminescence intensity of the PbS nanocrystals without a matrix grows at the temperature increase.

1. Introduction

Today PbS is the one of the most common narrow band gap semiconductor used in infra-red detectors, solar cell sensors and other elements in the semiconductor industry. Therefore, there are numerous methods described on the preparation of PbS itself and low-dimensional structures on its basis in an attempt to reduce the price of PbS preparation and/or improvement of its properties [1-5]. One of the lowest-cost technologies of the preparation of PbS quantum dots is the Langmuir-Blodgett (LB) technique which does not require the expensive equipment. However, the presence of a solid matrix in this method means the presence of surface defects in PbS nanostructures that can significantly influence its optical properties. But the recombination properties of PbS nanocrystals (NCs) synthesized by the LB method have been poorly studied so far [3]. Consequently, this research has been focused on the temperature and power dependences of the photoluminescence (PL) of PbS NCs formed by the LB method before and after the matrix removal.

2. Experimental methods

The structure under study was obtained by the Langmuir-Blodgett technique on the sapphire and porous carbon substrates. Initially, the lead behenate was grown by the substrate with thickness from 4 to 100 monolayes (ML's) for different samples. Then a behenic acid – matrix ($C_{21}H_{43}COOH$) and lead sulfide particles were formed as a result of sulfidizing. Finally, to remove the LB matrix several samples were annealed in the ammonia atmosphere NH3 at the temperature of 125 °C.

The investigation of the sample structures was made by the high-resolution transmission electron microscopy (HRTEM). The studies were carried out by the electron microscope of high resolution -

JEM-4000EX (JEOL) equipped with the ultra-high-resolution pole piece UHP-40H with the high resolution of 0.16 nm.

The optical measurements were carried out by the photoluminescence method. To excite the PL the HeCd laser with the wavelength/maximum power of 325 nm/10 mW or the YAG with the wavelength/maximum power of 512 nm/140 mW were used, respectively. The registration of the PL was conducted by the double diffraction monochromator equipped with different photodetectors to register visible and infra-red (IR) photons. The luminescence in a visible range was measured by the Si CCD matrix, whereas the luminescence in a near IR range ((0.8-1.7 mkm)) was measured by the Ge detector.

3. Results and discussion

The typical HRTEM image of the LB film of the lead behenate after sulfiding is shown in figure 1. As one can see, the nanoparticles visualized on this image as brighter areas in comparison with the background demonstrate the crystallinity properties. According to the analysis of the interplanar spacing, the PbS nanoparticles have a cubic crystal structure with the Fm3m space group with the cell parameter a = 0.59 nm. Therefore, we can conclude that the nanocrystals of the lead sulfide with the average size of 2-3 nm and surface density of 10^{12} $cm^{-}2$ are formed as the result of the sulfiding of the lead begenate films.

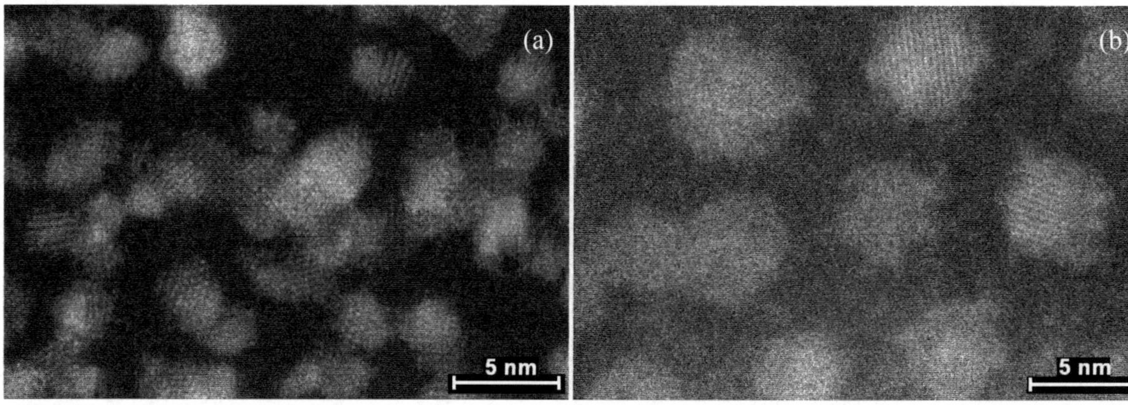

Figure 1. The TEM images of the (a) sample with PbS nanocrystalls in the matrix of 4 ML thickness and (b) sample with PbS nanocrystalls annealed at 125 C. The sample substrate is a carbon film.

The typical NC dispersion is 30 per cent. The annealing in the ammonia atmosphere results in the double increase of the average NC sizes (see Figure 1(b)). It should be also noted that the crystal conglomerates with the size of over 20 nm are formed as a result of the annealing but with a significantly smaller surface density.

The measurements of the optical properties of PbS nanocrystalls were carried out to confirm the quantization of the PbS nanocrystalls. The typical PL spectra of the samples under study at the room temperature are shown in figure 2. As one can see, the band with the maximum radiation of around 1.82 eV and a half width at the half-maximum (HWHM) of 110 meV dominates in the photoluminescence (PL) spectra of the non-annealed samples, whereas the band with the maximum radiation of around 0.8 eV and HWHM of 70 meV dominates in the PL spectra of the annealed samples. As these bands are absent in the PL spectra of the reference sample consisting of a single matrix, this allows us to refer them to the recombination of carriers in the PbS crystals. The shift of the PL band into a higher energy side can be explained only by the quantization effect as the band gap of the bulk PbS is equal to 0.43 eV [6]. According to the literature data [7], the energy of 1.8 eV corresponds to the QD of 2 nm thickness, whereas the energy of 0.8 eV corresponds to the PbS QDs of 5-6 nm diameter, which is in accordance with the microscopy data.

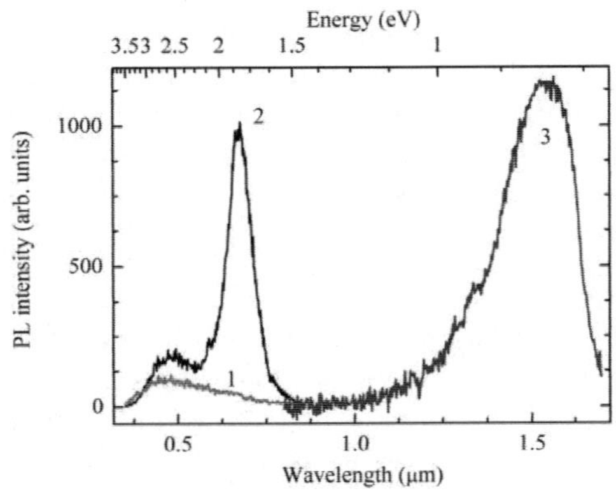

Figure 2. The PL spectra of the (1) LB matrix (2) sample with the PbS nanocrystals in the matrix and (3) the sample with the PbS nanocrystals after annealing. The measurement temperature is 300 K. The power of the HeCd laser is 10 mW.

The temperature dependencies of the PL intensity of the PbS QDs in the matrix and after the matrix removal are shown in figure 3. As one can see, the luminescence intensity of the PbS QDs in the matrix decreases at the temperature increase. The temperature dependence of the PL intensity of the PbS QDs in the matrix is well described by the following equation:

$$I(T) = \frac{I(0)}{1 + A\cdot\exp(-\frac{E_A}{kT}) + B\cdot\exp(-\frac{E_B}{kT})} \quad (1),$$

where the activation energy of the PL quenching is E_A=100±20 and E_B=14±3 meV. As the dependence of the PL intensity upon the excitation power remains linear at the temperature change (see figure 4) we suppose that the PL intensity decrease is caused by the transfer of the carriers from the bright non-defect QDs to the dark defect ones. In this case the number of the bright QDs and, consequently, the PL intensity decrease but the mechanisms of the carrier recombination in a QD do not change. Thus, the obtained energy can correspond to the average energy of the carrier delocalization from the QDs and the exciton binding energy, respectively.

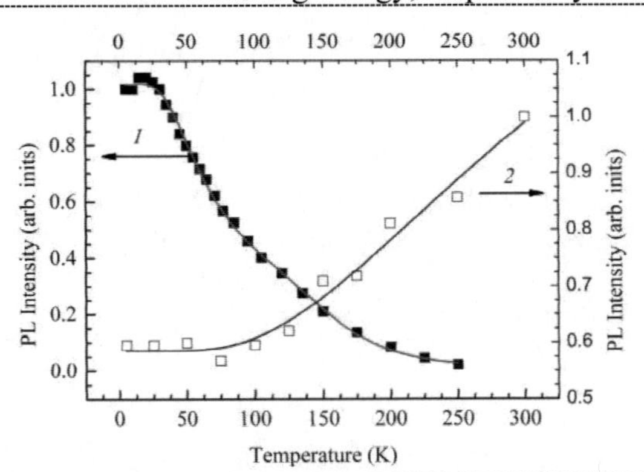

Figure 3. The temperature dependencies of the PL intensity of the PbS QDs in the matrix (1, ■) and without the matrix (2, □). The solid lines show the curves approximated by the equations (1) and (2). The excitation power of the YAG laser is 100 mW.

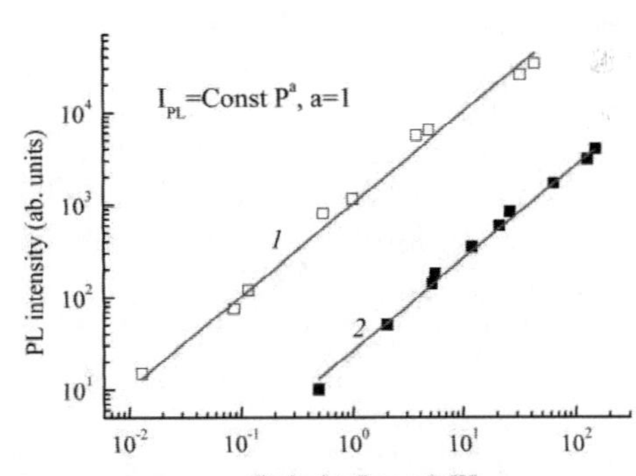

Figure 4. The dependence of the PL intensity of the PbS QDs without a LB matrix on the excitation power (the PL band maximum is 0.8 eV) at the (1) 5 and (2) 300 K. T=300 K. The PL was excited by the YAG laser.

IOP Conf. Series: Journal of Physics: Conf. Series **864** (2017) 012074 doi:10.1088/1742-6596/864/1/012074

The removal of the matrix in the ammonia atmosphere leads to the cardinal change in the temperature dependence of the PbS QDs, namely, the PL intensity starts increasing at the temperature increase. Therefore, we can conclude that the non-radiative recombination in the QDs is caused by the defects on the matrix/nanocrystal boundary. The temperature dependence of the PL intensity of the PbS QDs without a matrix is well described by the following equation:

$$I(T) = I(0) \cdot (1 + \exp(-\frac{E_C}{kT})) \qquad (2),$$

The activation energy of the PL enlargement is equal to E_C=40±5 meV that is close to the levels of the Pb vacancy in PbS [8]. It should be noted that the intensity of the PL of the PbS QDs without a matrix keeps increasing at the delay increase after the sample annealing, which can be explained as by the self-purification in semiconductor nanocrystals [9] as by the passivation of residual surface states by the oxygen.

4. Conclusion

The optical and structure properties of the PbS nanoclusters in the Langmuir-Blodgett matrix as well as their properties after the matrix removal in the ammonia atmosphere have been studied. The analysis of the obtained results has confirmed the formation of PbS QDs of 2-3 nm size in the LB matrix. The temperature and power dependences of QD PL indicate the possibility of the carrier transfer between the non-defect QDs and QDs with the surface states on the matrix boundary. The matrix removal in the ammonia atmosphere results in the enlargement of PbS QDs up to 5-6 nm accompanied by the passivation of surface defect states.

References

[1] H. Zhao, Z. Fan, H. Liang, G. S. Selopal, B. A. Gonfa, L. Jin, A. Soudi, D. Cui, F. Enrichi, M. M. Natile, I. Concina, D. Ma, A. O. Govorov, F. Rosei and A. Vomiero 2014 *Nanoscale* **6** 7004.

[2] Y. Yang, T. Lian 2014 *Coordination Chemistry Reviews* **263– 264** 229.

[3] Zdenek Remes, Tomas Novak, Jiri Stuchlik, The-ha Stuchlikova,Vladislav Drinek, Radek Fajgar, Konstantin Zhuravlev 2014 *Acta Polytechnica* **54** 426.

[4] Zhen Fang, Xiu Lin, Yueting Fan, Yufeng Liu, Yonghong Ni, Xianwen Wei 2010 *Journal of Alloys and Compounds* **493** L25.

[5] Liangfeng Sun, Lei Bao, Byung-Ryool Hyun, Adam C. Bartnik, Yu-Wu Zhong, Jason C Reed, Dai-Wen Pang, Hector D. Abrun˜a, George G. Malliaras, and Frank W. Wise 2009 *Nanoletters* **9** 789**.**

[6] J.N. Zemmel, J.D. Jensen, R.B. Schoolar 1965 *Phys. Rev. A* **140** 330.

[7] Giovanni Pellegrini, Giovanni Mattei, and Paolo Mazzoldi 2005 *J. Appl. Phys.* **97** 073706.

[8] Danylo Zherebetskyy, Yingjie Zhang, Miquel Salmeron, and Lin-Wang Wang 2015 *J. Phys. Chem. Lett.* **6** 4711.

[9] Gustavo M. Dalpian and James R. Chelikowsky 2006 *PRL* **96** 226802.

Defect-related transitions in luminescence of InAlAs on InP

A M Gilinsky[1], D V Dmitriev[1], A I Toropov[1], D V Gulyaev[1], O V Kibis[2] and K S Zhuravlev[1]

[1] A.V. Rzhanov Institute of Semiconductor Physics, pr. Ac. Lavrentieva, 13, 630090 Novosibirsk, Russia
[2] Novosibirsk State Technical University, pr. K. Marxa, 20, 630073 Novosibirsk, Russia

E-mail: gilinsky@isp.nsc.ru

Abstract. A study of photoluminescence (PL) of InAlAs grown on InP by molecular beam epitaxy was performed in a wide range of temperatures and excitation intensities. A novel defect-related transition has been observed for the first time by 120–180 meV below the near band edge PL line of InAlAs. The novel band appears in the spectra only in a limited range of temperatures around 50–160 K. In that temperature range the band may dominate in the PL spectrum. The characteristics of the band show that it is related to recombination via deep centres located in potential wells created by the alloy clustering. We show that by establishing quasi-stoichiometric conditions on the growth surface we were able to grow layers which show virtually zero intensity of the defect-related transitions and luminescence efficiencies by 1–2 orders of magnitude greater than that in the samples grown under standard conditions.

The ternary $In_xAl_{1-x}As$ semiconductor alloy is a well-known companion material for $In_yGa_{1-y}As$ used in a wide range of optoelectronics applications in the wavelength range above 1 μm, including both light emitters and photodetectors, as well as field-effect and bipolar transistors (see e.g. [1]). The major role that layers of InAlAs play is creation of high-quality lattice-matched heterostructure barriers for $In_{0.53}Ga_{0.47}As$ grown on InP without the need for quaternary phosphorus-based InGaAsP alloys. The recent proposal of a new field of application for the InAlAs material in avalanche photodiodes (APDs) can extend its use to new areas. As it has been calculated, the use of a charge multiplication layer made of InAlAs instead of InGaAsP should result in an improved temperature stability of the parameters of a photon-counting APD and a reduction of its dark count level [2,3]. However, attempts of realization of APD structures utilizing the charge multiplication layers of InAlAs have failed because of very large dark currents generated in such APD's that rendered their use impractical [4, 5]. The possible cause for the high dark current values found in InAlAs-based structures is carrier tunnelling in InAlAs via deep levels of point defects. This shows that our understanding of the spectrum of defects created during epitaxial growth of the material and their role in carrier transport and recombination is very limited, and requires a study of defect contents in InAlAs and optimization of the growth technology.

In this paper we present the results of a study of photoluminescence (PL) of InAlAs that allows us to reassess the properties of the recombination centres in the material. We have observed a novel long-wavelength luminescence transition that is related to defects in InAlAs for the first time, and consider ways of improving the defect contents in the material. We show that optimization of the conditions on the growth surface allows one to grow layers which do not display the defect-related long-wavelength transitions and have a luminescence efficiency by 1–2 orders of magnitude greater than that in the layers grown under standard conditions.

The samples studied in this work were 1 μm undoped epitaxial layers of $In_xAl_{1-x}As$ grown on semi-insulating (100) InP by molecular beam epitaxy. The samples were grown at a growth rate of 0.5 μm/h using various substrate temperatures and fluxes of constituent materials (In, Al, and As), and were not intentionally doped. Before the start of the growth of epitaxial layers the substrates were annealed in the flux of As_4 at a temperature of 540°C for oxide removal. The samples are listed in Table 1.

Table 1. Samples of $In_xAl_{1-x}As$.

Sample	x	T_S	BEP_{As} (Torr)	BEP_{As} / $BEP_{(4\times2)}$
A	0.492	520°C	1.50×10^{-5}	2.15
B	0.520	520°C	1.58×10^{-5}	2.25
C	0.524	535°C	1.50×10^{-5}	1.07

Here T_S is the substrate temperature and BEP_{As} is the beam equivalent pressure of the As_4 source used during the growth, and the BEP_{As} / $BEP_{(4\times2)}$ ratio shows by what amount the arsenic pressure exceeded the value that corresponded to switching from arsenic-rich to the metal-rich state of the growth surface, as determined using observation of the appearance of the (4×2) reconstruction in the reflection high energy electron diffraction patterns prior to the layer growth. The InAs contents x were determined by X-ray diffraction measurements. The PL measurements were performed in a continuous-flow helium cryostat using a 632.8 nm He-Ne or a 405 nm diode lasers for PL excitation.

Most of the experimental data presented here were taken on sample A that has the largest bandgap value that facilitates the assignment of unidentified PL bands. The PL spectra of sample A at various temperatures are shown in Figure 1. Two groups of transitions are present in the low-temperature spectra. The dominating line at 1.567 eV is the near band edge (NBE) PL of the InAlAs layer, while transitions at 1.419, 1.379 and 1.336 eV originate from the excitonic band, shallow-impurity related band and its phonon replica of the InP substrate, respectively. Also a weaker shoulder of the InAlAs line is seen around 1.515 eV that is related to an LO-phonon replica of the NBE transition.

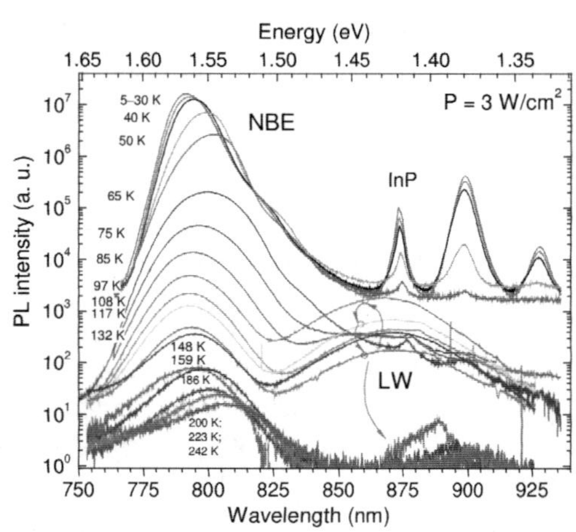

Figure 1. The PL spectra of sample A at various temperatures. The excitation was done by the He-Ne laser at a power density of 3 W/cm^2. Arrows indicate the direction of the evolution of LW emission with temperature.

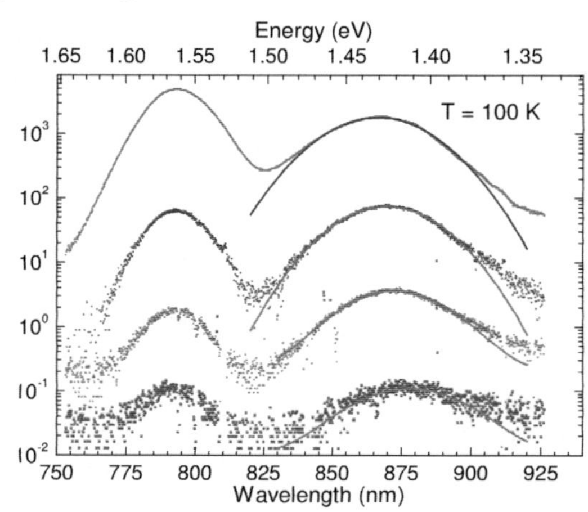

Figure 2. The PL spectra of sample A at a temperature of 100 K (dots) for excitation densities of 3, 0.3, 0.03 W/cm^2 and 3 mW/cm^2 (from top to bottom). The solid lines are Gaussian fits of the LW band.

As the sample temperature T is increased, the 'inverted S-shape' behaviour of the InAlAs NBE line reported earlier [6, 7] is observed in the spectrum: the NBE line exhibits first a red shift followed by a blue shift, which is accompanied with a monotonous decrease of its intensity caused by a redistribution of photogenerated charge carriers in favour of nonradiative recombination channels. An important new feature appears in the spectrum as the temperature is increased above ≈75 K: a novel wide band around 1.43 eV appears below the NBE line, referred to as 'LW' hereafter. The LW band intensity first increases as the sample temperature is raised to 100 K, and then decreases until T=150 K after which temperature the band is completely quenched. The remaining feature in that spectral region at T=160 K and above is the weak line of band to band recombination in InP at 1.394 eV. Figure 2 shows the PL spectra taken at a temperature of 100 K when the LW band attains its maximum intensity. It is seen from the fits of the experimental spectra (shown by solid lines) that the LW band has a Gaussian-like shape. A reduction in the excitation intensity leads to a red shift of the LW band at a rate of approximately 5.0±0.5 meV per decade of excitation power, while the NBE line does not show any change in its position. The LW band linewidth does not vary significantly with the excitation intensity neither at T=100 K nor above that temperature, the averaged linewidth remaining at 69 meV (FWHM) until the band disappears at T > 150 K. In contrast, the position of the band varies considerably with temperature, as is shown in Figure 3. The LW band exhibits a red shift that amounts to ≈50 meV with an increase in sample temperature. The NBE line, however, displays a blue shift in the temperature range from 50 K to 110 K, which shows that the mechanisms of the shifts of these two transitions differ. Additionally, the dependences of the NBE and LW transition intensities of the excitation power density P were studied. Both transitions displayed power-law dependences $I \sim P^{\gamma}$ without visible saturation in the range of intensities used. At low temperatures the NBE line has the exponent γ of 1.1 which rises to 1.5 and 1.7 at temperatures of 40 K and 100 K, respectively, while the LW band demonstrates somewhat lower vales of γ which amounts to 1.4 at T=100 K.

The origin of the LW band cannot be attributed to recombination in the substrate since it occupies the energy range that partly lies above the bandgap energy of InP. The band also cannot be related to the spatially indirect radiative transitions at the type-II heterointerface between InAlAs and InP which manifest themselves in the photon energy range below 1.25 eV [8]. The Gaussian line shape of the LW band and its position in the spectra allow us to suggest that the LW band is caused by recombination via a single deep centre located in InAlAs and characterized by a strong electron-phonon coupling [9]. From the linewidth of the LW band we then can estimate the vibrational mode energy of the excited state of the centre to be greater than 30 meV. The most intriguing behaviour of this band is its thermally-activated appearance in the spectrum and the temperature- and excitation

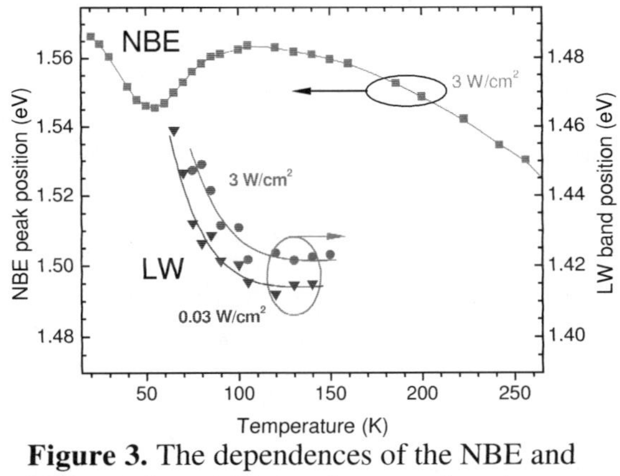

Figure 3. The dependences of the NBE and LW peak positions in sample A on temperature. Solid lines are guides for the eye. Note the different vertical scales used for the two transitions.

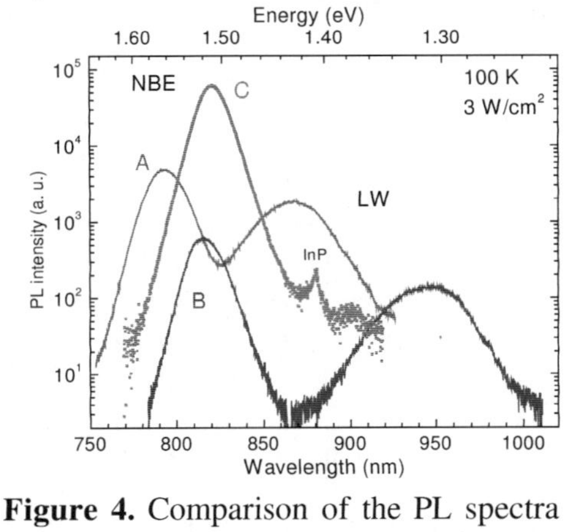

Figure 4. Comparison of the PL spectra of samples A to C at a temperature of 100 K.

intensity-induced shifts not commonly observed in deep centre PL of uniform materials. The amplitude of the shifts strongly exceeds the values observed for the donor-acceptor PL involving shallow impurity states in comparable materials such as GaAs or InP, therefore this mechanism can be ruled out. The emergence of the LW band above a certain temperature points to the existence of potential barriers that limit carrier transport to/from the participating centres at low temperatures. Both these peculiarities can be explained by taking into account the alloy clustering in InAlAs that is the cause of the 'inverted S-shape' behaviour of the NBE line illustrated in Figures 1 and 3 [6, 7]. The clustering results in modulations of the band edges with amplitudes much greater than the 8 meV value calculated for a perfectly random alloy [7]. A recombination centre located within a region separated by potential barriers due to clustering then should display the recombination features found for the LW transition. The centre's PL will be seen only at temperatures high enough to allow the carriers (presumably electrons) to reach the centre, while the spatial separation of the carriers of the second type, presumably holes, will result in shifts of the PL energy with changes in carrier concentration caused by variations in excitation power or temperature. We therefore conclude that the LW band is caused by deep centres localized in potential wells created due to the alloy clustering.

Figure 4 shows a comparison of the PL spectra of samples A to C taken at T=100 K. It is seen that sample B grown in conditions similar to that of sample A also demonstrates the LW band which is centered at 1.32 eV in this case. In contrast, sample C that was grown at a slightly higher substrate temperature using As pressure close to switching of the growth surface to the metal-rich state displays no LW emission, while its NBE PL intensity is by 12–100 times greater than that of samples A and B (by 100 times at T=300 K). This shows that the use of quasi-stoichiometric growth conditions allows one to improve the quality and reduce the deep centre concentration in the layers.

To summarize, the PL of InAlAs grown on InP by molecular beam epitaxy was studied in a wide range of temperatures and excitation intensities. A novel defect-related emission has been observed that lies by 120–180 meV below the near band edge PL of InAlAs. The novel band appears in the spectra only in a limited range of temperatures around 50–160 K, and can be observed neither at liquid helium nor at room temperatures. The Gaussian shape and thermally-activated appearance of the band and its large shifts in the spectrum show that it is related to deep centres located in potential wells created by the alloy clustering. Using quasi-stoichiometric growth conditions we grew layers which showed no defect-related long-wavelength transitions and had a luminescence efficiency by 1–2 orders of magnitude greater than that of the samples grown under standard conditions.

The authors gratefully acknowledge the support of this work by the Russian Foundation for Basic Research under grants nos. 14-29-08124 and 14-02-00033.

References

[1] Won T, Agarwala S and Morkoç H 1988 *Appl. Phys. Lett.* **53** 2311–3
[2] Mun S C L T, Tan C H, Dimler S J, Tan L J J, Ng J S, Goh Y L and David J P R 2009 *IEEE J. Quantum Electronics* **45** 566–71
[3] Tan L J J, Ong D S G, Ng J S, Tan C H, Jones S K, Qian Y and David J P R 2010 *IEEE J. Quantum Electronics* **46** 1153–7
[4] Meng X, Tan C H, Dimler S, David J P R and Ng J S 2014 *Optics Express* **22** 22608–15
[5] Karve G, Wang S, Ma F, Li X, Campbell J C, Ispasoiu R G, Bethune D S, Risk W P, Kinsey G S, Boisvert J C, Isshiki T D and Sudharsanan R 2005 *Appl. Phys. Lett.* **86** 063505
[6] Olsthoorn S M, Driessen F A J M, Eijkelenboom A P A M and Giling L J 1993 *J. Appl. Phys.* **73** 7798–803
[7] Ferguson I T, Cheng T S, Sotomayor Torres C M and Murray R 1994 *J. Vac. Sci. Technol.* B **12** 1319–27
[8] Brasil M J S P, Nahory R E, Quinn W E, Tamargo M C and Farrell H H 1992 *Appl. Phys. Lett.* **60** 1981–3
[9] Huang K and Rhys A 1950 *Proc. Royal Soc.* **204** 406–23.

IOP Conf. Series: Journal of Physics: Conf. Series **864** (2017) 012076 doi:10.1088/1742-6596/864/1/012076

Electronic structure and properties of highly ordered C_{60} nano arrays on Au (111): STM & DFT study

Zaw-Myo Win, Chao HUANG and Rui-Qin ZHANG*

Department of Physics and Materials Science, City University of Hong Kong, Tat Chee Avenue, Kowloon, Hong Kong SAR, China

E-mail: *aprqz@cityu.edu.hk

Abstract. Template assisted assembly of molecular nano arrays is one of the key steps towards molecular electronics and fullerene is one of the potential structural building blocks in fabrication of identical molecular nano arrays for miniature devices such as photovoltaic devices and single-molecule transistors. In this report, the reconstructed Au (111) with defect areas (steps) has been used as a template to assemble the highly ordered C_{60} nano array at low coverage studied with scanning tunnelling microscopy (STM) in conjunction with density functional theory (DFT). The interaction between the substrate and C_{60} nano arrays is strong enough to change the geometrical shape of C_{60}. As a result of strong interaction, the C_{60} molecule appears to be deformed into ellipsoidal shape which causes the reduction of C_{60} nano arrays on step sites of Au (111).

1. Introduction

Fullerene is one of the potential structural building blocks in fabrication of identical nano scale structures for miniature devices, such as photovoltaic devices and single-molecule transistors.[1] In addition to be a flexible building-block material, fullerenes are also a kind of electron acceptors that can offer an ease of fabrication of donor-acceptor nanojunctions.

Until now, the investigation of self-assembled structures of C_{60} on flat metal surfaces has been focused on monolayer structures by many research groups. [2] To our knowledge, a few reports have been done on the investigation of adsorption behavior of C_{60} residing on the step of vicinal metal surfaces which have been used as novel templates to prepare controllable nanostructures.[3] One big advantage of this method is that it can avoid introducing of impurities. Since no more experimental procedures are required, using vicinal gold templates saves much time than other methods as well.

Vicinal gold surfaces are useful for controlling the nano scale structure in specific energetic sites. One of the templates is Au (111) which has been studied extensively for the formation of 2 D arrays of metal islands. Adsorption of molecules occurs preferentially at defects such as step edges and kinks at room temperature. [4] For the adsorption of C_{60} on Au (111) at room temperature, the molecules quickly move to the preferential adsorption sites (steps).

In this work, we found that there is an apparent shortening of C_{60} arrays residing on linear steps of Au (111) by analyzing the line profile of STM topography. Our calculations using Density functional Theory (DFT) suggest that the shortening is due to the molecular distortions in the middle of arrays caused by the elongation of the C-C bond length and pit formation on the gold layer.

2. Experimental details

All experiments were carried out in an UHV chamber (base pressure < 10^{-10} mbar) equipped with an Omicron LT-STM operated at 77K. A tungsten tip was used, and all STM measurements were performed in a constant-current mode. The Au (111) single-crystal substrate was cleaned by several cycles of Ar+ sputtering (U=1.0kV) and subsequent annealing (T=300°C for 5 minutes, measured by a thermocouple directly attached to the crystal). C_{60} was purchased from Sigma Aldrich and used without further purification. The C_{60} was sublimed at 650 K from crucible to Au (111) surface that was held at room temperature for 1 minute. STM images of interested surface areas are obtained at~77 K.

3. Results and discussion

It is necessary to determine the crystallographic features of Au (111) substrate used as a template for C_{60} growth before the study of the adsorption behavior of it. The Au (111) surface we choose is of not only large area of terrace (50 %) but also linear steps (45 %) and kinks site (5 %). Even though, the substrate we choose is not a well prepared vicinal surface, some part of it is a good candidate as a natural template for growing of nano arrays. As a result of contraction in the [$\underline{1}$10] direction, the herringbone reconstruction forms. The reconstruction pattern can be seen in STM topography image of clean Au (111) surface in Fig. 1 (a). It occurs due to the two different surface atom packings, which are separated by two bright corrugation lines, alternatively. HCP (hexagonal closed packing) exists between two neighboring lines, whereas the wider-distance line area composed FCC (face centered cubic packing) lattices as shown in Fig. 1(b).

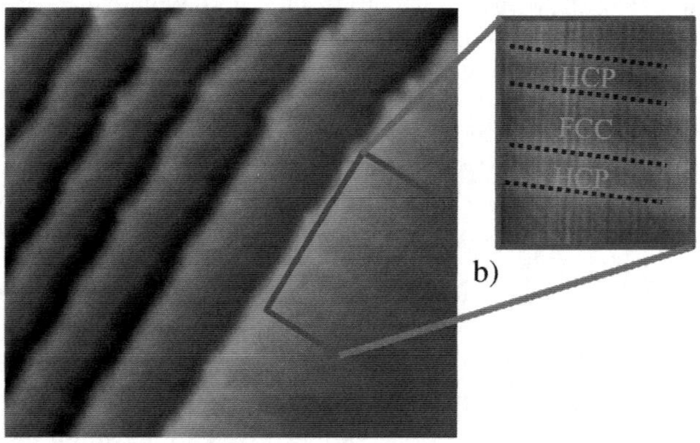

Figure 1. (a) STM topography image of Au (111) at I_t = 20 nA, V_{bias}= 1 V; and (b) Magnified STM image of HCP domains and FCC domains of Au (111)).

After deposition of ~0.05 Mono Layer (ML) (1 ML corresponds to a closed pack full coverage of molecules on the surface) molecules on the Au (111) surface. All the molecules organized to form nano arrays at kink or linear step edge as shown in Fig. 2 (a). This is because the C_{60} molecules are mobile on the terrace until they reach the preferred adsorption site (steps or kinks). It is found that these C_{60} molecules decorate the FCC stacking region at a low coverage. This shows that the C_{60} molecules resides the kinks or steps selectively.

Our analysis reveals that 90% of C_{60} nano arrays consist of four or five molecules which are decorating at the step area of Au (111). The intermolecular distance or the nearest neighbor distance (NND) of two C_{60} is ~1 nm (in crystal).[5] Line profile analysis of C_{60} nano arrays (from center to center) were carried for several hundreds of STM topography images in order to reveal the actual length of C_{60} molecular nano arrays. Clear reductions of C_{60} array lengths were observed. The result clearly state that the spacing of two C_{60} molecules in three, four and five molecular arrays is less than 1 nm as shown in Table 1. These results uncover the fact that, as the amount of molecules increase in an array, the intermolecular nearest neighbor distance is reduced compared with the reported value of

NND. We believe that some C_{60} in the middle of an array may elongate which results in shortening of the array length.

We further increased the coverage of C_{60} in order to achieve the monolayer deposition of C_{60} for comparing with the result of nano arrays with short lengths. However, we obtained C_{60} islands instead of nano arrays. The nearest neighbor distance between two C_{60} in the newly deposited island is ~1nm. The three lobe structure of C_{60} island can be seen in Figure 2(a) for comparison of line profile analysis. The average results of line profile analysis from our experiment has a ~ 13% reduction of the nearest neighbor distance comparing with the C_{60} monolayer island on the flat terrace.

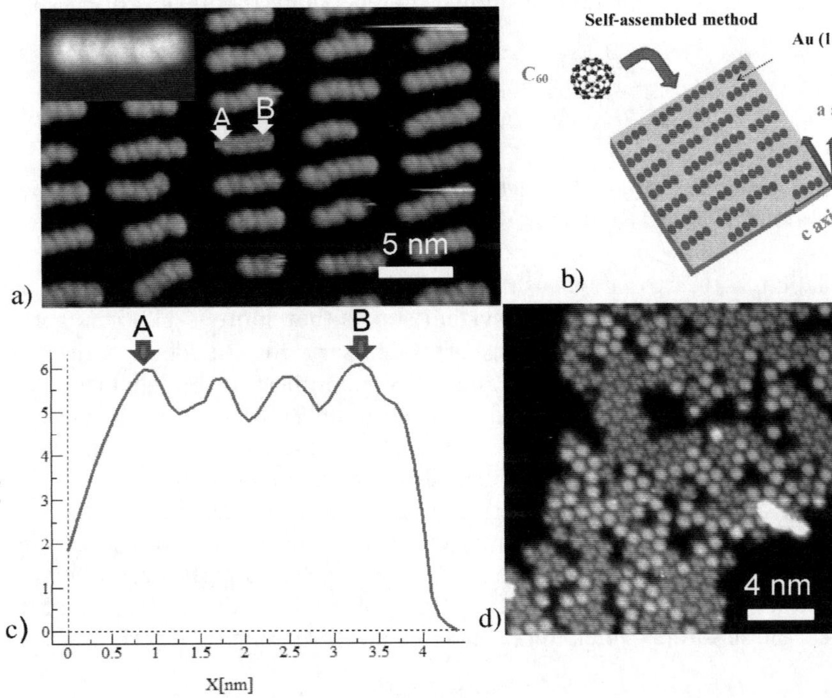

Figure 2. (a) STM image of periodic arrays of C_{60} selectively residing on the step edge of Au (111), It = 20 nA, Vbias= 3.5 V. The inset (I_t = 20 nA, Vbias= 3 V) shows the intermolecular resolution image of one C_{60} nano array (b) Schematic representation C_{60} deposited by self-assembly method on Au(111); (c) Line profile along the green line in (a); and (d) STM topography image of C_{60} island obtained at 77 K using I_t = 20 nA , V_{bias}= 2 V.

Table 1. Summary of lengths, nearest neighbour distance (NND) or intermolecular distance of C_{60} molecular nano arrays.

	Array 1 (Two C_{60}) (nm)	Array 2 (Three C_{60}) (nm)	Array 3 (Four C_{60}) (nm)	Array 4 (Five C_{60}) (nm)
Length[a]	0.8620	1.7049	2.5601	3.38136
Intermolecular distance[b]	0.8620	0.8520	0.8503	0.8454

[a] The length from one end to the other end of the nano array.
[b] The center to center distance between two C_{60} molecules.

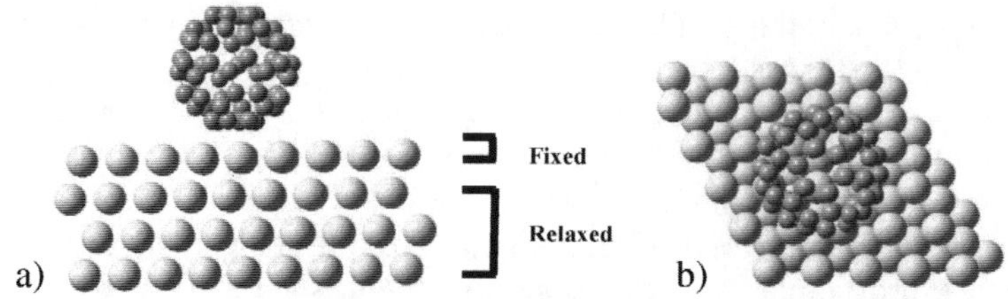

Figure 3. (a) Side view of adopted model for C_{60} sitting on the four atomic layers of gold. The bottom three layers are fixed and only the top layer is allowed to relax; and (b) top view of C_{60} on Au (111).

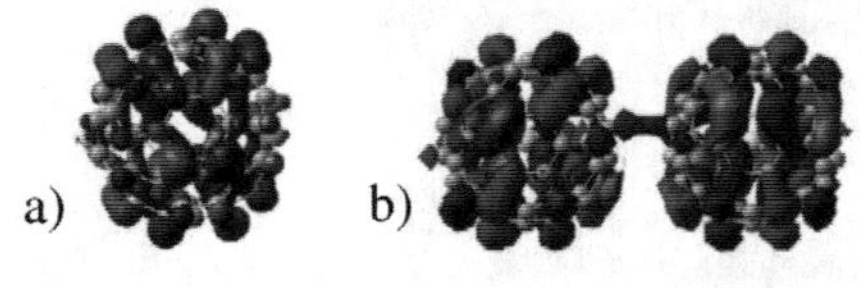

Figure 4. Computed lowest unoccupied molecular orbital of C_{60} on Au (111) with their respective relaxed structures: (a) one C_{60}; and (b) two C_{60}.

The DFT calculations were carried out using SIESTA [6] in order to provide better understanding of the STM results. The relaxed structure of system shows that some C-C bonds are elongated to certain extents. More theoretical calculations are necessary for the deformation trend in C_{60}. Moreover, the charge transfer from substrate to molecule could lead to the Jahn-Teller distortion due to covalent bond formation between the molecule and substrate.[7] To examine the electronic structure of C_{60} effectively at reasonable computational cost, we performed series of geometry optimizations with a certain number of gold atoms. The orbital distributions of the lowest unoccupied molecular orbital (LUMO) of C_{60} on Au (111) are plotted in Fig 4. These results indicate that the LUMO state of single C_{60} is not the same with other case like molecular array consists of two C_{60} as computed in Fig 4. The main reason is that the distribution of π orbital of two C_{60} is higher than the single C_{60}. It is obvious that the charge distribution increase with the number of C_{60} due to the π electron migrating between two C_{60} and the substrate, which might lead to the deformation of the orbitals.

4. Conclusion

To conclude, we investigated for the first time the reduction of nearest neighbor distance (NND) of C_{60} molecular arrays residing on the step sites of Au (111) which is less than 1 nm combining UHV STM experiments and DFT calculations. We proved that the shortening of nearest neighbor distance of nano molecular arrays is somehow related to the deformation of orbital as revealed by our DFT calculation based detail analysis. Our finding can be used to improve the identical nano scale structures for miniature devices such as photovoltaic devices and single-molecular transistors.

Acknowledgements

The work described in this paper was supported by a grant from the Research Grants Council of the Hong Kong SAR (Project Nos. CityU 103913).

References

[1] Yu L H and Natelson D 2004 *Nano Lett.* **4** 79–83
[2] Paßens M, Waser R and Karthäuser S 2015 *Beilstein J. Nanotechnol.* **6** 1421–31
[3] Néel N, Kröger J and Berndt R 2006 *Adv. Mater.* **18** 174–7
[4] Saywell A, Schwarz J, Hecht S and Grill L 2012 *Angew. Chemie - Int. Ed.* **51** 5096–5100
[5] Shi X-Q, Van Hove M A. and Zhang R-Q 2012 *J. Mater. Sci.* **47** 7341–7355
[6] Soler J M, Artacho E, Gale J D, Garcia A, Junquera J, Ordejon P and Sanchez-Portal D 2002 *J. physics. Condens. matter* **14** 2745-2779
[7] Hands I D, Dunn J L and Bates C A. 2010 *Phys. Rev. B.* **82** 155425

Investigation of Deep-level Defects in $CuGaSe_2$ Thin-film Solar Cells by Transient Photo-capacitance Spectroscopy

X B Hu[1], G E Weng[1], S Q Chen[1] and K Akimoto[2]
[1] Key Laboratory of Polar Materials and Devices, Ministry of Education, East China Normal University, Shanghai 200241, China
[2] Institute of Applied Physics, University of Tsukuba, Tsukuba 305-8577, Japan
E-mail: huxiaobo@ee.ecnu.edu.cn

Abstract. Properties of deep-level defects in $CuGaSe_2$ thin-film solar cells were investigated using transient photo-capacitance (TPC) spectroscopy. Two Gaussian-shaped deep-level defects centered at around 0.8 eV and 1.54 eV above the valence band were identified. The electronic structure of the two defects was illustrated by a configuration coordinate model to explain the thermal quenching effect in the two defects, which considered a large lattice distortion for the 0.8 eV defect while no distortion for the 1.54 eV defect.

1. Introduction

Chalcopyrite $Cu(In_{1-x},Ga_x)Se_2$ (CIGS) is one of the most promising materials for use in thin-film solar cells. CIGS-based solar cells have attained the highest efficiency (>20%) among all photovoltaic thin film technologies [1]. However, these high-efficiency solar cells are typically made from CIGS alloys with low Ga content (x~0.3), corresponding to bandgap energy of about 1.15 eV [2]. The developing of CIGS films with higher bandgap energy is also desirable since the optimum bandgap for the highest efficiency of solar cell is around 1.4 eV as theoretically predicted which means there is still much space to improve the efficiency [3], and also they are promising candidates as intermediate or top layers for application in the tandem solar cells which should attain higher efficiency than single-junction solar cells [4]. Especially, $CuGaSe_2$ (x =1.0) with a band gap of 1.68 eV would represent an ideal partner for $CuInSe_2$ in an all-chalcopyrite tandem structure [4]. However, the present efficiency of $CuGaSe_2$ is still much lower than theoretical predictions and needs further improved [5].

It is commonly recognized that formation of various point defects and defect phases in the films is considered as important factor limiting the efficiency since they are good candidates as effective carrier recombination centers and could directly induce the higher open-circuit voltage loss and poor carrier life-time compared to the CIGS solar cells with low bandgap energy [6]. Thus, a complete understanding of the deep-level defects is very useful to improve the performance of $CuGaSe_2$ solar cells. Transient photo-capacitance (TPC) spectroscopy is a powerful tool for the analysis of deep levels in the photo-absorber layer because of the very low detection limit [7]. Much has been done in studying CIGS solar cells by employing TPC spectroscopy in our previous work [8,9], however, these work were studied based on CIGS solar cells with Ga content 0.3~0.8, and studies on CIGS with higher Ga content such as $CuGaSe_2$ has not been reported.

In this study, transient photocapacitance (TPC) spectroscopy were used to investigate the properties of optically active deep-level defects in $CuGaSe_2$ solar cells.

2. Experimental

Polycrystalline $CuGaSe_2$ thin films were grown on Mo-coated soda-lime glass substrates by a three-stage process using a molecular beam epitaxy (MBE) system. The detailed deposition conditions and processes for the $CuGaSe_2$ solar cell have been described elsewhere [10]. The device parameters of $CuGaSe_2$ thin-film solar cells used in this study were measured under AM1.5 illumination at 25 °C, with short-circuit current of 14.49 mA/cm^2, open-circuit voltage of 0.818 V, fill factor of 0.643 and efficiency of 7.624%, respectively.

The TPC measurements were carried out from 100 K to 300 K by setting the samples in a cryostat chamber. The experimental conditions were the same as our previous work described [8,9]. During the photo-capacitance measurements, a halogen lamp was used as the monochromatic probe light source (700~1800 nm). The detail timing chart for the measurements is described in ref [11]. The transient response, which arises after the pulse, was integrated over a fixed time window (0.75 s) by using an analog-to-digital converter with the rate window fixed at 2s^{-1}.

3. Results and discussion

The TPC spectrum of the $CuGaSe_2$ sample measured at 100 K was shown in figure 1 (a). The spectrum shows three optical absorption features: one kind of interband absorption for photon energies ranging from about 1.6 to 1.8 eV, and two kinds of sub-bandgap absorption from 0.8 to 1.2 eV and from 1.2 to 1.6 eV caused by deep-level defects in the $CuGaSe_2$ cell. Regarding the interband absorption, we observed a steep increase in the intensity of the TPC signal with increasing photon energy due to the large cross section of the direct interband transition, this could be fitted by an exponential band tail as figure 1 (a) shows. With regard to the sub-bandgap absorption, the intensity of the TPC signal was lower than that of the interband absorption by one to two orders of magnitude, and two Gaussian shaped bands were used to fit the defect signal which considered two defects locating at around 0.8 eV and 1.54 eV above valence band respectively.

Figure 1 (b) shows the temperature dependence of the TPC spectrum measured in the temperature range from 100 to 300 K. The intensity of sub-bandgap TPC signal decreased with increasing temperature while significant variation in the inter-band signal was not observed with temperature. The decreasing of TPC intensity with temperature can be attributed to the thermal de-trapping of charges in defect levels with temperature. The inset shows the relation between the temperature and logarithm of TPC signal measured under monochromatic band gap light irradiation (at 0.8 eV and 1.54 eV as the inset shows). The variation of the TPC signal $S(T)$, with the temperature T was described by the thermal quenching model commonly used for the photoluminescence study [12], $S(T)=S(0)/[1+C\exp(-E_a/kT)]$ where C is a constant and E_a is the activation energy of thermal quenching

IOP Conf. Series: Journal of Physics: Conf. Series **864** (2017) 012077 doi:10.1088/1742-6596/864/1/012077

effect. The activation energy was obtained as 0.20 eV for the 0.8 eV defect and 0.13 eV for the 1.54 eV defect, respectively, as the inset shows.

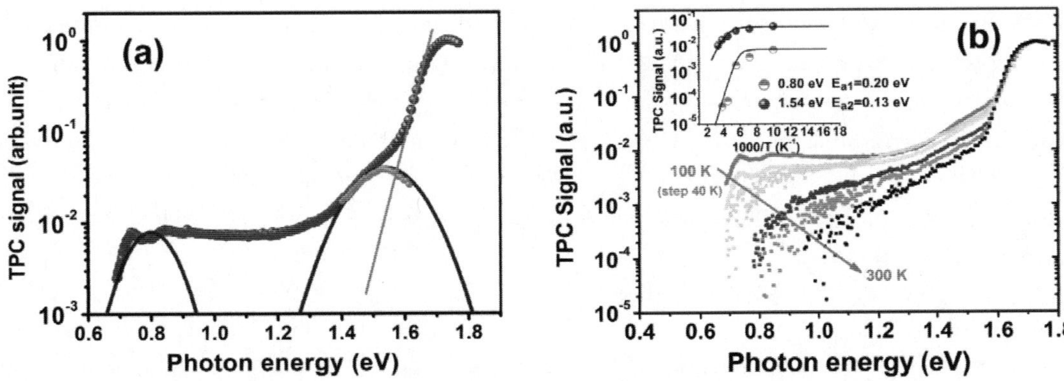

Figure 1. (a) TPC spectrum of the $CuGaSe_2$ sample measured at 100 K; (b) Temperature dependent of TPC spectrum of the $CuGaSe_2$ sample from 100 K to 300 K.

The mechanism of thermal quenching has been interpreted by using a discrete energy model [6]. In this model, the thermal excitation of the electrons trapped in deep defects to the conduction band leads to the quenching of the TPC signal. By using this model, the thermal quenching activation energy E_a is the difference between the bandgap energy and the optical transition energy. Thus, the activation energies of quenching are expected to be approximately 0.88 eV and 0.14 eV for the 0.8 eV defect and 1.54 eV defect (E_g=1.68 eV). For the 1.54 eV defect, the discrete model could explain the thermal activation energy properly since the experimentally obtained E_a=0.13 eV coincides well with the predicted 0.14 eV; however, for 0.80 eV defect, the value is considerably larger than that observed in the experiment as figure 1 (b) shows.

This could be interpreted using a configuration coordinate (CC) diagram, as shown in figure 2.

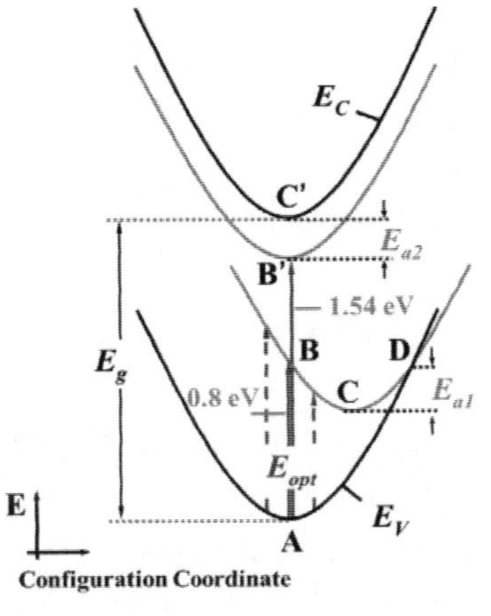

Figure 2. Configuration coordinate diagram describing the optical transition process for the defect 0.8 eV and 1.54 eV.

For the 0.8 eV defect, the optical process and thermal quenching process can be explained based on the CC model as follows: A→B: optical transition without changing configuration coordinate; B→C:

lattice relaxation with multi-phonon emissions; C→D: atomic vibrations by changing configuration coordinate, at low temperatures, the charged defects retain their configuration at position C due to lower vibration energy, however, when temperatures increases, atomic vibrations can shift their configuration from C to D, the energy difference between C and D is defined as the activation energy E_{a1}. D→A: non-radiative transition by shifting configuration coordinate. For the 1.54 eV defect, the thermal quenching process can be easily explained by the direct transition of electrons from B' to C' without atomic vibration, with the energy difference between B' to C' is defined as the activation energy E_{a2}. The CC model could also explain the spectrum broadening which was thought to be caused by the strong electron–phonon interaction between the valence band and the deep defect as the dash arrow shows in figure 2.

4. Conclusion

The deep-level defects in $CuGaSe_2$ thin film were investigated through transient photo-capacitance (TPC) technique. A 0.8 eV defect and a 1.54 eV above valence band maximum (VBM) were identified, and a configuration coordinate model was proposed to explain the thermal quenching effect for the two defects.

References

[1] Green M A, Emery K, Hishikawa Y, Warta W, and Dunlop E D 2016 *Prog. Photovolt: Res. Appl.* **24** 3–11

[2] Repins I, Contreras M A, Egaas B, DeHart C, Scharf J, Perkins C L, To B, and Noufi R 2008 *Prog. Photovolt: Res. Appl.* **16** 235–9

[3] Werner J H, Mattheis J, and Rau U 2005 *Thin Solid Films* **480** 399–409

[4] Nishiwaki S, Siebentritt S, Walk P, and Lux-Steiner M C 2003 *Prog. Photovolt: Res. Appl.* 11243–8

[5] Ishizuka S, Yamada A, Fons P J, Shibata H and Niki S, 2014 *Prog. Photovolt: Res. Appl.* **22** 821–9

[6] Sakurai T, Uehigashi H, Islam M M, Miyazaki T, Ishizuka S, Sakurai K, Yamada A, Matsubara K, Niki S, and Akimoto K 2009 *Thin Solid Films* **517** 2403–6

[7] Cohen J D, Unold T and Gelatos A V 1992 *J. Non-Cryst. Solids* **141** 142–54

[8] Hu X B, Sakurai T, Yamada A, Ishizuka S, Niki S, and Akimoto K, *Appl. Phys. Lett.* 2013 **103** 163905

[9] Hu X B, Sakurai T, Yamada A, Ishizuka S, Niki S, and Akimoto K 2014 *Jpn. J. Appl. Phys.* **53** 068008

[10] Islam M M, Yamada A, Sakurai T, Kubota M, Ishizuka S, Matsubara K, Niki S and Akimoto K 2012 *J. Appl. Phys.* **112** 103707

[11] Siebentritt S and Rau U 2006 *Wide-Gap Chalcopyrites* (Berlin: Springer Verlag) p 69

[12] Pankove J I 1971 *Optical Processes in Semiconductors* (New York: Dover)

Acknowledgments

The work was sponsored by Shanghai Pujiang Program (grant No. 16PJ1402600). The authors would also like to thank Mr Niki S from the AIST in Japan for providing the samples in this study.

Strong quantum-confined Stark effect in a lattice-matched GeSiSn/GeSn multi-quantum-well structure

RuizhiPeng, Chunfuzhang#, Genquan Han, and YueHao
State Key Discipline Laboratory of Wide Band Gap Semiconductor Technology, School of Microelectronics, Xidian University, Xi'an 710071, China
#Email: cfzhang@xidian.edu.cn

Abstract

This paper presents modeling and simulation of a multiple quantum well structure formed with $Ge_{0.95}Sn_{0.05}$ quantum wells separated by $Ge_{0.51}Si_{0.35}Sn_{0.14}$ barriers for the applications. These alloy compositions are chosen to satisfy two conditions simultaneously: type-I band alignment between $Ge_{0.95}Sn_{0.05}/Ge_{0.51}Si_{0.35}Sn_{0.14}$ and a lattice match between wells and barriers. This lattice match ensures that the strain-free structure can be grown upon a relaxed $Ge_{0.51}Si_{0.35}Sn_{0.14}$ buffer on a silicon substrate – a CMOS compatible process.

A electro-absorption modulator with the $Ge_{0.95}Sn_{0.05}/Ge_{0.51}Si_{0.35}Sn_{0.14}$ multiple quantum well structure based on quantum-confined Stark effect(QCSE) is demonstrated in theory. The energy band diagrams of the GeSiSn/GeSn multi-quantum-well structure at 0 and 0.5V bias are calculated, respectively. And the corresponding absorption coefficients as a function of cut-off energy for this multiple quantum well structure at 0 and 0.5Vbias are also obtained, respectively. The reduction of cut-off energy is observed with the applying of the external electric field, indicating a strong QCSE in the structure.

1. Introduction

The quantum-confined Stark effect (QCSE) [1] is a strong, electric field dependent change in optical absorption that has been seen in quantum well materials. It is used extensively for high-speed, low power dissipation optical modulators, for example, in telecommunications, and has also been used in large arrays of low power devices [2].

In this paper, we report strong QCSE in $Ge_{0.95}Sn_{0.05}$ quantum wells grown on Si substrates with $Ge_{0.51}Si_{0.35}Sn_{0.14}$ barriers and $Ge_{0.51}Si_{0.35}Sn_{0.14}$ buffer. These observations show clear QCSE whose performance is comparable to, or possibly better than, III–V QCSE effects at similar wavelengths. Here, we give an extended discussion of this work, and the absorption of $Ge_{0.95}Sn_{0.05}$ quantum wells is calculated by using an 8-band k.p solver. The lattice-matched GeSiSn/GeSn multi-quantum-well structure simulated in this paper has emission energy of 0.7eV. It advances applications in chemical and biological sensing, medical therapy, free space communication, spectroscopy-on-a-chip and it aids the development on-chip laser radar transmitters.

2. Band structure and sample design

This section illustrates the sample design in this paper. The design of the lattice-matched

IOP Conf. Series: Journal of Physics: Conf. Series **864** (2017) 012078 doi:10.1088/1742-6596/864/1/012078

GeSiSn/GeSn multi-quantum-well structure is based on the fact that electric field plays an active role in QCSE, which results in red shift of the cut-off wavelength for absorption spectra.

Fig. 1 shows the basic structure used for our experiments. $Ge_{0.95}Sn_{0.05}$ quantum wells with $Ge_{0.51}Si_{0.35}Sn_{0.14}$ barriers between them are grown on the top of a lattice-matched $Ge_{0.51}Si_{0.35}Sn_{0.14}$ layer on a Si substrate. The compositions of GeSn and GeSiSn are chosen to provide type-I band alignment at the Γ point as well as lattice matching. The structure would be situated on a relaxed buffer layer of GeSiSn-upon-silicon, a buffer whose lattice parameter is the same as that of the GeSn/GeSiSn structure, hence the entire structure is unstrained. The structure also allows photocurrent to be collected, from which the effective optical absorption coefficient of the quantum well region can be deduced.

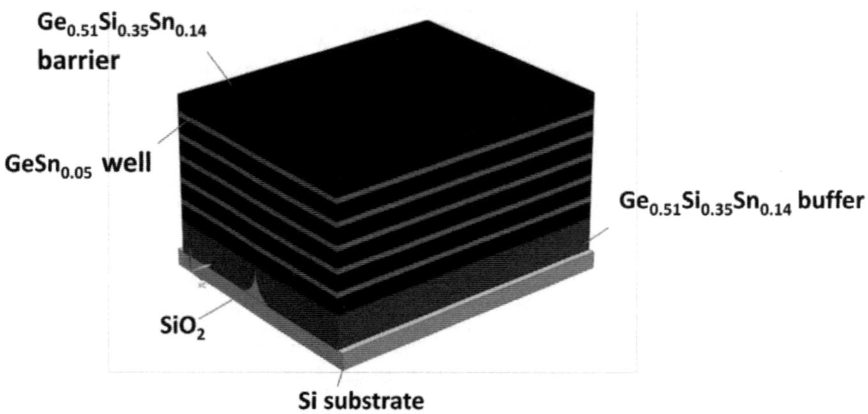

Fig.1 3D schematic of lattice-matched GeSiSn/GeSn multi-quantum-well structure

Since the band offsets between ternary Sn-containing alloys and Si or Ge are not known experimentally, we calculate the conduction-band minima for the lattice-matched heterostructure consisting of $Ge_{1-z}Sn_z$ and a ternary $Ge_{1-x-y}Si_xSn_y$. We used Jaros' band offset theory [3] which gives results in good agreement with experiment for many heterojunction systems[4]. The major input parameters of $GeSn_{0.05}$ and $Ge_{0.51}Si_{0.35}Sn_{0.14}$ for the calculations are listed in Table I, lattice constant was evaluated by linear interpolation based on the values of Ge and Sn given in Ref.5.

Table I. The major input parameters of $GeSn_{0.05}$ and $Ge_{0.51}Si_{0.35}Sn_{0.14}$

Parameter	GeSn0.05	Ge0.51Si0.35Sn0.14
Lattice constant, *a* (nm)	5.688283500	5.688283500
Elastic constant, c_{11} (GPa)	125.55	133.23
Elastic constant, c_{12} (GPa)	47.31	51.09
Elastic constant, c_{44} (GPa)	65.27	67.00
L	-31.34	-34.897
M	-5.90	-5.689
N	-34.14	-12.238

Fig. 2 shows what we believe to be the form of the line-up of various relevant bands in an unstrained GeSiSn/GeSn multi-quantum-well structure at 0V and 0.5V. We show the conduction band edge at the Γ point , as well as the valence band edges corresponding to the heavy hole , light hole and SO bands. Note that we expect type-I alignment (electron minimum energy and hole maximum energy in the same material) at the zone center, just as in typical III–V quantum well materials used for the QCSE.

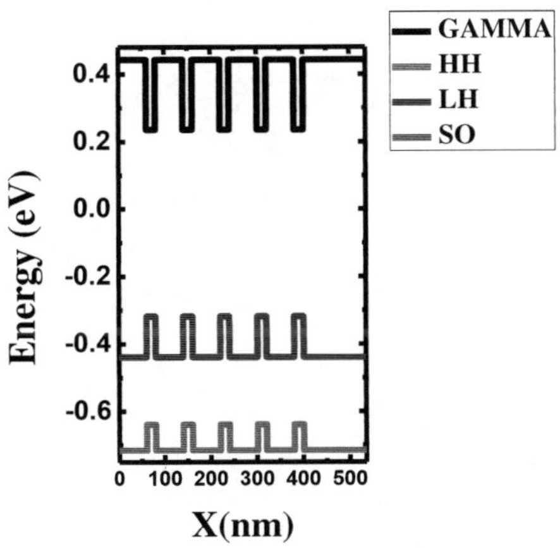

Fig. 2(a) Energy band diagrams of a multiple quantum well structure at 0V

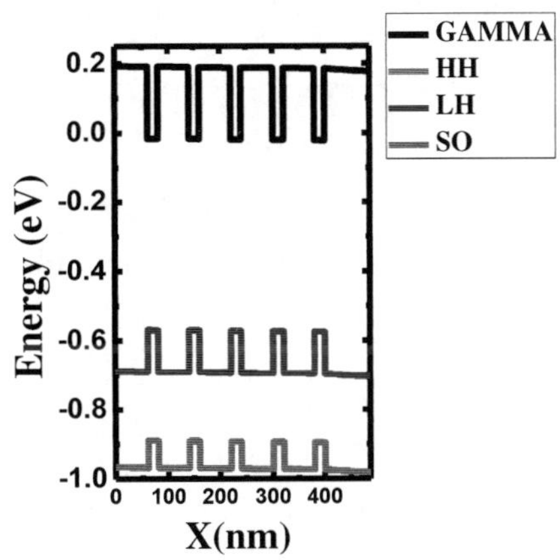

Fig. 2(b) Energy band diagrams of a multiple quantum well structure at 0.5V

3. Calculations of absorption coefficient in GeSiSn/GeSn multi-quantum-well structure

Absorption coefficient α is the most key parameter for determining the electrical and optical performance. In this work, the absorption was calculated by using an 8-band k.p solver. The technique is similar to the envelope function technique derived by Bastard[6] and a full derivation of the absorption coefficient: derivation of the absorption coefficient: α(ω) derivation of the absorption coefficient: αderivation of the absorption coefficient: absorption coefficient: α (ωderivation of the absorption coefficient: α (ω) as a function of angular frequency, ω can be found in Ref. 7. The absorption result is from the projection of the susceptibility tensor χij onto the polarization vector ε_j of the incident light to produce[7]

$$\alpha(\omega) = \frac{4\pi\omega}{nc}\varepsilon_i^{*}\chi_{ij}(\omega)\,\varepsilon_j$$

where *n* is the refractive index and *c* is the speed of light. The susceptibility is given in Ref.7.

Figure 5 shows the simulated optical absorption spectra for a multiple quantum well structure at 0V and 0.5V. A multiple quantum well structures at 0V and 0.5V demonstrate the cut-off energy of 1.35eV and 1.2eV, respectively. The absorption peak of multiple quantum well structures at 0V and 0.5V are 1.2eV and 0.8eV, Significant red shift of absorption edge is achieved in multiple quantum well structure at 0.5V due to QCSE. The absorption energy can be further decreased to low energy as voltage continues to increase, which will increase the electric field in the structure.

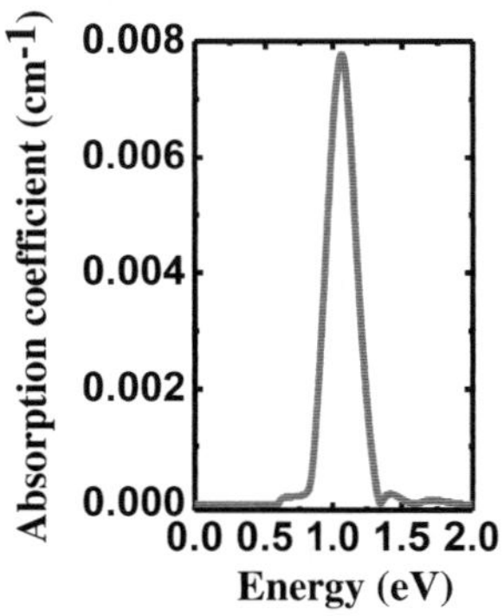

Fig. 3(a) The absorption coefficient as a function of cut-off energy for a multiple quantum well structure at 0V

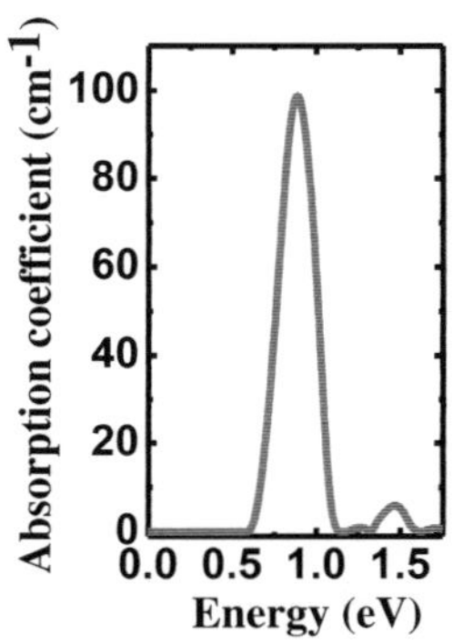

Fig. 3(b) The absorption coefficient as a function of cut-off energy for a multiple quantum well structure at 0.5V

4. Conclusion

In summary, a lattice-matched GeSiSn/GeSn multi-quantum-well structure is investigated by simulation. 8-band k·p method is utilized to calculate the energy band structure and optical transmission properties in the multi-quantum-well structure. The electric field is induced by the voltage that perpendicular to the well, thus resulting in decreased absorption energy . QCSE in a lattice-matched GeSiSn/GeSn multi-quantum-well structure is theoretically studied and the absorption coefficient can be modulated effectively by external electric field.

Reference

[1] D.A. B.Miller,D. S. Chemla, T. C. Damen, A. C. Gossard,W.Wiegmann, T. H. Wood, and C. A. Burrus, "Band-edge electroabsorption in quantum well structures: The quantum-confined Stark effect," Phys. Rev. Lett., vol. 53, no. 22, pp. 2173–2177, Nov. 1984.

[2] U. Arad, E. Redmard, M. Shamay, A. Averboukh, S. Levit, and U. Efron, "Development of a large high-performance 2-D array of GaAs-AlGaAs multiple quantum-well modulators," IEEE Photon. Technol. Lett., vol. 15, no. 11, pp. 1531–1533, Nov. 2003.

[3] M. Jaros, "Simple analytic model for heterojunction band offsets," Phys. Rev. B Condens. Matter 37(12), 7112–7114 (1988).

[4] Sun G, Soref R A, Cheng H H. Design of a Si-based lattice-matched room-temperature GeSn/GeSiSn multi-quantum-well mid-infrared laser diode.[J]. Optics Express, 2010, 18(19):19957-65.

[5] S. Gupta, V. Moroz, L. Smith, Q. Lu, and K. C. Saraswat, "7 nm FinFET CMOS design enabled by stress engineering using Si, Ge and Sn," IEEE Trans. Electron. Dev. **61**(5), 1222 – 1230 (2014).

[6] G. Bastard, Wave Mechanics Applied to Semiconductor Heterostructures (Monograph de Physique, Paris, 1988).

[7] Paul D J. 8-band k.p modeling of the quantum confined Stark effect in Ge quantum wells on Si substrates[J]. Physical Review B Condensed Matter, 2008, 77(77):761-768.

Quantum Generation Dynamics of Coherent Phonons: Analysis of Transient Fano Resonance

Yohei Watanabe,[1] Ken-ichi Hino,[2,3], Muneaki Hase,[4] and Nobuya Maeshima[2,3]

[1]Doctoral Program in Materials Science, Graduate School of Pure and Applied Sciences, University of Tsukuba, Tsukuba, Ibaraki 305-8573, Japan
[2]Division of Materials Science, Faculty of Pure and Applied Sciences, University of Tsukuba, Tsukuba 305-8573, Japan
[3]Center for Computational Sciences, University of Tsukuba, Tsukuba 305-8577, Japan
[4]Division of Applied Physics, Faculty of Pure and Applied Sciences, University of Tsukuba, Tsukuba 305-8573, Japan

E-mail: hino@ims.tsukuba.ac.jp

Abstract. We study the transient Fano resonance of a semiconductor Si observed in the early time region of coherent phonon generation induced by an ultrafast pump laser. We particularly examine effects of the detuning on the transient Fano resonance, where the detuning is defined by the difference between the central frequency of the pump laser and the band gap. It is clarified that asymmetric profiles of transient induced photoemission spectra, implying the Fano resonance, strongly depend on the detuning. This is attributed to energetically adjacent bosonic states, whose energy levels are strongly influenced by the detuning.

1. Introduction

Coherent phonon (CP) generation in semiconductors is one of the representative phenomena induced by irradiation of an ultrafast pulse laser [1]. The mechanism of the CP generation has been discussed with the phenomenological models such as the impulsive stimulated Raman scattering model [2] and the displacive excitation of CP model [3], where the initial phase of the CP oscillation is considered to be a key parameter. Although several theoretical works focusing on the initial phase have been carried out [4, 5, 6], the CP generation mechanism remains a point of controversy. Furthermore, recent time-resolved spectroscopy techniques unveil the early time dynamics of the CP, where a vestige of the Fano resonance (FR) [7] is transiently observed in a lightly n-doped Si [8, 9]. This FR is manifested only in about 100fs after the laser irradiation, and is different from the conventional FR in equilibrium, for example, observed in heavily-doped Si [10, 11]. In order to clarify the CP generation dynamics accompanying the transient FR, a microscopic theory based on a quantum mechanical model is required.

The present authors have presented a comprehensive quantum theory of the transient FR and the subsequent CP dynamics in semiconductors based on a polaronic-quasiparticle (PQ) picture, where the transient induced photoemission spectra have asymmetric profiles in non-polar crystals while symmetric in polar crystals [12]. The difference is mainly ascribed to a time-dependent phase factor of electron-phonon interactions; the phase factor is real in an optical-phonon deformation potential interaction, whereas pure-imaginary in a Frölich interaction. In

IOP Conf. Series: Journal of Physics: Conf. Series **864** (2017) 012079 doi:10.1088/1742-6596/864/1/012079

the present paper, we examine effects of a detuning on the transient induced photoemission spectra, where the detuning is defined by the difference between the central frequency of the pump laser and the band gap of the semiconductor. We have found that the asymmetric profiles of the spectra strongly depend on the detuning, which is attributed to energetically adjacent bosonic states.

2. Theory

In this chapter, we briefly summarize our theory of Ref. [12]. The total Hamiltonian of the system is represented by $\hat{H} = \hat{H}_e + \hat{H}'(t) + \hat{H}_p + \hat{H}_{e-p}$, where $\hat{H}_e = \sum_{b\boldsymbol{k}} \varepsilon_{b\boldsymbol{k}} a^\dagger_{b\boldsymbol{k}} a_{b\boldsymbol{k}} + \frac{1}{2}\sum_{\boldsymbol{q}\neq 0} V^{(C)}_{\boldsymbol{q}} \sum_{bb'\boldsymbol{k}\boldsymbol{k}'} a^\dagger_{b\boldsymbol{k}+\boldsymbol{q}} a^\dagger_{b'\boldsymbol{k}'-\boldsymbol{q}} a_{b'\boldsymbol{k}'} a_{b\boldsymbol{k}}$ is the two-band Hamiltonian of electrons with the Coulomb term, and $\hat{H}'(t) = -\sum_{\boldsymbol{k}}[\Omega_{cv}(t)\, a^\dagger_{c\boldsymbol{k}} a_{v\boldsymbol{k}} + \Omega_{vc}(t)\, a^\dagger_{v\boldsymbol{k}} a_{c\boldsymbol{k}}]$ is the electron-light interaction. Here $a^\dagger_{b\boldsymbol{k}}$ ($a_{b\boldsymbol{k}}$) represents a creation (destruction) operator of an electron with the Bloch momentum $\boldsymbol{k}$ in the b (=c for conduction or v for valence)-band. The electric dipole moment between b and b'-bands is given by $d_{bb'}$ and $\Omega_{bb'}(t) = d_{bb'}F(t)$, where $F(t)$ is the electric field of a pump laser. The Hamiltonian of the LO-phonon is given by $\hat{H}_p = \sum_{\boldsymbol{q}} \omega^{(LO)}_{\boldsymbol{q}}\, c^\dagger_{\boldsymbol{q}} c_{\boldsymbol{q}}$ and the electron-phonon interaction is represented by $\hat{H}_{e-p} = \sum_{b,\boldsymbol{q},\boldsymbol{k}}(g_{b\boldsymbol{q}} c_{\boldsymbol{q}} a^\dagger_{b\boldsymbol{k}+\boldsymbol{q}} a_{b\boldsymbol{k}} + g^*_{b\boldsymbol{q}} c^\dagger_{\boldsymbol{q}} a^\dagger_{b\boldsymbol{k}} a_{b\boldsymbol{k}+\boldsymbol{q}})$, where $c^\dagger_{\boldsymbol{q}}(c_{\boldsymbol{q}})$ is a creation (destruction) operator of an LO-phonon. $g_{b\boldsymbol{q}}$ is a coupling constant of the b-band electron with the LO-phonon; the coupling corresponds to the deformation potential interaction in non-polar crystals and to the Frölich interaction in polar crystals, respectively. Here, it should be noted that we use the atomic units in this paper unless otherwise stated.

In our theory, we consider the equation of motion for a composite operator $A^\dagger_{\boldsymbol{q}}(\boldsymbol{k}bb') \equiv a^\dagger_{b\boldsymbol{k}+\boldsymbol{q}} a_{b'\boldsymbol{k}}$. We need to treat the commutator expressed by

$$[\hat{H}_e + \hat{H}'(t), \bar{A}^\dagger_{\boldsymbol{q}}(\boldsymbol{k}bb')] \approx \sum_{\tilde{\boldsymbol{k}}\tilde{b}\tilde{b}'} \bar{A}^\dagger_{\boldsymbol{q}}(\tilde{\boldsymbol{k}}\tilde{b}\tilde{b}')\bar{Z}_{\boldsymbol{q}}(\tilde{\boldsymbol{k}}\tilde{b}\tilde{b}', \boldsymbol{k}bb'), \tag{1}$$

where $\bar{A}^\dagger_{\boldsymbol{q}}$ represents that the rotational wave approximation is employed [12]. We diagonalize the non-hermitian matrix $\bar{Z}_{\boldsymbol{q}}$ as $U^{L\dagger}_{\boldsymbol{q}}\bar{Z}_{\boldsymbol{q}} = \mathcal{E}_{\boldsymbol{q}} U^{L\dagger}_{\boldsymbol{q}}$ and $\bar{Z}_{\boldsymbol{q}} U^R_{\boldsymbol{q}} = U^R_{\boldsymbol{q}}\mathcal{E}_{\boldsymbol{q}}$. Here $\mathcal{E}_{\boldsymbol{q}}$ is the eigenvalue diagonal-matrix and $\{U^L_{\boldsymbol{q}}, U^R_{\boldsymbol{q}}\}$ are the associated biorthogonal set of eigenvectors. Then we introduce an α-th adiabatic quasi-boson operator with time fixed as an adiabatic parameter defined as

$$B^\dagger_{\boldsymbol{q}\alpha} = \sum_{\boldsymbol{k}bb'} \bar{A}^\dagger_{\boldsymbol{q}}(\boldsymbol{k}bb') U^R_{\boldsymbol{q}\alpha}(\boldsymbol{k}bb'), \tag{2}$$

and finally obtain the adiabatic coupled equations represented by

$$-i\frac{dB^\dagger_{\boldsymbol{q}\alpha}}{dt} = B^\dagger_{\boldsymbol{q}\alpha}\mathcal{E}_{\boldsymbol{q}\alpha} + c^\dagger_{\boldsymbol{q}} M'^*_{\boldsymbol{q}\alpha} + i\sum_{\alpha'} B^\dagger_{\boldsymbol{q}\alpha'} W_{\boldsymbol{q}\alpha'\alpha} + M''_{-\boldsymbol{q}\alpha} c_{-\boldsymbol{q}} \tag{3}$$

and

$$-i\frac{dc^\dagger_{\boldsymbol{q}}}{dt} = \sum_{\alpha} B^\dagger_{\boldsymbol{q}\alpha} M_{\boldsymbol{q}\alpha} + c^\dagger_{\boldsymbol{q}}\omega^{(LO)}_{\boldsymbol{q}}, \tag{4}$$

where $W_{\boldsymbol{q}\alpha'\alpha}$ is a non-adiabatic coupling between the α-th quasi-boson and the α'-th one [12]. $M_{\boldsymbol{q}\alpha}$, $M''_{-\boldsymbol{q}\alpha}$ and $M'^*_{\boldsymbol{q}\alpha}$ are effective couplings between the α-th quasi-boson and the LO-phonon. Almost all of the quasi-boson states are electron-hole pair excitations, resulting in the quasi-boson continuum. Embedding the LO-phonon level $\omega^{(LO)}_{\boldsymbol{q}}$ in the continuum $\mathcal{E}_{\boldsymbol{q}\alpha}$ is likely to induce the FR [12].

IOP Conf. Series: Journal of Physics: Conf. Series **864** (2017) 012079 doi:10.1088/1742-6596/864/1/012079

The coupled equations of motion (3) and (4) are considered as a scattering problem [7], and a new PQ operator $F_{\boldsymbol{q}}^\dagger = [B_{\boldsymbol{q}}^\dagger, c_{\boldsymbol{q}}^\dagger]V_{\boldsymbol{q}}^R$ is introduced, where $V_{\boldsymbol{q}}^R$ is obtained by solving a matrix equation [13]. Then, the retarded Green's function of $F_{\boldsymbol{q}}^\dagger$ and relevant physical quantities including the dielectric function and transient induced photoemission spectrum $\bar{A}_{\boldsymbol{q}}(t_p;\omega)$ are also calculated in our theoretical framework [12].

3. Results and Discussion

In this work, we have carried out calculations for Si, and before showing obtained results we introduce a key parameter, the detuning

$$\Delta\omega = \omega_0 - E_g, \tag{5}$$

where ω_0 is the central frequency of the pump laser and E_g is the band gap of Si. In addition, it should be noted that we employ a squared wave for the pump laser and the pulse width is set to 15fs, which ranges from $t = -7.5$ fs to 7.5 fs [12]. Then Ω_{cv} is set to 29 meV.

Figure 1 shows calculated results of $\bar{A}_{\boldsymbol{q}}(t_p;\omega)$ at $t_p = 7.5$ fs for $\Delta\omega = -27$ meV and $\Delta\omega = 82$ meV, respectively. Although asymmetric spectral profiles characterizing the FR appear for both cases, positions of the spectral peaks and the dips are clearly different; for $\Delta\omega = -27$ eV the peak is in the higher energy side of the dip while for $\Delta\omega = 82$ meV the peak is in the lower energy side of the dip. This difference is found to result from the behavior of the effective coupling $M_{\boldsymbol{q}\alpha}$. Figure 2(a) shows $M_{\boldsymbol{q}\alpha}$ for $\Delta\omega = 82$ meV as a function of time. We can see that changes its sign from positive to negative at $t = 26.8$ fs and at $t = 71.3$ fs, and negative to positive at $t = 70.2$ fs, which implies that the phase factor $M_{\boldsymbol{q}\alpha}$ abruptly changes between 0 and π. At $t_p = 7.5$ fs, $M_{\boldsymbol{q}\alpha}$ for $\Delta\omega = 82$ meV is positive and thus the phase factor is 0. By contrast, the phase factor for $\Delta\omega = -27$ meV is equal to π, because $M_{\boldsymbol{q}\alpha}$ for $\Delta\omega = -27$ meV is negative for any t. As is already discussed in Ref. [12], the spectral profiles strongly depend on the phase factor of $M_{\boldsymbol{q}\alpha}$. Thus the spectral profile for $\Delta\omega = 82$ meV is different from that for $\Delta\omega = -27$ meV.

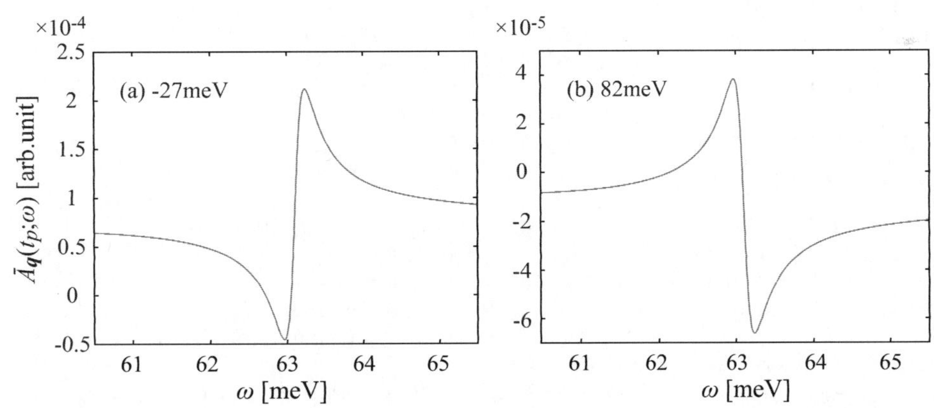

Figure 1. (a) Transient induced photoemission spectra $\bar{A}_{\boldsymbol{q}}(t_p;\omega)$ at $t_p = 7.5$ fs for $\Delta\omega = -27$ meV and (b) that for $\Delta\omega = 82$ meV, respectively.

Now, we focus on the the singular behavior of $M_{\boldsymbol{q}\alpha}$ for $\Delta\omega = 82$ meV; it seems to diverge at $t = 24.4$ fs and at 71.0 fs. In particular, the latter diverging behavior results in the change of the sign of $M_{\boldsymbol{q}\alpha}$ at $t = 70.2$ fs and at 71.3 fs. We have found that these singularities are attributed to the coupling between energetically adjacent adiabatic quasi-boson states. Figure 2 (b) shows

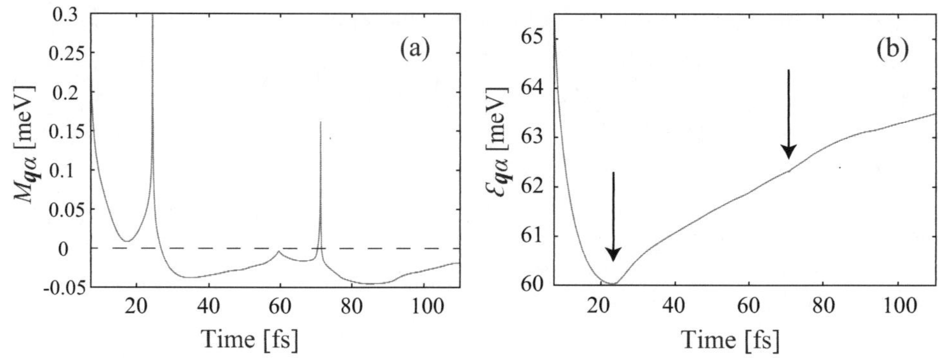

Figure 2. (a) The coupling constant $M_{q\alpha}$ and (b) associated adiabatic energy $\mathcal{E}_{q\alpha}$ for the detuning parameter $\Delta\omega = 82$ meV as a function of time.

the associated adiabatic energy $\mathcal{E}_{q\alpha}$ for $\Delta\omega = 82$ meV as a function of time. The abrupt energy changes occur at $t = 24.4$ fs and at 71.0 fs represented by arrows. At these points, $\mathcal{E}_{q\alpha}$ and the energy of the adjacent state labeled α' show the level crossing, resulting in the singularity of $M_{q\alpha}$. Here it should be noted that the level crossing shown here is considered to be closely related to the exceptional point (EP) [14, 15, 16], which is widely observed in non-hermitian problems in physics.

Acknowledgments

This work was supported by Grants-in-Aids for Scientific Research (C) (Grant No. 23540360 and Grant No. 15K05121) and for Young Scientists (B) (Grant No. 26800163) of the Ministry of Education, Culture, Sports, Science and Technology (MEXT), Japan.

References

[1] Kuznetsov A V and Stanton C J 1994 *Phys. Rev. Lett.* **73** 3243
[2] Yan Y-X, Gamble E B. and Nelson K A 1985 *J. Chem. Phys.* **83** 5391
[3] Zeiger H J, Vidal J, Cheng T K, Ippen E P, Dresselhaus G and Dresselhaus M S 1992 *Phys. Rev. B* **45** 768
[4] Stevens D M, Kuhl J and Merlin R 2002 *Phys. Rev. B* **65** 144304
[5] Riffe D M and Sabbah A J 2007 *Phys. Rev. B* **76** 085207
[6] Nakamura K G, Shikano Y and Kayanuma Y 2015 *Phys. Rev. B* **92** 144304
[7] Fano U 1961 *Phys. Rev.* **124** 1866
[8] Hase M, Kitajima M, Constantinescu A M and Petek H 2003 *Nature* **426**, 51-54
[9] Lee J D, Inoue J and Hase M 2006 *Phys. Rev. Lett.* **97** 157405
[10] Cerderia F, Fjeldly T A and Cardona M 1973 *Phys. Rev. B* **8** 4734
[11] Chandrasekhar M, Renucci J B and Cardona M 1978 *Phys. Rev. B* **17** 1623
[12] Watanabe Y, Hino K, Hase M and Maeshima N *Phys. Rev. B* (submitted), arXiv: 1510.00263.
[13] This matrix equation refers to Eq. (57) of Ref. [12].
[14] Moiseyev N 2011 *Non-Hermitian Quantum Mechanics* (New York: Cambridge) chapter 7-9.
[15] Haritan I, Gilary I, Amitay Z and Moiseyev N 2015 *J. Chem. Phys* **143** 154308
[16] Alfassi B, Peleg O, Moiseyev N and Segev M 2011 *Phys. Rev. Lett.* **106** 073901

IOP Conf. Series: Journal of Physics: Conf. Series **864** (2017) 012080 doi:10.1088/1742-6596/864/1/012080

Decoherence and fluctuation dynamics of the quantum dot nuclear spin bath probed by nuclear magnetic resonance

Evgeny A. Chekhovich

Department of Physics and Astronomy, University of Sheffield, Sheffield, S3 7RH, United Kingdom

E-mail: e.chekhovich@sheffield.ac.uk

Abstract. Dynamics of nuclear spin decoherence and nuclear spin flip-flops in self-assembled InGaAs/GaAs quantum dots are studied experimentally using optically detected nuclear magnetic resonance (NMR). Nuclear spin-echo decay times are found to be in the range 1-4 ms. This is a factor of ~3 longer than in strain-free GaAs/AlGaAs structures and is shown to result from strain-induced quadrupolar effects that suppress nuclear spin flip-flops. The correlation times of the flip-flops are examined using a novel frequency-comb NMR technique and are found to exceed 1 s, a factor of ~1000 longer than in strain-free structures. These findings complement recent studies of electron spin coherence and reveal the paradoxical dual role of the quadrupolar effects in self-assembled quantum dots: large increase of the nuclear spin bath coherence and at the same time significant reduction of the electron spin-qubit coherence. Approaches to increasing electron spin coherence are discussed. In particular the nanohole filled GaAs/AlGaAs quantum dots are an attractive option: while their optical quality matches the self-assembled dots the quadrupolar effects measured in NMR spectra are a factor of 1000 smaller.

1. Introduction

Quantum dots in III-V semiconductors are actively studied for applications in quantum information technologies. The advantages of III-V structures include excellent optical properties and advanced fabrication techniques available for these materials. However, all group III and V atoms have non-zero nuclear magnet moments. Interaction with nuclear spins is a major source of decoherence and information loss in electron and hole spin quantum dot qubits [1]. Nuclear quadrupolar effects and dipolar interactions between nuclei complicate the problem significantly, leading to complex dynamics of the electron-nuclear spin system which lacks complete understanding. Experimental probing of the nuclear spin dynamics is key to building a comprehensive picture of the electron-nuclear spin physics in quantum dots. In what follows measurements of the different timescales characterizing the nuclear spin bath are presented and discussed.

2. Nuclear spin dynamics in semiconductor quantum dots

2.1. *Experimental procedure*

We use optically detected nuclear magnetic resonance (ODNMR) to study a variety of quantum dot systems: self-assembled InGaAs/GaAs and InP/GaInP dots [2], monolayer fluctuation GaAs/AlGaAs dots [3] as well as GaAs/AlGaAs quantum dots grown by nanohole etching and in-filling [4]. All experiments are performed at T=4.2 K in magnetic fields up to B_z=10 T along the growth direction of the sample. Key to ODNMR measurement is the hyperfine interaction between electron and nuclear spins [1], which plays a dual role. Firstly, when quantum dot is excited with circularly polarized light, spin polarized electrons are generated repeatedly and hyperpolarize the nuclei in turn. Secondly, polarized nuclear spins produce an effective (Overhauser) magnetic field which manifests itself in the hyperfine shift E_{hf} of the electron spin levels and can be detected in the luminescence spectra – this allows ODNMR sensitivity down to individual quantum dot level.

2.2. *Longitudinal nuclear spin relaxation ($T_{1,nucl}$)*

Longitudinal relaxation time $T_{1,nucl}$ of the nuclear spin magnetization in the dark (without optical excitation) can be probed with optical "pump-probe" spectroscopy. Extremely long relaxation times exceeding 1000 s have been observed in self-assembled quantum dots [5, 6] and attributed to suppression of the dipolar nuclear flip-flops arising from large inhomogeneous strain induced quadrupolar effects. In strain-free GaAs/AlGaAs quantum dots the relaxation times are shorter (~60 s) but are still longer than in bulk semiconductors due to spin diffusion suppression [7]. Such $T_{1,nucl}$ times are much longer than any electron spin timescales in quantum dots implying that longitudinal nuclear spin relaxation does not limit electron spin coherence.

2.3. *Transverse nuclear spin relaxation ($T_{2,nucl}$)*

Unlike the longitudinal component, nuclear magnetization perpendicular to the external field is more prone to relaxation, since such process does not require energy exchange with environment. The transverse relaxation time of the nuclear magnetization $T_{2,nucl}$ is measured using Hahn echo pulse sequence [8, 9] shown in the inset of Fig. 1. The role of the π-pulse is to refocus the dephasing caused by inhomogeneous NMR spectral broadening, so that $T_{2,nucl}$ is a measure of the nuclear dipolar interactions only. The curves in Fig. 1 show the amplitude of the spin echo as a function of the evolution time 2τ measured for ^{69}Ga and ^{115}In nuclei of a single self-assembled InGaAs quantum dot. The echo decay times $T_{2,nucl}$ derived from fitting are noticeably larger than in strain-free structures (e.g. for ^{69}Ga nuclei we find $T_{2,nucl}$ =2.1 ms compared to $T_{2,nucl}$ =0.36 ms in strain-free dots [8]). Such increase in $T_{2,nucl}$ in self-assembled dots is due to the large inhomogeneous quadrupolar effects which shift the nuclear spin levels. Such shifts result in suppression of the dipolar flip-flops (flip-flop is a process where two nuclei exchange nuclear magnetization parallel to the magnetic field) as they become energetically forbidden.

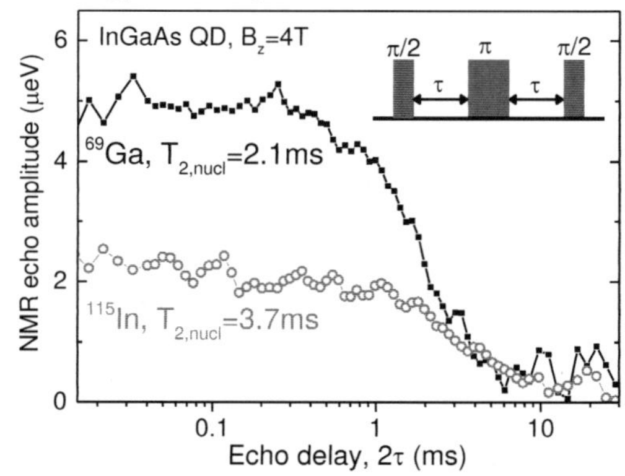

Figure 1. Nuclear spin echo decay in a single self-assembled quantum dot. Inset shows the Hahn echo pulse sequence.

On the other hand, upon closer examination, one finds that even complete freezing of the nuclear spin flip-flops does not lead to infinite nuclear spin coherence. Instead there is a maximum value of $T_{2,nucl}$ that can be calculated using the method of moments. For example, for ^{69}Ga the maximum $T_{2,nucl}$ is 3.0ms [9], thus experimental observation of a close to the limit $T_{2,nucl}$ =2.1 ms suggests that dipolar flip-flops are strongly suppressed in self-assembled dots. However, the degree of such suppression is difficult to gauge – this is an intrinsic feature of the spin echo and pulsed NMR in general. The upper limit on $T_{2,nucl}$ can be viewed as arising from parasitic spin dynamics induced by the "hard" pulses inducing the spin echo (an effect known as "instantaneous diffusion"). Probing equilibrium nuclear spin dynamics, which are of most interest, requires alternative NMR techniques discussed below.

2.4. *Using frequency comb NMR to measure homogeneous NMR linewidth ($\Delta\nu_{hom}$)*

Here we consider an alternative to pulsed NMR where low power radio frequency (rf) signals inducing non-coherent dynamics (without Rabi oscillations) are used. In the case of self-assembled quantum dots the inhomogeneous NMR spectrum of each isotope with linewidth $\Delta\nu_{inh}$ consists of a large number of individual NMR transitions with homogeneous linewidth $\Delta\nu_{hom} << \Delta\nu_{inh}$ (sketched with thick lines in Fig. 2(a)). Radio frequency signals with frequency comb profiles are used (thin lines in Fig. 2(a)). Depending on the mode spacing of the comb f_{MS} two opposite cases arise: (i) For $f_{MS} << \Delta\nu_{hom}$ all NMR transitions are uniformly excited and fast depolarization is expected, (ii) for $f_{MS} >> \Delta\nu_{hom}$ some of the NMR transitions are out of resonance with the comb, so that a slowdown in depolarization is expected. Thus non-coherent nuclear spin depolarization is sensitive to $\Delta\nu_{hom}$.

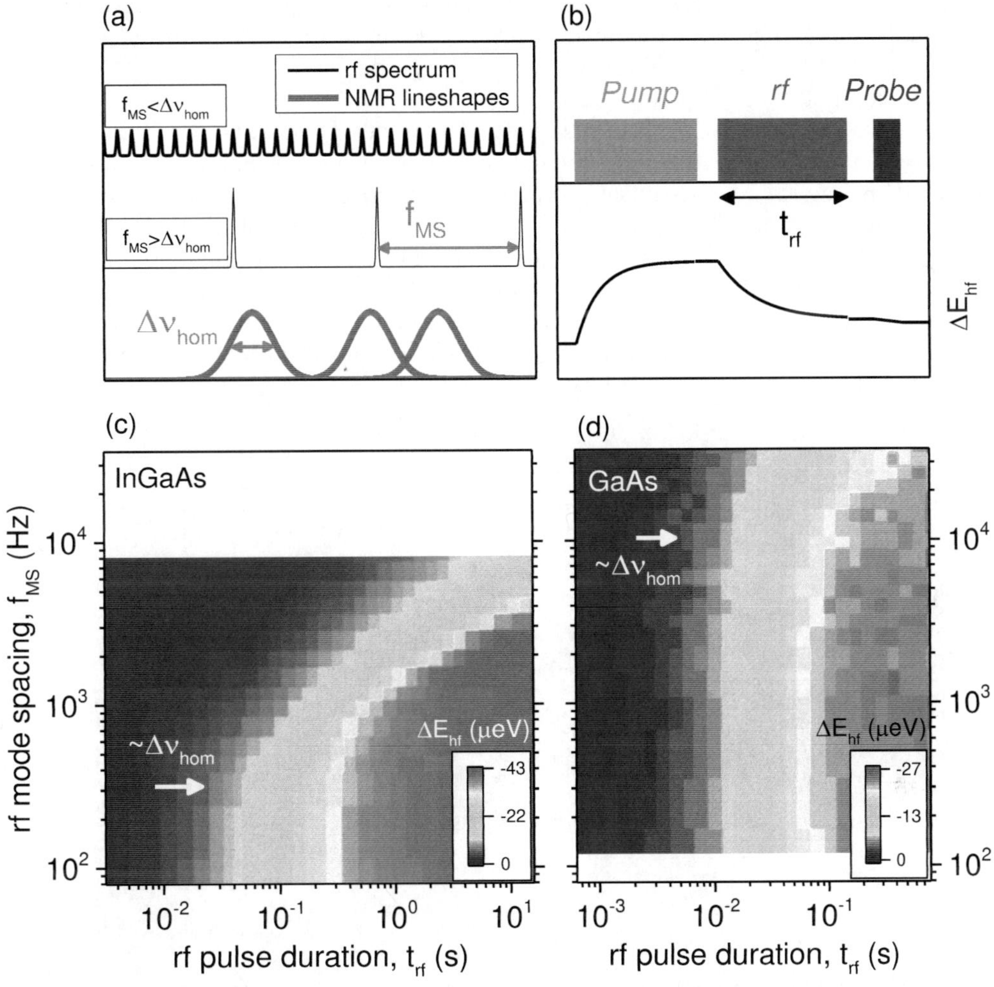

Figure 2. (a) Schematic spectra of NMR transitions (thick lines) and frequency combs (thin lines). (b) Time diagram of the measurement with a sketch of the corresponding change in the hyperfine shift ΔE_{hf}. (c, d) Experimental depolarization of ^{75}As spins in an InGaAs/GaAs (c) and GaAs/AlGaAs (d) quantum dot.

In experiment (Fig. 2(b)) the nuclei are first polarized optically. This is followed by depolarization with a frequency comb rf pulse in the dark (duration t_{rf}). Finally nuclear magnetization is probed by measuring the change in the hyperfine shift ΔE_{hf} in the photoluminescence spectrum. The dependence $\Delta E_{hf}(t_{rf}, f_{MS})$ is shown in Fig. 2(c) for the ^{75}As nuclei of an InGaAs quantum dot. As expected, at small

f_{MS} depolarization is fast. But above a certain threshold value of f_{MS} a pronounced slowdown is observed – this threshold in f_{MS} gives a few hundred Hz estimate of $\Delta\nu_{hom}$. A more accurate value of the homogeneous linewidth $\Delta\nu_{hom}\approx210$ Hz is obtained from the rate equation modelling [10]. The corresponding $T_{2,nucl}\approx1/(\pi\Delta\nu_{hom})\approx1.5$ ms is in reasonable agreement with the spin echo measurement [9]. A significantly different result is observed in Fig. 2(d) for a nearly strain-free monolayer fluctuation GaAs/AlGaAs quantum dot: the slowdown in nuclear depolarization is observed only for the comb mode spacing f_{MS} >10kHz, with fitting revealing much larger $\Delta\nu_{hom}\approx3200$ Hz and hence smaller $T_{2,nucl}\approx0.1$ s. Such homogeneous broadening is due to energetically allowed nuclear dipolar flip-flops and possibly due to some additional effect of charge fluctuations.

It is worth noting that even with completely frozen dipolar flip-flops, the homogeneous linewidth $\Delta\nu_{hom}\approx210$ Hz obtained for ^{75}As in an InGaAs quantum dot is smaller than the homogeneous broadening expected for all possible states of the nuclear spin environment surrounding each ^{75}As spin. This important feature of the frequency comb technique arises from the fact that the nuclear spin environment is sampled over a finite timescale, comparable to the exponential depolarization time achieved at $f_{MS} << \Delta\nu_{hom}$. Thus for sufficiently fast frequency comb rf depolarization the environment of each depolarized spin is effectively static ("frozen") and the dipolar field produced by such environment is eliminated as any other source of inhomogeneous broadening. As described below this property of the frequency comb NMR method opens a way for direct probing of the equilibrium flip-flop dynamics in self-assembled quantum dots.

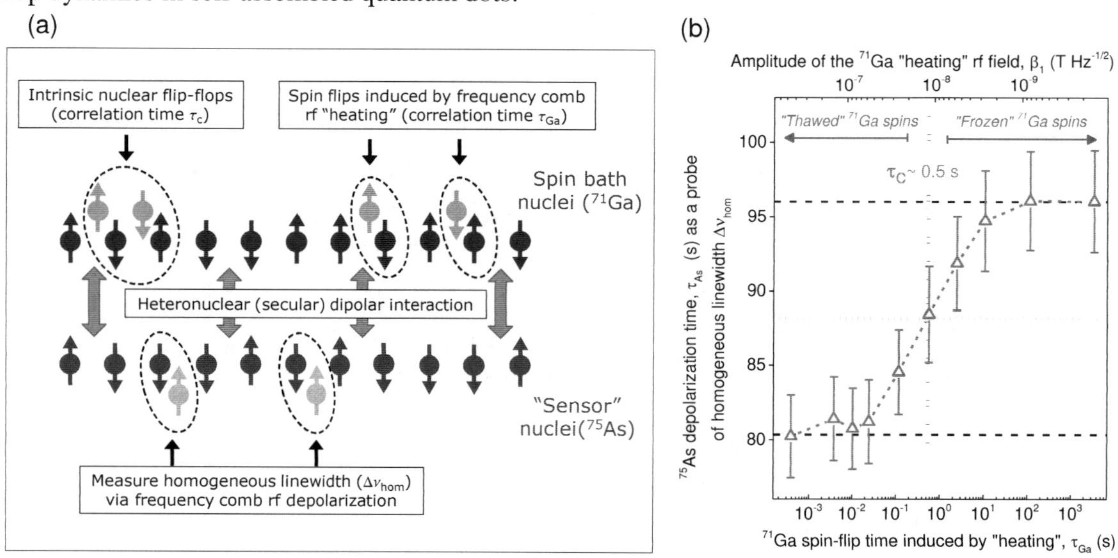

Figure 3. (a) Measurement protocol to determine correlation time τ_c of the nuclear dipolar flip-flops. (b) Experimental dependence of the ^{75}As rf-induced depolarization time on the amplitude of the ^{71}Ga rf heating amplitude. The half-range point (dotted lines) corresponds to the situation where natural flip-flops and rf heating induced spin flips of ^{71}Ga have the same effect on ^{75}As, revealing the correlation time τ_c of the ^{71}Ga flip-flops.

2.5. Nuclear spin flip-flop correlation times ($\tau_{c,nucl}$)

The strategy for probing the correlation time $\tau_{c,nucl}$ of the nuclear spin exchange flip-flops is shown in Fig 3(a). The ensemble of ^{71}Ga spins is studied, while ^{75}As nuclei are used as a "sensor". Heteronuclear (^{71}Ga-^{75}As) dipolar interaction is secular: the ^{71}Ga-^{75}As flip-flops are energetically forbidden but random configurations of the ^{71}Ga spins cause homogeneous broadening of the ^{75}As nuclear spins. The ^{71}Ga spins undergo random flip-flops with unknown correlation time $\tau_{c,nucl}$. This fluctuation correlation time can be shortened by an additional "heating" with a frequency comb rf (resonant with ^{71}Ga) which produces spin flips with a known correlation time τ_{Ga}. A pronounced speed up of the ^{71}Ga fluctuations occurs for $\tau_{Ga} \approx \tau_{c,nucl}$ and is detected via increased $\Delta\nu_{hom}$ of the ^{75}As spins. This allows for $\tau_{c,nucl}$ to be derived. Measuring the entire $\Delta E_{hf}(t_{rf}, f_{MS})$ dependence of ^{75}As is time consuming, however, in order to characterize the changes in $\Delta\nu_{hom}$ it is sufficient to measure the characteristic depolarization time τ_{As} at a fixed $f_{MS} \approx \Delta\nu_{hom}$. The resulting $\tau_{As}(\tau_{Ga})$ dependence is shown in Fig. 3(b): the transition point $\tau_{Ga} \approx \tau_{c,nucl}$ is approximately at the middle of the τ_{As} variation range (dotted lines) providing an estimate $\tau_{c,nucl} \approx 0.5$ s, with a more accurate $\tau_{c,nucl} \approx 1$ s obtained from model fitting [10].

The flip-flop correlation times $\tau_{c,nucl} \geq 1$ s derived from frequency comb NMR of neutral self-assembled quantum dots are very long and are a sign of extreme freezing of the flip-flops induced by inhomogeneous quadrupolar shifts. Nuclear dipolar flip-flops are responsible for the "spectral diffusion" process [11] which sets an ultimate upper limit on the coherence of the central (electron or hole) spin. It is thus evident that intrinsic nuclear spin dynamics of a bare quantum dot do not limit the electron spin coherence time $T_{2,elec}$ up to sub second regime. A more complex picture is expected when electron-nuclear interactions are present. This is a subject for a further investigation using frequency comb NMR on electron charged quantum dots.

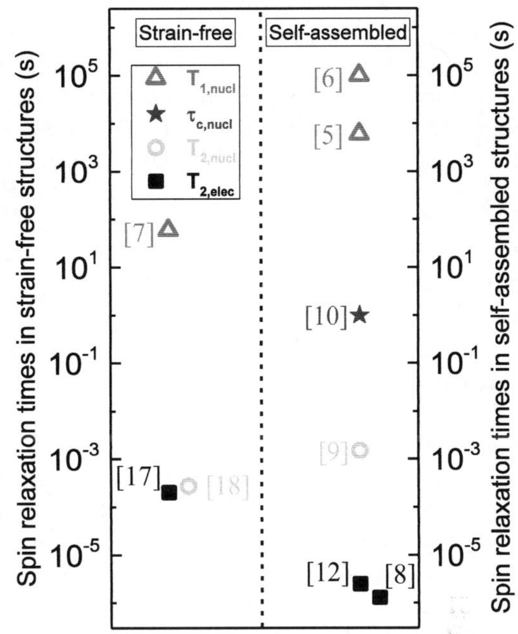

Figure 4. Electron and nuclear spin relaxation times in strain-free GaAs and self-assembled InGaAs quantum dots. Numbers show corresponding references.

3. Discussion and outlook

3.1. Comparison of electron and nuclear spin relaxation timescales

The various experimentally measured timescales characterizing the nuclear spin as well as electron spin dynamics are shown in Fig. 4 summarizing the recent studies for both strain-free lattice matched GaAs/AlGaAs and self-assembled strained InGaAs/GaAs quantum dots.

In strain-free structures electron spin coherence times $T_{2,elec}$ and nuclear spin echo times $T_{2,nucl}$ of few hundred microseconds are reported. Although not measured directly the flip-flop correlation times are expected to be comparable to the coherence times $\tau_{c,nucl} \approx T_{2,nucl}$ in structures with negligible quadrupolar effects. Thus one concludes that in the strain-free GaAs quantum dots the electron spin qubit coherence is limited by the nuclear spin fluctuation dynamics.

A very different and somewhat paradoxical picture is observed for self-assembled dots. Large quadrupolar effects result in a dramatic increase of all nuclear spin relaxation timescales. On the other hand, the electron spin coherence times are found to be much shorter than in strain-free quantum dots. There is a striking six orders of magnitude gap between electron spin coherence time $T_{2,nucl}$ and nuclear spin correlation time $\tau_{c,nucl}$. While initial studies considered extrinsic limitations to electron spin coherence (such as charge fluctuations in quantum dot environment [12]), more recent reports point to intrinsic limitations. The proposed mechanisms include electron mediated nuclear-nuclear interactions [13] as well as broadband nuclear spin noise arising from large inhomogeneous quadrupolar effects [14]. Building a complete picture of the factors limiting the electron spin coherence in self-assembled dots requires further studies: NMR techniques reported here are expected to be a valuable tool in such work. Meanwhile it is important to consider the available approaches to increasing the quantum dot spin qubit coherence as discussed below.

3.2. Approaches to increase quantum dot spin coherence

Recent proposals to improve qubit coherence in self-assembled dots include switching from electrons to valence band holes which have much smaller hyperfine interaction [15], as well as initializing the nuclear spin bath in an eigen state by achieving nuclear spin polarization close to 100% [16].

One more approach is to eliminate quadrupolar effects altogether: This is indeed possible in strain-free quantum dots in 2DEG in GaAs, however, such structures lack the optical access to the spin of the electron. On the other hand, a recently developed growth method based on in situ etching and filling of the nanoholes yields GaAs/AlGaAs quantum dots that combine the excellent optical properties of the self-assembled dots but with very small quadrupolar effects. Recent NMR measurements have shown that quadrupolar shifts in such dots are within ±20 kHz, which is three orders of magnitude smaller than in InGaAs dots [4]. Furthermore, application of external strain cancels the average quadrupolar shifts leaving only the effects of inhomogeneous NMR broadening that does not exceed ±10 kHz. A direct measurement of electron spin coherence in such dots could be very revealing in achieving deeper understanding of the electron-nuclear spin dynamics and the role of the quadrupolar interactions.

References

[1] Urbaszek B *et al* 2013 *Rev. Mod. Phys.* **85**, 79
[2] Chekhovich E A *et al* 2012 *Nature Nanotechnol.* **7**, 646
[3] Chekhovich E A *et al* 2013 *Nature Phys.* **9**, 74
[4] Ulhaq A *et al* 2016 *Phys. Rev. B* **93**, 165306
[5] Chekhovich E A *et al* 2010 *Phys. Rev. B* **81**, 245308
[6] Latta C *et al* 2011 *Phys. Rev. Lett.* **107**, 167401
[7] Nikolaenko A E *et al* 2009 *Phys. Rev. B* **79**, 081303(R)
[8] Makhonin M N *et al* 2011 *Nature. Mater.* **10**, 848
[9] Chekhovich E A *et al* 2015 *Nature Commun.* **6**, 6348
[10] Waeber A M *et al* 2016 *Nature Phys.* **12**, 688
[11] Merkulov I A *et al* 2002 *Phys. Rev. B* **65**, 205309
[12] Press D *et al* 2010 *Nature Photon.* **4**, 367
[13] Wüst G *et al* 2016 *Nature Nanotechnol.* doi:10.1038/nnano.2016.114
[14] Bechtold A *et al* 2015 *Nature Phys.* **11**, 1005
[15] Prechtel J H *et al* 2016 *Nature. Mater.* **15**, 981
[16] Schliemann J *et al* 2002 *Phys. Rev. B* **66**, 245303
[17] Bluhm H *et al* 2010 *Nature Phys.* **7**, 109
[18] Sanada H *et al* 2006 *Phys. Rev. Lett.* **96**, 067602

IOP Conf. Series: Journal of Physics: Conf. Series **864** (2017) 012081 doi:10.1088/1742-6596/864/1/012081

Efficiently tunable photon emission from an optically driven artificial molecule

Jonathan Cotrino-Lemus[1] and Hanz Y. Ramírez[1]

[1] Grupo de Física Teórica y Computacional, Escuela de Física, Universidad Pedagógica y Tecnológica de Colombia (UPTC), Tunja 150003, Boyacá, Colombia

E-mail: hanz.ramirez@uptc.edu.co

Abstract. In this work, the emission properties of double quantum dots driven by an intense monochromatic electromagnetic field, while undergoing resonant tunnelling, are investigated. We find the optically active energy transitions and their corresponding emission intensity, and compute resonance fluorescence spectra for different detunings between the direct and indirect exciton energies. The simulated emission exhibit either three, five, or seven peaks, tunable on demand. On the basis of the obtained results, our proposal offers efficient control of the resonance fluorescence of an artificial molecule, suitable for optoelectronic applications.

1. Introduction

Quantum dots (QDs) are semiconductor heterostructures at the nanometric scale, in which electrons and holes are confined in all three special dimensions and energy states are fully discretized [1]. This discretization yields optical properties well known in atoms such as single photon emission and Rabi oscillations [2,3]. In particular, the distinctive feature of a strongly radiation-matter system so-called Mollow triplet [4], has been successfully observed in QDs [5].

On the other hand, controllable coupling between neighboring QDs has been experimentally achieved by means of electric fields [6]; Hence, interacting double dots are often referred as quantum dot molecules (QDM) [7].

In this work, we propose to combine the advantages of solid emitters for realizing strong radiation matter coupling, with the tunability provided by the additional degree of freedom in double DQs, to form a photonic system in which the number of optical channels can be efficiently adjusted.

2. System and basis

The studied system consist of two QDs coupled by resonant tunneling, interacting with intense radiation from a monochromatic coherent source. This is schematically represented for an *InAs/GaAs* QDM in figure 1(a), where Δ stands for the energy difference between direct and indirect exciton states (Coulomb interaction neglected, $E_{XD} - E_{XI} = \Delta$).

In the non-interacting QDM-radiation basis (bare states), three states form the n-th rung of the Jaynes-Cummings (JC) ladder: the ground QDM state with n photons ($|n,\ g\rangle$), direct exciton state with $n-1$ photons ($|n-1,\ XD\rangle$) and indirect exciton state with $n-1$ photons ($|n-1,\ XI\rangle$), whose energies are $E_{n,g}$, $E_{n-1,XD}$, and $E_{n-1,XI}$, respectively. The energies of the bare and dressed states are schematically shown in figure 2(a), for the case in which the monochromatic mode (frequency ω_L) is out of resonance respect to the direct and indirect excitons ($\hbar\omega_L - E_{XD} = \Delta_{L-D}$).

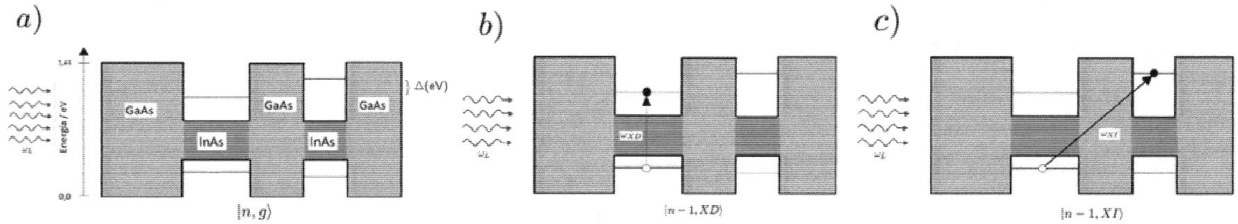

Figure 1. Considered basis. a) Ground state of the system with n photons; b) Direct exciton and $n-1$ photons; c) Indirect exciton and $n-1$ photons.

Then, the total Hamiltonian is given by

$$\widehat{H}_T = \widehat{H}_M + \widehat{H}_R + \widehat{H}_{M-R} + \widehat{H}_t\,,$$

where for each rung, the free QDM Hamiltonian is

$$\widehat{H}_M^n = E_{XD}|n-1,XD\rangle\langle n-1,XD| + E_{XI}|n-1,XI\rangle\langle n-1,XI| + E_0|n,g\rangle\langle n,g|\,,$$

the free-radiation Hamiltonian is

$$\widehat{H}_R^n = (n-1)\hbar\omega_L|n-1,XD\rangle\langle n-1,XD| + (n-1)\hbar\omega_L|n-1,XI\rangle\langle n-1,XI| + n\hbar\omega_L|n,g\rangle\langle n,g|\,,$$

the QDM-radiation interaction term is

$$\widehat{H}_{M-R}^n = g\sqrt{n}\,|n,g\rangle\langle n-1,XD| + g\sqrt{n}|n-1,XI\rangle\langle n,g|\,,$$

and the interdot tunneling term is

$$\widehat{H}_t^n = t|n-1,XD\rangle\langle n-1,XI| + t|n-1,XI\rangle\langle n-1,XD|\,.$$

Thus, the Hamiltonian for each of the subspaces is analytically diagonalizable and the dressed states can be obtained. These new eigenvectors are superpositions of the bare states, and the oscillator strengths for the allowed transitions are found in terms of the superposition coefficients

$$C_{i,j}^n = \langle (n-1), i|(n-1), j\rangle\,,$$

where $i = g, XD, XI$ and $j = 1,2,3$.

For a given average number of photons m in the mode ($m = \langle\widehat{N}\rangle$), set by the driving laser power; the transition energies $\Delta E_{j,j'}^m(g,t,\Delta_{L-D},\Delta) \equiv E_j^m - E_{j'}^{m-1}$, and the superposition coefficients $C_{i,j}^m(g,t,\Delta_{L-D},\Delta)$, depend on the four open parameters of our model, providing a promising scenario for flexible control of emission frequencies.

IOP Conf. Series: Journal of Physics: Conf. Series **864** (2017) 012081 doi:10.1088/1742-6596/864/1/012081

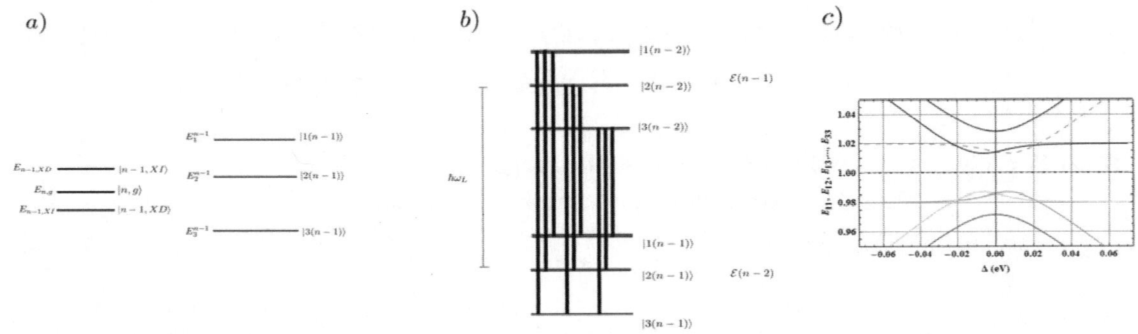

Figure 2. a) Bare (left) and dressed (right) states for the n-th rung of the JC ladder. b) two consecutive rungs and the allowed transitions. c) Transition energies as functions of the tuning parameter Δ.

3. Resonance fluorescence spectra

Setting constant values for the parameters $g\sqrt{m}, t, \Delta_{L-D}$ (10 meV, 10 meV, and 0 , respectively) [8,9], we simulate the resonance fluorescence (RF) spectra from the QDM for different values of the tunable parameter Δ. In figures 3 a), b), and c) the emission spectra for different values of Δ are shown, and an interesting on-demand triplet-pentaplet-heptaplet optical transition can be clearly observed.

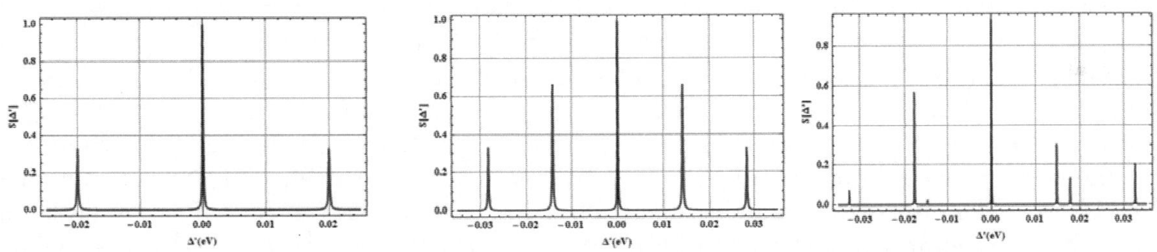

Figure 3. RF spectra from a driven QDM for a) $\Delta > 1\text{eV}$, b) $\Delta = 0$, and $\Delta = 0.014\ eV$.

In this model, effective tuning of Δ may be accomplished by increasing the indirect exciton energy through an external bias field, applied in the coupling direction of the QDM [6].
In conclusion, the resonance fluorescence of a quantum dot molecule driven by an intense laser, have been studied, and the predicted triplet-pentaplet-heptaplet transition appears as a convenient scenario for efficient control of optically active frequencies.

Financial support from the research division of UPTC is acknowledged.

References

[1] Alivisatos A P 1996 *Science* **271** 933
[2] Htoon H, Takagahara T, Kulik D, Baklenov O, Holmes A L Jr., and Shih C K 2002 *Phys. Rev. Lett.* **22** 087401
[3] Muller A, Flagg E B, Bianucci P, Wang X Y, Deppe D G, Ma W, Zhang J, Salamo G J, Xiao M, and Shih C K 2007 *Phys. Rev. Lett.* **99** 187402
[4] Mollow B R 1969 *Phys. Rev.* **188**
[5] Vamivakas A N, Zhao Y, Lu C Y, and Atatüre M 2009 *Nat. Phys.* **5** 198
[6] Stinaff E A et al. 2006 Science **311** 636
[7] Ramirez H Y, Camacho A S, and Lew Yan Voon L C 2007 *J Physics Cond. Matt.* **19** 346216
[8] He Y et al. 2015 *Phys. Rev. Lett.* **114** 097402
[9] A direct exciton radiative linewidth of 200 μeV, was used in the calculations.

IOP Conf. Series: Journal of Physics: Conf. Series **864** (2017) 012082 doi:10.1088/1742-6596/864/1/012082

Light and heavy hole exciton polariton Faraday rotation in a single GaAs microcavity

D C T Ferreira , A C S P Pimenta, and F M Matinaga[1]

Depto de Fisica, ICEX, UFMG, Belo Horizonte, MG, Brasil

[1] E-mail: matinaga@fisica.ufmg.br

Abstract. Faraday rotation (FR) can also be observed in a resonantly excited system without an external magnetic field. In a microcavity, polaritons spin population imbalance generates an effective magnetic field across the cavity mode, which results in an amplified FR on the emitted photon. Also, when the cavity is tuned to light hole (LH) exciton energy, an additional field momentum will be added to the effective magnetic field parallel to cavity mode, resulting in a aditional FR.

1. Introducion

Polaritons on a microcavity have a dynamical scattering process between states on the lower or upper polariton branch. It can generate an unbalanced spin plus/minus population, i.e. a spin splitting doublet, which results in a system with a effective magnetic field normal to the quantum well even without an external field.[1,2] Such field enhances the Faraday rotation for the emitted light, since it travel back and forth hundreds times in a Fabry Perot cavity. By tuning the cavity resonance for higher energy, light hole polaritons (LHP) can be generated in the same way as the heavy hole polaritons (HHP).[3] In this case, the effective magnetic field has an additional field due to the LH exciton magnetic momentum z – component (p_z). Such field would be an additional reason to use polaritons in a quasi resonant excitation as an all optical switching device. [4]

In this work we tune the cavity resonance from heavy hole to light hole exciton energy, measuring the Faraday rotation (FR) on the polariton emission as a function of the pump excitation, just by scanning the sample position. We discuss this effect in terms of the effective magnetic field in the cavity, without external field, i.e. the optically induced FR.

2. Experimental Results

An usual photoluminescence (PL) setup was used to measure the FR on the polaritons emission (fig.1a). The microcavity is formed by two DBR mirror with 26.5/24 $Al_{0.2}Ga_{0,8}As/AlAs$ pairs and a GaAs - SQW gain medium in the center of lambda cavity as described on the literature.[5] Reflectance measurement shows a cavity mode splitting at 798 nm (fig.1b) for the HH exciton (Δ=3.4meV) and a broadening due to the LH (792 nm) coupling (fig.1c). That exciton – cavity mode interaction is weaker than for HH exciton due to smaller LHP oscillator strength. However, it is not a simple change on the reflectance deepness due to LH absorption, showing a smaller coupling of the LH exciton with the cavity field.

We analyzed the linear polarization (LP) angle (θ) direction behavior of the polariton emission intensity for a resonant Ti:Sapphire continuous wave laser excitation as a function of the emission angle (we use $\lambda/2$ plate to rotate the emission polarization angle), as illustrated by the

IOP Conf. Series: Journal of Physics: Conf. Series **864** (2017) 012082 doi:10.1088/1742-6596/864/1/012082

red dashed line (fig.2a). The LP Polarization angle rotation, i.e. the FR effect was observed by measuring the θ (fig. 2b) for each pump power.

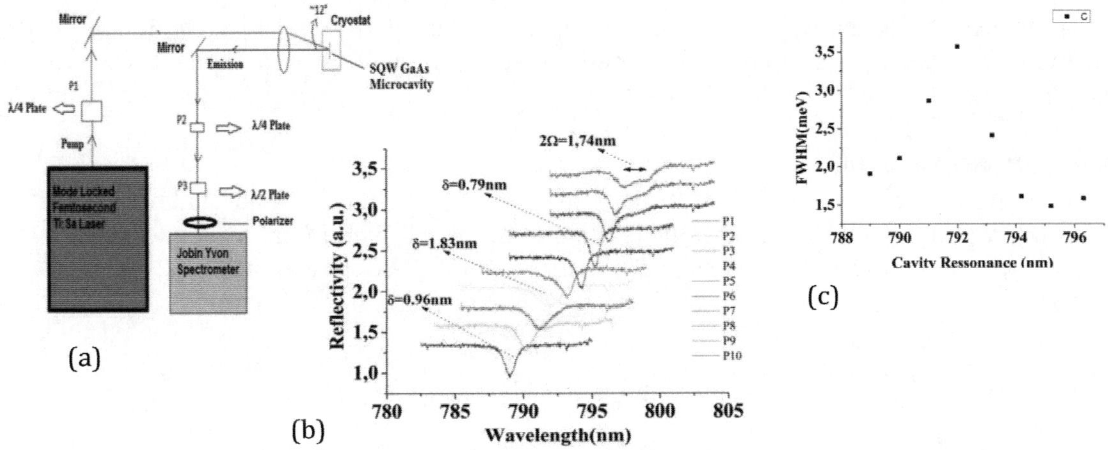

Figure 1: **a)** A single GaAs microcavity photo-luminescence characterization system. **b)** White light reflectance spectrum as a function of cavity resonance along of the sample positions (P_i). **c)** Cavity resonance deep FWHM for LHP.

Firstly, we took that data for a non-resonant excitation (fig.2c). For all these FR data, we took the emission LP degree ρ as:

$$\rho = {(I_{max} - I_{min})}/{(I_{max} + I_{min})}$$

Where $I_{max,\ min}$ are the maximum/minimum PL integrated intensity. Here we see a huge change on θ (~80°), fixing around 130° for higher pump power. This behavior, a pinning of the LP light emission in a fixed direction is somewhat previewed for high purity microcavity,[6] and the huge change determine the θ behavior around off resonant exciton laser pump threshold.

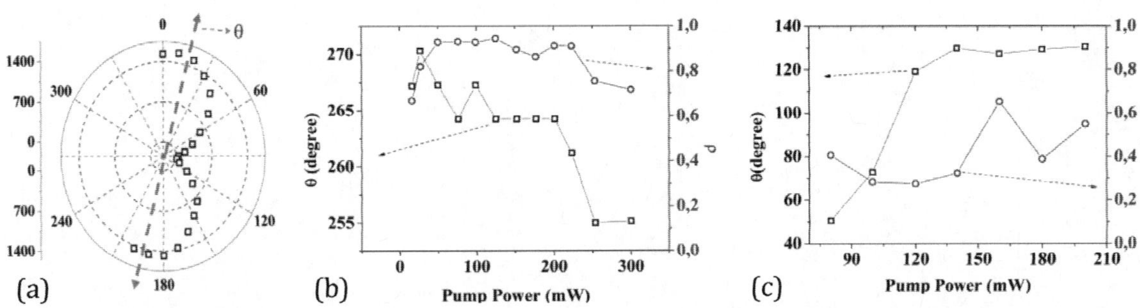

Figure 2: **a)** Polarization polar view: Polariton emission intensity as a function of the LP angle emission for a fixed 200 mW pump power. **b)** LP θ measurement for 793 nm, together with the LP degree (ρ) for a LP pump. **c)** θ measurement for off resonant excitation at 798 nm.

The LP angle emission changes (FR) for the polariton emission as a function of pump power show in this case a FR of 13°, this also represents a polarization change in relation to the pump excitation LP angle. Fig.2c shows a saturation angle change above pump threshold for the nonresonant excitation, indicating a fixed LP angle for the microcavity laser emission. This polarization data were measured above pump threshold for this nonresonant excitation intensity relation.

3. FR Results & Discussion

The FR in a microcavity is proporcional to the effective magnetic field ($\boldsymbol{B}_i$) parallel to the cavity mode, so we can have for a HHP emission, a field due to the population imbalance with spin up and down (eq.1) and for LHP, there is an additional z magnetic field momentum (eq.2,3).

$$g\mu_B B_{in} = 2(V_1 - V_2)(N_1 - N_2) \quad \text{(eq.1)}$$

$$\Phi_{\frac{3}{2}\frac{1}{2}} = -\frac{1}{\sqrt{6}}\left[(p_x + ip_y)\downarrow -2p_z \uparrow\right] \quad \text{(eq.2)}$$

$$\Phi_{\frac{3}{2}\frac{1}{2}} = -\frac{1}{\sqrt{6}}\left[(p_x + ip_y)\downarrow -2p_z \uparrow\right] \quad \text{(eq.3)}$$

Where g – exciton factor, μ_B - Bohr magnetic momentum, V_i – the triplet and singlet energy dispersion and N_i – spin up and down population. $\Phi_{m,n}$ represents the light hole momentum, with each spatial components p_i, indicating the spin up (↑) and down (↓) field components. The eq. 2,3 show an effective magnetic field parallel to the cavity mode(z), which also depends on the excitation pump resonant light polarization, CP or LP. [7]

Considering the $\boldsymbol{B}_i$, we measured the FR for LHP emission using LP and circular polarization (CP) excitation light as shown on fig. 3, together with the emission LP degree ρ. The cavity tuning was done by moving the pump spot along the sample position.

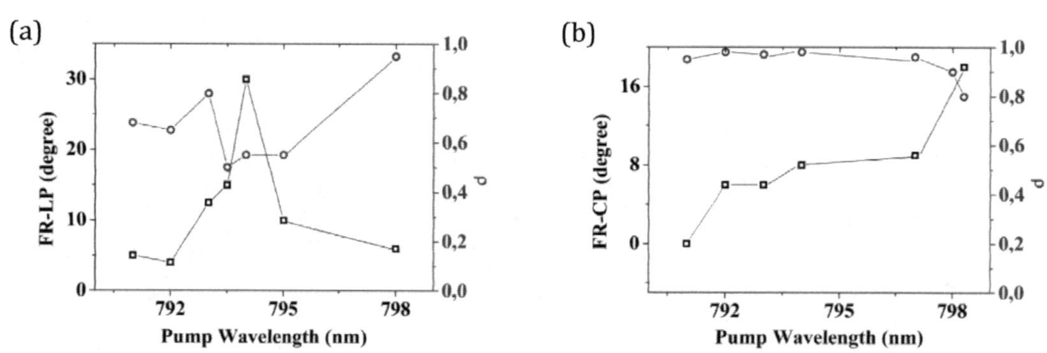

Figure 3: FR measurements for polaritons together with the emission LP degree ρ. **a)** FR for LP light excitation. **b)** FR for CP light excitation.

Since the FR for LHP results from two magnetic fields, it explain the higher FR (FR=30°) close to LH exciton resonance (793nm), compared to the HHP (FR=6°) for a LP pump excitation. On the other side, for the CP light excitation, we see a larger value for the HHP (FR=16°) in comparison to the LHP (FR=8°). For the LHP emission, we note a polariton spin ($\sigma^{+,-}$) population difference higher for the LP pump excitation, compared to the CP excitation. On the other side, for CP excitation, the scattering dynamic favor the LP light emission (fig.4). For the HHP emission, we saw a smaller FR for LP pump excitation and higher value for CP.

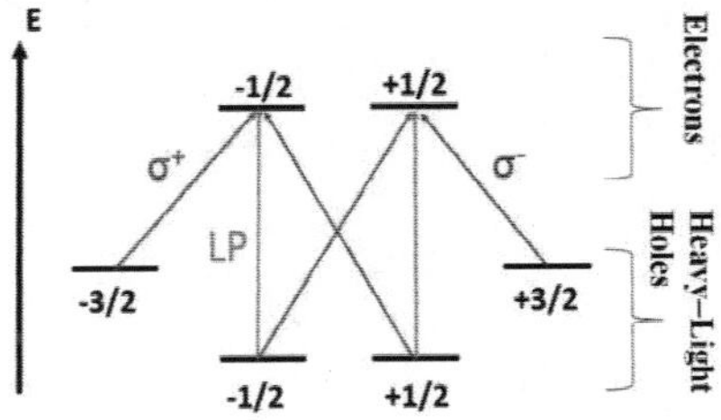

Figure 4: LP or a CP ($\sigma^{+,-}$) excitation, generating polaritons states, with balanced or not populations.

Those results reflect the $\sigma^{+,-}$ spin population imbalance produced by a CW pump (LP,CP) due to not only the dynamic behavior, but also to the spin exchange field, which is discussed somewhere.[8] Any way, we see a strong contribution on the cavity effective field when we excite the LHP, if we compare the much higher HHP emission (higher oscillator strength compared to the LH exciton) intensity for the same excitation intensity with a single quantum well.

4. Conclusion

The measured FR was 30 degree to LH LP excitation and 16 degree for HH CP excitation. Such huge FR results illustrate the enhancement of the cavity effective magnetic field due to z magnetic field component of the LHP emission parallel to cavity mode.

5. Aknowledgments

This work was supported by Fapemig and CNPq.

6. References

[1] Kavokin A et al 2003 Phys. Rev. B **67** 195321
[2] Giri R et al 2012 Phys. Rev. B **85** 195313
[3] Comitti V S et al 2014 Sol. Stat. Comm. **193** 37
[4] Amo A 2010 Nature Phot. **4** 361
[5] Cotta E A et al 2007 Phys. Rev. B **76** 073308
[6] Klopotowski L et al 2006 Sol. Stat. Comm. **139** 511
[7] Singh et al 1993 Physic of Semiconductors and their Heterostructure (New York Mc Graw-Hill)
[8] Renucci P et al 2005 Phys. Rev. B **72** 075317

IOP Conf. Series: Journal of Physics: Conf. Series **864** (2017) 012083 doi:10.1088/1742-6596/864/1/012083

Emission characteristics of high-gain GaN-based Vertical-Cavity Surface-Emitting Lasers

Guoen Weng[1], Shaoqiang Chen[1, *], Takashi Ito[2], Hidefumi Akiyama[2], Xiaobo Hu[1], and Baoping Zhang[3, *]

[1]Department of Electronic Engineering, East China Normal University, 500 Dongchuan Road, Shanghai 200241, China
[2]Institute for Solid State Physics, The University of Tokyo, 5-1-5 Kashiwanoha, Kashiwa, Chiba 277-8581, Japan
[3]Department of Electronic Engineering, Xiamen University, 422 South Siming Road, Xiamen 361005, Fujian, China

[*]sqchen@ee.ecnu.edu.cn, bzhang@xmu.edu.cn

Abstract: GaN-based vertical-cavity surface-emitting lasers (VCSELs) with high optical gain and short cavity lifetime are favorable for the generation of ultra-short pulses in the blue and green regions. In our previous works, 6 and 2 picosecond short-pulses have been generated from gain-switched InGaN VCSELs with 3- and 10-period InGaN/GaN quantum wells (QWs) in the active layers by using an up-conversion measurement system. To further increase the gain of the VCSEL for the generation of even shorter pulses, 20-period InGaN/GaN QWs samples were fabricated. The emission characteristics of these high-gain VCSELs were investigated and analyzed under the optical pumping at room temperature.

1. Introduction

Low-cost, compact and pulsed semiconductor lasers with ultra-short output pulses are on demand in many applications such as high-density optical storage, high-speed optical communications, optical signal processing, and time-resolved spectroscopy [1, 2]. Vertical-cavity surface-emitting laser (VCSEL), as a novel semiconductor laser, has many particular advantages over conventional edge-emitting lasers including single-longitudinal-mode operation, low laser threshold, circular output beam with low divergence, and densely-packed two-dimensional arrays. Moreover, the high optical gain and short cavity lifetime in the VCSELs due to their short cavity lengths make them favorable for the generation of ultra-short pulses via gain-switching. Actually, picosecond pulses have been widely reported in both optically and electrically pumped gain-switched GaAs-based VCSELs over the past few decades [3, 4]. However, the increasing requirements for next-generation of optical storages with higher storage density and capacity, and laser printers with higher graphical resolution have promoted the shortening of the laser wavelengths, which makes nitride-based VCSELs with the laser emission in blue and green regions into very promising devices. In recent years, enormous efforts have been devoted to obtain ultra-short blue pulses using the InGaN VCSELs via a simple gain-switching technique. Optical pulses as short as 6 and 2 ps have been generated from gain-switched InGaN VCSELs with 3- and 10-period InGaN/GaNquantum wells (QWs) in the active regions in our previous works[5, 6] by using an up-conversion measurement system. Nevertheless, the gain-switched blue pulses should be further shortened, on the basis of our rate-equation analysis [7], by increasing optical gain in the active regions of InGaN VCSELs; for example, by increasing the QW numbers in the cavity.

In this paper, high-gain InGaN VCSELs with 20-period InGaN/GaN QWs were fabricated. Distributed Bragg reflectors (DBRs) with high reflectivity of 97.5% centered at about 420 nm and lower reflectivity below 400 nm were designed and fabricated for optical pumping. To avoid the effect of dopants, undoped GaN layers were grown in the active region. Optical properties of the 20-period InGaN/GaN QW VCSELs were studied with femtosecond pulse laser excitation. The results show that the InGaN VCSELs demonstrate single-longitudinal-mode lasing at 411.6 nm with a linewidth of 0.3 nm under impulsive optical pumping with a threshold pumping energy of about 271 nJ. The success in the fabrication of the high-gain InGaN QW VCSELs should be useful for further gain-switching investigation towards ultrashort pulse generations.

2. Experimental methods

The VCSEL epitaxial structure grown on a (0001)-oriented sapphire substrate by metal-organic chemical vapor deposition includes a 30-nm GaN nucleation layer, a 3-μm GaN bulk layer, 20 period quantum wells consisting of 15-nm GaN barriers and 2-nm InGaN wells, and a 100-nm GaN cap layer. 12.5 pairs of Ta_2O_5 and SiO_2 were evaporated on the top of the grown VCSEL structure to form the bottom distributed Bragg reflector (DBR). Then, the structure was bonded onto a quartz substrate, followed by a laser lift-off (LLO) process using a KrF excimer laser to remove the sapphire substrate. Next, the GaN surface after LLO was thinned and polished through inductively coupled plasma (ICP) etching and chemical mechanical polishing (CMP) techniques. Finally, a second dielectric DBR consisting of 8 pairs of Ta_2O_5 and SiO_2 was deposited. The Ta_2O_5/SiO_2 dielectric DBRs exhibit a broad stopband width of about 85 nm, with peak reflectivities of 97.5% and 99.8% centered at about 420 nm for the top and bottom sides, respectively. The schematic diagram of the fabricated GaN-based VCSEL is shown in Fig. 1(a). The root mean square (RMS) roughness of the GaN surface measured by atomic force microscopy (AFM) was 0.7 nm in a scan area of 2 μm × 2 μm after CMP. The sub-nanometer roughness smooth surface provide favorable condition for high reflectivity DBRs evaporation [8]. Figure 1(b) shows the experimental apparatus for the optical measurements of the VCSEL. The GaN-based VCSEL was optically pumped at room temperature by a fs impulsive optical excitation at 370 nm, with a repetition rate of 1 kHz and a pulse width of 150 fs.

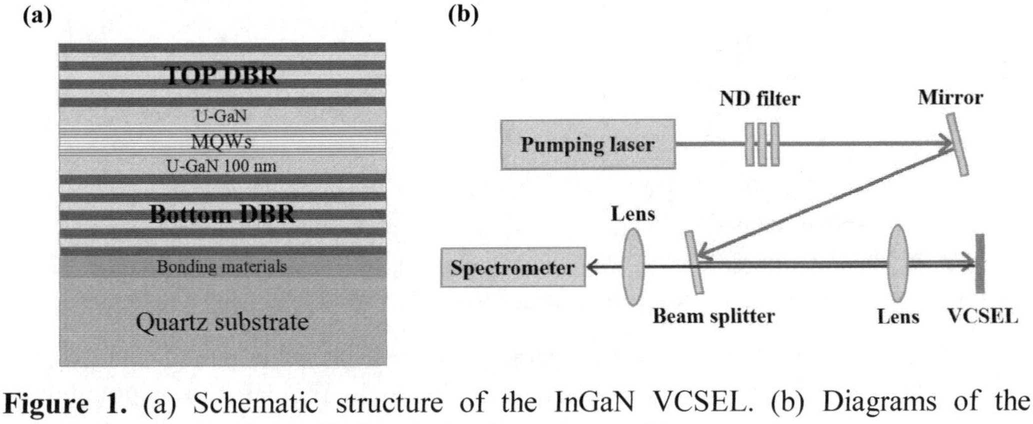

Figure 1. (a) Schematic structure of the InGaN VCSEL. (b) Diagrams of the experimental setup for optical pumping measurements.

3. Results

Figure 2(a) shows the variation of emission spectrum with increasing the pumping energy. As shown in the figure, these spectra obviously show the transition from spontaneous emission to stimulated emission with a single dominant peak. Above the threshold pumping energy, a sharp emission peak appears at 411.6 nm with a linewidth of 0.3 nm (resolution limit). The inset shows the laser emission intensity versus the pumping time at the pumping energy of 1.8 E_{th}, indicating highly stable and reliable performance of the VCSEL device. Figure 2(b) shows the light emission intensity from the VCSEL as a function of the pumping energy. A distinct threshold characteristic was observed at the threshold pumping energy of about 271 nJ corresponding to an energy density of 13.8 mJ/cm^2. Note that the reflectivity of the top DBR is 88% at the pumping wavelength of 370 nm, it is respected that the threshold pumping energy should be much more lower since only 12% of the pumping energy can be absorbed by the active region of the VCSEL.

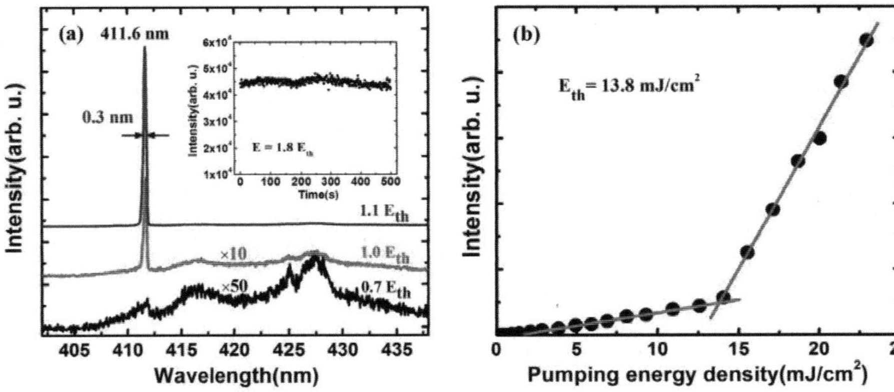

Figure 2. (a) The variation of emission spectrum with increase of the pumping energy. The inset is the laser emission intensity vs the pumping time at the pumping energy of 1.8 E_{th}. (b) Emission intensity as a function of the pumping energy at room temperature.

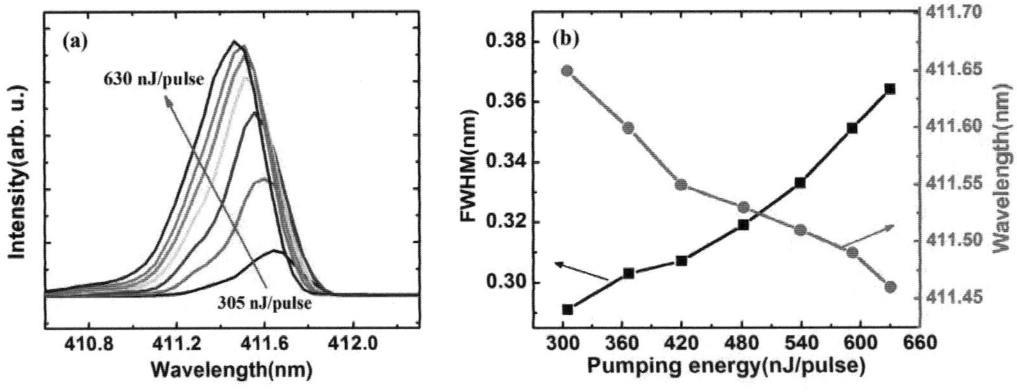

Figure 3. Lasing spectra (a) and lasing wavelengths and FWHM of the lasing spectra (b) as a function of pumping energy from 305 nJ/pulse to 630 nJ/pulse.

The variation of the laser emission spectra with increasing pumping energy from 305 nJ/pulse (1.1 E_{th}) to 630 nJ/pulse (2.3 E_{th}) were shown in Fig. 3(a). The pumping-energy dependence of the FWHM and peak wavelength of the lasing emission were shown in Fig.

IOP Conf. Series: Journal of Physics: Conf. Series **864** (2017) 012083 doi:10.1088/1742-6596/864/1/012083

3(b). The blue shift of the lasing wavelength and the increase of the linewidth with increasing pump energy are typical behaviors of gain-switched semiconductor lasers during pulse lasing, which can be explained by the initial transient spectral broadening (or wavelength chirping) on the short-wavelength side of the gain-switched pulses at elevated pumping energy [9].

Figure 4(a) is a replot of the laser emission intensity as a function of pumping energy in a logarithmic scale. The spontaneous emission coupling factor β, estimated from the difference between the heights of the emission intensities before and after lasing, was ~0.16, which is nearly one order of magnitude higher than those of previous reported [10], indicating a higher quality resonant cavity of the fabricated VCSEL in present study. Figure 4(b) depicts the polarization characteristics of the VCSEL. The degree of the polarization, given by $(L_{max}-L_{min})/(L_{max}+L_{min})$, where L_{max} and L_{min} are respectively the maximum and minimum relative light intensities, was measured to be about 65% at the pumping energy of 1.8 E_{th}.

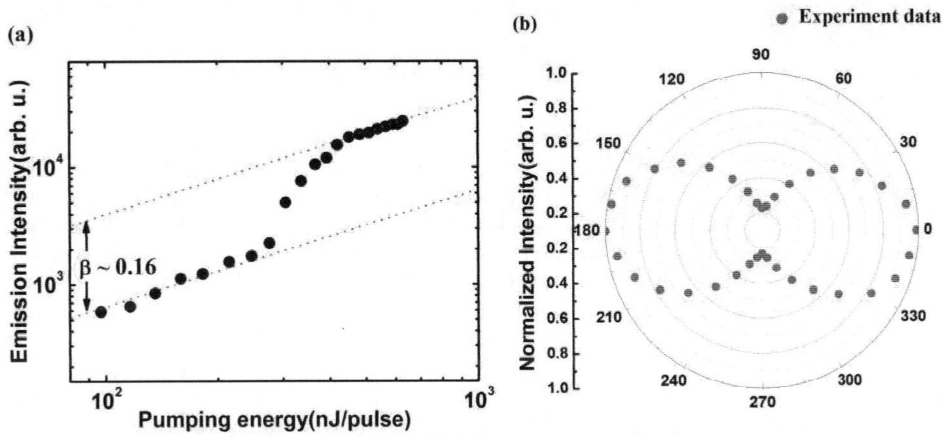

Figure 4. (a) Emission intensity vs pumping energy in a logarithmic cale. The dash lines are guides for the eye. (b) The polarization characteristics of the laser emission from the InGaN VCSEL at the pumping energy of 1.8 E_{th}.

4. Conclusion

High-gain GaN-based VCSEL with 20-period InGaN/GaN QWs were design and fabricated. The laser emission characteristics of the VCSEL were investigated under optical pumping at room temperature. The laser emitted at a wavelength of 411.6 nm with a linewidth of 0.3 nm and a threshold pumping energy of about 271 nJ. Further investigations on the gain-switching dynamics of this high-gain VCSEL are in preparation.

References

[1] Parthenopoulos D A and Rentzepis P M 1989 *Science* **245** 843.
[2] Kawakami R *et al* 2013 *Sci. Rep.* **3** 1014.
[3] Karin J R *et al* 1990 *Appl. Phys. Lett.* **57** 963.
[4] Nakazawa M, Hasegawa H and Oikawa Y 2007 *IEEE Photon. Tech. Lett.* **19** 1251.
[5] Chen S Q *et al* 2012 *Appl. Phys. Lett.* **101** 191108.
[6] Asahara A *et al* 2014 *Sci. Rep.* **4** 6401.
[7] Chen S Q *et al* 2014 *Opt. Express* **22** 4196.
[8] Liu W J *et al* 2013 *IEEE Photon. Tech. Lett.* **25** 2014.
[9] Chen S Q *et al* 2014 *Sci. Rep.* **4** 4325.
[10] Chu J T *et al* 2006 *Appl. Phys. Lett.* **89** 121112.

Oxygen polyvacancy in anatas as a filament: first principle investigation

T V Perevalov[1,2], D R Islamov[1,2]

[1] Rzhanov Institute of Semiconductor Physics SB RAS, Novosibirsk 630090, Russian Federation
[2] Novosibirsk State University, Novosibirsk 630090, Russian Federation

E-mail: timson@isp.nsc.ru

Abstract. We investigate oxygen-deficient anatase using quantum-chemical simulation within the density functional theory. It was found that the most energetically favorable spatial configuration of an oxygen polyvacancy is a three-dimensional chain in crystallographic direction [100] or [010]. The ability of oxygen polyvacancy in the form of a chain to act as a conductive filament and to participate in the resistive switching is discussed.

1. Introduction

The currently, titanium oxide (TiO_2) is being intensively studied as an active medium for nonvolatile memory devices, based on the reversible switching of dielectric from a high resistance state to a low resistance state under the current pulse in the metal-insulator-metal (MIM) structure [1, 2]. The most discussed model of the resistive switching is based on the formation and dissociation of a filament - a conductive nanowire with a diameter of 1-10 nm in the dielectric [1, 3]. It is a widespread belief that oxygen vacancies are responsible for such dielectric layer structure transformation [1, 3]. According to the filament model, a conductive channel in the resistive memory element is formed by the oxygen atoms release and migration from the lattice sites in some region of the dielectric film. Thus, the study of the atomic and electronic structure of TiO_2 with the vacancy complexes of closely spaced oxygen vacancies (that is, polyvacancies) is very important for determining their ability to form ordered structures and participate in resistive switching.

Oxygen divacancies and the electronic structure of highly oxygen-deficient rutile (r-TiO_2) with a concentration up to 5% were calculated in [4]. When the vacancy concentration is less than 1%, the defective zone remains rather narrow, which does not agree with experiment and suggests that, perhaps, a defects clustering, i.e. the formation of polyvacancies, occurs. The analysis of the oxygen vacancies location in r-TiO_2 in the (110) plane showed that it is energetically favorable for vacancies to line up in chains [5]: a one-dimensional chain in the direction [110] or two-dimensional in the direction [001]. Titanium ions, separating vacancies in these chains, form a conductive filament. We can assume that similar conclusions are true for anatase (a-TiO_2). However, the weak point of this model is that the selection of the plane (110), in which oxygen vacancies are created, is arbitrary.

The goal of the present work is to study the atomic and electronic structure of vacancy complexes (oxygen polyvacancies) in a-TiO_2 by means of quantum- chemical simulation and X-ray photoelectron spectroscopy.

IOP Conf. Series: Journal of Physics: Conf. Series **864** (2017) 012084 doi:10.1088/1742-6596/864/1/012084

2. Computational details

The simulation was carried out using a Quantum ESPRESSO software package within the density functional theory (DFT) in a periodic supercell model. Two types of exchange-correlation functional were used: the generalized gradient approximation in the calculations of the structural parameters optimization and another, correctly describing the value of the band gap of TiO_2, hybrid functional B3LYP in the calculations of the density of electronic states. We used *a*-TiO_2 supercells of 96, 144, and 192 atoms. To simulate the atomic structure of oxygen divacancy, after removal of the first oxygen atom, followed by relaxation, the second oxygen atom was removed according to the principle of the least energy required for this: all possible options for the location of the second vacancy were tested, and the energies of its formation were compared. The structure of oxygen trivacancy was determined similarly, but for a pair of fixed vacancies, and so on. X-ray photoelectron spectra (XPS) were calculated by summing the partial densities of filled states with weight factors equal to the corresponding values of the photoionization cross section and by smoothing using a Gaussian function with a half-width of 0.8 eV.

3. Results and discussion

First of all, the total density of states (TDOS) and partial density of states (PDOS) spectra of *a*-TiO_2 with oxygen monovacancy in 96 atoms supercell were calculated (figure 1). For clarity, PDOS is presented for Ti and O atoms closest to the vacancy, and only part of the spectrum is shown which includes the band gap and the conduction band. The obtained spectra are in qualitative agreement with the results obtained in the framework of close techniques in Ref. [6], as well as with experimental data [7]. The presence of oxygen vacancies leads to the two filled spin-polarized states with the one spin orientation (up) are appearance in *a*-TiO_2 band gap. These defective levels are formed mainly by Ti 3*d* states closest to the vacancy atoms. It was found that the oxygen vacancy ground state is the magnetic (triplet) only in the case of simulation of relatively small supercell size (no more than 108 atoms). The oxygen vacancy ground state is nonmagnetic (singlet) for larger than 108 atoms supercells. The transition from triplet state to singlet state is due to the increase in the distance between oxygen vacancies in neighbouring supercells.

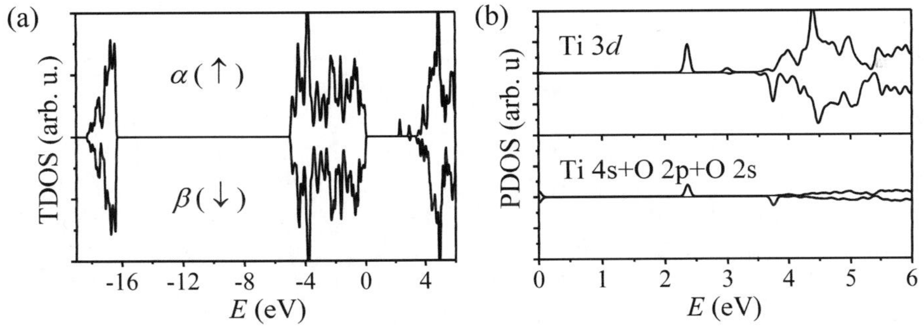

Figure 1. (a) TDOS spectra for electrons with spin up (α) and down (β) calculated for 96 atom *a*-TiO_2 supercell with a neutral O vacancy. (b) PDOS spectra of *a*-TiO_2, calculated for the closest to the vacancy titanium and oxygen atoms. The zero of energy is taken the position of valence band top.

An ideal supercell of 96 atoms *a*-TiO_2 with the four oxygen atoms that undergo removal as well as the removal sequence is represented in figure 2a. For each subsequent vacancy, formation near the existing vacancies is energetically favorable. However, it should not be as close as possible, but at the diametrically opposite end of the octahedron. The indicated O atoms are arranged in rectangle tops.

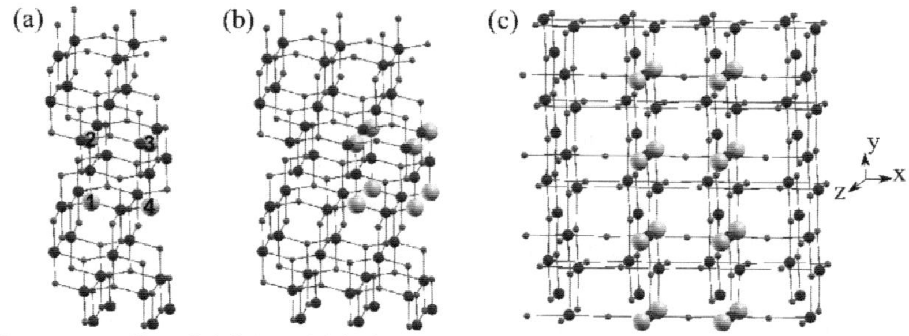

Figure 2. *a*-TiO_2 supercells of 96 (a), 144 (b) and 192 (c) atoms. The O atoms are red, Ti atoms are blue, and O atoms removable for vacancies creating are yellow. Numbers is the sequence of removal.

The calculated valence band XPS for a perfect *a*-TiO_2 and with 1, 2, 3, and 4 O vacancies as well as experimental XPS (from [8]) for *a*-TiO_2 before and after Ar^+ ions etching are shown in figure 3. The experimental XPS for stoichiometric *a*-TiO_2 is in good agreement with published data [9] and is well described by the theoretical spectrum. This demonstrates the correctness of our theoretical model.

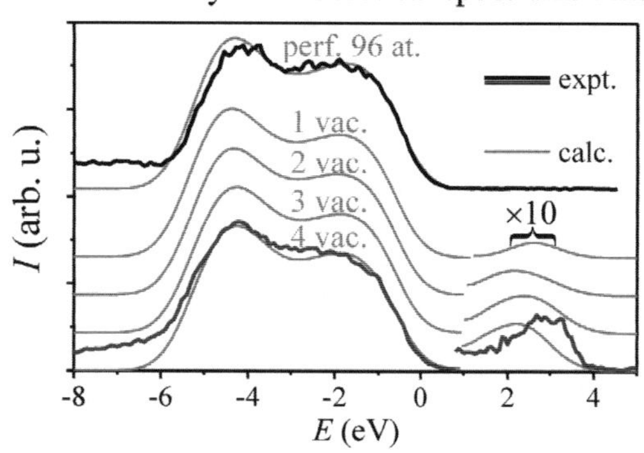

Figure 3. Experimental X-ray photoelectron spectra of the *a*-TiO_2 valence band before (black curve) and after (blue curve) Ar^+ etching and the calculated spectra for *a*-TiO_2 without defects and with one, two, three, and four oxygen vacancies (red curves). The zero energy corresponds to the valence band top.

The Ar^+ ions etching of *a*-TiO_2 leads to the formation of a broad peak with a maximum 2.9 eV above the valence band top. The calculated spectra of *a*-TiO_2 with oxygen polyvacancies show a similar peak, and its intensity is greater with a higher degree of oxygen deficiency. The position of the maximum of the calculated peak is 2.3-2.6 eV, depending on the size of polyvacancy. Thus the Ar^+ ions etching leads to the formation of an oxygen-deficient surface layer in *a*-TiO_2. The experimental peak in the band gap is best described by the theoretical one obtained by the calculations for a 96-atom supercell with 4 O vacancies. So the vacancies concentration in the actual structure is about 10^{20} cm^{-3}.

To determine the potential of an O polyvacancy in *a*-TiO_2 for further growing in size, supercells of 144 and 192 atoms were used (figures 1b and 1c). Eight O atoms that are easiest to remove from the 144-atom supercell form a rectangular parallelepiped with a square base, extending along the crystallographic axis [001]. Subsequent O atoms removal is the least energy cost in the [100] or [010] direction. With an arbitrary selection of one of these directions, further increase of a polyvacancy in size will take place in this direction (Fig. 1c). Thus, the formation of O polyvacancies as a directional three-dimensional chain with a cross section diameter of 0.6 nm is energetically favorable for anatase. This calls into question the structure of the chain of oxygen vacancies in r-TiO_2 predicted in [5].

The energetic efficiency of the formation of oxygen vacancies chain in *a*-TiO_2 is consistent with the filament model of the resistive switching of dielectric. However, these chains have an orientation in the absence of an external field. The directional character of growth of a polyvacancy in size in *a*-TiO_2 can be naturally explained by the presence of selected high symmetry directions. In real MIM structures, TiO_2 layers are amorphous or polycrystalline. However, since the electronic structure of solids is mainly determined by a short range order, we can assume that amorphous TiO_2, as well as rutile and anatase, are characterized by energetic efficiency of the formation of each successive vacancy namely in the proximity of already existing ones, and the external electric field sets the direction of growth of a polyvacancy in the form of a certain chain (filament).

The TDOS spectra in the energy range from the valence band top (E_V) to the bottom of the conduction band (E_C), for O polyvacancies in a 192-atom a-TiO_2 supercell are shown in figure 4. For ease of interpretation and visualization of the results, calculations were carried out only for the Γ-point. Each removal of O atom adds level in the band gap. These levels are filled with electrons and are distributed on the band gap preferentially near the E_C. The growth of an O polyvacancy in size in the form of a chain yields a broad defective band in the energy spectrum of a-TiO_2 filled with electrons, which eventually intersects with the conduction band. Thus, a-TiO_2 can acquire conductivity that is implemented through the channel of titanium ions separating vacancies in this chain.

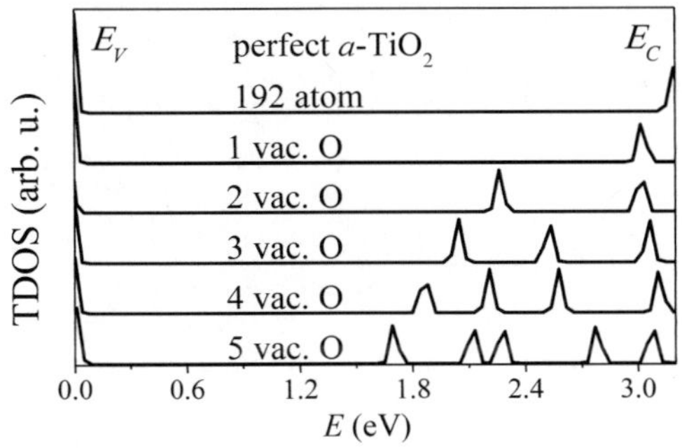

Figure 4. Fragments of the total density of states spectra, showing the defect states in the band gap, caused by the presence of one, two, three, four, and five oxygen vacancies in a 192-atom a-TiO_2 supercell. Zero energy corresponds to E_V, the upper limit of energy corresponds to E_C.

4. Summary

In the framework of quantum-chemical simulation, the atomic and electronic structure of oxygen polyvacancies in anatase is studied. It is demonstrated that the calculation method used is correct. We used an original procedure of searching the positions of oxygen atoms in a-TiO_2 supercells, which are removed for polyvacancies simulating. It is found that the polyvacancies formation in the form of a three-dimensional chain with a diagonal of 0.6 nm is energetically favorable for a-TiO_2. These chains of oxygen vacancies can act as a conductive shunt.

Acknowledgments

This work was supported by the Russian Foundation for Basic Research, project no. 16-32-00119 mol_a. Simulation was performed using a computing cluster of the Rzhanov Institute of Semiconductor Physics, Siberian Branch Russian Academy of Sciences.

References

[1] Kwon D H, Kim K M, Jang J H, Jeon J H, Lee M H, Kim G H, Li X S, Park G S, Lee B, Han S, Kim M and Hwang C S 2010 *Nature Nanotechnology* **5** 148-153
[2] Hu C Q, McDaniel M D, Posadas A, Demkov A A, Ekerdt J G and Yu E T 2014 *Nano Letters* **14** 4360-4367
[3] Bersuker G, Gilmer D C, Veksler D, Kirsch P, Velli L, Padovani A, Larcher L, McKenna K, Shluger A, Iglesias V, Porti M and Nafria M 2011 *Journal of Applied Physics* **110** 124518
[4] Yu N C and Halley J W 1995 *Physical Review* B **51** 4768-4776
[5] Park S G, Magyari-Kope B and Nishi Y 2011 *IEEE Electron Device Letters* **32** 197-199
[6] Finazzi E, Valentin C Di, Pacchioni G and Selloni A 2008 *The Journal of Chemical Physics* **129** 154113
[7] Henrich V E, Dresselhaus G and Zeiger H J 1976 Physical Review Letters **36** 1335-1339
[8] Perevalov T V, Islamov D R and Saraev A A 2016 *Technical Physics Letters* **42** 601-604
[9] Sanjines R, Tang H, Berger H, Gozzo F, Margaritondo G and Levy F 1994 *Journal of Applied Physics* **75** 2945-2951